# INTEGRATED MANAGEMENT OF INSECTS IN STORED PRODUCTS

# INTEGRATED MANAGEMENT OF INSECTS IN STORED PRODUCTS

EDITED BY

**BHADRIRAJU SUBRAMANYAM**

*University of Minnesota*
*St. Paul, Minnesota*

**DAVID W. HAGSTRUM**

*U.S. Department of Agriculture*
*Manhattan, Kansas*

CRC Press
Taylor & Francis Group
Boca Raton London New York

CRC Press is an imprint of the
Taylor & Francis Group, an **informa** business

CRC Press
Taylor & Francis Group
6000 Broken Sound Parkway NW, Suite 300
Boca Raton, FL 33487-2742

First issued in paperback 2019



ISBN-13: 978-0-8247-9522-1 (hbk)
ISBN-13: 978-0-367-40159-7 (pbk)



**Library of Congress Cataloging-in-Publication Data**

Integrated Management of insects in stored products / edited by
Bhadriraju Subramanyam, David W. Hagstrum.
p. cm.
Includes bibliographical references and index.
ISBN 0-8247-9522-9 (hardcover : alk. paper)
1. Food storage pests. 2. Food storage pests—Integrated control.
I. Subramanyam, Bhadriraju. II. Hagstrum, David W.
SB937.I585 1995
632'.7—dc20 95-31158
CIP

**Visit the Taylor & Francis Web site at**
**http://www.taylorandfrancis.com**

**and the CRC Press Web site at**
**http://www.crcpress.com**

Publisher's Note The publisher has gone to great lengths to ensure the quality of this reprint but points out that some imperfections in the original copies may be apparent

**To our parents, wives, and children,**
**and especially to our mentors,**
**Phillip K. Harein, Laurence K. Cutkomp,**
**and William H. Ewart**

# Preface

The management of insects in stored products is in transition from a dependence on regular applications of chemical insecticides to the use of integrated pest management (IPM). IPM involves understanding interactions between stored-product environment and insects associated with stored products, and replacing all or most of the chemical applications with cost-effective nonchemical alternatives. The increased emphasis on adopting IPM strategies for stored-product insects has been brought about primarily by the development of resistance in insects to insecticides, consumers' demand for food free of insecticide residues, and the loss of existing insecticides due to federal regulations. The transition from a heavy reliance on chemicals to little or no chemical input and increased use of nonchemical approaches requires a thorough understanding of stored-product insects, their identification, monitoring, biology, ecology, and response to chemical and non-chemical management options.

This book summarizes the current knowledge of insect biology, ecology, and sampling; the current status of chemical, physical, and biological control methods; and approaches for developing, implementing, and evaluating insect sampling, resistance management, and IPM programs. It is intended for all people involved in the protection of stored products, especially those involved in developing and implementing IPM programs for insects, and we hope that these individuals will use the information in this book and adapt it to their specific situation.

The audience for this book is broader than just those involved in the management of insects in raw and processed food products. This is an invaluable resource to educators, pest control operators, extension specialists, researchers, students, consultants, regulatory personnel, attorneys, IPM advocates, and citizens involved in maintaining the quality of our food supply.

The chapter sequence is arranged logically. Insect identification, biology, and ecology are covered first, because correct identification and understanding pests' interaction with the environment are initial steps in the development of a pest management program. These chapters also provide insights into the aspects of insect biology and ecology that need to be considered in developing a sampling program. Information on sampling tools and statistics is provided in enough detail to facilitate development, implementation, and evaluation of a sampling program. The chapters on physical, biological, and chemical control provide an overview of available insect-control options that are of value in the management of insects. Resistance measurement and management are covered because they will be important as long as chemical and biological controls are components of an IPM program. The last chapter reviews the literature on the use of IPM against insects. IPM approaches have been developed and utilized more extensively for field and orchard crop insect pests than for stored-product insects. Therefore, the applicability of these methods in the management of stored-product insects is discussed in the IPM chapter. The extensive list of references in each chapter is an excellent source of additional information for readers interested in specific subjects or topics.

We would like to thank the excellent staff at Marcel Dekker, Inc., especially Anita Lekhwani, Associate Acquisitions Editor; Jodie McCune, Assistant Production Editor; Marilyn Ludzki, Promotion Manager; and John McGarrell, Copywriter, for their constant interest and invaluable assistance. Celia Hartmann did an outstanding job of copyediting. Susan Norwood and Michael Tufte assisted in verifying the book index. We are grateful to all the anonymous reviewers for their constructive criticism of the chapter outlines and individual chapters. Frank Arthur, Robert Barney, Richard Beeman, Gerrit Cuperus, Barry Dover, Alan Dowdy, Scott Fargo, Jerry Heaps, William Hutchison, Phil Kenkel, Robert Meagher, Thomas Phillips, Carl Reed, David Sauer, and Paul Wileyto deserve special mention for reviewing drafts of various chapters and for their intellectual input. Their contributions significantly improved the quality of the book. We would also like to thank all contributing authors.

*Bhadriraju Subramanyam*
*David W. Hagstrum*

# Contents

# Contributors

**Robert J. Barney, Ph.D.** Associate Research Director, Community Research Service, Kentucky State University, Frankfort, Kentucky

**John H. Brower, Ph.D.** Research Entomologist, U.S. Grain Marketing Research Laboratory, Agriculture Research Service, U.S. Department of Agriculture, Manhattan, Kansas

**Paul G. Fields, Ph.D.** Research Scientist, Research Branch, Agriculture and Agri-Food Canada, Winnipeg, Manitoba, Canada

**Paul W. Flinn, Ph.D.** Research Biologist, U.S. Grain Marketing Research Laboratory, Agriculture Research Service, U.S. Department of Agriculture, Manhattan, Kansas

**David W. Hagstrum, Ph.D.** Research Entomologist, U.S. Grain Marketing Research Laboratory, Agriculture Research Service, U.S. Department of Agriculture, Manhattan, Kansas

**Ralph W. Howard, Ph.D.** Research Chemist, U.S. Grain Marketing Research Laboratory, Agriculture Research Service, U.S. Department of Agriculture, Manhattan, Kansas

**James G. Leesch, Ph.D.** Research Entomologist, Horticultural Crops Research Laboratory, Agriculture Research Service, U.S. Department of Agriculture, Fresno, California

**William E. Muir, Ph.D.** Professor, Department of Biosystems Engineering, The University of Manitoba, Winnipeg, Manitoba, Canada

**David P. Rees, B.Sc.*** Post Harvest Entomologist, Department of Grain Technology, Natural Resources Institute, Chatham, England

**John D. Sedlacek, Ph.D.** Principal Investigator, Community Research Service, Kentucky State University, Frankfort, Kentucky

**Lincoln Smith, Ph.D.** Cassava Program, Department of Entomology, CIAT, Cali, Colombia

**Bhadriraju Subramanyam, Ph.D.** Assistant Extension Entomologist, Department of Entomology, University of Minnesota, St. Paul, Minnesota

**Patrick V. Vail, Ph.D.** Laboratory Director, Horticultural Crops Research Laboratory, Agricultural Research Service, U.S. Department of Agriculture, Fresno, California

**Paul A. Weston, Ph.D.** Principal Investigator, Community Research Service, Kentucky State University, Frankfort, Kentucky

**Noel D. G. White, Ph.D.** Research Scientist, Research Branch, Agriculture and Agri-Food Canada, Winnipeg, Manitoba, Canada

**Current affiliation*: Division of Entomology, Institute of Plant Production and Processing, CSIRO, Canberra, Australian Capital Territory, Australia

# 1

# Coleoptera

**David P. Rees***
*Natural Resources Institute, Chatham, England*

The order Coleoptera (beetles) comprises some 250,000 known species, many of which are able to exploit human-made and human-modified habitats and, in doing so, are now important pests. Members of some 40 families of beetles have been recorded in stores worldwide (Halstead 1986). However, almost all the species of major importance as storage pests belong to one of seven families: Bostrichidae, Bruchidae, Cucujidae, Curculionidae, Dermestidae, Silvanidae, and Tenebrionidae.

The Coleoptera undergo complete metamorphosis. Adults are typically heavily sclerotized and can be easily recognized by the presence of elytra, or highly modified hardened front wings. These conceal the usually functional hind wings, the dorsal, or upper, surface of the metathorax, and most of the dorsal surface of the mesothorax. In most species the elytra also conceal all or most of the dorsal surface of the abdomen (Fig. 1). The head of most species projects directly forwards but in some, notably the bostrichids, it is deflexed (projects downwards) under the prothorax.

Larvae, which have well-developed biting mouthparts, vary in form (Fig. 1a–d). They all lack the prolegs or false legs found on most lepidopteran larvae. Campodeiform larvae are elongate active creatures with well-developed legs and antennae. These are often obligate or facultative carnivores, for example, larvae of the Carabidae, Histeridae, and Tenebrionidae. Most secondary pests have larvae of this type whereas larvae of primary pests are usually scarabaeiform. Scarabaeiform larvae are typically herbivores and spend all or most of their life within a seed or solid food. They are crescent-shaped and have less well developed legs and antennae than campodeiform larvae. Examples include larvae of the Ano-

**Current affiliation*: Division of Entomology, Institute of Plant Production and Processing, CSIRO, Canberra, Australian Capital Territory, Australia

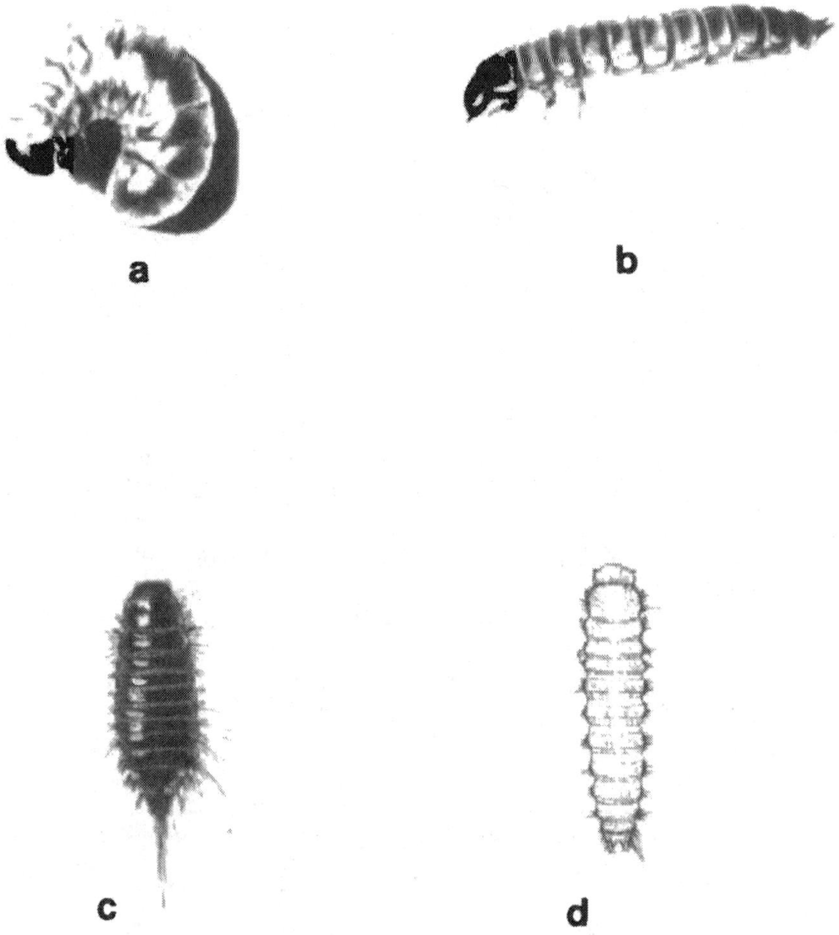


**Figure 1** (a) Larva of *Ptinus tectus* (Ptinidae) (MAFF/CSL). (b) Larva of *Tribolium castaneum* (Tenebrionidae) (MAFF/CSL). (c) Larva of *Trogoderma granarium* (Dermestidae) (MAFF/CSL). (d) Larva of *Dermestes maculatus* (Dermestidae) (NRI).

biidae, Bostrichidae, Bruchidae, Curculionidae, and Ptinidae. Larvae of the Curculionidae are apodous (legless) and spend all their life within grain kernels.

Nearly all ecological niches found in stores are exploited by beetles. Members of the Bostrichidae, Bruchidae, and Curculionidae can attack undamaged pulse and cereal grains as primary pests. Damage caused makes the grain much more susceptible to attack by other insects, the secondary pests, which are not generally capable of breeding rapidly or at all on intact grains. These species can, nevertheless, cause heavy additional damage to previously attacked material and can also attack processed commodities such as flour and milled rice and material damaged by processing, bad handling, threshing, drying, or other processes. Most of these species belong to the Cucujidae, Silvanidae, and Tenebrionidae.

Several families, notably the Cleridae and Dermestidae, contain species that attack material of animal origin such as dried skins, dried fish, and fishmeal.

Many beetles of the families Cryptophagidae, Mycetophagidae, and Ptinidae are mold feeders or scavengers. Others, for example, many nitidulids and some members of the Bruchidae, are frequent inhabitants of ripening crops but can only persist in storage on poorly dried commodities.

Wood-feeding members of the Bostrichidae, Lyctidae, and Scolytidae can damage wooden storage structures. Other beetles, including the Trogossitidae and some Dermestidae, will feed elsewhere but their larvae will often burrow into wood, soft plaster, and other materials in search of a site to pupate. Several families of predaceous beetles occur in stores, notably the Carabidae, Histeridae, and Staphylinidae. While most are casual strays from the local fauna and are of little consequence, some, notably several histerids, are strongly associated with the particular storage pests on which they prey.

Of all of the beetle species known, probably fewer than 100 could be called regular inhabitants of food stores. Many of these, while rare or little known in nature, have been associated with human activities since at least the time of ancient Egypt, Greece, or Rome (Buckland 1981; Panagiotakopulu and Buckland 1991). As a result of millennia of trade, most important species have a cosmopolitan or a pantropical distribution. However some, such as the dermestid *Trogoderma granarium* Everts and most notably the bostrichid *Prostephanus truncatus* (Horn), appear to have colonized only a proportion of their potential range and are still spreading. There also remains the possibility of new species adapting to the storage environment and becoming pests or of known storage pests developing a taste for new commodities.

A number of publications provide identification of storage Coleoptera in more detail than is possible here. These include works by Collier (1981a), Gorham (1987), Mound (1989), Haines (1991), Halstead (1986), and Kingsolver (1987a), and keys for larvae by Anderson (1987) and Collier (1981b). Other works give additional information on the distribution and biology of pest species together with data on environmental requirements. These include works by Aitken (1975), Evans (1981), Haines (1974, 1981, 1991), and Gorham (1991).

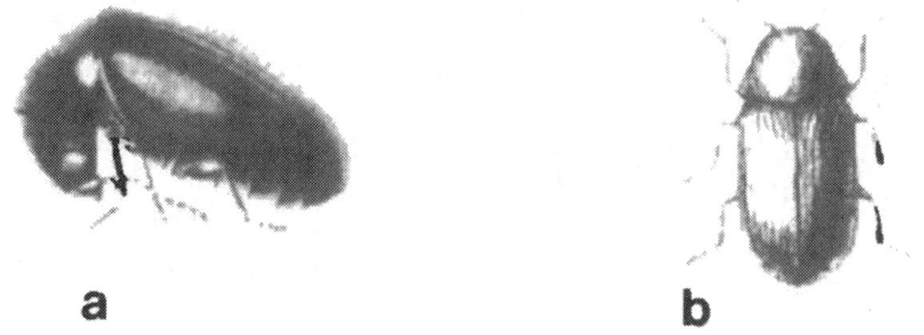

**Figure 2** (a) *Lasioderma serricorne* (Anobiidae): lateral view (2–2.5 mm) (MAFF/CSL). (b) *Stegobium paniceum* (Anobiidae): dorsal view (2–3 mm) (MAFF/CSL). (c) *Araecerus fasciculatus* (Anthribiidae): dorsal view (3–5 mm) (R. E. White). (d) *Araecerus fasciculatus*: lateral view (R. E. White). (e) *Rhyzopertha dominica* (Bostrichidae): dorsal view (2–3 mm) (MAFF/CSL). (f) *Rhyzopertha dominica*: dorsal view of prothorax (NRI). (g) *Rhyzopertha dominica*: lateral view shows gently convex declivity of elytra (NRI).

## I. ECOLOGY, DISTRIBUTION, AND IDENTIFICATION

### 1. Anobiidae

*Description*

The Anobiidae are small oval or cylindrical beetles with the head strongly deflexed under the prothorax. Most species are wood borers and some of these will attack wooden storage structures. Only two species are important as pests of stored products: *Lasioderma serricorne* (F.) (Fig. 2a) and *Stegobium paniceum* (L.) (Fig. 2b). *L. serricorne* is an oval beetle 2–3 mm long and light brown. The elytra are smooth and covered with fine hairs. The antennae of *L. serricorne* are serrate (sawlike); those of *S. paniceum* have a loose club formed from the last three segments. The elytra of *S. paniceum* have longitudinal striae, or grooves, which are not present in *L. serricorne*.

*Life Cycle*

Eggs are laid loosely on the commodity. On hatching, the scarabaeiform larvae burrow into the commodity, feeding as they go. After four to six instars, the larva pupates in a cell made out of fragments of food bound up with silk. The adults, which do not feed, live for 2–6 weeks. They are active and strong fliers and can be attracted to light. Development takes place most rapidly for both species at 30°C, 60–80% relative humidity (RH) and is possible at 20–37°C and 13–34°C for *L. serricorne* and *S. paniceum*, respectively. Biological data on *L. serricorne* have been published by Howe (1957) and for *S. paniceum* by Lefkovitch (1967).

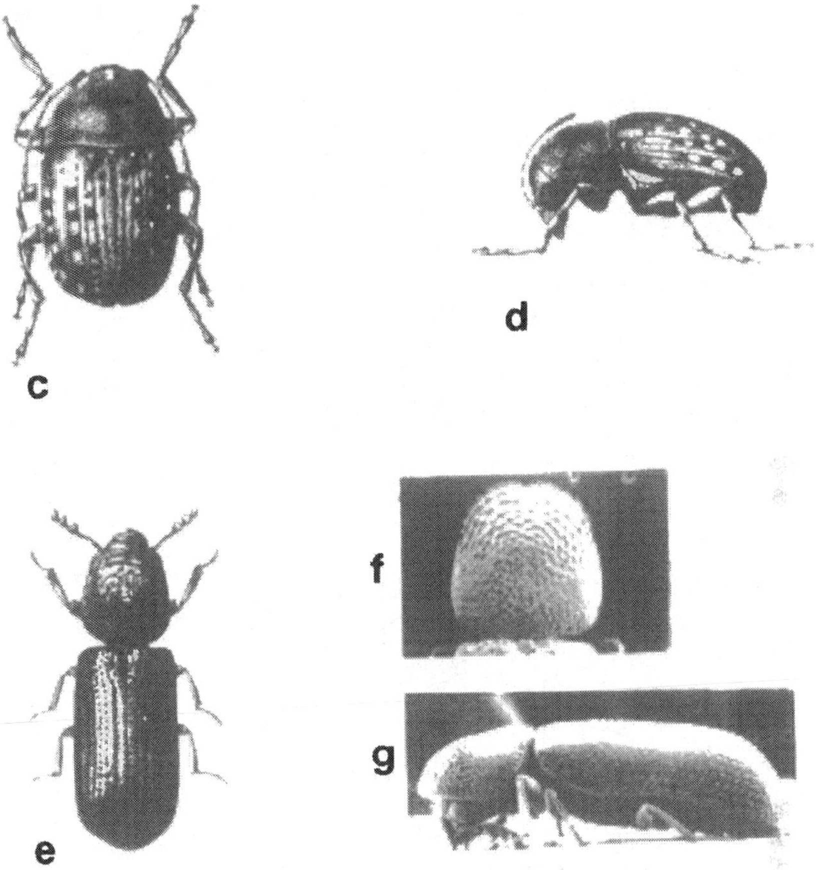


*Distribution and Habitat*

Both species are cosmopolitan; in temperate areas *S. paniceum* is more common. Both attack a wide range of commodities, from cereal grains to processed foods, and are most serious as pests when they attack high-value processed commodities, for example, tobacco products, chocolate, herbs, and spices.

## 2. Anthicidae

Adults of this family bear a passing resemblance to ants. Their larvae typically feed on decaying vegetation. Members of the genus *Anthicus* have been recorded from a wide range of stored products. They are cosmopolitan in distribution but are not important as pests. Storage species can be identified by the key provided by Mound (1989).

## 3. Anthribidae

Members of the Anthribidae are typically found on fungi and dead wood. Only one species, *Araecerus fasciculatus* (Degeer), is of economic importance on stored products. The adult, 3–5 mm long, resembles a bruchid. The elytra, which are mottled in appearance, leave the last abdominal segment exposed. The last three segments of the relatively long antennae form a loose club (Fig. 2c,d).

*A. fasciculatus* is found throughout the tropics and subtropics. It is best known as a pest of stored coffee and cocoa beans. It is also a serious pest of drying cassava and will attack other commodities, for example, maize, nuts, spices, and roots. Mphuru (1974) has reviewed the literature on this species.

## 4. Apionidae

This family is closely related to the Curculionidae and its members are similar in appearance. Many apionids attack growing crops but are not often found in stored commodities. *Piezotrachelus* spp, which infest the ripening pods of *Vigna* spp (cowpeas and grams), sometimes emerge in large numbers from the recently stored seeds of these pulses.

## 5. Bostrichidae

### *Description*

The Bostrichidae are principally wood-borers. However, three genera, *Rhyzopertha*, *Prostephanus*, and *Dinoderus*, are very important as primary pests of cereal grains and dried cassava. Many wood-boring species, especially in the tropics, attack wooden storage structures. Adult bostrichids are cylindrical (circular in cross-section) with a large pronotum and deflexed head. Both the pronotum and the apical end of the elytra (often called the declivity) are often adorned with horns, hooks, and spikes, although in the storage species these are modest or absent.

Adult *Rhyzopertha dominica* (F.) are 2–3 mm long and typically bostrichid in shape (Fig. 2c). The pronotum is rounded at the front where there are transverse rows of flattened tubercles (Fig. 2f). The elytra have regular rows of punctures and short setae and the elytral declivity is gently convex without any ornamentation (Fig. 2g).

Most *Dinoderus* species (a notable exception is *Dinoderus distinctus* Lesne) have a pair of slight depressions at the base of the pronotum (Fig. 3a). These are absent in *Prostephanus truncatus* (Fig. 3c) and *R. dominica* (Fig. 2f). Elytral hairs are short and erect and the declivity is strongly convex (Fig. 3b) and lacks any ornamentation or ridges. In *P. truncatus*, unlike the previous species, the elytral declivity is flattened and steep (Fig. 3d). The tip and sides of the declivity are marked by a ridge (Fig. 3c).

Species of Bostrichidae found in stores can mostly be identified with the key of Fisher (1950) and Haines (1991).

*Life Cycle and Environmental Requirements*

Adults readily burrow into their preferred commodities, producing characteristically large quantities of flour. Eggs are laid either singly or in clumps, in the tunnels and in crevices and cracks in the commodity. The scarabaeiform larvae then feed on the flour or burrow into the commodity at right angles to existing tunnels. Pupation takes place within the foodstuff in a chamber made by the larva. Development is fastest and survival to adulthood better in well-stabilized commodities than in loose grains or in flour. Adults live for several months, the several hundred eggs produced by each female being laid throughout her life.

Optimal conditions for development are in the region of 32°C, 80% RH, typical of major storage pests. However, bostrichids can survive and breed in much drier grain (with a moisture content as low as 8%) and slightly hotter conditions than can *Sitophilus* spp. This gives them a competitive advantage over *Sitophilus* spp in areas with a hot, dry climate.

*Distribution and Habitat*

*Rhyzopertha dominica* is a very important primary pest of whole cereal grains, especially of small grains such as wheat, sorghum, millet, and rice. It is one of the few beetles that is a significant pest on paddy rice. *R. dominica* is cosmopolitan, being especially serious as a pest in hot, dry conditions throughout the tropics and subtropics and in sheltered situations in temperate regions.

*Dinoderus* spp are most frequently found on bamboo and wood products and are often intercepted in temperate regions on imports of these materials. In tropical Africa they are important pests of dried cassava and, occasionally, farm-stored maize. Most species are pantropical, although their exact distributions are poorly known. *Dinoderus minutus* (F.) is the species most often intercepted in temperate regions.

*Prostephanus truncatus* is endemic to Mexico, Central America, northern South America, and the extreme south of the United States. In Mexico it occurs from coastal tropical regions to temperate areas at 2000 m altitude, where it is a pest of cob maize, especially that stored on-farm by subsistence producers. *P. truncatus* was accidentally introduced into Tanzania, East Africa, in the late 1970s and has spread throughout that country and into neighboring ones. A second outbreak occurred in Togo, West Africa, in the mid-1980s, and the species can now be found in Benin, Ghana, Nigeria, Burkina Faso, and Guinea. In Africa it is a pest of major importance of farm-stored maize cobs and dried cassava. On both

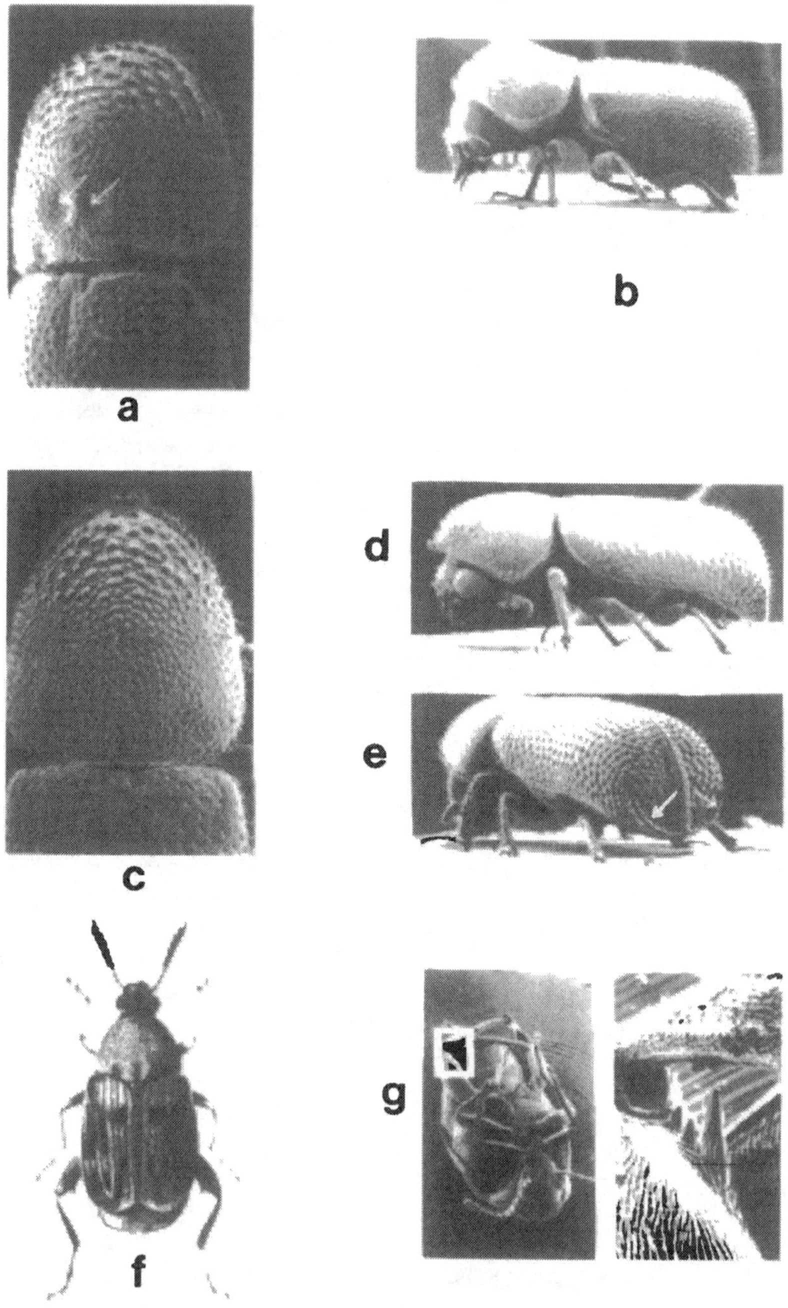
a
b
c
d
e
f
g

continents it is not a significant pest in modern mechanized storage and transport systems; however, care needs to be taken to ensure that this pest is not transported to as-yet-uninfested regions. Large populations of *P. truncatus* are known to occur in natural forest and scrub habitats well away from human activity (Rees et al. 1990), making its control difficult and eradication impossible. As part of an integrated control package against this pest, a Central American histerid beetle, *Teretriosoma nigrescens* (Lewis), has been introduced into Togo and Kenya as a biocontrol agent.

## 6. Bruchidae

*Description*

Members of the Bruchidae feed on plant seeds, especially those of the Leguminaceae or pea family. Adults are very easy to recognize (Fig. 3f): the body, which is covered in short hairs, is compact and globular. The elytra, which are often patterned, are short and expose the last abdominal segment (called the pygidium). The antennae are long and, in some species, pectinate (comblike).

Those species important as pests of stored products are primary pests of legume seeds (Fig. 2). The most important genera as pests of stored legumes are *Acanthoscelides* (Fig. 3f), *Callosobruchus* (Fig. 4a), *Zabrotes* (Fig. 4d), and *Caryedon* (Fig. 4f). Others, such as *Bruchus*, *Bruchidius*, and *Specularius*, while important field pests, do not survive long on well-dried pulses and usually die out in storage. However, in many countries their presence at harvest may cause the rejection of beans when submitted for intake into storage.

Members of the main genera can be easily identified by characteristics on the hind legs of the adult. The hind femur of *Acanthoscelides* has three teethlike structures in a row (Fig. 3g). *Zabrotes* has two movable spurs on the far end of the hind tibia (Fig. 4e). In *Callosobruchus* spp there is a blunt outer and a sharp inner tooth on the ventral apex of the hind femur. These teeth vary somewhat according to species (Fig. 4b,c). In *Caryedon* the hind femur is broad and flattened (Fig. 4g). Adult *Caryedon*, at 4–7 mm long, are bigger than the other species, which are 2–4

---

**Figure 3** (a) *Dinoderus minutus* (Bostrichidae): dorsal view of prothorax shows two depressions or foveae (arrow) (NRI). (b) *Dinoderus minutus*: lateral view shows strongly convex elytral declivity (2.5–3.5 mm) (NRI). (c) *Prostephanus truncatus* (Bostrichidae): dorsal view of prothorax (3.0–4.5 mm) (NRI). (d) *Prostephanus truncatus*: lateral view shows flattened elytral declivity (NRI). (e) *Prostephanus truncatus*: hind view shows carinae (arrow) at apex of elytral declivity (NRI). (f) *Acanthoscelides obtectus* (Bruchidae): dorsal view (3–4.5 mm) (MAFF/CSL). (g) *Acanthoscelides obtectus*: ventral teeth of the hind femur: location in ventral view (left) and detail (right) (NRI).

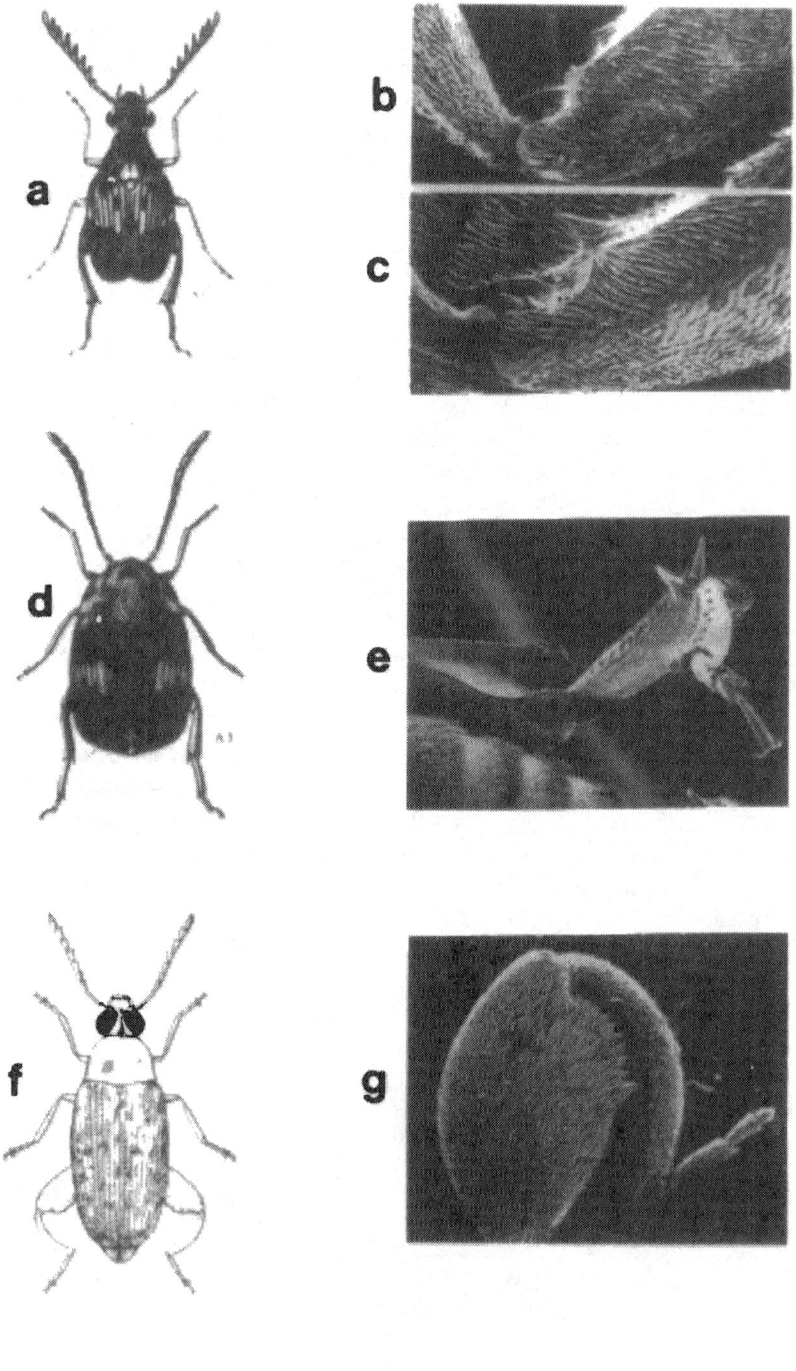
a
b
c
d
e
f
g

mm in length. Keys to the species found in stores include those provided by Kingsolver (1987d), Haines (1991), and (for *Callosobruchus* spp only) Haines (1989).

*Life Cycle and Environmental Requirements*

Eggs are laid onto seedpods or directly on legume seeds. *Callosobruchus*, *Caryedon*, and *Zabrotes* firmly attach their eggs to the pod or seed, while *Acanthoscelides* lays its eggs loosely in cracks and crevices of the seed or pod. On hatching, the scarabaeiform larvae immediately bore into the seed. They usually spend their larval life within one seed, excavating a cavity as they feed. Normally pupation takes place inside the seed; however, *Caryedon* larvae often construct a thin paperlike cocoon attached to the outside of the seed or pod. Larvae that pupate within the seed usually prepare, prior to pupation, a route through which the adult will escape. At this point they eat away the cotyledon until only the seed coat or testa remains, leaving what appears as a circular window on the seed surface. On emergence the short-lived, nonfeeding adult bites and pushes its way out through this escape route. Adults are very active and can run well and fly readily and are often very visible running over the surface of infested seeds.

The major pest genera are found throughout the regions of the world where their host crops are grown and stored. They are especially abundant in tropical and warm temperate areas. In tropical developing countries, where legume seeds are often the main source of protein in the human diet, the losses caused by these insects are of major significance.

The life cycle of the most important genera is relatively short. Under optimal conditions (30–35°C, 70–90% RH) complete development takes place in as little as 22–25 days for *Acanthoscelides obtectus* (Say), *Callosobruchus* spp, and *Zabrotes subfasciatus* (Boheman). Minimum temperatures for development for these species are roughly 15, 18, and 20°C, respectively.

---

**Figure 4** (a) *Callosobruchus chinensis* (Bruchidae): dorsal view of male (2.5–3.5 mm) (MAFF/CSL). (b) *Callosobruchus analis* (Bruchidae): outer view of ventral apex of hind femur shows large blunt outer tooth and small inner tooth (NRI). (c) *Callosobruchus maculatus* (Bruchidae): outer view of ventral apex of hind femur shows large blunt outer tooth and large sharp inner tooth (NRI). (d) *Zabrotes subfasciatus* (Bruchidae): dorsal view of female (2.0–2.5 mm) (MAFF/CSL). (e) *Zabrotes subfasciatus*: ventral view of apical half of hind leg shows the two equal movable spurs on the tibial apex (and the absence of ventral teeth on the femur) (NRI). (f) *Caryedon serratus* (Bruchidae): dorsal view (3.5–6.5 mm) (BM/NH). (g) *Caryedon serratus*: hind leg shows the broad femur with its comb of teeth (NRI).

*Distribution and Habitat*

***Acanthoscelides obtectus.*** Other species of this genus are found mostly in central and southern America. However, *A. obtectus* is now widely distributed in most temperate and tropical regions. Of the bruchids of major importance, it is the one most likely to be found in temperate areas.

*A. obtectus* is a very important pest of *Phaseolus* beans, *P. vulgaris* (kidney/haricot bean) and the lima bean, *P. lunatus*. It has also been recorded attacking *Vicia* spp (field bean, broad bean) and the bambarra groundnut, *Vicia* (= *Voandzeia*) *subterranea*. In the Americas, other species of *Acanthoscelides* occur as storage pests. In Central America and the West Indies, *A. zeteki* Kingsolver can be a serious pest of pigeon peas (*Cajanus cajan*).

***Callosobruchus analis* (F.).** *C. analis* is an Asian species, although it is sometimes accidentally exported to other areas. In India and parts of Southeast Asia it is an important pest of *Vigna unguiculata* (cowpea), *Vigna radiata* (green gram), and sometimes *Glycine max* (soya bean).

***Callosobruchus chinensis* (L.).** *C. chinensis* originated in tropical Asia and is now cosmopolitan in all tropical and subtropical regions. It is a very important pest of cowpea, green gram, chickpea (*Cicer arietinum*), and the lentil (*Lens culinaris*).

***Callosobruchus maculatus* (F.).** *C. maculatus* is widely distributed in all tropical and subtropical regions and appears to be the dominant *Callosobruchus* species in Africa, where it appears to have originated. It is a very important pest of cowpeas, green gram, and lentils.

Several workers describe an "active" or "flight" form of *C. maculatus*, which is apparently more active and more heavily marked. The exact role of this form, which appears in populations as a result of environmental and genetic changes, is not clearly understood.

**Other *Callosobruchus* Species.** Other species known to attack stored pulses include *C. phaseoli* (Gyllenhal), *C. rhodesianus* (Pic), *C. subinnotatus* (Pic), and *C. theobromae* (L.). *C. phaseoli* is found in Central and South America and tropical Africa, attacking dolichos beans (*Lablab purpureus*) as well as cowpeas and green gram. *C. rhodesianus* is restricted to South, East, and West Africa, and attacks cowpeas. *C. subinnotatus* appears to be established in West Africa, infesting bambarra groundnuts. *C. theobromae* is known from Africa and South and Southeast Asia. It has been recorded on groundnuts (*Arachis hypogaea*) in Nigeria, pigeon peas in India, and soya beans in Indonesia.

***Zabrotes subfasciatus.*** This species originated from, and is particularly important in, tropical and subtropical Central and South America. It is also found in other warm regions of the world, notably central and eastern Africa, Madagascar, countries bordering the Mediterranean, and India. It mostly attacks kidney and lima beans but, in Uganda and parts of West Africa, it has also been recorded attacking cowpeas.

***Caryedon serratus* (Olivier).** *Caryedon serratus* is almost entirely associated in commerce with groundnuts and tamarind (*Tamarindus indica*). It is a serious pest of groundnuts in West Africa and also occurs in subtropical and tropical areas of Asia, the West Indies, Hawaii, and Latin America north to Mexico.

## 7. Carabidae

The Carabidae or ground beetles are mostly nocturnal predators (Fig. 9e). Many species have been recorded from stores in very small numbers. These are usually members of the local fauna that strayed accidentally into the store.

## 8. Cerylonidae

Adults of this family are small shiny oval beetles some 1.5 mm long. The antennae characteristically fit into dorsal cavities on the front angles of the pronotum. Only one genus, *Murmidius*, in particular *M. ovalis* (Beck) (Fig. 7e), has been found on a regular basis in stored products and usually only if they are contaminated with mold.

## 9. Cleridae

Some 2000 species of Cleridae are known, most of which are predatory and live in the tropics. However, some members of the genus *Necrobia* feed on stored products. Clerids are lightly covered with distinctive blunt stiff hairs, especially at the margins of the prothorax and elytra. In some species the posterior of the prothorax narrows to form a "neck" with the abdomen. Adults are often metallic blue, green, or bronze.

*Necrobia rufipes* (Degeer) is a shiny metallic greenish-blue beetle 4–5 mm long (Fig. 5a). Its legs and basal antennal segments are red. *Necrobia ruficollis* (Fab.) is a similar color but the pronotum and basal part of the elytra are rust red. Both species are widespread in the tropics, subtropics, and warm temperate regions. They will attack animal products such as ham and cheese. *N. rufipes* is a common pest of underdried copra and is often found in association with *Dermestes* spp on dried fish, where it both feeds on fish flesh and preys on other insects present.

Another clerid, the predatory *Thaneroclerus buqueti* (Lefevre) (Fig. 5b), is found in association with the anobiids *Stegobium paniceum*, *Lasioderma serricorne*, and *Anobium punctatum*. It is 4–5 mm long and dull brown.

## 10. Cryptophagidae

The Cryptophagidae are small beetles, about 1.5–4 mm long. Some are reddish-brown while others are patterned but all are pubescent (hairy) (Fig. 5c). They are mold-feeders, their presence in stores generally indicating damp unhygienic conditions; genera recorded include *Cryptophagus* and *Henoticus*. Species found in

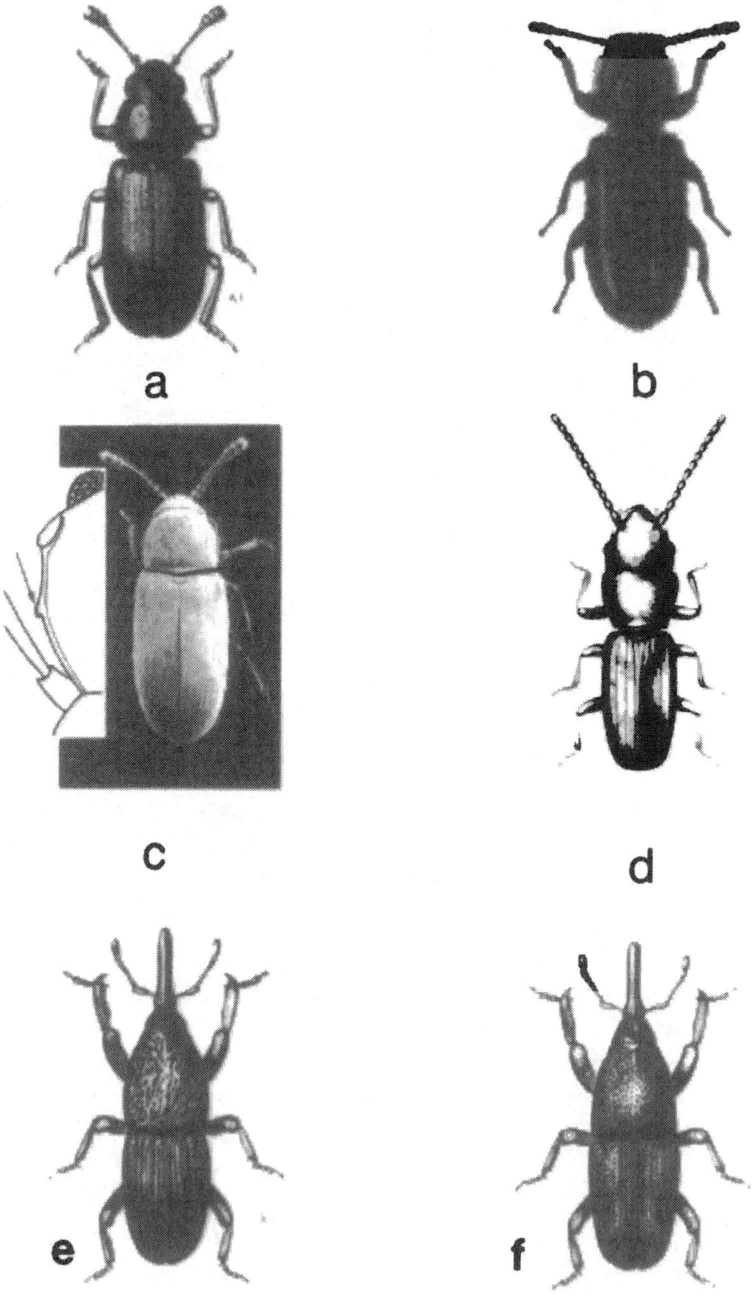
a
b
c
d
e
f

the United States can be identified using keys provided by Woodroffe and Coombs (1961) and Kingsolver (1987c).

## 11. Cucujidae

### *Description*

Adults of this family are characteristically flattened and are generally about 1.5–2 mm long. The antennae are usually long, often half or more of the body length. Members of the genus *Cryptolestes* are common secondary pests of cereals and cereal products. Adults are small (1.5–2 mm long), oblong, flattened, light-brown beetles. The head and prothorax are comparatively large and account for nearly half of the body length. A ridge or carina runs from behind each eye and then in parallel down each side of the pronotum (Fig. 5d).

Eight species of the genus *Cryptolestes* are known to infest stored products. All are externally very similar and microscopic examination of genitalia of cleared mounted specimens is the most reliable means to identify species. Banks (1979) describes how this may be done for the six most common species. *C. klapperichi* Lefkovitch is described by Green (1979). All species can be identified by the keys of Halstead (1993).

### *Life History and Behavior*

Eggs (up to about 200 per female) are laid among the commodity. After four instars the campodeiform larva pupates in a silk cocoon. The life cycle of *C. ferrugineus* (Stephens) takes 103 to 17 days at temperatures from 21–38°C, 75% RH. Optimal conditions are 33°C, 70% RH, when the life cycle is completed in 23 days. *C. ferrugineus* can survive exposure to winter conditions in temperate climates. However, *C. pusillus* (Schönherr) is less capable of surviving conditions of low temperature and humidity and *C. pusilloides* (Steel & Howe) is particularly sensitive to low humidities.

*Cryptolestes* species are some of the most common secondary pests of cereals, cereal products, and a variety of other materials in both temperate and tropical regions. They can attack grains that have only very slight imperfections. Being highly flattened they are able to enter very small cracks and crevices in grains and enter otherwise well packaged processed food.

---

**Figure 5** (a) *Necrobia rufipes* (Cleridae): dorsal view (4–5 mm) (MAFF/CSL). (b) *Thaneroclerus buqueti* (Cleridae): dorsal view (5.0–6.5 mm) (Ohio Biol. Survey). (c) *Cryptophagus* sp (Cryptophagidae): dorsal view and insert of side of prothorax shows anterolateral swelling and lateral tooth (arrow) (1.5–3.3 mm) (NRI). (d) *Cryptolestes ferrugineus* (Cucujidae): dorsal view (1.5–2.0 mm) (MAFF/CSL). (e) *Sitophilus granarius* (Curculionidae): dorsal view (2.4–4.5 mm) (MAFF/CSL). (f) *Sitophilus oryzae* (Curculionidae): dorsal view (2.4–4.5 mm) (MAFF/CSL).

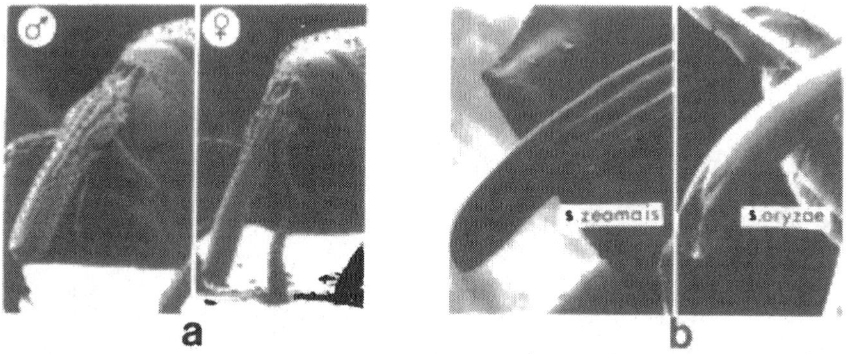


**Figure 6** (a) *Sitophilus zeamais* and *S. oryzae* (Curculionidae): rostra of male (left) and female (right) show the differences in shape and surface texture (NRI). (b) Apex of aedaegus of male *Sitophilus* on *S. zeamais* (left) the two longitudinal depressions are shown. On *S. oryzae* (right) the smooth convex surface is shown (NRI). (c) *Trogoderma granarium* (Dermestidae): dorsal view (2.0–3.0 mm) (MAFF/CSL). (d) Median ocellus of a dermestid (*Trogoderma* sp) shows front of head plus enlargement with ocellus (arrow), which appears as a slight, hairless bump usually shining in fresh specimens (NRI). (e) *Dermestes frischii* (Dermestidae): dorsal view (6–10 mm) (BM/NH). (f) *Dermestes maculatus*: apex of elytra shows apical teeth (NRI).

*Distribution*

*C. ferrugineus* is the most widely distributed species, being cosmopolitan. It is a common pest in all temperate regions. *C. pusillus* is also cosmopolitan but is most abundant under warm, humid conditions, such as, for example, in Southeast Asia. *C. pusilloides* has been recorded from tropical and subtropical regions of the southern hemisphere. *C. turcicus* (Grouvelle) has been found in most temperate areas of the world, especially under humid conditions, but apparently not from Australasia. *C. capensis* (Waltl) occurs in Europe and North Africa and is probably established in southern Africa. *C. ugandae* Steel & Howe appears restricted to tropical Africa. *C. klapperichi* is a little-known species initially described in Afghanistan but also found in South and Southeast Asia. *C. cornutus* Thomas & Zimmerman is a recently described species intercepted in California on an import of dried chili peppers from Thailand.

## 12. Curculionidae

*Description*

The Curculionidae are characterized by the presence in adults of a forwardly pointing, snoutlike projection to the head (Figs. 5e,f, 6a). At the end of this structure, known as the rostrum, are the mouthparts. The legless (apodous) larvae

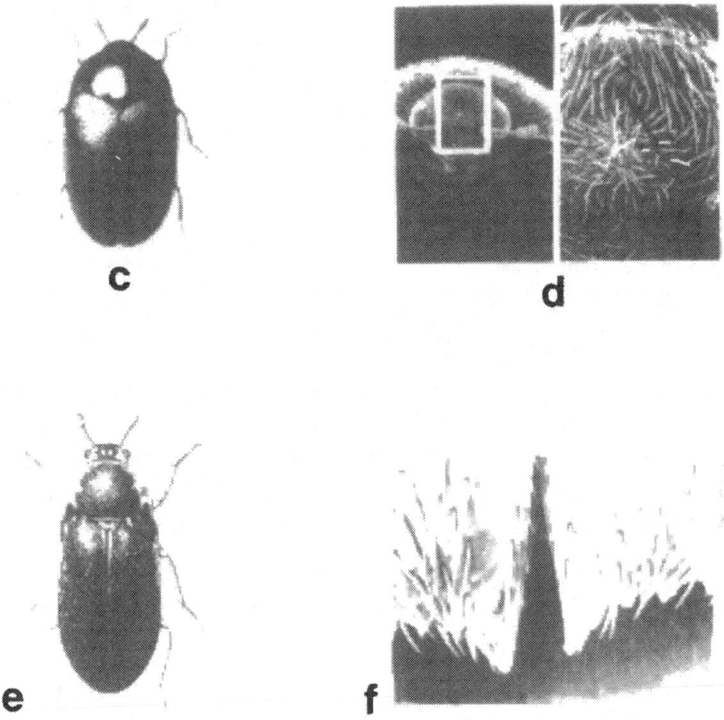


feed by boring into roots, stems, or seeds of plants. The family contains many very destructive agricultural pests, among which only *Sitophilus* spp are important as storage pests.

The three species, *S. zeamais* Motschulsky, *S. oryzae* (L.), and *S. granarius* (L.), are probably the most widespread and destructive primary pests of stored cereals. Another species, *S. linearis* (Herbst), is of minor importance as a pest of stored tamarind pods. The taxonomy of this genus, in particular the relationship between *S. zeamais* and *S. oryzae*, was, until recently, very confused; Haines (1991) gives an account.

*S. granarius* is similar in appearance to *S. zeamais* and *S. oryzae*, but can be differentiated from them by the absence of hind (flight) wings and by the oval shape of punctures on the prothorax; *S. zeamais* and *S. oryzae* have round punctures. The elytra of *S. granarius* are unmarked and brown, while those of *S. zeamais* and *S. oryzae* have four reddish-yellow spots (Fig. 5e,f).

*S. zeamais* and *S. oryzae* are indistinguishable by external characteristics. To differentiate between them requires the dissection and examination of genitalia.

Either gender can be used but the diagnostic characteristics are most easily seen in males, which can be identified by their shorter, more pitted rostra (Fig. 6a). In *S. zeamais* the outer surface of the penis or aedeagus has a central ridge between two longitudinal depressions, while in *S. oryzae* it is completely smooth and convex (Fig. 6b).

*Life Cycle*

The adults are long-lived (several months to 1 year) and feed on the commodity. Up to about 150 eggs are laid by each female throughout her life. Eggs are inserted individually into small cavities chewed in the grain by the female. The cavity is then covered with a waxy secretion, sealing the egg into the grain. Eggs hatch in about 6 days at 25°C, the larva developing within the grain, excavating a cavity as it grows. Larvae are cannibalistic on smaller or weaker individuals; as a result, rarely will more than one adult emerge from a single grain of wheat or rice, although two or three may emerge from a maize grain. After four instars, pupation takes place within the grain. Upon hatching the adult bites its way out of the grain, leaving a circular exit hole. Complete development is possible at temperatures between about 15 and 35°C, and takes 35 days under optimal conditions, which are 27°C, 70% RH for the tropical species and slightly cooler for *S. granarius*. Mortality of juveniles increases in grain with moisture contents (MC) below 13%, and eggs are usually not laid at all on commodities below 10% MC. Development takes place most rapidly on grain with a moisture content of 14–16%.

*Distribution and Habitat*

*Sitophilus* spp are among the most widespread and destructive primary pests of stored cereals in the world. *S. zeamais* and *S. oryzae* are cosmopolitan, but are especially abundant as pests in warm temperate to tropical regions. In cool temperate areas, such as northwest Europe, these species are largely replaced by *S. granarius* which is also found in cool highland areas in the tropics.

*S. granarius* is most often found on wheat and barley but will breed on other cereals. *S. zeamais* is a common pest of stored maize but is frequently found on other cereals including milled and paddy rice; *S. oryzae* is most often found attacking small-grained cereals (rice, wheat, and sorghum) (Haines 1981). Both *S. zeamais* and *S. oryzae* will attack solid processed products such as pasta and dried cassava. A few strains of both species are able to feed on legumes such as peas, cowpeas, and grams (Coombs et al. 1977; Haines 1981).

Although both *S. zeamais* and *S. oryzae* can fly, *S. zeamais* appears to do so more often. When grain is grown and stored in close proximity, *S. zeamais* is likely to be found flying from the store into a ripening crop and becoming established prior to harvest. *S. oryzae* is generally more of a pest of commerce and large-scale storage than of the farm. In contrast, *S. granarius* has only vestigial flight wings and is thus flightless.

A few other curculionids are known as minor storage pests, including *Caulophilus oryzae* (Gyllenhal), differentiated from *Sitophilus* spp by its short rostrum. It is found on soft or damaged maize in the southern states of the United States and in Mexico and Central America, and on dry ginger root in Jamaica. Several species of *Catolethrus* have been found in Africa, Mexico, and Central America infesting farm-stored maize cobs stored by subsistence producers.

## 13. Dermestidae

### *Description*

There are about 700 known species within the Dermestidae. Pest species are known from a wide range of commodities, including dried fish, skins, woollen articles, museum specimens, and cereal grains. Dermestid beetles are ovoid and between 1.5 and 12 mm in length (Fig. 6c,e). They are usually covered with hairs or colored scales. When the latter are present, as in *Anthrenus* spp, they give a colorful patterned appearance (Fig. 7d). Dermestid adults (except *Dermestes* and *Thorictodes* species) have a simple third eye (median ocellus) between the two compound eyes (Fig. 6d). The campodeiform larvae are very hairy (Fig. 1c,d). Setae can become detached from both cast skins and live larvae and can be highly irritant on the skin or when ingested or inhaled.

Members of the Dermestidae can be identified using the keys provided by Banks (1994), Haines (1991), Kingsolver (1987b), and Peacock (1993).

***Trogoderma* Species.** *Trogoderma* species are oval, hairy reddish-brown to black beetles 2–5 mm long. The elytra are either one color or have a weak pattern. Adults have a median ocellus and can be differentiated from the otherwise similar *Anthrenus* species by not being covered with colored scales.

***Trogoderma granarium* Everts.** Adult *T. granarium* are oval, reddish-brown hairy beetles 2–3 mm long, females being somewhat larger than the males (Fig. 6c). Adults rarely, if ever, feed or drink and, although possessing flight wings, are not known to fly. The larvae are very hairy and grow to about 5 mm (Fig. 1c).

Larval development does not occur at temperatures below 21°C, but can occur at humidities as low as 2%. However, development is most rapid in hot humid conditions, taking 18 days at 35°C, 73% RH. These optimal conditions do not correspond with its known distribution, which is mainly confined to hot dry areas (Banks 1977). Under moister conditions, *T. granarium* appears not to compete well against other, faster breeding, species such as *Sitophilus* spp or *R. dominica*.

Some larvae can undergo a form of suspended animation or facultative diapause for months or even years, usually in response to adverse conditions, during which metabolism falls to a very low level, making them highly tolerant to fumigants and insecticides.

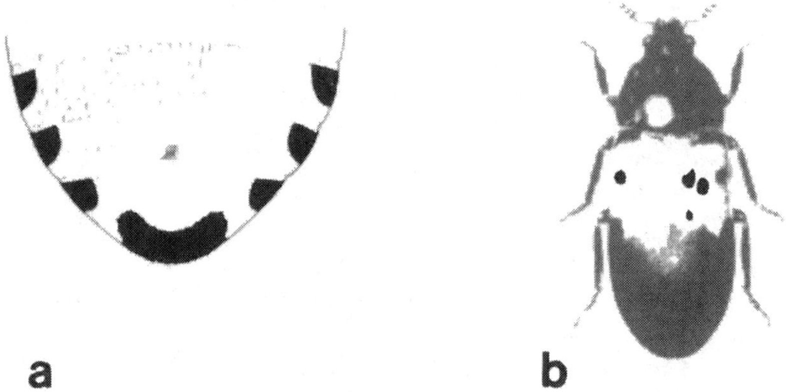


**Figure 7** (a) *Dermestes frischii* (Dermestidae): ventral apex of abdomen shows black and white pattern (BM/NH). (b) *Dermestes lardarius* (Dermestidae): dorsal view (6.0–10.0 mm) (MAFF/CSL). (c) *Attagenus* sp (Dermestidae): dorsal view (3.0–6.0 mm) (BM/NH). (d) *Anthrenus verbasci* (Dermestidae): dorsal view (1.7–3.2 mm) (BM/NH). (e) *Murmidius ovalis* (Cerylonidae): dorsal view (1.5 mm or less) (D.G.H. Halstead). (f) *Carcinops pumilio* (Histeridae): dorsal view (1.6–2.7 mm) (BM/NH).

*T. granarium* is an important pest of cereals and oilseeds but will also attack pulses. It is most serious as a pest in hot dry regions, in particular in areas inside a broad band stretching from the Sahel in western Africa to northern regions of the Indian subcontinent. It has also been found in certain specialized warm, dry habitats in temperate regions, for example, in maltings in Britain. *T. granarium* appears not to be established in the Americas, Australia, and parts of Southeast Asia. It is the subject of strict quarantine regulations to prevent its entry to a number of countries, for example, the United States and Australia. Its distribution, current and potential, is reviewed by Banks (1977).

**Other *Trogoderma* Species.** A number of other *Trogoderma* species have been found in association with stored products, but none has the pest status of *T. granarium*. The presence of these species often creates interest because of their similarity in appearance to *T. granarium*. Species that commonly occur in stores can mostly be identified by examination of a number of characteristics including mouthparts, antennae, antennal cavity, elytral pattern, and genitalia (Banks 1994; Beal 1954, 1956; Kingsolver 1987b). Keys for larvae also exist (Banks 1994; Beal 1960; Spangler 1961; Kingsolver 1987b).

*T. glabrum* (Herbst) is known from Europe and North America. It can breed on cereal and cereal products alone and heavy infestations have been reported in stored wheat and maize in the United States. *T. variabile* Ballion has been reported on cereal grains and pulses in many parts of the United States. It is also established

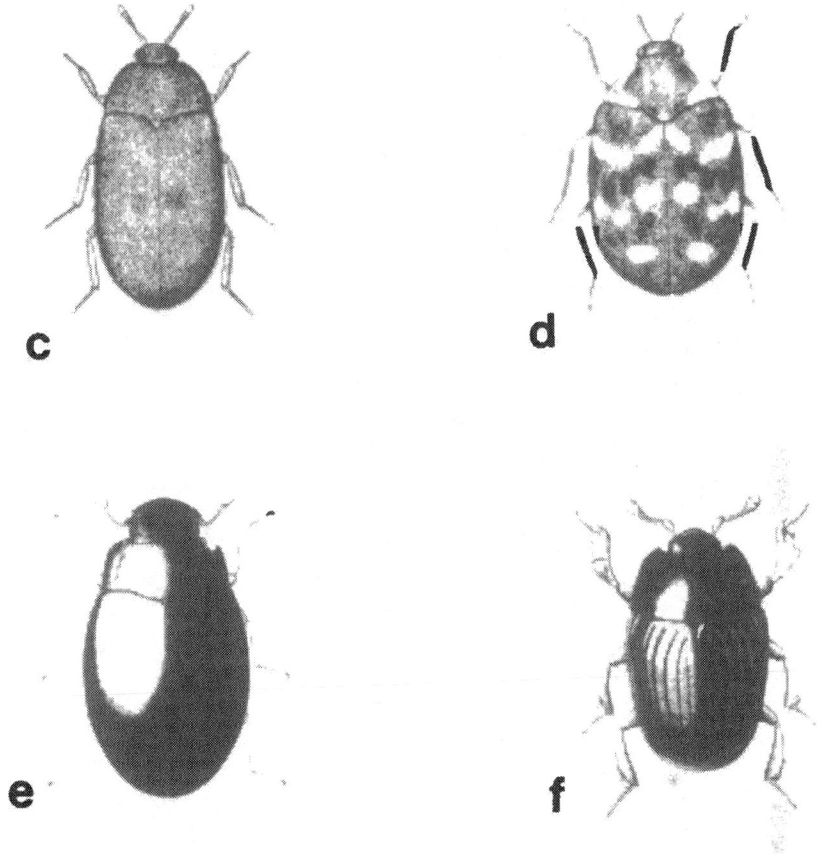


in Australia. Other species from stores include *T. inclusum* LeConte and *T. sternale* Jayne.

***Dermestes* spp.** *Dermestes* species are elongate oval beetles, 6–10 mm long. Adults do not have a median ocellus. The upper surface of the adult is black or dark brown, covered with black, grey, brown, or yellowish hairs (Figs. 6e,f, 7a). The underside is covered with either black and white or brown and golden hairs. *Dermestes* species can be identified using the keys referred to under *Trogoderma*, or that of Haines and Rees (1989) and Peacock (1993).

***Dermestes maculatus* Degeer and *D. frischii* Kugelann.** These species can be differentiated from other *Dermestes* spp by the white underside of the abdomen with black spots at the sides and tip (Fig. 7a). *D. maculatus* has a spine at the end of each elytron, which is absent in *D. frischii* (Fig. 6f).

Both species are important cosmopolitan pests of animal products, especially of hides and dried fish in both tropical and temperate regions. They have also been found in animal accommodation, for example, chicken houses. Dried fish of freshwater origin is most frequently attacked by *D. maculatus* and salt-water fish by *D. frischii*. The latter species appears more tolerant of the higher salt content.

About 100–150 eggs are laid by a female over 2 months. Egg laying is increased if the female has water to drink. The larvae (Fig. 1d) burrow into the foodstuff as they feed, passing through five or more instars, the number of which increases as conditions become more unfavorable. Before pupation, larvae burrow into a solid material to make a pupal cell. This may be the infested commodity, another material stored nearby, or the wood of a drying rack or storage structure. Such tunnelling may, over time, so weaken drying racks as to necessitate their replacement.

Complete development takes place in about 5–7 weeks under optimal conditions of 30–35°C, 75% RH. Adults are strong fliers and can quickly find new sources of food.

**Other *Dermestes* spp.** The third most widespread species attacking dried fish in tropical climates is *D. ater* Degeer. It is also an important pest of copra, as is *D. maculatus*. *D. carnivorus* F. is an important pest of dried fish in South and Southeast Asia and is also known from Europe and the Americas.

*D. lardarius* L. is a cosmopolitan pest of a wide range of animal products. It is also a scavenger, feeding on the remains of dead insects and other animals, but can breed on cereal products such as wheat germ. It is often an inhabitant of domestic premises in temperate countries. The elytra in this species are distinctive: the basal half of each elytron has a large area of pale brown-grey hairs (Fig. 7b) that is absent in other species.

Two other species, *D. haemorrhoidalis* Kuster and *D. peruvianus* Laporte de Castelnau, believed to be of South American origin, have spread to North America and Europe. In the United Kingdom, for example, both species have become domestic pests.

***Attagenus* spp.** A number of species of *Attagenus* (2–5 mm long) are common and destructive domestic pests in temperate regions, feeding on wool, fur, skins, and other materials (Fig. 7c). They also occur in stores, usually as scavengers feeding on dead insects. Species include *A. unicolor* (Brahm), *A. fasciatus* (Thunberg), *A. cyphonoides* Wrestler, and *A. belle* (L.). *Attagenus* species can be identified using keys provided by Beal (1970) and Peacock (1979 and 1993).

***Anthrenus* spp.** *Anthrenus* are small (1.5 to 4 mm) oval to round beetles easily recognized by their dorsal surfaces covered with colorful scales arranged in distinct patterns (Fig. 7d). Many are widely distributed or cosmopolitan and are well known as domestic pests and are often encountered in stores. Species include *A. verbasci* (L.), *A. flavipes* LeConte, *A. coloratus* Wrestler, and *A. museorum* (L.). Many species can be identified using the keys of Hinton (1945a) and Peacock (1993).

The larvae are common household pests on wool and are a serious problem in museums housing dried biological specimens. They are also common inhabitants of bird nests. The adults feed on the nectar and pollen from flowers.

## 14. Endomychidae

Only one species, *Mycetaea hirta* (Marsham) (1.5–1.8 mm long), is known to be associated with stored products. Both adults and larvae feed on molds and are found on damp moldy produce as well as in a range of domestic and natural habitats in Europe, Asia, North America, Africa, the Caribbean, and some Pacific islands.

## 15. Histeridae

About 3200 species are known, of which fewer than 20 have been found in stores (Hinton 1945b). They are compact, heavily sclerotized oval beetles. The elytra are short and do not cover the last one or two segments of the abdomen. All species known from storage habitats are either glossy black or metallic (Fig. 7f). They are good fliers and can feign death very effectively when disturbed. The larvae are campodeiform and have large sickle-shaped mandibles.

Both adults and larvae are predatory, feeding on insects and other arthropods. In temperate and tropical areas *Carcinops pumilio* (Erichson) and *C. troglodytes* (Paykull), respectively, are found in store residues, old birds' nests, and dead animals in which they feed on fly larvae and other materials. *Saprinus* spp prey on *Dermestes* spp and *Teretriosoma* and *Teretrius* are usually predators of wood-boring beetles. The Central American histerid *Teretriosoma nigrescens* (Lewis) is found to be closely associated with the bostrichid *Prostephanus truncatus* (Rees et al. 1990). *T. nigrescens* has been introduced into the African states of Togo and Kenya as a biocontrol agent against this pest. Species of *Carcinops* can be identified using the key of Halstead (1969).

## 16. Languriidae

Members of this family are infrequently found in stores and none is an important pest. Genera recorded in stores include *Cryptophilus* and *Pharaxontha*. For the identity of species in store, consult Hinton (1945a).

## 17. Lathridiidae

About 35 species of this family have been found in stored products. All are fungus feeders and their presence may indicate problems with moisture content or the presence of damp residues. Adults are very small: 1–3 mm long. Genera found in storage include *Lathridius* (Fig. 8a) and *Dienerella*. Hinton (1941) and Kingsolver and Andrews (1987) describe many of the species associated with stored food.

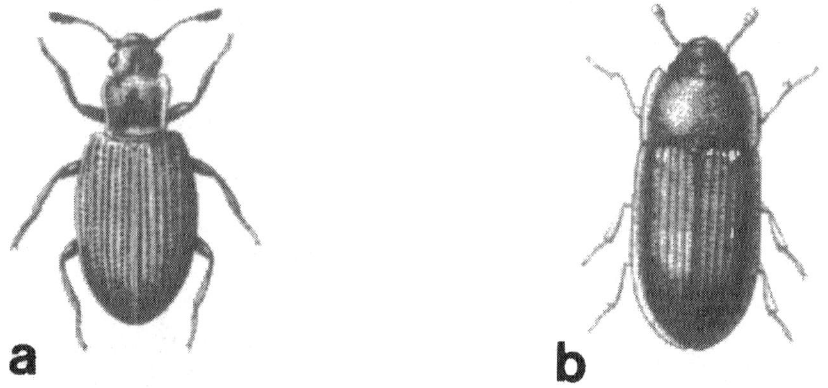

**Figure 8** (a) *Lathridius pseudominutus* (Lathridiidae): dorsal view (2.3–3.0 mm) (BM/NH). (b) *Lophocateres pusillus* (Lophocateridae): dorsal view (2.7–3.2 mm) (BM/NH). (c) *Lyctus brunneus* (Lyctidae): dorsal view (3.0–7.0 mm) (BM/NH). (d) *Typhaea stercorea* (Mycetophagidae): dorsal view (2.2–3.0 mm) (BM/NH). (e) *Carpophilus hemipterus* (Nitidulidae): dorsal view (2–4 mm) (MAFF/CSL). (f) *Niptus hololeucus* (Ptinidae): dorsal view (3.5–4.5 mm) (MAFF/CSL).

## 18. Lophocateridae

Formerly classified as a part of the Trogossitidae, the Lophocateridae includes *Lophocateres pusillus* (Klug), the only species regularly found in stored products (Fig. 8b). This is a minor secondary pest often found on paddy and milled rice in Southeast Asia and also on other cereals, pulses, cassava, and groundnuts.

## 19. Lyctidae

Members of this family are wood-borers that are able to damage wooden storage structures and will sometimes attack stored crops, especially dried cassava. Adults are elongate, parallel-sided beetles with a distinct two-segmented antennal club (Fig. 8c). The most frequently recorded species from stores is the cosmopolitan *Lyctus brunneus* (Stephens). It is often found in farm stores in tropical countries.

## 20. Merophysidae

The only members of this family of fungivores belong to the genus *Holoparamecus*. These are small beetles: 1.0–1.5 mm long. The most common species found in stores, at least in the tropics, is *H. depressus* Curtis (Hinton 1941).

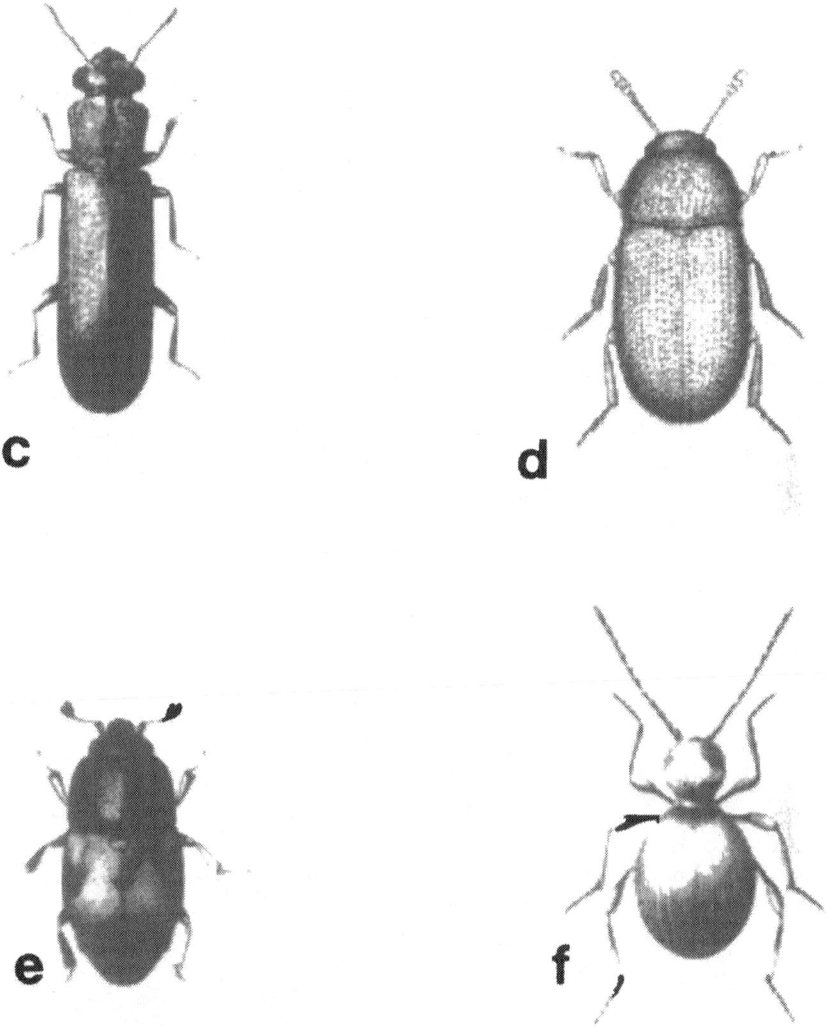
c
d
e
f

## 21. Mycetophagidae

Members of this fungivorous family typically live under tree bark and in rotting vegetation. The species most often associated with stored products is *Typhaea stercorea* (L.). It is a hairy, oblong beetle about 2.5–3.0 mm long (Fig. 8d). It has been found on a wide range of commodities in all parts of the world, but is most common in the humid tropics. Hinton (1945a) and Aitken (1975) give additional information.

## 22. Nitidulidae

Some 2000 species of nitidulids are known, most of which feed on the sap of trees and the juice of fruit. *Carpophilus* and *Urophorus* are the genera most often found in stores. The adults are slightly flattened ovate to oblong beetles, 2–5 mm in length and light brown to black. The elytra of *Carpophilus* are shortened, leaving two or three segments of the abdomen exposed; in some species, they are marked with characteristic yellow or red patches (Fig. 8e).

More than a dozen species of *Carpophilus* have been recorded from stores on a wide range of commodities. Many are field pests of growing and ripening crops that do not persist in well-dried commodities. Some species survive on cereal grains stored in humid areas, especially in the tropics, and are frequently found on maize stored on-farm by subsistence producers. Elsewhere, they may be found in store residues or on commodities with an intrinsically high moisture content, such as dried fruit.

Apart from *C. hemipterus* (L.), a distinctive species known as an important pest of dried fruit, most other species of storage *Carpophilus* are very similar in appearance, making precise identification difficult (see Dobson 1954a,b, 1960; Connell 1987). *C. dimidiatus* (F.) is cosmopolitan on cereals, oilseeds, cocoa, nuts, and several other commodities. In the tropics it is particularly associated with maize under conditions of rural storage (Haines 1981). *C. pilosellus* Motschulsky is found throughout the tropics; on cereals it is most often found on milled rice (Haines 1981). Other species found on stored products include *C. freemani* Dobson, *C. fumatus* Boheman, *C. maculatus* Murray, and *C. obsoletus* Erichson.

In the genus *Urophorus*, *U. humeralis* (F.) has been recorded on damaged maize, copra, dates, and dried fruit in tropical and subtropical regions. It is better known as an important field pest of pineapples.

## 23. Ptinidae

Adults of this family bear a superficial resemblance to spiders, hence their common name of "spider beetles." Most are covered with brown or golden hair (Fig. 8f) and are scavengers usually found under bark, in rotting logs, and in animal nests. In stores, ptinids are most often associated with poor conditions of hygiene, feeding on damp residues, corpses and excreta of insects and vertebrates. As pests they are most important in temperate regions.

A number of genera have been recorded in stores; however, the nomenclature of these insects is confused. The most important species is the cosmopolitan *Ptinus tectus* Boieldieu, which is most abundant in northwest Europe, but is still a relatively minor pest. Several other species of *Ptinus* are known from stores. Other widely distributed species, *Mezium americanum* (Laporte de Castelnau) and *M. attine* Boieldieu, are known as scavengers in stores. *Trigonogenius globulus* Solier and *T. particularis* Pic are known from temperate regions and highland areas in the tropics. Species of *Niptus* and other genera may occur in domestic premises. *Gibbium psylloides* (de Czenpinski) and *G. aequinoctiale* Boieldieu are unlike other ptinids, being dark reddish brown, hairless, and strongly shining. The former species is known from Europe, the latter from Asia and the Americas. Other genera known from stores include *Pseudeurostus*, *Stethomezium*, and *Tipnus*. For details of biology and identification of ptinids, consult Mound (1989), Howe (1991), and Spilman (1987a).

## 24. Scolytidae

Only a few species of this predominantly wood-boring family are associated with stored products. Many are superficially similar in appearance and habit to the Bostrichidae. Many have been reported attacking the timbers of wooden storage structures. Although these usually do not feed on the commodity, they may accidentally find their way into it. Only one species, *Pagiocerus frontalis* (F.), is known to breed regularly on stored grain, but several species of *Hypothenemus* have also been found in small numbers on stored products. *H. hampei* (Ferrari), is a well-known pest of coffee cherries and is frequently associated with coffee beans.

*P. frontalis* is a pest of farm-stored soft maize varieties grown and stored on-farm in the high Andes of South America. Elsewhere in South America and in the Caribbean it is known to breed on seeds of the avocado.

## 25. Silvanidae

### *Description*

Members of this family are generally 2–4 mm long, flattened, and parallel-sided (Fig. 9a,c,d). Most species have projections on the prothorax in the form of teeth or swellings at the front corners or several teeth along the sides.

Most species feed on plant material. Some live under bark and others are associated with wood-boring beetles. A few are associated with stored products as secondary pests; of these *Oryzaephilus surinamensis* (L.) and *O. mercator* (Fauvel) are the most important. Others, for example, *Ahasverus advena* (Waltl) and

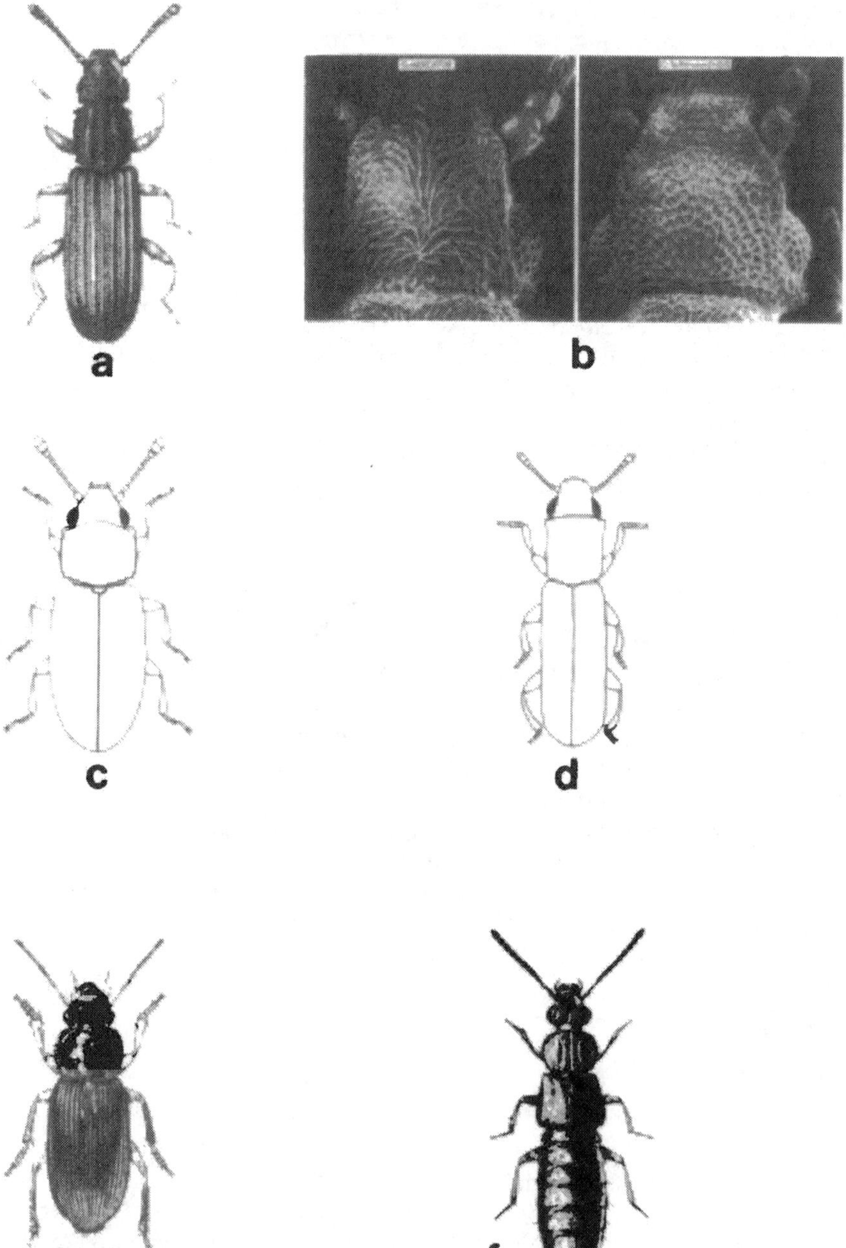
a
b
c
d
e
f

*Cathartus quadricollis* (Guerin), are frequently found in storage but are generally less important. Species found in stores can be identified using the keys of Halstead (1986 and 1993).

***Ahasverus advena.*** *A. advena* is usually slightly shorter (2–3 mm) but more robust than *Oryzaephilus*. On each front corner of the prothorax there is a blunt tooth (Fig. 9c). *A. advena* has a cosmopolitan distribution and is known from a wide range of foodstuffs, usually under conditions of high humidity, or when mold growth has occurred. It is, for example, a common inhabitant of milled rice in Southeast Asia.

***Cathartus quadricollis.*** *C. quadricollis* is a slim, light-brown parallel-sided beetle 2–3 mm long. The pronotum is square to rectangular in shape (Fig. 9d). It is a common minor pest of a wide range of commodities, including cereals, dried fruit, and cacao. In warm temperate and tropical areas, such as the southern United States, Central America, and West Africa, it attacks maize in the field and is very common in the tropics as a pest of maize stored on-farm by subsistence producers.

***Oryzaephilus* Species.** Adults of both species are slender, flattened, parallel-sided dark-brown beetles 2.5–3.5 mm long (Fig. 9a). The prothorax has six toothlike projections along each side (Fig. 9a). Both species are very alike; however, the side of the head behind the eye is much shorter in *O. mercator* than in *O. surinamensis* (Fig. 9b).

*Life Cycle*

Life cycles of these species are similar. Eggs are laid loosely on foodstuffs or in crevices when available. The campodeiform larvae pass through two to four larval instars. When feeding on grain, the germ is attacked with preference. The life cycle of *O. surinamensis* takes between 20 and 80 days at 17.5–37.5°C, 10–90% RH. The optimal conditions for development are 30–35°C, 70–90% RH. *O. surinamensis* is more tolerant than *O. mercator* to extremes of temperature and humidity (Howe 1956a) and can survive short periods at temperatures below 0°C.

*Distribution and Habitat*

Both species are very common cosmopolitan secondary pests of stored grain, cereal products, dried fruit, oilseeds, and other foodstuffs. *O. surinamensis* is more

---

**Figure 9** (a) *Oryzaephilus surinamensis* (Silvanidae): dorsal view (2.5–3.5 mm) (MAFF/CSL). (b) *Oryzaephilus* species: dorsal view of heads of *O. mercator* (left) and *O. surinamensis* (right) shows relative dorsal lengths of eye and temple (NRI). (c) *Ahasverus advena* (Silvanidae): dorsal view (2.0–3.0 mm) (MAFF/CSL). (d) *Cathartus quadricollis* (Silvanidae): dorsal view (2.5–3.0 mm) (MAFF/CSL). (e) *Harpalus rufipes* (Carabidae): dorsal view of typical carabid beetle (BM/NH). (f) *Oxytelus sculptus* (Staphylinidae): dorsal view of typical staphylinid (BM/NH).

often associated with cereals and their products and *O. mercator* with oilseeds and materials with a high oil content.

As well as being common in tropical regions both species, but especially *O. surinamensis*, are important pests in cool temperate areas. In the United Kingdom, for example, *O. surinamensis* is widely regarded as the most serious beetle pest of stored grain and is able to survive winter conditions without protection.

The long-lived adults often wander from foodstuffs into crevices, making them difficult to eradicate. They have also been found under bark on trees near stores. Their flattened shape makes it relatively easy for them to enter cracks in grains, nuts in shell, and products in packets and boxes (for example, biscuits, breakfast cereals, dried pet food) in commercial and domestic premises.

## 26. Staphylinidae

The Staphylinidae is a very large family of elongated parallel-sided beetles with characteristically very short elytra (Fig. 9f). Like the Carabidae, most are general predators and those found in stores are usually strays from the local fauna.

## 27. Tenebrionidae

*Description*

The Tenebrionidae is a very large family of beetles with some 10,000 known members, of which about 100 species have been found associated with stored products. Several of these are among the most widespread and important secondary pests of stored durable food products.

Adults of species found in stored products range from 3 to 10 mm long, are reddish-brown to black, flattened, and parallel-sided. Larvae are active, campodeiform, and well sclerotized (Fig. 1d).

Of the many tenebrionids associated with stored products, few are well adapted to life in well-dried commodities. By far the most important species are *Tribolium castaneum* (Herbst) and *T. confusum* J. du Val. Other species are, to a greater or lesser extent, more commonly associated with inadequately dried material or are little more than scavengers.

Infestation by tenebrionids often leads to persistent disagreeable odors remaining on the commodity as a result of excreted benzoquinones. Repeated consumption of such contaminated material probably constitutes a health risk to both humans and livestock.

Tenebrionidae found in stores can be identified using Halstead (1986), Haines (1991), and Spilman (1987b).

***Alphitobius* Species.** *Alphitobius* spp (Fig. 10a) are reddish brown to black beetles, 5–8 mm long, and more ovoid than *Tribolium* (Fig. 11a). Two species are typically found in stores. *A. diaperinus* (Panzer) is cosmopolitan and will feed on grain, cereal products, and animal feeds, especially damp, moldy residues. In many temperate countries it is also a serious pest in deep-litter poultry houses. The similar *A. laevigatus* (F.) is also widely found in moldy commodities but not in poultry houses. The two species can be separated by the eye characteristic shown in Fig. 10b and c.

***Latheticus oryzae* Waterhouse.** *L. oryzae* is typically tenebrionid in shape and 2.7–3.0 mm long. However, the terminal segment of the antenna is distinctly narrower than other segments (Fig. 10e); in other genera it is the same size or wider. The adult is slimmer in appearance than *Tribolium* and usually paler in color (Figs. 10d, 11a,b).

*L. oryzae* is found throughout the tropics and subtropics, but is especially common in warm, humid areas of South and Southeast Asia. Its apparent inability to breed at temperatures below 25°C probably confines it to such areas.

Compared to *Tribolium*, it is a minor secondary pest of cereals and cereal products, for example, rice and wheat. It thrives best on diets with a low oil content.

***Gnatocerus* Species.** *Gnatocerus* species resemble *Tribolium* in form and color (Figs. 10f, 11a,b). Males have obvious mandibular horns, large in *G. cornutus* (F.) and smaller in *G. maxillosus* (F.) (Fig. 10g); these are absent in females. Both genders can be differentiated from *Tribolium* spp by the shape of the prosternal process, between the front pair of legs (Fig. 11c,d).

*Gnatocerus* spp are common secondary pests of cereals, oilseeds, and sometimes other foodstuffs. Their optimal rate of increase is lower than that of *Tribolium* species, hence they are usually less serious pests. *G. cornutus* is found throughout warm temperate and tropical regions and is a common pest of flour mills in warm temperate regions, such as Australia and the southern United States. *G. maxillosus* is largely confined to the tropics and is a common pest of maize cobs under conditions of subsistence agriculture.

***Palorus* Species.** Adults of this genus are of typical tenebrionid shape, reddish-brown, and 2.5–3.0 mm long (Fig. 12c). They can be differentiated from the other major genera by their smaller size and undivided eye (Fig. 12a,b). A number of species have been recorded as relatively minor secondary pests, the most common of which are *P. subdepressus* (Wollaston) and *P. ratzeburgi* (Wissmann). The former has a mostly tropical distribution, the latter is cosmopolitan. *Palorus* spp are most often associated with grain and grain products stored under hot humid conditions and are common in the farm-stored grain of tropical subsistence producers.

***Tenebrio* Species.** Two species, *T. molitor* L. and *T. obscurus* F., are the most frequently recorded from stores. They are very resistant to cold conditions and are mostly found in temperate areas. They are black, parallel-sided beetles about 10 mm long, and feed on a wide range of dried animal and vegetable materials.

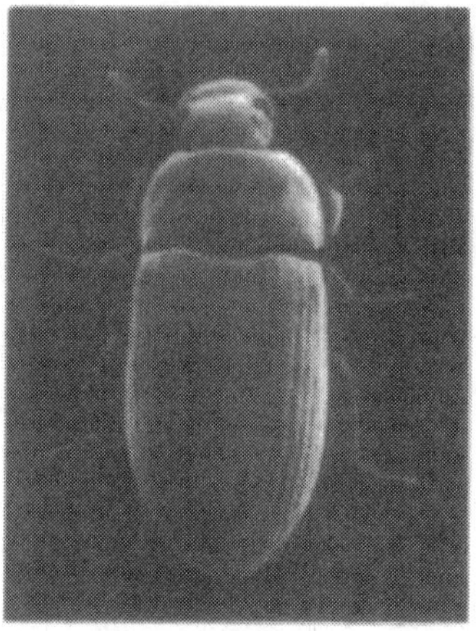

a

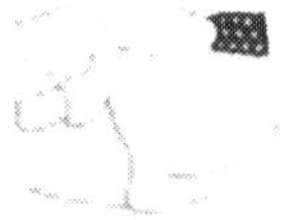

b

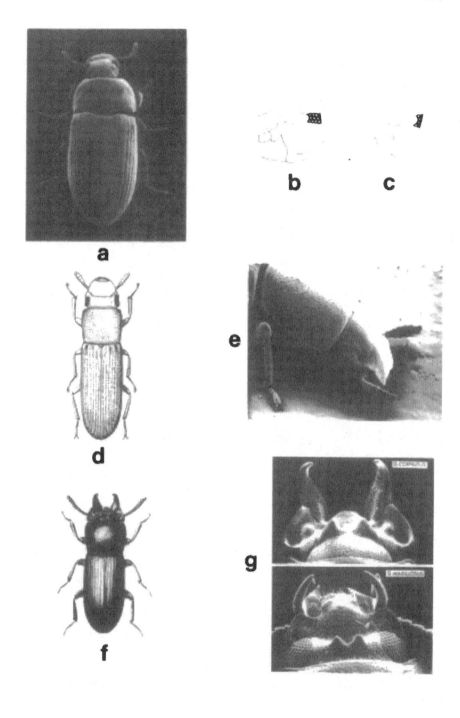

c

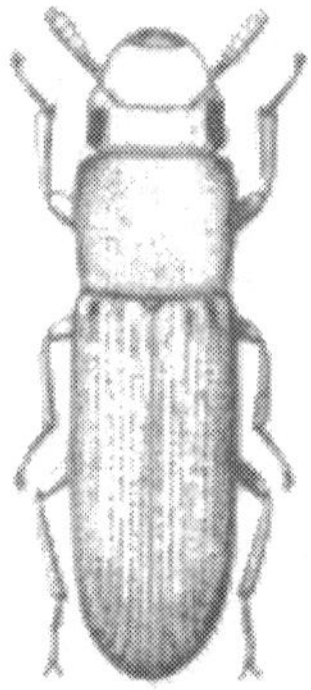

d

e

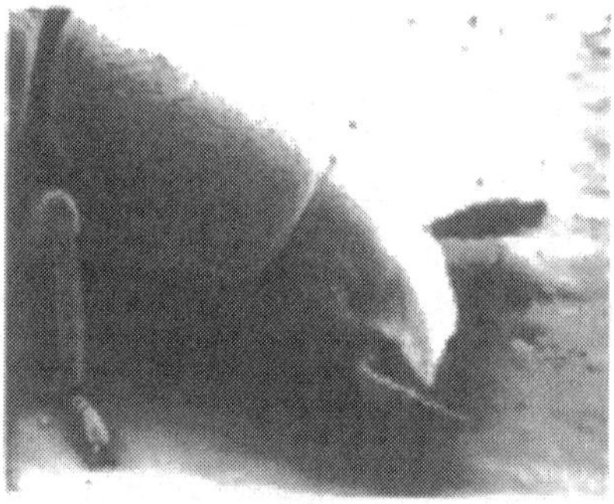

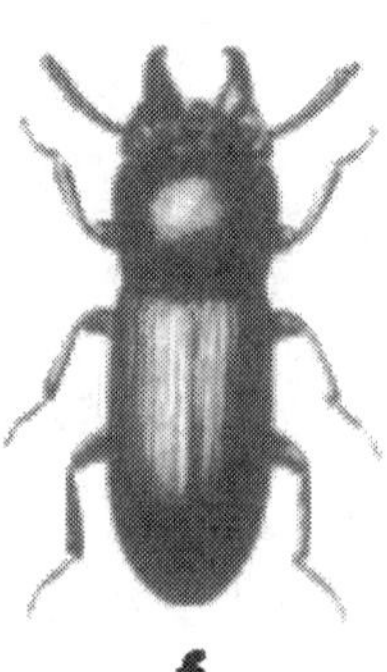

f

g

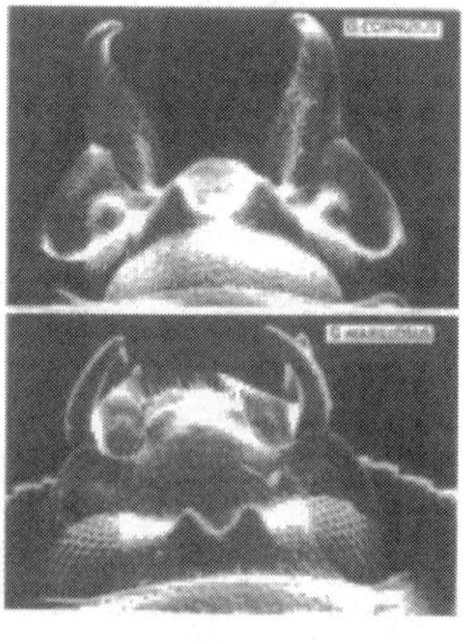

Usually they act as scavengers and are not serious pests because their optimal rate of increase is very low.

***Tribolium* Species.** Several species of *Tribolium* have been recorded in stores, but by far the most important are *T. castaneum* and *T. confusum*. Both are cosmopolitan, *T. castaneum* being most common in tropical to warm-temperate areas, while *T. confusum* is uncommon in tropical regions but common and widespread in temperate zones. Both are 3–4 mm long and reddish brown (Fig. 11a,b). They can be differentiated by the distance between the eyes, which is narrow in *T. castaneum* and wide in *T. confusum*, when viewed ventrally (Fig. 11e,f).

In most regions of the world one or the other species is a very important, or the most important, secondary pest of a wide range of stored products. These include almost every kind of dried animal and plant product, but especially cereals and cereal products. They do not breed rapidly on undamaged clean grain, but are very troublesome in flour, milled cereals, animal feeds, and processed foods.

Adult *T. castaneum* can live for many months, even several years, under temperate conditions. Females lay 2–10 eggs each day throughout most of their lives. Development can be very rapid; the life cycle can be completed in about 21 days under optimal conditions of 35°C, 75% RH, and is possible between 22 and 40°C. Optimal, maximum, and minimum temperatures for the development of *T. confusum* are about 2.5°C lower than those for *T. castaneum*. The effect of temperature and humidity on development was investigated by Howe (1956b, 1960, 1962). Cannibalism plays an important part in the diet of these insects and they will also prey on the eggs, young larvae, and pupae of other storage insects.

Under optimal conditions *T. castaneum* populations can increase at a rate of up to 70–100 times a month, faster than that recorded for any other storage pest. Both species can fly but *T. castaneum* is more ready to do so, especially under tropical conditions. Both can readily disperse and search for new food resources without human help. As a result, they are often one of the first insects to recolonize a commodity after fumigation.

There is a vast amount of literature on these species. Much of this is reviewed by Sokoloff (1972, 1974, 1977).

---

**Figure 10** (a) *Alphitobius diaperinus* (Tenebrionidae): dorsal view (5.5–7.0 mm) (NRI). (b) *Alphitobius diaperinus*: lateral view of head with some eye facets shown (NRI). (c) *Alphitobius laevigatus* (Tenebrionidae): lateral view of head with some eye facets shown (NRI). (d) *Latheticus oryzae* (Tenebrionidae): dorsal view (2.7–3.0 mm) (BM/NH). (e) *Latheticus oryzae*: lateral view of head (NRI). (f) *Gnatocerus cornutus* (Tenebrionidae): dorsal view of male (3.5–4.9 mm) (MAFF/CSL). (g) *Gnatocerus* species: dorsal view of heads of *G. cornutus* (top) and *G. maxillosus* (bottom) shows shape and proportions of mandibular horns (NRI).

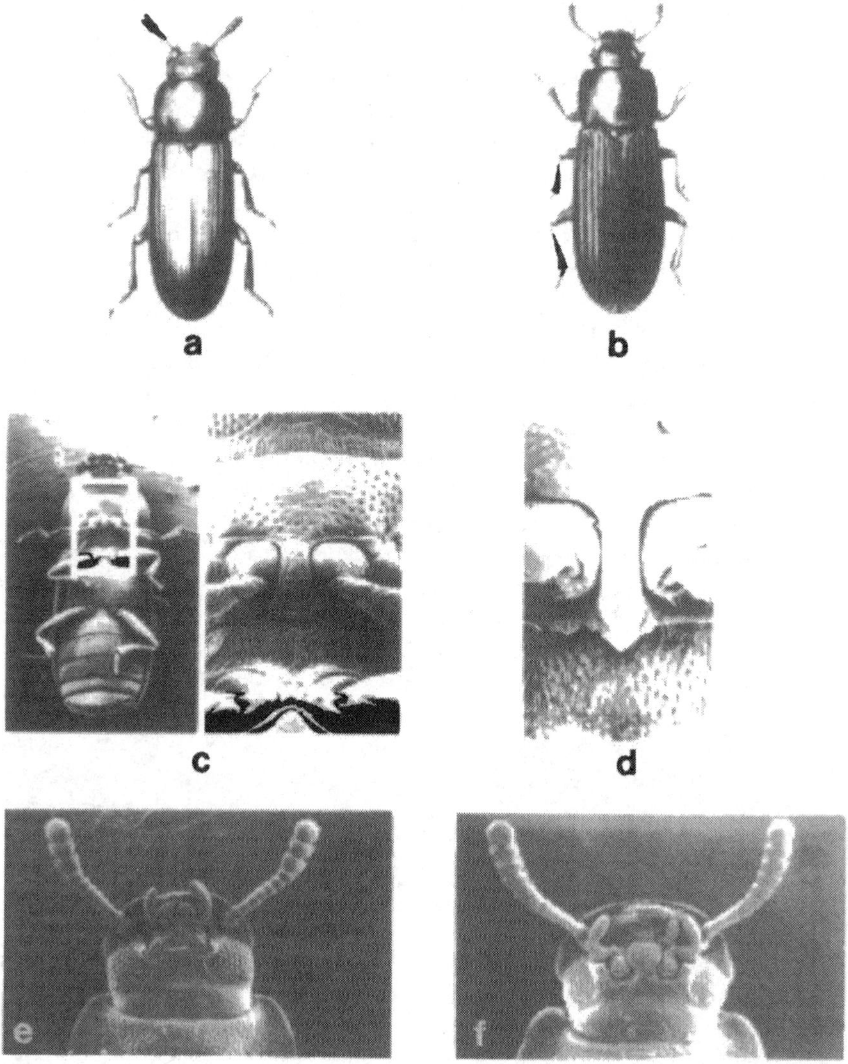


**Figure 11** (a) *Tribolium castaneum* (Tenebrionidae): dorsal view (2.3–4.4 mm) (MAFF/CSL). (b) *Tribolium confusum* (Tenebrionidae): dorsal view (2.3–4.4 mm) (MAFF/CSL). (c) *Tribolium* species: ventral view of body (left) with position of prosternal process shown in square and enlargement (right) (NRI). (d) *Gnatocerus* species (Tenebrionidae): prosternal process (NRI). (e) *Tribolium castaneum*: ventral view of head (NRI). (f) *Tribolium confusum*: ventral view of head (NRI).

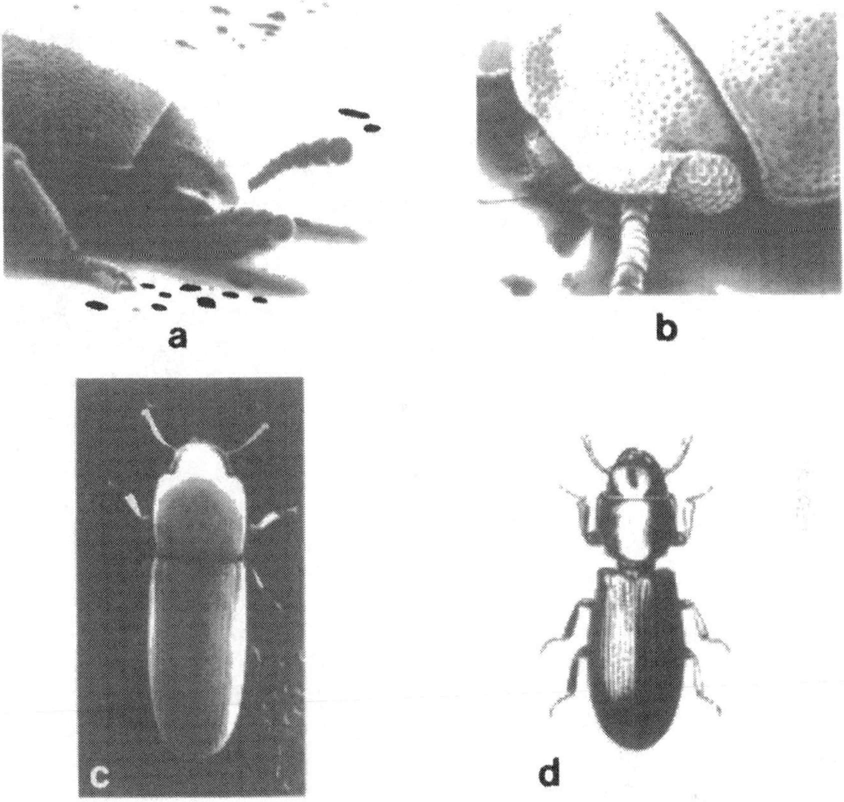


**Figure 12** (a) *Tribolium castaneum* (Tenebrionidae): lateral view of head (NRI). (b) *Palorus subdepressus* (Tenebrionidae): lateral view of head (NRI). (c) *Palorus subdepressus*: dorsal view (2.7–3.0 mm) (NRI). (d) *Tenebroides mauritanicus* (Trogossitidae): dorsal view (5.0–11.0 mm) (MAFF/CSL).

Several other species of *Tribolium* are associated with stored products. *T. destructor* Uyttenboogaart has been recorded from Europe and in cool highland areas in Africa and western Asia. An African species, *T. anaphe* Hinton, has been recorded mainly from cotton seed but also from cocoa beans and palm kernels. *T. madens* (Charpentier) has been found in grain stores in Europe and North Africa. The very similar species, *T. audax* Halstead, is apparently restricted to North America. There it is sometimes a serious pest in stored cereals and cereal products.

### 28. Trogossitidae

The only trogossitid known to occur frequently in stored products is *Tenebroides mauritanicus* (L.). It is a flattened, glossy black beetle, about 10 mm long and parallel sided with a "waist" or constriction between the prothorax and the abdomen (Fig. 12d). It has a cosmopolitan distribution and is a minor pest of stored products, especially cereals and oilseeds. It is most often found in residues and poorly dried commodities. Larvae will burrow into wooden structures in search of a site to pupate.

## ACKNOWLEDGMENTS

The author acknowledges the collaboration of colleagues and in particular, Dr. C. P. Haines, at the Natural Resources Institute (NRI), Chatham, England, and the inclusion of material drawn from the NRI publication by Haines (1991). The author is grateful to Dr. H. J. Banks for the opportunity to write this chapter as a visiting scientist in the CSIRO Division of Entomology, Canberra, Australia.

Permission to reproduce illustrations originating elsewhere has been given by the Trustees of the British Museum (Natural History) (BM/NH in the figure legends), the Ministry of Agriculture, Fisheries and Food's Central Science Laboratory (MAFF/CSL in the legends); the Ohio Biological Survey; Dr. D. G. H. Halstead; and Dr. R. E. White.

## REFERENCES

Aitken, A. D. (1975). Insect Travellers. Volume 1: Coleoptera. MAFF Technical Bulletin No. 31. London: HMSO.

Anderson, D. A. (1987). Larval beetles (Coleoptera). in Gorham, J. R. (ed.) (1987), Vol. 1:95–113.

Banks, H. J. (1977). Distribution and establishment of *Trogoderma granarium* Everts (Coleoptera: Dermestidae): Climatic and other influences. *J. Stored Prod. Res.*, *13*:183–202.

Banks, H. J. (1979). Identification of stored product *Cryptolestes species* (Coleoptera: Cucujidae): A rapid technique for preparation of suitable mounts. *J. Aust. Ent. Soc.*, *18*:217–222.

Banks, H. J. (1994). Illustrated identification keys for *Trogoderma granarium*, *T. glabrum*, *T. inclusum*, and *T. variable* (Coleoptera: Dermestidae) and other Trogoderma associated with stored products. CSIRO Division of Entomology Technical Paper No. 32, CSIRO, Canberra.

Beal, R. S. Jr. (1954). Biology and taxonomy of the nearctic species of *Trogoderma* (Coleoptera: Dermestidae). *Univ. California Pub. Ent.*, *10*(2):35–102.

Beal, R. S. Jr. (1956). Synopsis of the economic species of *Trogoderma* occurring in the United States with a description of a new species (Coleoptera: Dermestidae). *Ann. Ent. Soc. America*, *49*(6):559–566.

Beal, R. S. Jr. (1960). Description, biology and notes on the identification of some *Trogoderma* larvae (Col.: Dermestidae). Tech. Bull. 1228. Washington D.C.: US Dept. of Agriculture.

Beal, R. S. Jr. (1970). A taxonomic and biological study of species of Attegenini (Coleoptera: Dermestidae) in the United States and Canada. *Ent. Am.*, *45*(3):141–235.

Buckland, P. C. (1981). The early dispersal of insect pests of stored products as indicated by archeological records. *J. Stored Prod. Res.*, *17*(1):1–12.

Collier, D. J. (1981a). Identification of adult Coleoptera found in stored products. Proceedings of the Australian Development Assistance Course on the preservation of stored cereals, CSIRO, Canberra, Vol. 1: 70–95.

Collier, D. J. (1981b). Identification of larval Coleoptera found in stored products. Proceedings of the Australian Development Assistance Course on the preservation of stored cereals, CSIRO, Canberra, Vol. 1: 96–107.

Connell, W. A. (1987). Sap beetles (Nitidulidae, Coleoptera). in Gorham, J. R. (ed.) (1987), Vol. 1: 151–174.

Coombs, C. W., Billings, C. J., and Porter, J. E. (1977). The effect of yellow split peas (*Pisum sativum* L.) and other pulses on the productivity of *Sitophilus oryzae* (L.) (Col. Curculionidae) and the ability of other strains to breed thereon. *J. Stored Prod. Res.*, *13*:53–58.

Dobson, R. M. (1954a). The species of *Carpophilus* Stephens (Col., Nitidulidae) associated with stored products. *Bull. Ent. Res.*, *45*(2):389–402.

Dobson, R. M. (1954b). A new species of *Carpophilus* Stephens (Col., Nitidulidae) found on stored produce. *Entomol. Monthly Mag.*, *90*:299–300.

Dobson, R. M. (1960). Notes on the taxonomy and occurrence of *Carpophilus* Stephens (Col., Nitidulidae) associated with stored products, *Entomol. Monthly Mag.*, *95*: 156–158.

Evans, D. E. (1981). The biology of stored product Coleoptera. Proceedings of the Australian Development Assistance Course on the preservation of stored cereals, CSIRO, Canberra, Vol. 1: 149–185.

Fisher, W. S. (1950). A revision of the North American species of the beetles belonging to the family Bostrichidae. Miscellaneous publications of the United States Dept. of Agriculture, No. 698.

Gorham, J. R. (ed.) (1987). Insect and mite pests of food: An illustrated key. U.S. Department of Agriculture handbook, No. 655.

Gorham, J. R. (ed.) (1991). Ecology and management of food industry pests. FDA Technical bulletin No. 4, Association of Official Analytical Chemists, Arlington, Virginia.

Green, M. (1979). *Cryptolestes klapperichi* Lefkovitch in stored products and its identification (Coleoptera: Cucujidae). *J. Stored Prod. Res.*, *15*:71–72.

Haines, C. P. (1974). Insects and arachnids from stored products: a report on specimens received by the Tropical Stored Products Centre 1972–1973. Tropical Products Institute Publication No. L39, London, UK.

Haines, C. P. (1981). Insects and arachnids from stored products: A report on specimens received by the Tropical Stored Products Centre 1973–1977. Tropical Products Institute Publication No. L54, London, UK.

Haines, C. P. (1989). Observations on *Callosobruchus analis* (F.) in Indonesia, including a key to storage *Callosobruchus* spp. (Coleoptera, Bruchidae). *J. Stored Prod. Res.*, *25*:9–16.

Haines, C. P. (ed.) (1991). Insects and arachnids of tropical stored products: Their biology and identification, 2nd Edition. Natural Resources Institute, Chatham, Kent.

Haines, C. P., and Rees, D. P. (1989). A field guide to the types of insects and mites infesting cured fish. FAO Fisheries technical paper No. 303, FAO, Rome, Italy.

Halstead, D. G. H. (1969). A key to the species of *Carcinops* Warseul (Coleoptera, Histeridae) associated with stored products, including *C. troglodytes* (Paykull) new to this habitat. *J. Stored Prod. Res.*, *5*:83–85.

Halstead, D. G. H. (1986). Keys for the identification of beetles associated with stored products. I—Introduction and keys to families. *J. Stored Prod. Res.*, *22*(4):163–203.

Halstead, D. G. H. (1993). Keys for the identification of beetles associated with stored products. II–Laemophloeidae, Passandridae and Silvanidae. *J. Stored Prod. Res.*, *29*:99–197.

Hinton, H. E. (1941). The Lathridiidae of economic importance. *Bull. Ent. Res.*, *32*:191–247.

Hinton, H. E. (1945a). A monograph of the beetles associated with stored products. Volume 1. London: British Museum (Natural History). Reprinted in 1963 by Johnson Reprint Co., London.

Hinton, H. E. (1945b). The Histeridae associated with stored products. *Bull. Ent. Res.*, *35*:309–340.

Howe, R. W. (1956a). The biology of the two common storage species of *Oryzaephilus* (Coleoptera; Cucujidae). *Ann. App. Biol.*, *44*(2):341–355.

Howe, R. W. (1956b). The effect of temperature and humidity on the rate of development and the mortality of *Tribolium castaneum* (Herbst) (Coleoptera, Tenebrionidae). *Ann. App. Biol.*, *44*(20):356–368.

Howe, R. W. (1957). A laboratory study of the cigarette beetle *Lasioderma serricorne* (F.)(Col. Anobiidae) with a critical review of the literature. *Bull. Ent. Res.*, *48*:9–56.

Howe, R. W. (1960). The effect of temperature and humidity on the rate of development and the mortality of *Tribolium confusum* Duval (Coleoptera, Tenebrionidae). *Ann. Appl. Biol.*, *48*(2):363–376.

Howe, R. W. (1962). The effects of temperature and humidity on the oviposition rate of *Tribolium castaneum* (Hbst) (Coleoptera, Tenebrionidae). *Bull. Ent. Res.*, *53*:301–310.

Howe, R. W. (1991). Spider beetles: Ptinidae. in Gorham, J. R. (ed.) (1991), pp. 177–180.

Kingsolver, J. M. (1987a). Adult beetles (Coleoptera). in Gorham, J. R. (ed.) (1987), Vol. 1:75–94.

Kingsolver, J. M. (1987b). Dermestid beetles (Dermestidae, Coleoptera). in Gorham, J. R. (ed.) (1987), Vol. 1:115–117.

Kingsolver, J. M. (1987c). Cryptophagid beetles (Cryptophagidae, Coleoptera). in Gorham, J. R. (ed.) (1987), Vol. 1:175–179.

Kingsolver, J. M. (1987d). Seed Beetles (Bruchidae, Coleoptera). in Gorham, J. R. (ed.) (1987), Vol. 1:215–221.

Kingsolver, J. M., and Andrews, F. G. (1987). Minute brown scavenger beetles (Lathridiidae, Coleoptera). in Gorham, J. R. (ed.) (1987), Vol. 1:179–184.

Lefkovitch, L. P. (1967). A laboratory study of *Stegobium paniceum* (L.) (Coleoptera: Anobiidae). *J. Stored Prod. Res.*, *3*:235–249.

Mound, L. (1989). Common Insect pests of stored food products. Economic Series No. 15, 7th edition. London, British Museum (Natural History).

Mphuru, A. N. (1974). *Araecerus fasciculatus* Degeer (Coleoptera: Anthribidae): A review. *Tropical Stored Prod. Info.*, *26*:7–15.

Panagiotakopulu, E., and Buckland, P. C. (1991). Insect pests of stored products form late bronze age Santorini, Greece. *J. Stored Prod. Res.*, *27*(3):179–184.

Peacock, E. R. (1979). *Attagenus smirnovi* Zhantiev (Coleoptera: Dermestidae) a species new to Britain, with keys to the adults and larvae of the British *Attagenus*. *Entomol. Gazette*, *30*(2):131–136.

Peacock, E. R. (1993). Adults and larvae of hide, larder and carpet beetles and their relatives (Coleoptera: Dermestidae) and of derodontid beetles (Coleoptera: Derodontidae). Handbooks for the identification of British Insects, Vol. 5, Part 3. Royal Entomological Society of London, London.

Rees, D. P., Riveva, R. R., and Herrera, F. J. (1990). Observations on the ecology of *Teretriosoma nigrescens* (Col.: Histeridae) and its prey *Prostephanus truncatus* (Horn)(Col.: Bostrichidae) in Yucatan peninsula, Mexico. *Trop. Sci.*, *30*:153–165.

Sokoloff, A. (1972). The biology of *Tribolium* with special emphasis on genetic aspects. Volume 1. London: Oxford University Press.

Sokoloff, A. (1974). The biology of *Tribolium* with special emphasis on genetic aspects. Volume 2. London: Oxford University Press.

Sokoloff, A. (1977). The biology of *Tribolium* with special emphasis on genetic aspects. Volume 3. London: Oxford University Press.

Spangler, P. J. (1961). Notes and pictorial key for separating Khapra beetle (*Trogoderma granarium*) larvae from all other Neartic species of the genus. *Coop. Econ. Insect Rep.*, *11*(6):61–62.

Spilman, T. J. (1987a). Spider beetles (Ptinidae, Coleoptera). in Gorham, J. R. (ed.) (1987), Vol. 1:137–147.

Spilman, T. J. (1987b). Darkling beetles (Tenebrionidae, Coleoptera). in Gorham, J. R. (ed.) (1987), Vol. 1:185–214.

Woodroffe, G. E., and Coombs, C. W. (1961). A revision of the North American Cryptophagus Herbst (Coleoptera: Cryptophagidae). Miscellaneous Publication, Entomological Society of America, 2:179–211.

# 2

# Lepidoptera and Psocoptera

**John D. Sedlacek, Paul A. Weston, and Robert J. Barney**
*Kentucky State University, Frankfort, Kentucky*

Approximately 70 species of moths, primarily belonging in the families Pyralidae, Tineidae, Oecophoridae, and Gelechiidae, are associated with infestations of stored products (Cox and Bell 1991). However, only the almond moth (*Cadra cautella*, Pyralidae), Mediterranean flour moth (*Ephestia kuehniella*, Pyralidae), tobacco moth (*E. elutella*, Pyralidae), raisin moth (*C. figulilella*, Pyralidae), Indianmeal moth (*Plodia interpunctella*, Pyralidae) (i.e., flour moths), and the Angoumois grain moth (*Sitotroga cerealella*, Gelechiidae) (i.e., a grain moth) are considered widely distributed and major pests of stored foods (Cox and Bell 1991).

Grain and grain products occasionally are infested with very small insects known as Psocoptera or psocids. They are widely distributed in North America and Europe. Psocids feed on a variety of organic matter and are considered pests mainly because of their presence and not the damage they cause.

This chapter is organized into two sections. The first discusses stored-product moth pests and is further broken down into primary and secondary colonizing moths. The second section deals briefly with stored-product psocid pests. Our objective is to familiarize the reader with sight identification, biology, life history, and habits of stored-product Lepidoptera and Psocoptera. Confirmation of species must be accomplished by chaetotaxy for larvae and dissection and preparation of genitalia for adults. We refer those requiring detailed taxonomic information on the identification of families, genera, and species to works by Corbet and Tams (1943), Aitken (1963), Goater (1986), Mound (1989), Neunzig (1990), Gentry et al. (1991), and USDA (1991).

## I. LEPIDOPTERA

### 1. Basic Structure and Life Cycle

Stored-product moths are common insects characterized by small overlapping scales on two pairs of large overlapping wings. Adults are usually dull colored and small, with a wingspan of 2.5 cm or shorter. Their bodies and legs are also covered with scales, which rub off easily when touched. The biting and chewing mouthparts of adults have been lost during evolutionary time and replaced with a long coiled, suctorial proboscis.

The damaging stage is the larval or caterpillar stage. Larval mouthparts include mandibles that are used for chewing. In some groups, modified salivary glands produce silk used primarily for making the pupal cocoon. Larvae have well-developed head capsules, three pairs of thoracic legs terminating in claws, three to six pairs of abdominal prolegs, and a pair of terminal claspers. The prolegs bear rows of tiny hooks called crochets, which are used for gripping. There are typically five to six larval instars. Each body segment has a lateral spiracle, and the arrangement of setae and bristles is quite specific and can be used for identification. The larvae of stored-product moths are white to beige-orange and are concealed by the substrate in which they reside.

Moths are holometabolous, meaning that they undergo complete metamorphosis (i.e., egg, larval, pupal, and adult stages). Eggs are globular or spindle-shaped (i.e., broad at middle and narrow at ends), upright, and distinctively sculptured, or flattened and ovoid. They can be laid singly or in groups and are usually stuck firmly to the substrate. The pupal stage is usually concealed inside a cocoon attached to the substrate or to packaging material, especially the surface of grain sacks.

### 2. Pest Status/Nature of the Problem

In addition to causing damage by feeding directly on sound grain (Fig. 1), stored-product moth pests produce silk that may bind kernels together and cause mechanical damage to various types of food industry machinery. *Ephestia* spp, *Plodia interpunctella*, and *Corcyra cephalonica* produce especially dense webbing (Fig. 2). The almond moth, *C. cautella*, is second only to the coleopteran *Tribolium castaneum* in the amount of damage caused to stored products. *Plodia interpunctella* is the most widely distributed of all moths infesting stored foods. In tropical and subtropical regions of the Old World, the most devastating pest in mills is the rice moth, *C. cephalonica*. In temperate climates, the main pest of flour mills is the Mediterranean flour moth, *E. kuehniella*. The most serious moth pest of grain in the New World and Africa is *Sitotroga cerealella*. In fact, in central Kentucky, *S. cerealella* has recently emerged as the most significant moth pest of on-farm stored shelled maize (Sedlacek, unpublished data). Whenever grain

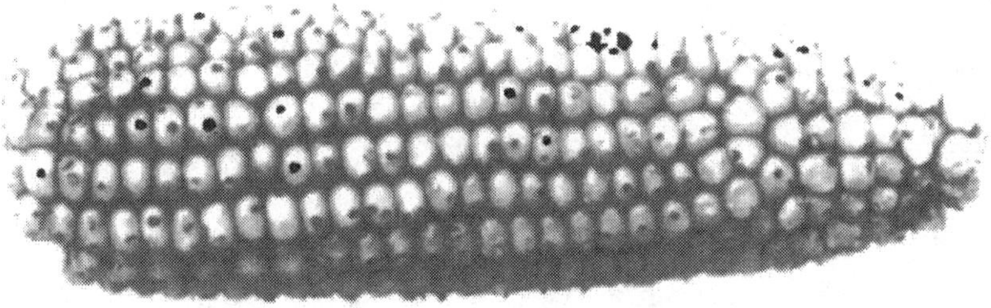

**Figure 1** *Sitotroga cerealella* damage to corn. (From Anonymous 1986.)

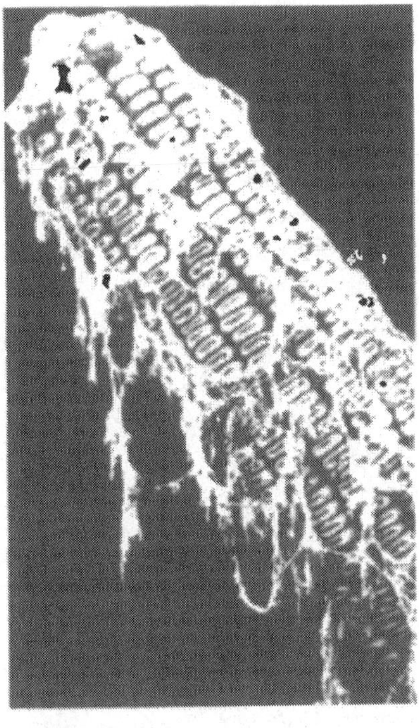

**Figure 2** *Plodia interpunctella* webbing on corn. Dense webbing such as this and that found on shelled grain and/or bin walls may cause clogging and damage to machinery. (From Anonymous 1986.)

is stored for extended periods, such as in Europe during the 1940s, the most serious moth pest is the tobacco moth, *E. elutella*.

The moth species described in this chapter have successfully adapted to storage environments. The wide range of food types for these moths and their adaptability to storage environments, where large concentrations of food are accumulated, are undoubtedly the reasons for their success.

## 3. Primary Colonizers

Several species of Lepidoptera are capable of infesting intact grain. The most widely distributed species is *Sitotroga cerealella* (Olivier), commonly known as the Angoumois grain moth. Because of the importance of *S. cerealella* as a pest of stored grains, most details will be provided for this insect, with notable exceptions for the other species detailed in the following sections.

### *Angoumois Grain Moth: Sitotroga cerealella* (Olivier)

**Description.** *Sitotroga cerealella* (family Gelechiidae) is a small (1.5 cm long) moth not easily confused with other stored-grain lepidoptera. The buff-colored wings are fringed and are sometimes lightly mottled (Fig. 3a). The larvae (Fig. 3b) are rarely seen, because they are strictly internal feeders.

**Distribution.** *Sitotroga cerealella* is found worldwide in a variety of stored grains including corn (maize), wheat, rice, sorghum, and millet. Although it has been reported from northern latitudes, this pest is more common in temperate to tropical regions. *Sitotroga cerealella* is particularly effective as a primary colonizer because it is highly mobile and has very flexible nutritional requirements. It has been found in sites far removed from grain storage facilities (Cogburn and Vick 1981) and is capable of infesting grains both prior to harvest and in storage.

**Ecology.** *Sitotroga cerealella* begins as a translucent, white egg approximately 2 mm long. Eggs are laid in crevices and initially conform to the contour of the oviposition site. Eggs turn pink and hatch after 4–8 days (Koone 1952). After eclosing, larvae usually spin a silken entrance cocoon and bore into the grain. Once inside, larvae excavate a cylindrical chamber that is enlarged as they grow. A larva usually completes development through its four instars within a single grain kernel, and larger grains may support the development of several larvae. Prior to pupating, the larva excavates seed tissue just up to the cuticle, leaving a transparent, circular "window" that is diagnostic of the pest (Fig. 1). Larvae then cut the cuticle partially around the circumference of the window, leaving a flap that is easily pushed aside by the emerging adult.

Successful development from egg to adult occurs under a wide variety of conditions. Virtually any stored grain will support growth of *S. cerealella* larvae. In addition, larvae successfully colonize many grains prior to harvest. Infestation in maize, for example, may occur when grain is in the "milk" stage, although

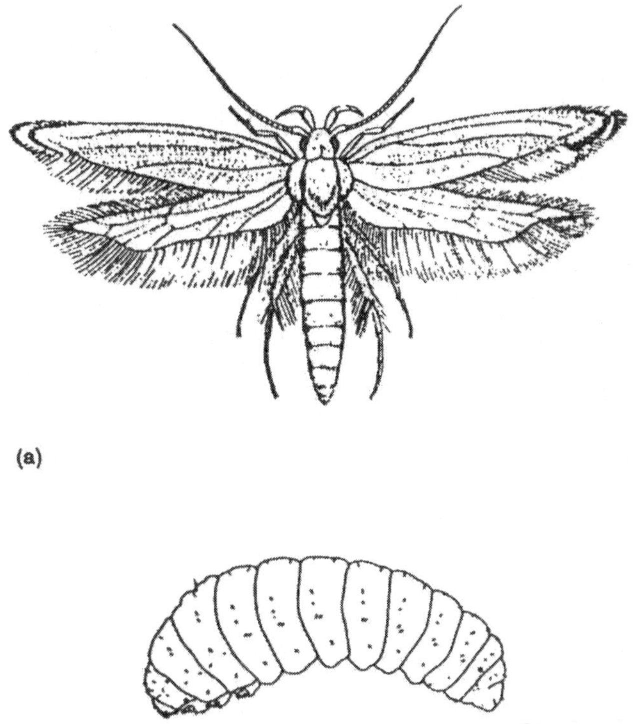

(a)

(b)

**Figure 3** (a) Adult *Sitotroga cerealella* are buff-colored with fringed wings. (b) Larvae are internal feeders. (a, From Hill 1990, reprinted with permission; b, From USDA 1991.)

survivorship is lower than that on mature corn. *Sitotroga cerealella* also completes development on a variety of wild grasses and other plants with seeds of sufficient size. The primary prerequisite for development appears to be a firm substrate into which larvae can burrow; Ayertey (1982) found that larvae did not survive on finely ground maize.

Developmental rate varies with grain type, temperature, and grain moisture content. Development takes approximately 35 days, but is faster on smaller grains. The fastest time reported for development from egg to adult is 21 days on sorghum (Shazali and Smith 1985). The shorter development time is probably related to the smaller size achieved on smaller grains. The optimal temperature for development is 28–30°C, with little growth occurring at temperatures below 17°C or above 36°C. Developmental rate typically decreases as moisture content of grain decreases. Development has been reported under extreme conditions to take as long as 367 days (Candura 1954).

Survivorship of immature and adult *S. cerealella* also is influenced by temperature and humidity (grain moisture content). Egg-to-adult survivorship on sorghum decreased as temperature increased over the range of 25–35°C and increased as relative humidity increased from 40–80% (Shazali and Smith 1985). These differences were mediated by differential egg hatch as well as differential ability of larvae to develop on sorghum under the various conditions. Adult longevity is prolonged markedly if water is available. Adults denied access to free water lived an average of 6–10 days; when water was provided, this duration was approximately doubled, with one female living for 52 days at approximately 16.5°C.

Temperature limits of *S. cerealella* no doubt account for its abundance at middle and low latitudes and its relative scarcity at higher latitudes. In the United States, for example, the insect is a serious pest of stored maize in the southern states, but is rarely a problem in northern states. In the Midwest, the dividing line bisects Indiana; Russell (1962) found *S. cerealella* infesting maize only in the southernmost counties of the state. Simmons and Ellington (1933) reported that only late instar larvae are capable of surviving average winter temperatures near freezing. It is assumed that survival of this life stage also decreases as winter temperatures decrease, limiting occurrence of this insect to lower and middle latitudes.

The nutritional requirements of *S. cerealella* have been fairly well established. Although larvae complete development on grains devoid of the germ, development is faster and female weight is greater when the germ is present (Mills 1965). When reared on artificial diets, larvae also grew optimally when wheat germ was included (Chippendale 1971). Besides nutritional factors, wheat germ also contains substances that elicit arrestment and feeding (Chippendale and Mann 1972).

Females mate soon after adult eclosion and may mate more than once. Oviposition begins shortly after the first mating, often within the first day of adult life. Most eggs are laid within 4 days after oviposition begins. Mating is not required for oviposition, although eggs laid by unmated females are not viable. Production of a sex pheromone, (Z-E)-7,11-hexadecadien-1-ol acetate, by females assists males in locating mates. Average fecundity is 100–150 eggs/female; the largest number of eggs recorded for a single female is 389 (Simmons and Ellington 1933).

Gravid females are not particularly choosy when depositing eggs. When ovipositing in grain, females typically deposit eggs in crevices formed between adjacent grains or between the grain and associated plant tissue (e.g., the glume of maize kernels). Virtually any crevice of the appropriate dimensions, however, will stimulate oviposition, as demonstrated in choice tests in the laboratory (Weston, unpublished data) and by the commonly noted observation that eggs can be collected by placing strips of tightly folded paper into cages housing gravid females.

Because selection of a site for oviposition is apparently very lax in this species, it is particularly important for females to locate accurately the appropriate habitat for oviposition before depositing eggs. Host location is not a major concern for insects emerging from grain in a storage structure, but might be important for females in field situations.

Although it is well documented that *S. cerealella* infests grain in the field prior to harvest, no literature exists documenting the processes by which gravid females locate appropriate host plants for oviposition. Pheromone trapping has revealed that male *S. cerealella*, whose flight activity in the field peaks near dusk, are more abundant in the vicinity of plots of maize than in neighboring open fields (Weston, unpublished data), but no studies have documented the movement of females in a field situation. Simmons and Ellington (1933) suggested that moths move from storage structures to fields of ripening wheat in summer, but provided no evidence that such migration actually occurs.

Besides infesting cultivated crops, *S. cerealella* successfully develops in uncultivated plants, such as wild grasses (Joubert 1966). Thus, control of this insect is likely to be a continuing struggle even if infestations in harvested grain are eliminated.

*Pink Scavenger Caterpillar: Pyroderces rileyi* (Walsingham)

**Description.** *Pyroderces rileyi* (family Cosmopterigidae) is easily differentiated from *S. cerealella*. It is smaller (wingspan of approximately 1 cm) and has very narrow hind wings with long fringes (Fig. 4a). The forewings are banded and mottled with brown, yellow, and black. The legs and antennae are gray ringed with white. The larvae are white upon hatching, but turn pink shortly thereafter. Full-grown larvae are about 8 mm long and are relatively slender (1.2 mm) (Fig. 4b). Prior to pupation, the ultimate instar spins a cocoon about 7 mm long. As with most stored-product Lepidoptera, particles of grain and frass are incorporated into the cocoon. The eggs are similar to those of *S. cerealella* but are smaller, pearly white, and have a strongly wrinkled surface and one truncated end.

**Distribution.** *Pyroderces rileyi* apparently has higher thermal requirements than *S. cerealella* because its distribution rarely extends into temperate regions. It has been recorded from the southern United States as well as Mexico, Hawaii, and Asia, where it is a pest.

**Ecology.** *Pyroderces rileyi* feeds on cotton and maize prior to harvest and on maize in storage, and has been reported from sorghum, cowpeas, citrus fruits, and leaves from a variety of plants. Larvae feeding on maize prior to harvest consume kernels, cobs, and husks, and often follow infestation by corn earworm (*Helicoverpa zea* [Boddie]). In addition, first instar larvae feed on maize silks or mold growing thereon. The irregular feeding damage is easily differentiated from that of *S. cerealella*.

Development is considerably faster than that of *S. cerealella*. Development

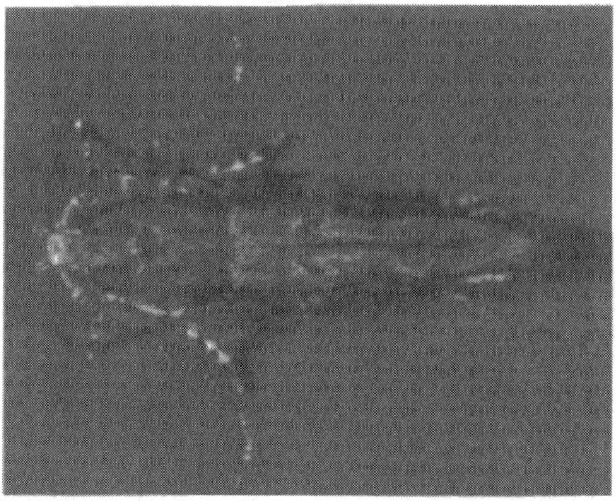

(a)

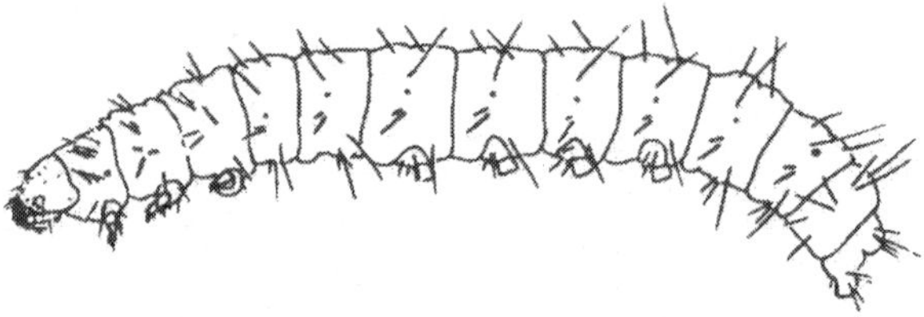

(b)

**Figure 4** (a) Adult *Pyroderces rileyi* have narrow hind wings and long wing fringes. (b) Larvae are white, but turn pink and are 8 mm long when full-grown. (a, From Anonymous 1986; b, From USDA 1991.)

from egg to adult can occur in as little as 23 days, but 25–28 days is more typical. As with most insects, development rate of *P. rileyi* increases with temperature. Although optimal, upper, and lower temperatures have not been published, the geographical distribution of this insect suggests that these values are higher than those of *S. cerealella*.

Emergence of adults is virtually continuous throughout the growing season. Douglas et al. (1962) reported that infestations of *P. rileyi* in storage are due primarily to preharvest infestation. They reported that infestation by this insect in stored maize occurred only in damp or moldy grain, and speculated that other plants are utilized as a resource by females emerging early in the growing season.

Females have a life span similar to that of *S. cerealella*. Douglas et al. (1962)

reported that females lived for 5–18 days, and laid their first eggs after 4–9 days, although some females began laying eggs as soon as 2 days after emergence. Up to 147 eggs are laid per female, but the average number is closer to 55. Eggs are laid singly or in groups of up to four. No information regarding ovipositional stimuli has been published.

Because *P. rileyi* infests primarily grain previously colonized by other insects or fungi, and rarely infests grain in storage, control of this insect is probably of little concern to managers of stored grains or their products.

*Rice Moth: Corcyra cephalonica* (Stainton)

**Description.** *Corcyra cephalonica* (family Pyralidae, subfamily Galleriinae) is more variable in size than *S. cerealella*, but typically has a wingspan of 1.7 cm (males) to 1.9 cm (females). The wings are uniformly dark gray, sometimes streaked with darker lines along the veins. The head is snoutlike and blunter in males (Fig. 5a). The moth is often difficult to spot because of its general inactivity during the day and its ability to blend in with many backgrounds.

Unlike *S. cerealella*, *C. cephalonica* is an external feeder. The larva is initially creamy white and gradually darkens to a dirty white with a reddish-brown head capsule (Fig. 5b). After completing development through seven or eight instars, it weaves a cocoon incorporating food grains, debris, and other materials. Cocoons are often found in clusters and matted together in sheets. The cocoon is thinner but stronger and more closely woven than that of other stored-product Lepidoptera. The pupa is leathery brown and measures about 8 mm. Eggs are pearly white ellipsoids with a rough, irregularly sculpted surface and a small nipple at one end. Eggs turn yellowish as they develop.

**Distribution.** *Corcyra cephalonica* is widely distributed but is most common in tropical regions. It is the most significant pest of stored products in many tropical locations and survives on a wider range of food materials than *S. cerealella*; it reproduces not only on most stored cereals but also on dried fruits, cocoa, groundnuts, and processed forms of the same. It has been reported from temperate locations, but these instances have usually been associated with shipments from tropical locales.

**Ecology.** *Corcyra cephalonica* differs considerably from the preceding two moths in its ecological role. Although it is capable of feeding on intact grains, it performs better on broken and processed grains. Thus, it is more like a secondary lepidopteran colonizer of stored products. It usually completes development after seven or eight instars, but this number may increase on poor foodstuffs or under adverse environmental conditions.

Its primarily tropical distribution reflects higher temperature optima than *S. cerealella*; its optimal temperature for survival and development is 30–32°C, with no development occurring below 17°C or above 35°C. Egg durations increase as

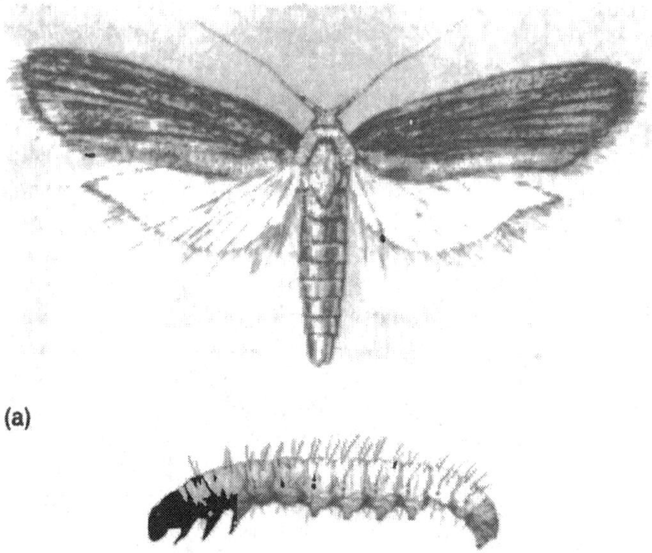

(a)

(b)

**Figure 5** (a) *Corcyra cephalonica* adults have uniformly dark wings and, at times, darker streaked wing veins. (b) Larvae have a dark head capsule and a creamy white body that darkens to dirty white as they mature. (a and b, From Anonymous 1986.)

temperature decreases, reaching a maximum of 17 days at 17.5°C. Egg hatchability decreases as temperature and relative humidity decrease. Adult longevity decreases as temperature increases, with males outliving females at all temperatures.

Ovipositional habits are also different from those of *S. cerealella*. Eggs are laid singly, usually on rough surfaces, and are attached to the substrate with a sticky substance.

Despite these differences, many similarities with *S. cerealella* exist. Females mate soon after emergence, and most eggs are laid within 5 days of adult emergence. Females lay about 160 eggs during this time, and neonate larvae appear after approximately 4 days. Larvae complete development within about 40 days, but the actual duration varies with grain type, relative humidity (moisture content), and temperature. Development is fastest on maize. Adults do not seem to require any nourishment.

## 4. Secondary Colonizers

Secondary colonizers or flour moths, so called because they seldom attack sound grain, are among the most common and serious pests of stored grains. They prefer broken kernels (e.g., those injured mechanically by combines or augers) or those fed on by primary insect pests. Especially preferred are milled cereal products such as flour, breakfast foods, and meals. *Plodia interpunctella* and the meal moth, *Pyralis farinalis*, are the two most frequently reported secondary moth pests. However, *P. interpunctella* has the widest distribution, infesting stored foods globally. *Ephestia kuehniella* is the primary pest of flour mills in temperate zones, and *E. elutella* is a severe pest where grain is stored for several seasons. Other species included in this group are *C. cautella* and *C. figulilella*.

All secondary moth colonizers described are in the family Pyralidae. *Plodia interpunctella* will be dealt with primarily because it is important on a global scale and feeds on a vast array of foods. Exceptions to its life history will be addressed in the review of the other species.

*Indianmeal Moth: Plodia interpunctella* (Hübner)

**Description.** *Plodia interpunctella* is a stricking moth with a wingspan of approximately 1.9 cm. The adult is easily differentiated from other stored-product moths by the markings on its forewings, which are dark reddish brown on the distal two-thirds and whitish gray on the proximal one-third (Fig. 6a).

The fully grown larvae are 1.25 cm long and dirty white in color, with occasional variation into the pink and green hues (Fig. 6b). The larvae spin silken cocoons and metamorphose into light-brown pupae, from which adults moths emerge.

**Distribution.** *Plodia interpunctella* has a global distribution, having been found in North, Central, and South America; Europe; North, West, and South Africa; Asia; Japan; Australia; and New Zealand (Tzanakakis 1959). It feeds on many grains (e.g., barley, buckwheat, maize, oats, rice, rye, sorghum, wheat, and milo) or on grain products such as various grain brans and meals. Dried fruits such as raisins, apricots, dates, figs, peaches, and prunes, and nuts such as almonds, peanuts, hazelnuts, and walnuts are also consumed along with peas, beans, lentils, various spices, and other stored commodities (Williams 1964). Thus, *P. interpunctella* has a wide range of hosts and is a good colonizer because it is highly mobile with apparently a wide variety of hosts and variable nutritional requirements.

**Ecology.** *Plodia interpunctella* eggs are white, grayish white, or yellowish white and are 0.3–0.5 mm long, oval, and have a reticulated chorion (Tzanakakis 1959). Eggs are laid singly or in groups directly on the food material or on its packaging. Hatching occurs after 4 days at 30°C and 70% relative humidity (RH), and has been reported to be most dependent upon temperature. Upon hatching, larvae disperse in search of food. Their small size enables them to enter containers that appear to be well sealed. The larvae feed inside grain kernels or on meal and always leave silken threads or webbing that may inhibit fumigation. Larvae are

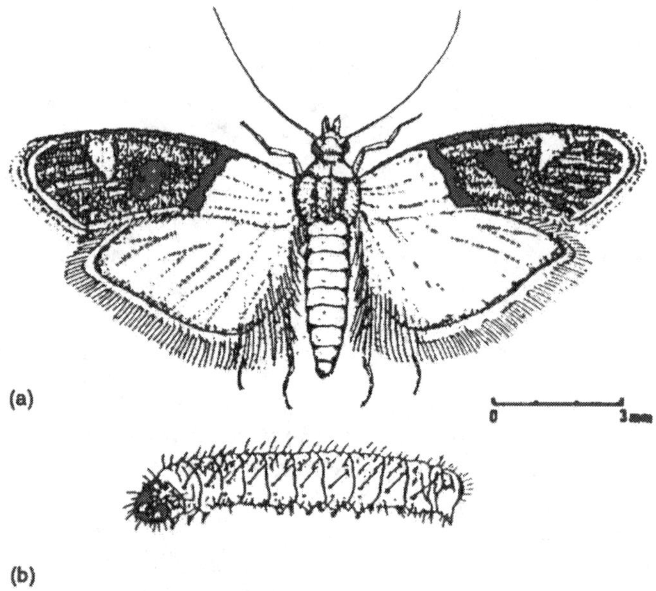


**Figure 6** (a) *Plodia interpunctella* adults are easily differentiated from other stored-product moths. The forewings are dark reddish-brown on the distal two-thirds and whitish-gray on the proximal one-third. (b) *P. interpunctella* larvae are 1.25 cm long and dirty white. (a and b, From Hill 1990, reprinted with permission.)

notoriously polyphagus. However, the germs of grains they attack appear to be favored.

Most mature larvae leave the food medium and search for a suitable place to spin a cocoon in which to pupate. Usually a crack or other protected place, typically in dark locations, is preferred. Some larvae spin their cocoons in the food medium just below the surface. Cocoons spun by the pupating larvae can be differentiated from those made by the hibernating or diapausing larvae. The hibernaculum (i.e., hibernation cocoon) is dense and completely closed, whereas the pupal cocoon is flimsy, loose fitting, tapering, and opens anteriorly to permit exit of the adult.

Following hibernation, the larva opens a hole in the hibernaculum and either spins its pupal cocoon inside, or comes out and constructs the pupal cocoon nearby. It appears that larvae spin pupal cocoons outside the hibernaculum only when the hibernaculum is not large enough to include the pupal cocoon. Duration of the pupal stage seems to depend primarily on the temperature during that stage, and to be independent of the duration of larval stage (Fraenkel and Blewette 1946). The duration of the adult stage is dependent upon environmental factors such as temperature and humidity, the occurrence of mating, the opportunity for oviposition, and the presence or absence of water or other liquids for consumption.

Successful development from egg to adult can occur under a variety of conditions. Ability to complete development is dependent upon type of food (e.g., different grains or grain hybrids), temperature, and grain moisture content (Mbata and Osuji 1983). Developmental rate varies on different maize hybrids (Abdel-Rahman et al. 1968) and with degree of kernel damage. Temperature and grain moisture content also affect development (Tzanakakis 1959). *Plodia interpunctella* ceases to develop below 40% RH (Howe 1965). Optimal temperatures for development range from 28°C to 32°C (Howe 1965), with little development occurring below 18°C (Howe 1965) or above 35°C (Cox and Bell 1991). Development is longer and mortality is higher at low (25%) rather than high (70%) relative humidity.

The length of egg incubation varies with temperature, with no hatch occurring at 15°C (Bell 1975). Young eggs of *P. interpunctella* are less tolerant of cold than any of the *Ephestia* or *Cadra* species. Duration of egg development decreases from 7.2 days at 22°C to 4.1 days at 25°C and 3.0 days at 30°C (Bell 1975). At room temperature, eclosion occurs at 4–8 days, whereas at 25°C and 53% RH it occurs in 4–5 days. Size of larvae varies with temperature, moisture content, and type of food. There are five to seven larval instars. On stored groundnuts, development from first larval instar to adult was fastest at 30°C and 80% RH and slowest at 25°C and 60% RH. Size and weight of larvae vary with food and temperature, whereas size and weight of pupae vary with gender, age of pupae, and with food (Tzanakakis 1959). The pupal stage lasts 15–20 days at 20°C, 8–11 days at 25°C, and 7–8 days at 30°C (Bell 1975). Typically, there are one to two generations per year in more temperate climates, whereas in warmer climates there may be up to eight generations per year (Tzanakakis 1959).

Survivorship is dependent upon temperature, relative humidity, and food type. Young larvae may survive temperatures as low as 10°C (Cox and Bell 1991). Most larval mortality occurs during the first instar. Survivorship of immatures and adults increases as temperature and moisture content increase to the upper thresholds of 32°C and 80% RH (Abdel-Rahman et al. 1968; Bell 1975); above this threshold survivorship decreases. Size and hardness of grain kernels also affect mortality and development of *P. interpunctella* (Abdel-Rahman et al. 1968). Mortality of *P. interpunctella* eggs and larvae ranged from 4% on tough wheat to 28% on yellow maize; duration of development ranged from 35–41 days on wheat to 36–327 days on yellow maize (Williams 1964).

Females mate very soon after adult eclosion. They produce a sex pheromone, *cis*-9, *trans*-12-tetradecadienyl acetate (Kuwahara et al. 1971), that is very attractive to male *P. interpunctella*. In addition, males produce a sex pheromone that causes females to become receptive to male mating attempts (McLaughlin 1982). Females lay an average of 150–200 eggs, but this value may range from 40 to 400 eggs per individual (Tzanakakis 1959; Lum and Flaherty 1969). The number of eggs laid depends upon temperature, larval food, and on the size of individuals (Tzanakakis 1959). Generally speaking, larger moths have larger ovaries and lay more eggs. The ovipositional period has been reported to be 1–18 days, but the majority of viable eggs are laid during the first 4 days (Silhacek and Miller 1972). Maximum egg production occurs in individuals 2–3 days old and is affected by the presence or absence of food (Mullen and Arbogast 1977). Full potential fecundity is realized only if a stimulus from a suitable food is present. Furthermore, fecundity appears to increase when adult females oviposit on the host on which they were reared (Mullen and Arbogast 1977). Stress caused by fluctuating or changing light regimes also reduces egg production (Lum and Flaherty 1969).

The wide host range and temperatures under which this species is capable of surviving and reproducing account for its status as a cosmopolitan pest. Diapause, which greatly extends the developmental period, may be induced by short photoperiod, low temperature, or high population pressure (Williams 1964; Bell 1975; Cox and Bell 1991). Diapause provides a means by which the species may overwinter or survive periods of adverse environmental conditions. These same attributes make control of this pest difficult.

*Almond Moth: Cadra (= Ephestia) cautella* (Walker)

**Description.** Adult *C. cautella* are not particularly distinctive looking. The forewings are reddish brown with faintly colored white cross-lines and the hindwings are pale gray (Fig. 7a). Wingspan is 1.4–2.2 cm and wing fringes are short.

Larvae are pale gray with many setae and small dark spots that may be used to confirm identification. Head capsule is dark, and full grown larvae are 1.2–1.5 cm long (Fig. 7b).

**Distribution.** *Cadra cautella* (also known as the tropical warehouse moth and dried-currant moth in places other than the United States) is a pest predominantly in the tropics and warmer regions of the world. However, it can be a pest in heated storage facilities in more temperate regions. Dried fruits are the preferred food but the insect also feeds on many types of stored vegetable matter including flours, grains, dates, cocoa beans, nuts, and seeds.

**Ecology.** *Cadra cautella* eggs are globular and white when laid and gradually turn orange as they develop. The eggs are laid directly in the commodity. Consumption by larvae is the main type of damage, but frass and silk-filled galleries cause significant contamination of the commodity. Larvae preferentially feed on seed germs.

Like *P. interpunctella*, *C. cautella* does not complete development at 15°C. Development is affected most by temperature and relative humidity during larval

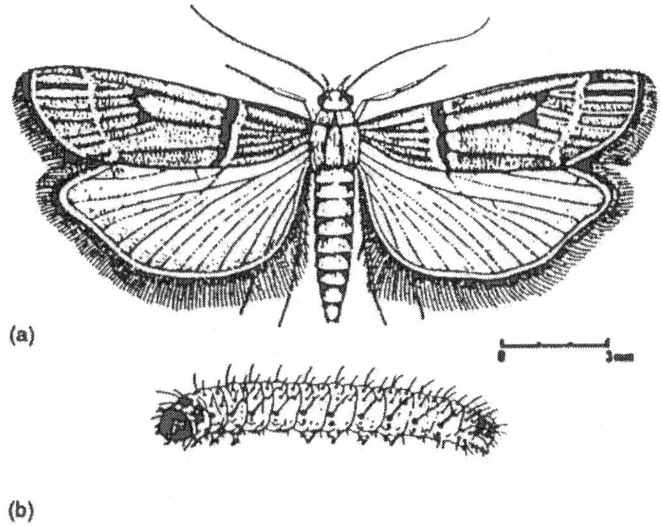


**Figure 7** (a) *Cadra cautella* adults are reddish-brown with wings folded. (b) *C. cautella* larvae are pale gray with many setae. (a and b, From Hill 1990, reprinted with permission.)

stages. Egg and pupal stages seem unaffected by relative humidity except at extremely low (40%) or high (>90%) relative humidities. Eggs hatch in an average of 17.5 days at 15°C, 7.4 days at 20°C, 4.7 days at 25°C, and 3.4 days at 30°C and 70% RH (Bell 1975). Duration of the pupal stage averages 17.5 days at 20°C, 8.9 days at 25°C, and 7.0 days at 30°C and 70% RH.

Successful development from egg to adult is dependent upon temperature, moisture content, and type of food (Burges and Haskins 1965). Optimal development occurs at 30–32°C and 70–80% RH. Lower development threshold is around 15°C while the upper threshold is about 36°C. Duration of egg and pupal stages is not influenced by relative humidity but by temperature. Low humidity retards larval growth, and very low humidity results in death. This most likely results from the low humidity reducing the amount of available water in food, making the food harder to eat and resulting in net water loss due to fecal elimination and active feeding.

Larval mortality, especially in very young larvae, is higher than egg and pupal mortality. Extremes of temperature and humidity are particularly likely to increase mortality. Eggs may hatch at the extremes but larvae invariably fail to complete development.

Survival of mated adults is lower than for virgin adults (Burges and Haskins 1965). Average longevity of males and females varied from 3.1 days at 35°C to

12.3 days at 15°C. Average longevity is longest at 15°C (193.7 days) and shortest (41.3 days) at 35°C.

Oviposition by *C. cautella* occurs in response to the onset of darkness, falling temperature, or both. The presence of food also has a significant effect on oviposition; like *P. interpunctella*, full potential fecundity is not realized unless a stimulus from a suitable food supply is provided (Mullen and Arbogast 1977). Availability of free moisture is also important; female *C. cautella* provided with water laid about 46% more eggs than those not given water (Hagstrum and Tomblin 1975).

Females lay an average of 200 eggs (Hill 1990). The preovipositional period is significantly longer at lower temperatures than at moderate to higher temperatures at all humidities (Tuli et al. 1966). The ovipositional period is likewise longer at higher humidities for all temperatures up to 35°C. Fecundity increases with RH over the range of 30%–90% at temperatures ranging from 15–35°C. Exposure to high temperatures, however, results in male impotence, which, in turn, lowers egg productivity and fertility. Photoperiod also apparently affects mating. An average of 33% more *C. cautella* remained unmated during a 24 h exposure to absolute dark than when kept under dim illumination.

*Cadra cautella* females also use a sex pheromone to attract potential mates. Air permeated with synthetic sex pheromone at various moth densities decreased the average number of mated female moths after 24 h by 51% (McLaughlin and Hagstrum 1976). When combined, darkness and pheromone effects appeared to be additive, suppressing mating by 85% at 24 h and 71% after 48 h. Both photoperiod and pheromone effects appear to be dependent on moth density. Combining a sex pheromone-permeated atmosphere and sterile males reduces reproduction of *C. cautella* more than either of the two methods alone (Hagstrum et al. 1978).

*Mediterranean Flour Moth: Ephestia kuehniella* Zeller

**Description.** *Ephestia kuehniella*, with a wingspan of 2.0–2.5 cm, is the largest moth in this genus. The forewings are gray, speckled brown and white with transverse, wavy black markings; the hind wings are white with brownish-gray veins (Fig. 8a).

Larvae spin silken threads wherever they go and bind with webbing particles of food on which they have been feeding. Most damage is done when larvae interfere with flour production by spinning copious amounts of webbing that clogs machinery and by chewing holes in screening. When fully grown, the larvae are about 1.25 cm long and are whitish or pinkish, with a few small black spots on their backs (Fig. 8b).

**Distribution.** *Ephestia kuehniella* is found worldwide but is not abundant in the tropics and cannot survive well in winter or for extended periods of cold unless in heated storage facilities or other buildings (Cox 1987). Thus, it is essentially a temperate or Mediterranean species (Bell 1975; Jacob and Cox 1977). Although

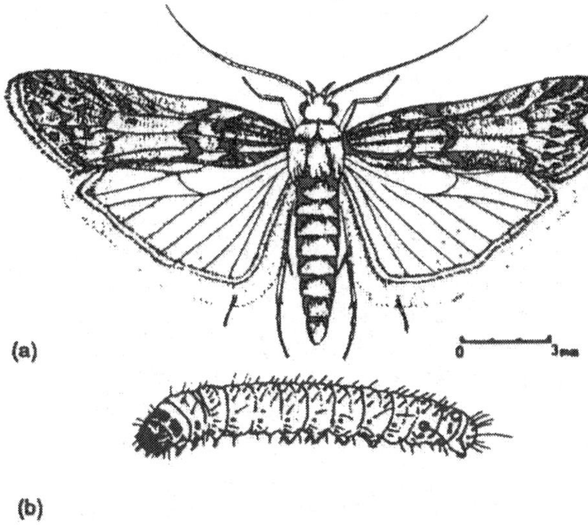


**Figure 8** (a) *Ephestia kuehniella* adults are the largest in the genus. Forewings are gray, speckled brown and white with wavy black markings. The hindwings are white with brownish-gray veins. (b) Larvae are 1.25 cm long and white or pink with few small black spots on their backs. They spin copious amounts of webbing. (a and b, From Hill 1990, reprinted with permission.)

*E. kuehniella* prefers flour and meal, it also attacks grain, bran, cereal products, soybeans, rolled rice, peanuts, almonds, cocoa beans, sesame seeds, and other types of food (Anonymous 1986; Cox and Bell 1991).

**Ecology.** Female *E. kuehniella* lay small white eggs in accumulated flour, meal, or waste grain (Anonymous 1986). Eggs are elliptical, 0.51 × 0.29 mm. Most eggs exhibit folding of the chorion at the anterior end and some have a nipplelike projection. The surface of the chorion is marked by a reticulate pattern of wavy ridges joining weak tubercles. The pattern is distinctly heavier on the ends than the middle in most eggs (Arbogast et al. 1980).

Mating is unsuccessful at 12.5°C; however, oviposition occurs at 7.5°C but not at 5.0°C. Eggs hatch between 10°C and 31°C without any apparent influence of ambient humidity on duration of development or viability (Jacob and Cox 1977). Larvae fail to pupate at 12°C at humidities below 70% RH, while at 31°C no larvae pupate. Low humidity consistently extends duration of larval development. Humidity has no consistent effect on pupal development. However, at 18°C and 75% RH, duration of the pupal stage was considerably longer than at other humidities.

The lower developmental threshold temperature for complete development is 12°C while the upper threshold is 28°C (Jacob and Cox 1977). Development takes longer and survival is poorer at 40% RH than at 70% RH. However, survival is good at 20 and 25°C and 15% RH. A surprising 36% survival was recorded for individuals completing development at 25°C and 0% humidity. Development was fastest at 25°C and 75% RH, taking an average of 74 days from oviposition to adult emergence.

*Ephestia kuehniella* diapauses as last-instar larvae. Diapause of this species is influenced by temperature, photoperiod, and nutrition. Approximately 50% of the larvae of a strain from Scotland entered diapause when reared at 25°C in complete darkness and 30% exhibited the same phenomenon in short photoperiods (Cox et al. 1981). Diapause lasted 2–3 months in most photoperiods at 20 and 25°C. However, diapause lasted around 7 months in continuous darkness at 25°C and was terminated more rapidly after chilling to 7.5°C for 6 weeks. The incidence of diapause was about 50% on diets of whole maize or wheat flour and 12% on an artificial diet containing glycerol, yeast, and wheatfeed (Cox et al. 1984a).

These observations have important implications for control of this pest. Lower rates of metabolism and locomotion associated with larval diapause cause them to be less susceptible to insecticides and fumigants (i.e., methyl-bromide and phosphine) (Cox et al. 1984b). Entering diapause at 25°C would reduce the rate of population increase during summer months when conditions are ideal for rapid development. However, in temperate regions, this is the time of year when most chemical pest control is used. Thus, larvae in diapause during this time will be more likely to survive than nondiapausing larvae.

Female *E. kuehniella* are attractive to males only when assuming their characteristic calling posture. In this position the abdomen is flexed upward, abdominal segments 8–10 are extruded, and a mating pheromone is liberated from the membranous cuticle between abdominal segments 8 and 9 (Dickins 1936). Females at 50 h of age were more attractive to males than females aged less than 22 h. Traynier (1970) found that 72-h-old females yielded the most potent pheromone extracts. Furthermore, extracts were of equal potency during different times of the day. According to Edwards (1962) *E. kuehniella* exhibits a circadian cycle of flight. In natural light at a constant 24°C, one period of flight occurs in females at sunset and two occur with males; one following sunset and one preceding sunrise (Edwards 1962). The maximum response of males was at dawn (Traynier 1970).

Male *E. kuehniella* kept under continuous light from the pupal stage onward had a much lower reproductive capacity than males kept at a 12 h light/dark photoperiod (Riemann and Rudd 1974). Three factors apparently contribute to this phenomenon: decreases in mating, sperm production, and transfer of sperm from spermatophore to spermathecae.

Based on the effects of temperature and photoperiod on the life cycle and reproductive biology, it would seem that using these two physical parameters would enable commodity managers to control populations of this pest in warehouses merely by manipulating them. However, populations of insects can be rapidly selected for or against. Therefore, manipulation of warehouse environmental conditions may support control measures only in the short term.

*Raisin Moth: Cadra (= Ephestia) figulilella* Gregson

**Description.** *Cadra figulilella* is one of the smaller species in this genus with a wingspan of 1.5–2.0 cm. The forewings are pale yellow-gray with indistinct markings and the hind wings whitish with a yellow-gray border (Hill 1990; Richards and Thomson 1932).

Newly hatched larvae are pink and turn white as they age. The mature larva is about 1.3–1.5 cm long and is similar in appearance to *C. cautella*.

**Distribution.** *Cadra figulilella* is widespread in the Mediterranean region and in similar areas of climate in the Americas and Australia (Cox 1974). Its distribution increased when it was accidentally imported into Britain on dried fruits and carobs (Aitken 1963). This moth may attack ripening crops before harvest and dried fruits and carobs in storage (Cox 1975a). Thus, it may fly into stores or be carried there on the commodity itself. It is commonly associated with fallen fruit and freshly harvested carobs (Cox 1974) and in cottonseed cake, cocoa beans, cashew kernels, and other crops (Donohoe et al. 1949).

**Ecology.** *Cadra figulilella* eggs are ellipsoid to ovoid and 0.40 × 0.33 mm. They lack projections and folds at the anterior end. The chorionic sculpturing consists of tubercles joined by low-elevation sinuous ridges (Arbogast et al. 1980).

Eggs are laid loosely on and near the food source and do not adhere to the surface of the commodity. The average number of eggs produced by 25 pairs of adults kept on raisins at room temperature was 351, 75% of which hatched (Donohoe et al. 1949). Total oviposition ranged from 0 to 692. No hatching occurred at 10 or 12.5°C and only 10% hatched at 15°C (Cox 1974). The shortest mean incubation periods were at 30 and 32.5°C and 70% RH. Duration of development increased at 37.5°C and at temperatures ranging from 27.5 to 17.5°C.

Larval mortality is 100% at 15 and 37.5°C. However, development is influenced by humidity as well as by temperature. For example, the lowest mortality and shortest development time of *C. figulilella* was at 30°C and between 70 and 90% RH (Cox 1974). This, along with longer larval wandering periods at low temperatures, is evidence that this is a semitropical species most likely having its origin away from the equator.

The critical photoperiod for diapause induction of *C. figulilella* is between 12 h light/12 h dark and 13 h light/11 h dark (Cox 1975b). *Cadra figulilella* enters diapause at photoperiods less than or equal to 13 h of light per day at 30°C and 70% RH. In continuous darkness for 60 days, few larvae exhibit developmental delay. However, in continuous darkness for 140 days, 91% of the experimental population diapauses at 20°C and 70% RH. Diapause delays development in this species by 74 days (Cox 1975b).

*Tobacco/Warehouse Moth: Ephestia elutella* (Hübner)

**Description.** *Ephestia elutella* is a relatively small member of this genus with a wingspan of 1.5–1.6 cm. The forewings are brownish gray, crossed with two light-colored bands, and the hindwings are uniformly gray (Ebeling 1975; Mallis 1982) (Fig. 9a).

Mature larvae vary in length from 1.0 to 1.5 cm and are creamy white shaded with yellow, brown, or pink (Fig. 9b). Pupae turn from light-brown to black immediately before adult emergence (Ebeling 1975).

**Distribution.** This moth is considered cosmopolitan but is uncommon in the tropics (Hill 1990). Thus, it is a warm temperate species where it may survive outside or in unheated buildings. *Ephestia elutella* can be a serious pest of stored tobacco and may consume entire leaves except for the largest veins (Mallis 1982). Tobacco leaves that are not consumed may still be damaged by webbing and excrement. This moth also infests cereals, chocolate, cocoa beans, coffee, cotton-

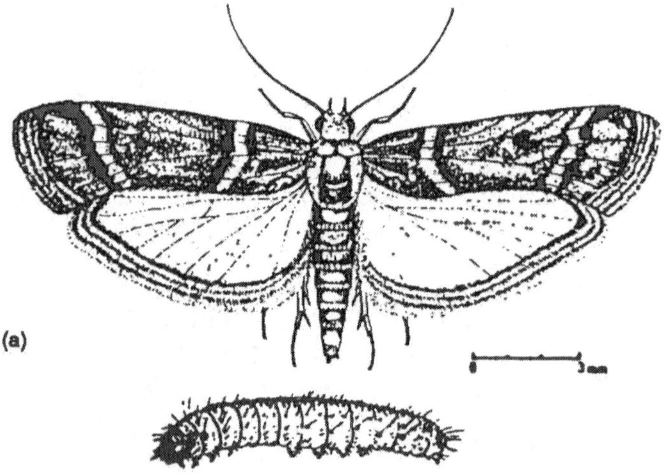

(a)

(b)

**Figure 9** (a) *Ephestia elutella* is a small member of this genus. The forewings are brownish gray crossed with two light-colored bands. Hindwings are uniformly gray. (b) Mature larvae vary from 1.0 to 1.5 cm in length and are creamy-white shaded with yellow, brown, or pink. (a and b, From Hill 1990, reprinted with permission.)

seed, dried fruits, flours, meals, and processed flours, nuts, seeds, and shelled maize (Mallis 1982).

**Ecology.** *Ephestia elutella* eggs are ellipsoid and 0.49 × 0.34 mm. Like *C. figulilella* eggs, they lack projections and folds at the anterior end. Chorionic sculpturing and texture are similar to those of *C. cautella* but with less prominent tubercles and fewer ridges (Arbogast et al. 1980).

Eggs are laid singly, in short rows along crevices, or in small clusters on or near stored tobacco or other commodity. As with the previously discussed species, fecundity is affected by differences in larval nutrition, temperature during larval development, duration of diapause, initial weight of adults, temperature during oviposition, and access to drinking water (Waloff et al. 1948). Under optimal conditions, average fecundity is 150–200 eggs per female; however, as many as 327 have been produced (Waloff et al. 1948; Cox and Bell 1991). Eggs are laid primarily at dusk or during the night through the first week of adulthood; at 25°C most eggs are laid during the first 4 days. In warehouses, adults live up to 3 weeks.

*Ephestia elutella* develops successfully at temperatures ranging from 15 to 30°C. However, no development occurs at 10°C and at 30°C individuals are infertile. On most foods, *E. elutella* takes longer to develop than related species of moths. It takes approximately 6–7 weeks at 25°C and 70% RH and 11–12 weeks at 20°C from oviposition to adult emergence (Bell 1975). *Ephestia elutella* exhibits a strong tendency to diapause at lower temperatures or short photoperiods, which prolongs the developmental period and renders the species univoltine in cool climates. The developmental ranges and cold hardiness of *E. elutella* suggest that this is a temperate species.

*Meal Moth: Pyralis farinalis* Linnaeus

**Description.** *Pyralis farinalis* is an attractive moth with a wingspan of 2.2–3.0 cm. The adult is easily identified by the reddish-brown color of the forewing base and apex, pale wing center bordered by a wavy white line, pale brown hindwing with white markings, and black spots posteriorly. Adult moths hold their wings flat, laterally extended, and the abdomen curled over the body (Fig. 10a).

Fully grown larvae are 2.5 cm long and gray with a dark head and shield (Fig. 10b). Larvae live in silken galleries attached to a solid substrate.

**Distribution.** *Pyralis farinalis* is cosmopolitan (Mallis 1982). It is an occasional pest of sound wheat and cereal products stored in moist places (Levinson and Buchelos 1981). Larvae of this insect are most commonly found in damp, spoiled grain and grain products in poor condition (Sinha et al. 1962; Cotton 1963; Davidson and Peairs 1966; Anonymous 1986). They are also found in damp straw, plant refuse, and moldy leaves (Berns 1958; Ebeling 1975) and on poultry farms (Bell and Daehnert 1962; Cox 1986).

According to Trematerra (1988), the distribution and development of larvae and pupae are limited by factors that reduce the growth of fungi, yeasts, and

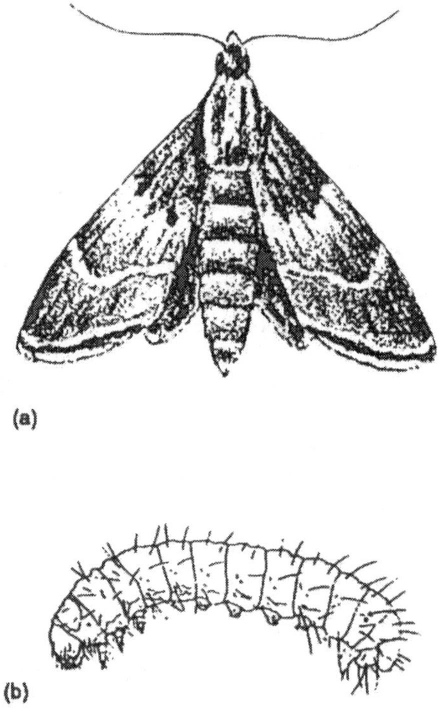


**Figure 10** (a) *Pyralis farinalis* adult with characteristic spread wings. (b) *P. farinalis* larvae are conspicuous because of the webbing they create. (a and b, From USDA 1991.)

bacteria. Trematerra and Galli (1985) demonstrated that larvae could develop on maize or wheat flours, meadow hay, or artificial diet contaminated with fungi or yeasts, but not cereal products without fungi. Arbogast and van Byrd (1981) reared *P. farinalis* collected from peanut warehouses on wheat and yeast with a water wick.

**Ecology.** *Pyralis farinalis* eggs are oval, 0.52 × 0.37 mm. The eggs are marked by a reticulate pattern of closely spaced wavy ridges that are crossed by elevated ridges forming a more or less continuous network over the entire surface of the egg (Arbogast and van Byrd 1981).

Meal moth larvae are usually conspicuous because of the webbing they produce, which, in turn, binds the seeds together. Larvae spin tubes of silk that contain particles of food. They rest in these tubes, which are very durable, and feed from the openings at the ends. Fully grown larvae leave the tubes, spin cocoons (also often covered with food particles), and metamorphose into pupae. The female moth lives about 1 week and lays 200–400 eggs (Anonymous 1986). Development from egg to adult takes 6–8 weeks in the summer.

*Pyralis farinalis* caused 99.4% kernel mass loss and consumed all parts (i.e., bran, germ, and endosperm) of damaged wheat kernels (Madrid and Sinha 1982). However, it was unclear whether this species completed development to the adult stage on this commodity.

Curtis and Landolt (1992) used a commercial formulation of *Helicoverpa zea* (Boddie) diet containing agar, ground pinto beans (54%), torula yeast (16.6%), ground wheat germ (26%), ascorbic acid (2%), methyl-p-hydroxy-benzoate (1%), and sorbic acid (0.5%) at 24°C and 50% RH to examine life table parameters of this moth.

Average egg hatch was 81.3% and their incubation period was 9 days (Curtis and Landolt 1992). The average survival rate for development from egg to adult was 62.4% at egg densities of 50/jar and 38.3% at egg densities of 250/jar. Development time from oviposition to adult emergence averaged 61.6 days for males and 64.5 for females for the above conditions. Mean generation time (egg to egg) was 65.4 days.

*Pyralis farinalis* had an average preovipositional period of 1.9 days and most females mated and began ovipositing within 2 days. The majority of eggs (88%) were laid loose, but 7% and 5% were laid on the walls of the vial or paper, respectively. Average adult lifespan for mated adults was 10.2 days for males and 9.8 days for females. Heavier females laid significantly more eggs and lived longer than lighter females.

Landolt and Curtis (1982) reported that males of *P. farinalis* and *Amyelois transitella* (Walker) (navel orangeworm), were cross-attracted to females of the other species and both were found in traps baited with females of either species. They also found that male meal moths court and attempt to copulate with female navel orangeworms. The cross-attraction between these two species could result in some interference or disruption of mating in situations where both species are abundant. Mating success of each species might be adversely affected if populations of the other species dominate.

## II. PSOCOPTERA: BASIC STRUCTURE, LIFE CYCLE, AND PEST STATUS

Psocoptera are a relatively small order of insects, approximately 6000 species worldwide, that are found in a variety of natural terrestrial ecosystems. Also known as psocids, these insects are generally thought to feed on microflora and organic debris, although some are known to be predators. Some psocids, also known as booklice, have adapted to living in food stores, granaries, warehouses, and bulk grain in transport. Because only a limited number of psocids are of economic importance, limited information is available on most species. However, New (1987) has published a comprehensive review of the order.

Psocids are small (0.7–6.0 mm) soft-bodied insects with long antennae, chewing mouthparts, and may be winged or wingless (Fig. 11). Although psocids are often overlooked and considered to be of minor economic importance due to their small size, heavy psocid infestations have been reported to cause significant damage to stored wheat and milled rice. Psocids also cause economic damage in the food processing industry and create possible health concerns by transferring micro-organisms and contaminating food products with feces and carcasses. Champ and Smithers (1965) identified 15 species of Psocoptera in stored products in Queensland, Australia, and speculated that the greatest economic losses from psocid infestations were from contamination of processed food, and that damage to bulk products was overshadowed by major pests.

Although there are three suborders of psocids, most known pests of domestic food products are found in two: Trogiomorpha (>20 antennal segments; *Lepinotus*, *Trogium*, and *Psillipsocus* species) and Troctomorpha (<20 antennal segments; *Liposcelis* species). The scientific literature on psocids that are domestic pests is limited. *Liposcelis bostrychophilus* Badonnel is a relatively common, cosmopolitan pest of stored products and will be used as a model in the following discussions. Notable exceptions to this model will be referenced.

*Liposcelis bostrychophilus* Badonnel

**Description.** This tiny insect (about 1 mm long) is pale brown with a soft rounded body, protruding eyes, and threadlike antennae. Sinha (1988) provides an

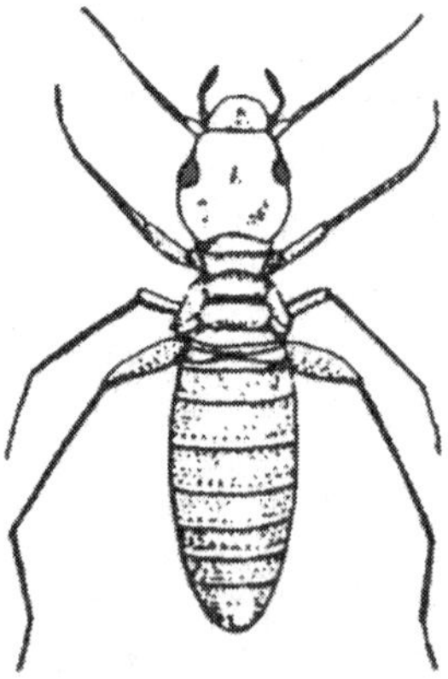

**Figure 11** Psocoptera adults are pale in color, winged or wingless, and soft bodied. Heads are relatively large, eyes are poorly developed, and antennae are long and slender. (From Hill 1990, reprinted with permission.)

excellent scanning electron micrograph of *L. bostrychophilus*. Psocids undergo simple metamorphosis with typically six instars, but this number is reduced to five in *Psyllipsocus* and four in Liposcelidae (New 1987). Psocids most resemble Mallophaga and Anoplura, which are ectoparasites of birds and mammals.

**Distribution.** Believed to be of African origin, *L. bostrychophilus* has been found worldwide in such diverse habitats as stored grains, fungal cultures, cliff swallow nests, and termite nests (Williams 1972). Turner and Maude-Roxby (1988) believe that this species, although of tropical origin, has been able to extend its range because it is parthenogenetic and capable of surviving long periods without food (up to 2 months).

As is true for many other stored-product insects and mites, the geographical origin of many domestic psocids is unknown. Inferred habitat ranges of many tropical psocids are often based on inadequate data. Psocids, like most small insects, are frequently incorporated in the aerial plankton after active take-off, and thus may be transported considerable distances. Distribution via commerce has undoubtedly occurred.

**Ecology and Behavior.** Although microscopic molds are its primary diet, *L. bostrychophilus* has omnivorous feeding habits like most Psocoptera (Williams 1972). Recorded as infesting over 50 types of foods in the United Kingdom, *L. bostrychophilus* is considered most troublesome in flours, rice, semolina, and breakfast cereals (Turner and Maude-Roxby 1988). Watt (1965) recorded observations of this species inflicting feeding damage on stored wheat and sorghum seed, and Sinha (1988) cited an infestation in rapeseed. *Liposcelis bostrychophilus* has been documented to prey upon eggs of *P. interpunctella* (Lovitt and Soderstrom 1968) and eggs of an anobiid, wood-infesting beetle (Williams 1972). *Liposcelis divinatorius* (Muller) was reported to prey upon eggs of *Sitotroga* sp (Finlayson 1933). In a controlled laboratory experiment, Mills et al. (1992) found several species of fungi, broken wheat kernels, screenings, and weed seeds as suitable food substrates for *L. bostrychophilus*.

Parthenogenesis (egg development without fertilization), coupled with obligatory thelytoky (all progeny female), permit *L. bostrychophilus* populations to grow rapidly under favorable conditions. Development from egg to adult is about 3 weeks at 27°C and 70% RH (Sinha and Watters 1985). Adult longevity is from 72 to 144 days depending upon environmental conditions. *Liposcelis bostrychophilus* is capable of surviving for considerable periods of time in adverse conditions without food. Knulle and Spadafora (1969) and Devine (1982) showed that this species maintained body water levels by absorbing atmospheric water vapor when humidity was 60% or above. Below this level, water loss leads to death. Domestic psocids are generally more tolerant of desiccation than other psocids. Most psocids live inside the grain bulk within the intergranular microclimate, which they have little difficulty penetrating.

**Economic Importance.** *Liposcelis* spp may cause measurable weight loss and quality deterioration in stored grain (Rees and Walker 1990). They reported visible damage to rice grains by selective feeding on the germ and softer parts of the exposed endosperm on broken grains. Rees and Walker (1990) concluded that *Liposcelis* spp are secondary pests of grain whose diet is supplemented by mold. Their pest status in tropical countries has often been underestimated due to the less than obvious nature of the damage compared to primary feeders.

In a 20-year study of bulk stored grain in Canada (Sinha 1988), only one serious psocid population explosion was observed: *Lepinotus reticulatus* Enderlein on wheat. An analysis of several biotic and abiotic factors found that the outbreak could be partially explained by a preceding rise in temperature and moisture content of the grain. Sinha (1988) postulated that these factors could have enhanced population growth directly or indirectly by promoting above-normal growth of seed-borne fungi.

*Liposcelis bostrychophilus* has become a cosmopolitan pest due to the diverse range of foodstuffs capable of supporting populations, rapid population growth potential possibly resulting from a parthenogenetic life history, and a tolerance to adverse environmental conditions.

## ACKNOWLEDGMENTS

We thank B. D. Price, P. L. Rattlingourd, and S. J. Hill, Jr. (Community Research Service, Kentucky State University) for reviewing an early draft of this manuscript. This manuscript was supported by grants from USDA–CSRS to Kentucky State University under agreements KY.X-10-91-18P and KY.X-10-90-15P.

## REFERENCES

Abdel-Rahman, H. A., Hodson A. C., and Christensen, C. M. (1968). Development of *Plodia interpunctella* (Hb.) (Lepidoptera, Phycitidae) on different varieties of corn at two levels of moisture. *J. Stored Prod. Res.*, *4*:127–133.

Aitken, A. D. (1963). A key to the larvae of some species of Phycitinae associated with stored products, and of some related species. *Bull. Entomol. Res.*, *54*:175–188.

Anonymous. (1986). Stored grain insects. USDA Handbook No. 500.

Arbogast, R. T., and van Byrd, R. (1981). External morphology of the eggs of the meal moth, *Pyralis farinalis* (L.), and the murky meal moth, *Aglossa caprealis* (Hübner) (Lepidoptera: Pyralidae). *Int. J. Insect Morphol. Embryol.*, *10*:419–423.

Arbogast, R. T., LeCato, G. L., and Byrd, R. V. (1980). External morphology of some eggs of stored-product moths (Lepidoptera: Pyralidae, Gelechiidae, Tineidae). *Int. J. Insect Morphol. Embryol.*, *9*:165–177.

Ayertey, J. N. (1982). Development of *Sitotroga cerealella* on whole, cracked, or ground maize. *Ent. Exp. Appl.*, *31*:165–169.

Bell, C. H. (1975). Effects of temperature and humidity on development of four pyralid moth pests of stored products. *J. Stored Prod. Res.*, *11*:167–175.

Bell, D., and Daehnert, R. H. (1962). Control of house flies on poultry ranches with anti-resistant DDT. *J. Econ. Ent.*, *55*:817–819.

Berns, R. E. (1958). Straw meal moths. *Pest Cont.*, *26*:16.

Burges, H. D., and Haskins, K. P. F. (1965). Life-cycle of the tropical warehouse moth, *Cadra cautella*, at controlled temperatures and humidities. *Bull. Ent. Res.*, *55*: 775–789.

Candura, G. S. (1954). Notes on *Sitotroga cerealella* in northern Italy and other moths infesting stored foodstuffs. *Rev. Appl. Entomol. A*, *42*:35–36.

Champ, B. R., and Smithers, C. N. (1965). Insects and mites associated with stored products in Queensland. 1. Psocoptera. *Queensland J. Agr. Anim. Sci.*, *22*:259–262.

Chippendale, G. M. (1971). Observations on the physical and chemical composition of diets for the Angoumois grain moth. *J. Ins. Physiol.*, *17*:1257–1266.

Chippendale, G. M., and Mann, R. A. (1972). Feeding behavior of Angoumois grain moth larvae. *J. Ins. Physiol.*, *18*:87–94.

Cogburn, R. R., and Vick, K. W. (1981). Distribution of Angoumois grain moth, almond moth, and Indian meal moth in rice fields and rice storages in Texas as indicated by pheromone-baited adhesive traps. *Environ. Entomol.*, *10*:1003–1007.

Corbet, S., and Tams, W. H. T. (1943). Keys for the identification of the Lepidoptera infesting stored food products. *Proc. Zool. Soc. London.*, *113*:55–148.

Cotton, R. T. (1963). Pests of stored grain and grain products. Minneapolis: Burgess.

Cox, P. D., and Bell, C. H. (1991). Biology and ecology of moth pests of stored foods. *In* Ecology and Management of Food-Industry Pests (ed. Gorham, J. R.). FDA Tech. Bull. 4. Association of Official Analytical Chemists, pp. 181–193.

Cox, P. D. (1987). Cold tolerance and factors affecting the duration of diapause in *Ephestia kuehniella* Zeller (Lepidoptera: Pyralidae). *J. Stored Prod. Res.*, *23*:163–168.

Cox, P. D. (1986). A survey of stored product Lepidoptera in New Zealand. *N. Z. J. Exp. Agric.*, *14*:71–76.

Cox, P. D. (1974). The influence of temperature and humidity on the life-cycles of *Ephestia figulilella* Gregson and *Ephestia calidella* (Guenée) (Lepidoptera: Phycitidae). *J. Stored Prod. Res.*, *10*:43–55.

Cox, P. D. (1975a). The suitability of dried fruits, almonds, and carobs for the development of *Ephestia figulilella* Gregson, *E. calidella* (Guenée) and *E. cautella* (Walker) (Lepidoptera: Phycitidae). *J. Stored Prod. Res.*, *11*:229–233.

Cox, P. D. (1975b). The influence of photoperiod on the life-cycles of *Ephestia calidella* (Guenée) and *Ephestia figulilella* Gregson (Lepidoptera: Phycitidae). *J. Stored Prod. Res.*, *11*:75–88.

Cox, P. D., Allen, L. P., Pearson, J., and Beirne, M. A. (1984a). The incidence of diapause in seventeen populations of the flour moth, *Ephestia kuehniella* Zeller (Lepidoptera: Pyralidae). *J. Stored Prod. Res.*, *20*:139–143.

Cox, P. D., Bell, C. H., Pearson, J., and Beirne, M. A. (1984b). The effect of diapause on the tolerance of larvae of *Ephestia kuehniella* to methyl bromide and phosphine. *J. Stored Prod. Res.*, *20*:139–143.

Cox, P. D., Mfon, M., Parkin, S., and Seaman, J. E. (1981). Diapause in a Glasgow strain of the flour moth, *Ephestia kuehniella*. *Physiol. Entomol.*, *6*:349–356.

Curtis, C. E., and Landolt, P. J. (1992). Development and life history of *Pyralis farinalis* L. (Lepidoptera: Pyralidae) on an artificial diet. *J. Stored Prod. Res.*, *28*:171–177.

Davidson, R. H., and Peairs, M. (1966). Insect pests of farm, garden, and orchard. New York: Wiley.

Devine, T. L. (1982). The dynamics of body water in the booklouse *Liposcelis bostrychophilus* (Badonnel). *J. Exp. Zool.*, *222*:335–347.

Dickins, G. R. (1936). The scent glands of certain Phycitidae (Lepidoptera). *Trans. R. Entomol. Soc. London*, *85*:331–361.

Donohoe, H. C., Simmons, P., Barnes, D. F., Kaloostian, G. H., Fisher, C. K., and Heinrich, C. (1949). Biology of the raisin moth. Tech. Bull. U.S. Dept. Agric. 994.

Douglas, W. A., Henderson, C. A., and Langston, J. M. (1962). Biology of the pink scavenger caterpillar and its control in corn. *J. Econ. Entomol.*, *55*:651–655.

Ebeling, W. (1975). Urban Entomology. Univ. of California Div. of Agricultural Sciences.

Edwards, D. K. (1962). Laboratory determinations of the daily flight times of separate sexes of some moths in naturally changing light. *Can. J. Zool.*, *40*:511–530.

Finlayson, L. R. (1933). Some notes on the biology and life-history of psocids. *Entomol. Soc. Ontario Rep.*, *63*:56–58.

Fraenkel, G., and Blewette, M. (1946). The dietetics of the caterpillars of three *Ephestia* species, *E. kuehniella*, *E. elutella* and *E. cautella* and of a closely related species, *Plodia interpunctella*. *J. Exp. Biol.*, *22*:162–171.

Gentry, J. W., Harris, K. L., and Gentry, J. W. Jr. (1991). Order Lepidoptera (moths and butterflies). *In* Microanalytical Entomology for Food Sanitation Control, Vol. 1. Gentry and Harris, Melbourne, FL.

Goater, B. (1986). British pyralid moths. Harley Books, Martins, England.

Hagstrum, D. W., and Tomblin, C. F. (1975). Relationship between water consumption and oviposition by *Cadra cautella* (Lepidoptera: Phycitidae). *J. Georgia Entomol. Soc.* *10*:357–358.

Hagstrum, D. W., McLaughlin, J. R., Smittle, B. J., and Coffelt, J. A. (1978). Sterile males in a sex pheromone permeated atmosphere to reduce reproduction of *Ephestia cautella*. *Environ. Entomol.*, *7*:759–762.

Hill, D. S. (1990). Pests: Class insecta. *In* Pests of Stored Products and Their Control. CRC Press, Boca Raton.

Howe, R. W. (1965). A summary of estimates of optimal and minimal conditions for population increase of some stored products insects. *J. Stored Prod. Res.*, *1*:177–184

Jacob, T. A., and Cox, P. D. (1977). The influence of temperature and humidity on the life-cycle of *Ephestia kuehniella* Zeller (Lepidoptera: Pyralidae). *J. Stored Prod. Res.*, *13*:107–118.

Joubert, P. C. (1966). Field infestation of stored product insects in South Africa. *J. Stored Prod. Res.*, *2*:159–161.

Knulle, W., and Spadafora, R. R. (1969). Water vapour sorption and humidity relationships in *Liposcelis* (Insecta: Psocoptera). *J. Stored Prod. Res.*, *5*:49–55.

Koone, H. D. (1952). Maturity of corn and life history of the Angoumois grain moth. *J. Kans. Ent. Soc.*, *25*:103–105.

Kuwahara, Y., Kitamura, C., Takahashi, S., Hara, H., Ishii, S., and Fukami, H. (1971). Sex pheromone of the almond moth and the Indianmeal moth: cis-9, trans-12-tetradecadienyl acetate. *Science, 171*:801–802.

Landolt, P. J., and Curtis, C. E. (1982). Interspecific sexual attraction between *Pyralis farinalis* L. and *Amyelois transitella* (Walker) (Lepidoptera: Pyralidae). *J. Kans. Entomol. Soc., 55*:248–252.

Levinson, H. Z., and Buchelos, C. Th. (1981). Surveillance of storage moth species (Pyralidae, Gelechiidae) in a flour mill by adhesive traps with notes on the pheromone-mediated flight behavior of male moths. *Z. Angew. Entomol., 92*:233–254.

Lovitt, A. E., and Soderstrom, E. L. (1968). Predation on Indian meal moth eggs by *Liposcelis bostrychophilus*. *J. Econ. Entomol., 61*:1444–1445.

Lum, P. T. M., and Flaherty, B. R. (1969). Effect of mating with males reared in continuous light or in light-dark cycles on fecundity in *Plodia interpunctella*. *J. Stored Prod. Res.*, 5:80–94.

Madrid, F. J., and Sinha, R. N. (1982). Feeding damage of three stored-product moths (Lepidoptera: Pyralidae) on wheat. *J. Econ. Entomol., 75*:1017–1020.

Mallis, A. (1982). Handbook of pest control. Cleveland: Franzak and Foster. 1101 pp.

Mbata, G. N., and Osuji, F. N. C. (1983). Some aspects of the biology of *Plodia interpunctella* (Hübner) (Lepidoptera: Pyralidae), a pest of stored groundnuts in Nigeria. *J. Stored Prod. Res., 19*:141–151.

McLaughlin, J. R. (1982). Behavioral effect of a sex pheromone extracted from forewings of male *Plodia interpunctella*. *Environ. Entomol., 11*:378–380.

McLaughlin, J. R., and Hagstrum, D. W. (1976). Effects of a dark environment and air permeation with synthetic sex pheromone on mating in the almond moth. *Environ. Entomol., 5*:1057–1058.

Mills, J. T., Sinha, R. N., and Demianyk, C. J. (1992). Feeding and multiplication of a psocid, *Liposcelis bostrychophilus* Badonnel (Psocoptera: Liposcelidae), on wheat, grain screenings, and fungi. *J. Econ. Entomol., 85*:1453–1462.

Mills, R. B. (1965). Early germ feeding and larval development of the Angoumois grain moth. *J. Econ. Entomol., 58*:220–223.

Mound, L. (1989). Common insect pests of stored food products. A guide to their identification. British Museum of Natural History.

Mullen, M. A., and Arbogast, R. T. (1977). Influence of substrate on oviposition by two species of stored-product moths. *Environ. Entomol., 6*:641–642.

Neunzig, H. H. (1990). Pyraloides, Pyralidae (part). *In* The Moths of America North of Mexico. Fasc. 15.3. (ed. Dominick, R. B. et al.). Allen Press, Lawrence, Kansas.

New, T. R. (1987). Biology of the Psocoptera. *Oriental Insects, 21*:1–109.

Rees, D. P., and Walker, A. J. (1990). The effect of temperature and relative humidity on population growth of three *Liposcelis* species (Psocoptera: Liposcelidae) infesting stored products in tropical countries. *Bull. Entomol. Res., 80*:353–358.

Richards, O. W., and Thomson, W. S. (1932). A contribution to the study of the genus *Ephestia* Gn. (including *Strymax* Dyar) and *Plodia* Gn. (Lepidoptera: Phycitidae) with notes on parasites of the larvae. *Trans. Ent. Soc. London, 80*:169–248.

Riemann, J. G., and Ruud, R. L. (1974). Mediterranean flour moth: effects of continuous light on the reproductive capacity. *Ann. Entomol. Soc. Am.*, 67:857–860.

Russell, M. P. (1962). Field infestation of corn in Indiana by the Angoumois grain moth and a rice weevil. *J. Econ. Entomol.*, *55*:814–815.

Shazali, M. E. H., and Smith, R. H. (1985). Life history studies of internally feeding pests of sorghum: *Sitotroga cerealella* (Ol.) and *Sitophilus oryzae* (L.). *J. Stored Prod. Res.*, *21*:171–178.

Silhacek, D. L., and Miller, G. L. (1972). Growth and development of the Indian meal moth, *Plodia interpunctella* (Lepidoptera: Phycitidae), under laboratory mass-rearing conditions. *Ann. Entomol. Soc. Am.*, *65*:1084–1087.

Simmons, P., and Ellington, G. W. (1933). Life history of the Angoumois grain moth in Maryland. *U.S. Dept. Agric. Tech. Bull.*, *351*:1–34.

Sinha, R. N. (1988). Population dynamics of Psocoptera in farm-stored grain and oilseed. *Can. J. Zool.*, *66*:2618–2627.

Sinha, R. N., and Watters, F. L. (1985). Insect pests of flour mills, grain elevators, and feed mills and their control. Agriculture Canada Publication 1776E. Agriculture Canada, Ottawa.

Sinha, R. N., Liscombe, E. A. R., and Wallace, H. A. H. (1962). Infestation of mites, insects, and microorganisms in a large wheat bulk after prolonged storage. *Can. Entomol.*, *94*:542–555.

Traynier, R. M. M. (1970). Sexual behavior of the Mediterranean flour moth, *Anagasta kuehniella*: Some influences of age, photoperiod, and light intensity. *Can. Entomol.*, *102*:534–540.

Trematerra, P. (1988). Anthropod antagonists of *Pyralis farinalis* (L.) (Lepidoptera: Pyralidae) and visitors of its larval shelters. *J. Appl. Entomol.*, *105*:525–528.

Trematerra, P., and Galli, A. (1985). On the feeding habits of the larvae of *Pyralis farinalis* (L.). *Acad. Nay. Ital. Ent.*, *14*:543–550.

Tuli, S., Mookherjee, P. B., and Sharma, G. C. (1966). Effect of temperature and humidity on the fecundity and development of *Cadra cautella* Wlk. in wheat. *Indian J. Entomol.*, 28:305–317.

Turner, B. D., and Maude-Roxby, H. (1988). Starvation survival of the stored product pest *Liposcelis bostrychophilus* Badonnel (Psocoptera, Liposcelidae). *J. Stored Prod. Res.*, *24*:23–28.

Tzanakakis, M. E. (1959). An ecological study of the Indian meal moth, *Plodia interpunctella*, with emphasis on diapause. *Hilgardia*, *29*:205–246.

USDA (1991). Insect and Mite Pests in Food: An Illustrated Key. (ed. Gorham, J. R.). USDA Handbook No. 655.

Waloff, N., Norris, M. J., and Broadhead, E. C. (1948). Fecundity and longevity of *Ephestia elutella*. *Trans. R. Entomol. Soc.*, *99*:245–268.

Watt, M. J. (1965). Notes on pests of stored grain *Liposcelis bostrychophilus* and *Sitophilus* spp. *Agr. Gazette*, *1965*:693–696.

Williams, G. C. (1964). The life-history of the Indian meal-moth, *Plodia interpunctella* (Hübner) (Lep. Phycitidae) in a warehouse in Britain and on different foods. *Ann. Appl. Bio.*, *53*:459–475.

Williams, L. H. (1972). Anobiid eggs consumed by a psocid (Psocoptera: Liposcelidae). *Ann. Entomol. Soc. Am.*, *65*:533–553.

# 3

# Ecology

**David W. Hagstrum, Paul W. Flinn, and Ralph W. Howard**
*U.S. Department of Agriculture, Manhattan, Kansas*

Insect ecology is the study of factors regulating insect distribution and abundance. Knowledge of insect ecology is particularly important in the development of integrated pest management (IPM) programs. The nonchemical control methods used in IPM programs are generally more dependent on an understanding of insect ecology than are chemical control methods. An overview of factors regulating insect populations in storage, transportation, and processing facilities is provided to help us design pest management programs that take advantage of these regulating factors. Computer models are discussed as a way of using this ecological knowledge to predict potential pest damage. The chapter on integrated pest management shows how these models can be used to design pest management programs.

Stored-product insects are well adapted to surviving on food residues in storage, transportation, and processing facilities. Adaptations include the ability to utilize a broad range of foods and the mobility to locate food, mates, and oviposition sites even when they are scarce. The need to utilize scarce resources makes dispersal a particularly important part of population dynamics. Their adaptations for utilizing scarce resources make stored-product insect pests difficult to manage and this must be considered in designing pest management programs.

Ecological studies done under conditions similar to those found in storage, transportation, and processing facilities are most likely to be useful in improving pest management. However, the majority of studies of stored-product insect ecology have been conducted at unrealistically high densities and under unnatural conditions (Bellows 1982; Gordon and Stewart 1988; Sokoloff 1974, and references cited therein). In this chapter, we will emphasize studies that were conducted at insect densities and under conditions typically encountered by insects infesting raw and processed commodities. In laboratory cultures, crowding becomes the primary factor regulating population growth, while in storage, transportation, and processing facilities, the effects of temperature, moisture, and food availability and quality are the main factors regulating population growth in the absence of pest control. Pest management as a factor regulating insect populations will be discussed in other chapters on biological, physical, and chemical control.

The objective of this chapter is to provide an understanding of the ecology of stored-product insects that can help us to develop better pest management programs. The chapter has two major subdivisions. The first deals with some of the most important factors regulating insect distribution and abundance, and the second examines the population dynamics of source populations, and populations in raw or processed commodities.

## I. FACTORS REGULATING INSECT DISTRIBUTION AND ABUNDANCE

Some of the primary factors regulating stored-product insect distribution and abundance are temperature, moisture, food availability and quality, commodity handing and processing practices, and behavior-eliciting chemicals. Environmental factors and food influence insect abundance through their effects on insect developmental times, survival and egg production, while other factors such as handling and processing of commodities and behavior-eliciting chemicals mainly affect survival and behavior. Insect behavior is important in determining insect distribution because insects tend to prefer certain environmental conditions and must locate resources such as food, mates, and oviposition sites.

### 1. Temperature, Moisture, and Diet

Insect development time, survival, and egg production depend upon the suitability of environment and diet. Insect population growth rates can increase as a result of shorter development times, increased survivals, or higher egg production.

#### *Development Time*

The temperature and moisture content of a commodity have a major influence on insect development time. Table 1 shows the effects of temperature on the total development times of 11 species of stored-product beetles and six species of stored-product moths. The majority of stored-product insect pests are either beetles or moths. The development time of beetles was generally affected more by temperature than by moisture or diet (Hagstrum and Milliken 1988). For moths, differences in development time caused by relative humidity (2.2–2.6-fold) and diet (1.1–3.3-fold) were smaller than those caused by temperature (2.8–13.2-fold (Subramanyam and Hagstrum 1993). For two species of moths, the effects of diet were similar across a broad range of temperatures. The development time of *Sitophilus granarius* (L.) reared on five different types of grain varied from 35.1 days on rice to 45.5 days on maize (Table 2). This table also shows the large

**Table 1** Effects of Temperature on the Predicted Egg-to-Adult Development Times (in Days) of Stored-Product Insects[a]

| | Temperature (°C) | | | | | | | | |
|---|---|---|---|---|---|---|---|---|---|
| Species | 17.5 | 20 | 22.5 | 25 | 27.5 | 30 | 32.5 | 35 | 37.5 |
| Moths | | | | | | | | | |
| *Ephestia kuehniella* | — | 69.1 | 56.0 | 46.5 | 40.6 | 39.2 | — | — | — |
| *Plodia interpunctella* | 150.9 | 99.3 | 67.3 | 48.1 | 37.9 | 34.9 | 38.4 | 49.1 | — |
| *Cadra cautella* | 108.9 | 76.7 | 57.1 | 45.3 | 38.3 | 34.4 | 32.5 | 31.8 | — |
| *Corcyra cephalonica* | 192.0 | 92.6 | 57.6 | 44.8 | 39.7 | 37.4 | 36.1 | 35.2 | — |
| *Ephestia calidella* | 94.2 | 62.7 | 43.6 | 32.9 | 28.2 | 28.7 | 34.2 | 45.8 | — |
| *Cadra figulilella* | 129.2 | 98.7 | 76.6 | 60.9 | 50.8 | 45.9 | 46.5 | 54.1 | — |
| Average | 135.0 | 83.2 | 59.7 | 46.4 | 39.3 | 36.7 | 37.5 | 46.1 | — |
| Long-lived beetles | | | | | | | | | |
| *Cryptolestes ferrugineus* | — | — | 53.4 | 37.0 | 28.1 | 23.2 | 20.6 | 19.0 | 18.2 |
| *Cryptolestes pusillus* | — | — | 53.1 | 45.1 | 38.5 | 32.9 | 28.4 | 25.1 | 24.5 |
| *Oryzaephilus surinamensis* | — | — | 48.5 | 36.4 | 27.9 | 22.4 | 19.8 | 20.8 | 27.0 |
| *Sitophilus oryzae* | — | 52.9 | 43.2 | 35.9 | 30.6 | 27.4 | 26.7 | 29.1 | 36.7 |
| *Tribolium castaneum* | — | — | — | 41.8 | 32.7 | 28.4 | 26.3 | 23.4 | 21.7 |
| *Tribolium confusum* | — | — | 56.2 | 44.6 | 35.6 | 28.5 | 23.0 | 20.0 | 34.1 |
| *Rhyzopertha dominica* | — | — | — | 58.8 | 49.9 | 42.4 | 36.1 | 31.0 | — |
| Average | — | — | 50.9 | 42.8 | 34.7 | 29.4 | 25.8 | 24.1 | 27.0 |
| Short-lived beetles | | | | | | | | | |
| *Acanthoscelides obtectus* | 82.0 | 60.4 | 45.4 | 35.7 | 30.2 | 28.9 | — | — | — |
| *Callosobruchus maculatus* | 167.4 | 72.5 | 42.0 | 31.5 | 27.3 | 25.3 | 23.9 | 22.8 | 21.9 |
| *Lasioderma serricorne* | | 94.8 | 62.1 | 43.1 | 32.9 | 28.3 | 27.9 | 30.7 | 36.5 |
| *Stegobium paniceum* | 153.5 | 105.4 | 73.4 | 52.9 | 41.9 | 41.6 | 58.4 | — | — |
| Average | 134.3 | 83.3 | 55.7 | 40.8 | 33.1 | 31.0 | 36.7 | 26.8 | 29.2 |

[a]Predicted development times of moths and long-lived beetles are based on equations from Subramanyam and Hagstrum (1993) and Hagstrum and Milliken (1988), respectively. The same equation is used to predict the development times of short-lived beetles using data from Menusan (1934) for *A. obtectus*; from El-Sawaf (1956), Giga and Smith (1983), and Mookherjee and Chawla (1964) for *C. maculatus*; from Howe (1957) for *L. serricorne*; and from Lefkovitch (1967) and Momoi and Sadamori (1982) for *S. paniceum*.

variation in development times between individual insects on the same type of grain. Similar results have been reported for *Tribolium castaneum* (Herbst) (Pant and Dang 1969) and *Sitophilus oryzae* (L.) (Baker 1988).

Average development times of six moth species were longer than those of seven species of long-lived beetles between 15 and 40°C (Table 1). Food eaten by

**Table 2** Development Time from Egg to Adult and Progeny Production of *S. granarius* on Five Types of Grain at 27.5°C and 75% Relative Humidity

| | Development Time (Days) | | | Progeny Production per 20 Weevils/4 Days | | |
|---|---|---|---|---|---|---|
| Grain | Mean | SD | Range | Mean | SD | Range |
| Maize | 45.5 | 1.5 | 32–66 | 21.0 | 7.7 | 10–33 |
| Barley | 41.1 | 1.5 | 30–62 | 96.7 | 17.2 | 73–126 |
| Oats | 40.2 | 2.0 | 34–52 | 9.4 | 3.8 | 4–17 |
| Wheat | 39.9 | 0.5 | 32–60 | 82.4 | 17.6 | 56–112 |
| Rice | 35.1 | 0.6 | 30–48 | 12.5 | 3.4 | 8–17 |

*Source*: Based upon data from Schwartz and Burkholder (1991).

moth larvae was used by adult moths for egg production, and adult moths are short-lived and generally do not feed. In contrast, adults of long-lived beetle species must feed to produce eggs and they reproduce throughout most of their lifetime. Most of the difference in development times between moths and long-lived beetles was due to moths spending a larger percentage of their total development time as feeding larvae (Table 3). Like moths, several species of short-lived beetles such as *Acanthoscelides obtectus* (Say), *Callosobruchus maculatus* (F.), *Stegobium paniceum* (L.), and *Lasioderma serricorne* (F.) tend to have long development times and produce eggs without feeding. Developmental times for short-lived beetles are not as long as for moths, but the former also produce fewer eggs.

**Table 3** Effect of Insect Order and Moisture Content of Commodity on Percentage of Total Development Time Spent in Each Stage

| | Stage | | |
|---|---|---|---|
| Insect order | Egg | Larva | Pupa |
| High moisture content | | | |
| Moths | 8 | 77 | 15 |
| Beetles | 15 | 66 | 19 |
| Low moisture content | | | |
| Beetles | 12 | 72 | 15 |

*Source*: Based on data from Hagstrum and Milliken (1988), and Subramanyam and Hagstrum (1993).

Food consumption rate increases with temperature (Baker and Loschiavo 1987) and may partially explain the effects of temperature on development time. The effects of temperature on the feeding rate of *C. maculatus* are shown in Figure 1. Larval feeding activity increased as temperature increased from 13 to 25°C, leveled off between 25 and 38°C, and then decreased as temperatures increased from 38 to 45°C. Day–night fluctuations in temperature also affect the development times of several species of stored-product insects (Hagstrum and Milliken 1991). Development times at constant temperatures are significantly longer at low temperatures and shorter at high temperatures than at fluctuating temperatures with the same mean. These differences are due to the nonlinear relationship between temperature and developmental time.

Typical changes in the age structure of a beetle population started with a single group of young adults are shown in Figure 2. With each generation, the number of *Rhyzopertha dominica* (F.) larvae increases, and larvae temporarily become a larger portion of the population. However, once these larvae begin to pupate and emerge as adults, larvae again become a smaller portion of the population. The amplitude of this fluctuation in age structure decreases with each generation as the population approaches a stable age structure of about four larvae per adult. Once the population reaches a stable age structure, the percentage of total population in each stage remains constant. Moisture affects the duration of larval stage, but not egg or pupal stages (Table 3). The time spent in the larval

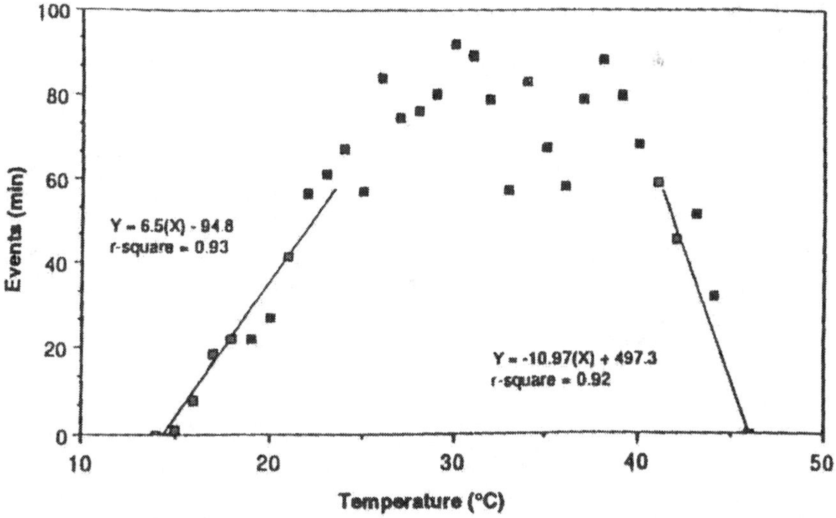


**Figure 1** Feeding activity of *C. maculatus* larvae over a range of temperatures. (From Shade et al. 1990).

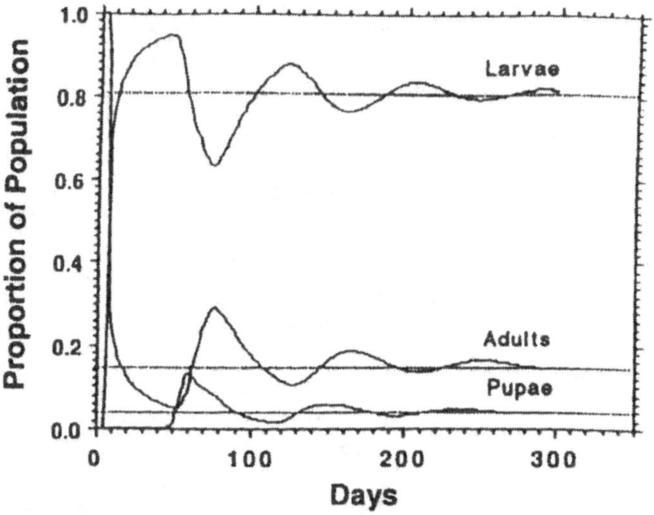


**Figure 2** Changes in the age structure of *R. dominica* populations at 27°C and 14% wheat moisture content as predicted by computer simulation model. The dashed line shows the stable age distribution. (From Hagstrum and Flinn 1992.)

stage is increased from 66 to 72% by decreasing moisture. Because the duration of a stage determines the number of insects in that stage when a population reaches a stable age structure, the age structure of a population will be affected by moisture. However, the percentage of total development time spent in each stage is the same over the 20–35°C temperature range, and thus the stable age structure of the population is less likely to be affected by temperature than moisture.

In addition to knowing the mean development time, the variation in developmental times between individuals must be considered to predict insect population growth accurately. Figure 3 shows typical emergence curves for adult *T. castaneum* between 22.5 and 35°C. Predictions of models incorporating variable adult emergence times are more realistic than those assuming that all adults emerge at the same time.

*Survival*

Insect population growth is influenced by immature and adult survivorship. For *T. castaneum*, the average percentage survival from egg to adult is above 80% between 25 and 37.5°C and decreases rapidly above and below this temperature range (Fig. 4). From the standard deviations, we can infer that, between 25 and 37.5°C, we are 65% confident that average survival was 71–100%. Immature survival for several other species of stored-product insects are within one standard deviation for 48% of the data and within two standard deviations for 86% of the

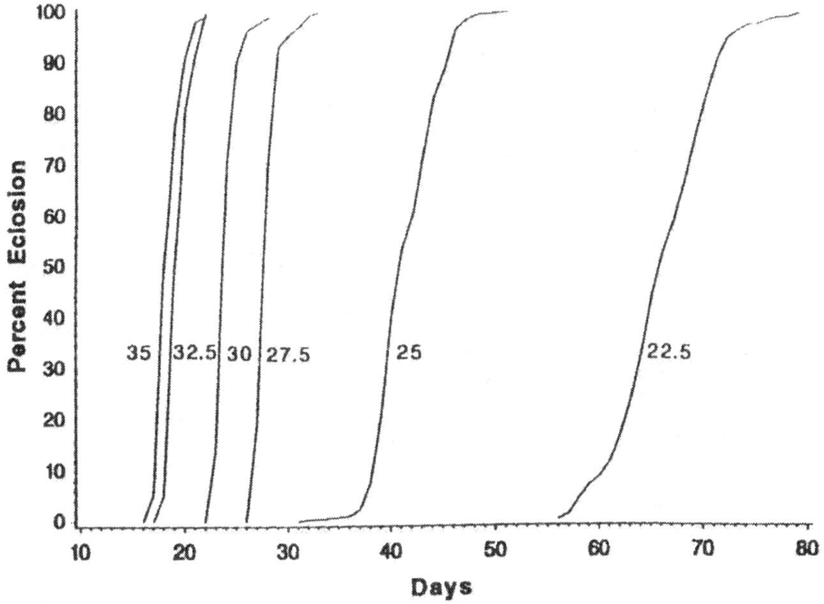


**Figure 3** Effects of temperature on adult eclosion curves for *T. castaneum*. (Based on unpublished portion of data from Hagstrum and Milliken 1991.)

data. If the relationship between temperature and survival were the same for all species, 65 and 95% of data would be expected within one and two standard deviations, respectively. The tendency for survival rates of other species to fall within the standard deviations for *T. castaneum* suggests clearly that the effects of temperature on immature survival are similar for these seven species. Studies have shown that the majority of immature mortality of *R. dominica* and *S. oryzae* occurs in the first instar (Birch 1945; Howe 1952). High mortality of young larvae is probably a consequence of the difficulty of metamorphosis from egg to larval stage and initiation of larval feeding, and may also be characteristic of many other species. As moisture decreases, survival of *R. dominica* immatures decreases (Fig. 5). The shape of the survival curve of this species also changes as a result of survival being reduced more at low temperatures than at high temperatures. In making control decisions, the graph suggests that when it is not possible to cool the grain sufficiently to reduce population growth, populations might also be reduced by drying the grain or keeping it at high temperatures.

Adult survivorship also is affected by temperature and moisture. Because reproduction can occur throughout much of adult life, adult survivorship affects total egg production. Figure 6 provides the survivorship curves of a beetle, *T.*

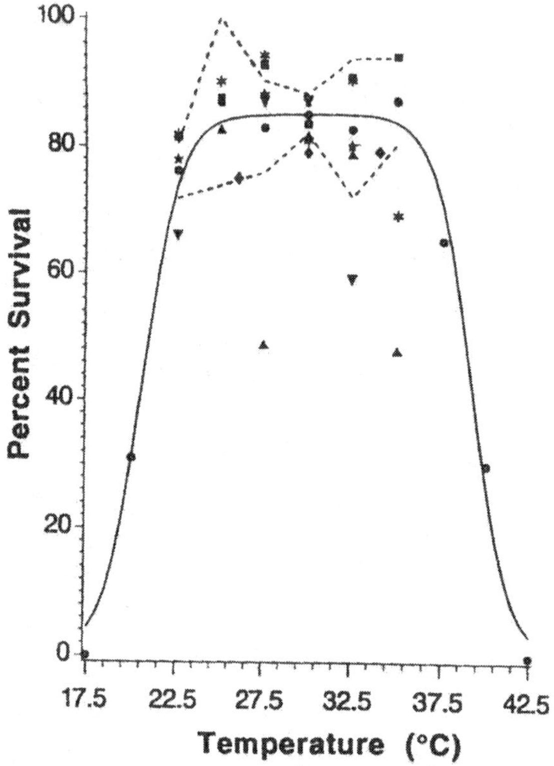


**Figure 4** Effect of temperature on observed (●) and predicted (——) percentage survival of *T. castaneum* (based on unpublished portion of data from Hagstrum and Milliken 1991). Using equations similar to those developed by White (1985), the survival of *T. castaneum* can be described as: survival = $0.85/[1 + e^{(19.73 - 0.96 \times \text{temperature})}]$ if temperature ≤ 29.85 or survival = $0.85/[1 + e^{(19.73 - 0.96 \times (59.7 - \text{temperature}))}]$ if temperature > 29.85. The dashed lines show ± 1 standard deviation. Survival data from Birch (1945), Arbogast (1976), Komson and Stewart (1968), Smith (1965), Currie (1967), and Howe (1960) are also given for *R. dominica* (◆), *O. mercator* (▼), *O. surinamensis* (★), *C. ferrugineus* (*), *C. pusillus* (▲), and *T. confusum* (■).

*castaneum*, over a range of temperatures and grain moistures. Adult survivorship increased as moisture increased and decreased as temperature increased. Figure 7 shows the effects of temperature on the survivorship of the moth, *Ephestia kuehniella* (Zeller). Adult survivorship decreased steadily as temperature increased from 20 to 27.5°C. At 20°C, adults all lived 5 days and rarely lived more than 11 days, while at 27.5°C, adults all lived 3 days and rarely lived more than 7 days.

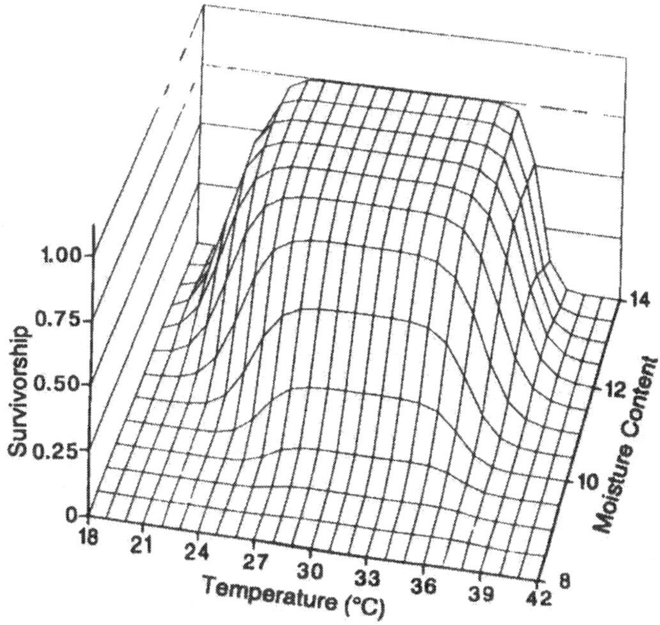


**Figure 5** Effects of temperature and wheat moisture content on the survival of *R. dominica*. (Redrawn from Birch 1945.)

Mortality at low temperatures is a function of cooling rate, exposure time, temperature, and insect species (Fields 1992). The slower the cooling rate, the better insects acclimate and survive exposure to cold temperatures. Figure 8 shows that the survivorship of immatures and adults exposed to 9°C is higher with, than without, acclimation. Also, survivorship decreased more rapidly at 9°C than at 13.5°C. An interesting finding is that survivorship of adults was greater at 13.5°C than at 32°C. This is probably due to low reproduction and activity at 13.5°C increasing adult longevity, and greater reproduction and activity at 32°C reducing longevity. Among five species, survivorship at cold temperatures tended to be highest for *Oryzaephilus surinamensis* (L.) and lowest for *T. castaneum* (Fig. 9).

*Egg Production*

The number of eggs produced over the lifetime of a female varies with insect species, temperature, moisture, and diet. The pattern of egg production differs between stored-product moths and beetles. Moths accumulate nutrition for egg production as larvae, produce large numbers of eggs soon after adult eclosion, and die shortly afterward (Fig. 7). In contrast, some adult beetles live much longer than adult moths and produce eggs over several months (Fig. 6). Egg production of *T.*

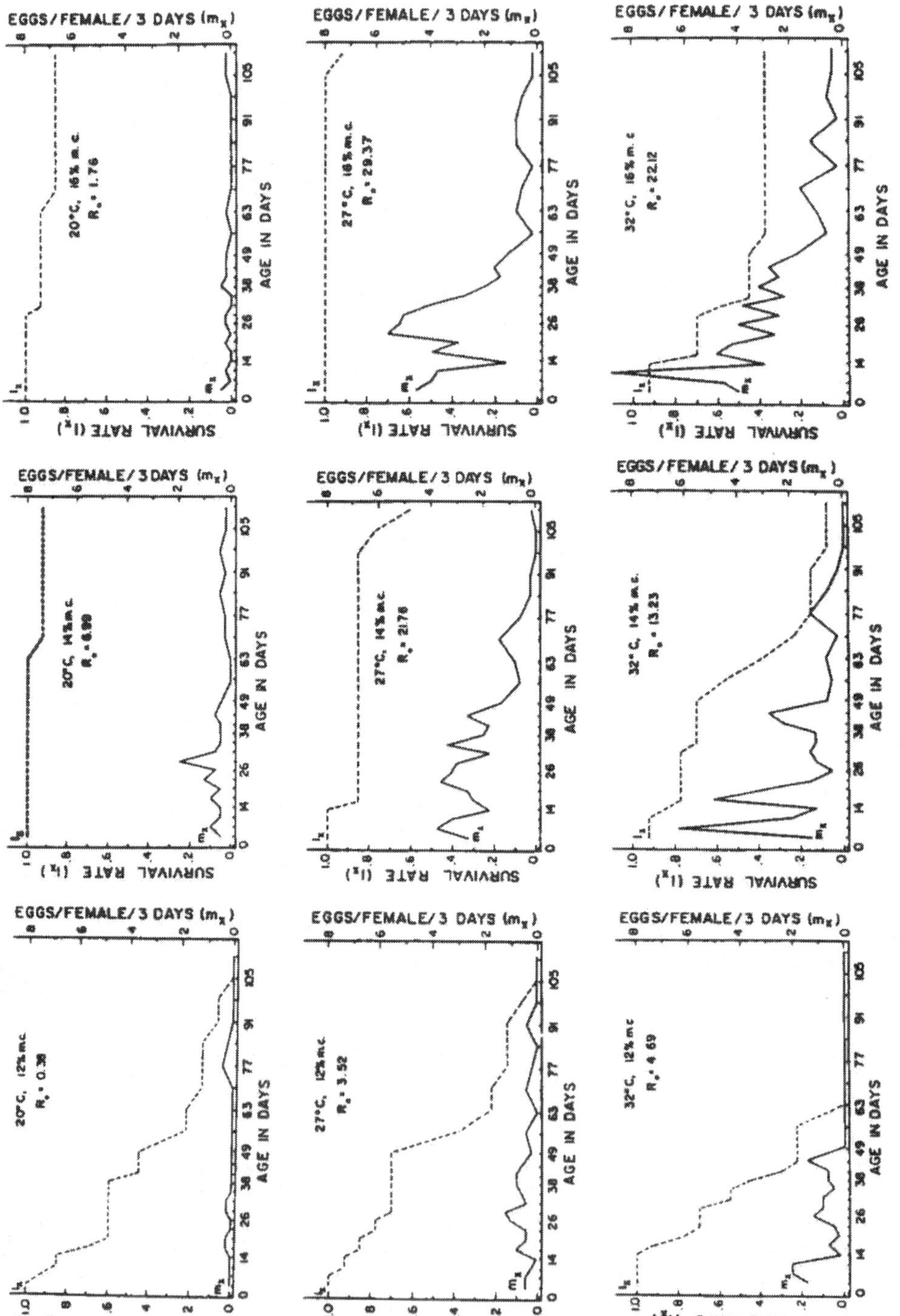

20°C, 12% m.c.
$R_o$ = 0.38
27°C, 12% m.c.
$R_o$ = 3.52
32°C, 12% m.c.
$R_o$ = 4.69
20°C, 14% m.c.
27°C, 14% m.c.
$R_o$ = 21.78
32°C, 14% m.c.
$R_o$ = 13.23
20°C, 16% m.c.
$R_o$ = 1.76
27°C, 16% m.c.
$R_o$ = 29.37
32°C, 16% m.c.
$R_o$ = 22.12
EGGS/FEMALE/3 DAYS ($m_x$)
SURVIVAL RATE ($l_x$)
AGE IN DAYS

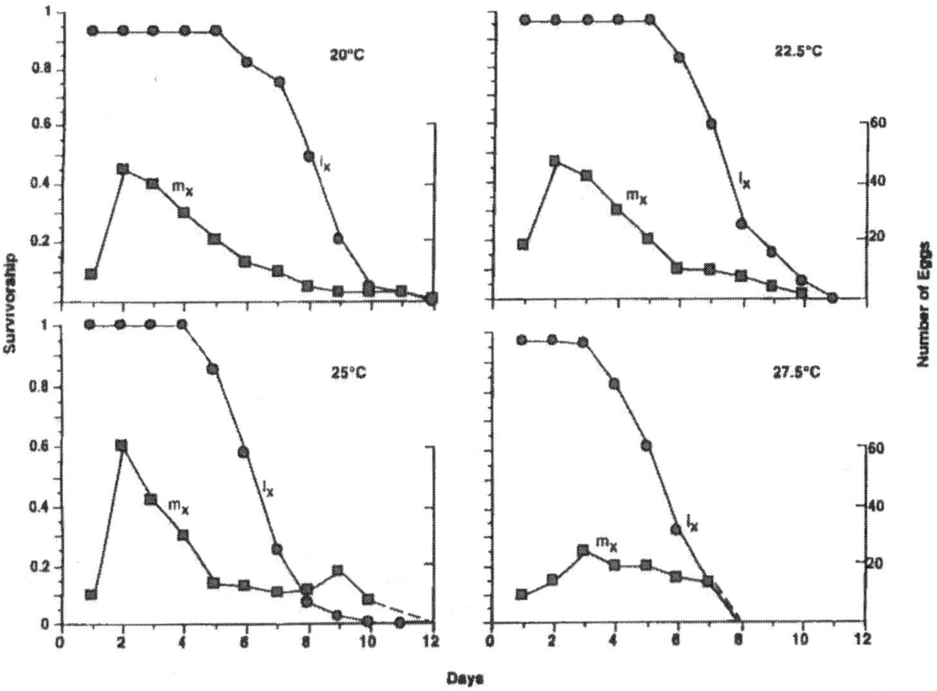


**Figure 7** Effect of temperature on adult longevity ($l_x$) and fecundity ($m_x$) of *E. kuehniella*. (Redrawn from Siddiqui and Barlow 1973.)

*castaneum* increased with increasing temperature and moisture. However, interaction of adult survival and egg production resulted in maximal reproduction at 27°C and 16% moisture. The progeny production of *S. granarius* reared on five different types of grain varied from 9.4 on oats to 96.7 on barley (Table 2). There is considerable variation in progeny production between groups of insects of one species on the same type of grain. The data in Table 2 suggest that the suitabilities of grain types for insect development do not predict their suitabilities for progeny production.

For pyralid moths, oviposition decreased when mating was delayed (Barrer 1976) or when females mated with males reared under continuous light (Lum and Flaherty 1970), and increased when females had access to free water (Hagstrum and Tomblin 1975). Delayed mating reduced oviposition because female moths

---

**Figure 6** Effect of temperature and wheat moisture content on adult longevity ($l_x$) and fecundity ($m_x$) of *T. castaneum*. (Reprinted with permission from Lhaloui et al. 1988.)

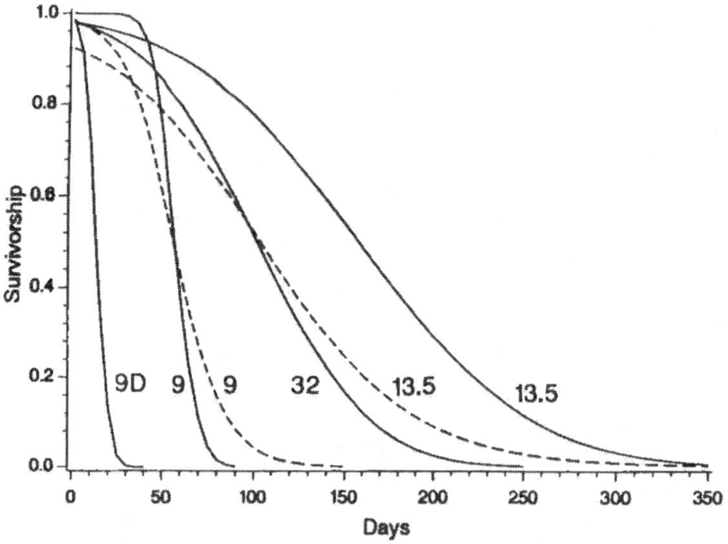


**Figure 8** Effect of temperature on the survivorship of immature (----) and adult (——) *R. dominica*. Cooling rate from rearing temperature of 32°C to 9 or 13.5°C was 4.5°C per week, except for one case in which insects were transferred directly from 32 to 9°C (9D). (Based on data from Evans 1983, 1987.)

did not live to lay all of their eggs. The increase in egg production with free water indicates that females could not produce as many eggs if energy had to be used to produce metabolic water. Free water also increased the fecundity and longevity of the beetle, *Dermestes lardarius* L. (Jacob and Fleming 1982). Feeding increased the oviposition of the beetle, *Callosobruchus chinensis* (L.) (Shinoda and Yoshida 1984).

Insect developmental time, survival, and egg production, and thus population growth, are affected by temperature and moisture. Therefore, reducing the suitability of environment can be an important part of an insect pest management program. Many species of stored-product insects can feed on a broad range of commodities, and insect developmental time, survival, and egg production vary with commodity. Even for a particular commodity, the susceptibility to insects may vary with the variety of the commodity, as has been shown for wheat (Baker et al. 1991; McGaughey et al. 1990), rice (Cogburn and Bollich 1990), maize (Dobie 1974), and cowpeas (Fitzner et al. 1985).

## 2. Interactions among Individuals and Species

Intraspecies interactions among individual insects such as crowding may be important in limiting the size of insect populations breeding on residual food in

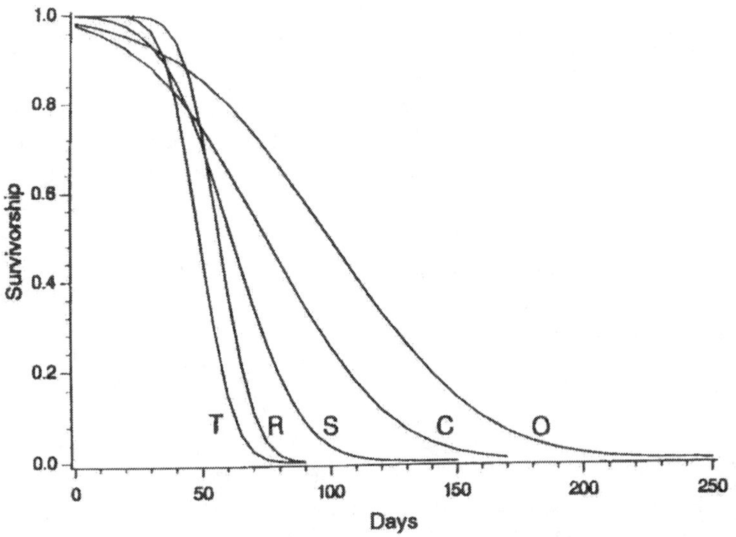


**Figure 9** Survivorship of adult *T. castaneum* (T), *R. dominica* (R), *S. oryzae* (S), *C. ferrugineus* (C), and *O. surinamensis* (O) at 9°C. Cooling rate from rearing temperature of 32°C to 9°C was 4.5°C per week. (Based on data from Evans 1983.)

empty storage, transportation, or processing facilities, but the importance of crowding is reduced by the tendency of adults to emigrate before crowding occurs (Hagstrum and Gilbert 1976). Solomon (1953) reviewed the effects of crowding on mortality and egg production of stored-product insects. The effects of crowding increased logarithmically with population density. For eight species, mortality reached 20% at densities of 1–32 insects/g of infested commodity and 80% at densities of 7–130/g. For many species, egg production was reduced 20% at densities of 0.3–16 insects/g and 80% reductions were observed above three insects/g. Overall, reproduction tended to be depressed more than mortality at a given density. These are extremely high densities. Wheat is considered infested if densities are ≥0.002 insects/g, and densities of 0.3 insects/g can cause heating of the grain (Hagstrum and Flinn 1992).

Interspecies interactions also can be important in regulating population dynamics (Lefkovitch 1968; LeCato 1975a). Under crowded conditions, many species of stored-product insects eat the inactive stages such as eggs or pupae of their own or other species. LeCato (1975b) found that dead eggs or adults of *Plodia interpunctella* (Hubner) were eaten by *T. castaneum*, and that this increased population growth and reduced mortality. Arbogast (1989) showed that feeding on dead moths increased the population growth of *Latheticus oryzae* Waterhouse on maize more than such feeding increased the population growth of *T. castaneum*. Another type of interspecies interaction is the production of hastisetae by *Trogoderma inclusum* LeConte and *Trogoderma variabile* Ballion (Kokubu and Mills 1980). Hastisetae attach to setae of other insects and can incapacitate them by entangling their appendages. Of 13 species tested, *S. oryzae* and *Cryptolestes pusillus* (Schonherr) were the most susceptible with 93 and 83% mortality, respectively, at the end of 72 h exposure. Many species were able to remove hastisetae by cleaning their body surface with their legs and mouthparts; however, this reduces the time available for feeding and reproduction.

## 3. Handling and Processing

Many of the practices used to harvest, dry, move, clean, and process commodities are part of the ecology of stored-product insects and may have a positive or negative influence on insect populations. Although stored-product insects are often thought to spend most of their time immersed in food, food can be limiting. In bulk storage, some species of insect may need to find wheat or maize kernels with cracks in the outer covering (pericarp) and rice or peanuts with cracked hulls. Commodities may be more susceptible to insects at some stages of processing than at others. For processed commodities, insect-resistant packaging is widely used to prevent insects from finding food.

### *Harvesting*

Combine threshing can result in 20–50% wheat kernel damage (Tuff and Telford 1964; White 1982; White and Bell 1990). Damaged kernels are more susceptible than whole kernels to insect attack by *O. surinamensis* on wheat (Fleming 1988), *T. castaneum* on wheat (White 1982), *Cryptolestes ferrugineus* (Stephens) on wheat (White and Bell 1990) or maize (Throne and Culik 1989), *C. pusillus* on maize (Cline 1991), and *S. oryzae* on wheat (Khare et al. 1979) or on resistant sorghum (Williams and Mills 1980). However, for wheat, *O. surinamensis* is probably the only major insect pest that requires broken kernels for development and reproduction (Flinn et al. 1992b). Setting the combine to minimize kernel damage can be an important part of a stored-grain insect pest management program.

### *Drying*

Drying affects insect populations directly by reducing survival or indirectly by causing cracks that increase the susceptibility of a commodity to insects. Fields (1992) reviews the effects of high temperatures on insect survival. The susceptibility of insects to heat increases as the moisture content of a commodity decreases. Farrar and Reed (1942) determined the lethal drying conditions for six species of insects on wheat. Survival rates depended upon whether lethal temperatures were maintained for enough time to kill insects. *R. dominica* was the most resistant, and *S. oryzae* and *S. granarius* were the least resistant. *Sitotroga cerealella* (Olivier) was more resistant than *O. surinamensis* and both species were more resistant than *C. pusillus* and *Tribolium confusum* Duval. They found that the tendency for air to follow channels through a mass of grain may result in differences of as much as 28°C in temperature within a distance of 5–10 cm. Such differences in temperatures could allow some insects to survive the drying process. Sun drying of barley by farmers in Japan killed some first to third instar *Sitophilus zeamais* Motschulsky within 10 min, but had no effect on fourth instars or pupae (Yoshida 1974). Drying rice with heated air produced splits in the husk that allowed *S. oryzae* and *S. zeamais* to feed and oviposit (Takahashi and Mizuno 1982). For maize infested by *S. zeamais* in the field in Georgia, threshing separated fewer than 20% of live weevils from seed, and drying killed roughly 90% of insects (Keever et al. 1988). Traditional exposure of harvested azuki beans to the sun for a period of 1–3 days did not affect the survival of *C. chinensis*, although the beans were exposed to temperatures of 33.6°C and their water content was reduced by 2.5% (Shinoda and Yoshida 1985).

### *Moving*

Insect populations in grain were reduced by augering (Muir et al. 1977; Watters and Bickis 1978) or by pneumatic conveying (Bryan and Elvidge 1977; Bahr 1973, 1975). Muir et al. (1977) reported mortalities of 61% of larvae and 83% of adults of *C. ferrugineus* as a result of augering. Bahr (1975) found that mortalities often exceeded 90%, and ranked the susceptibility of different species to mortality from pneumatic conveying as *S. granarius* $<$ *S. oryzae* $<$ *C. ferrugineus* $<$ *R. dominica* $<$ *T. castaneum* $<$ *O. surinamensis*. In these studies, pneumatic conveying reduce survival more than augering. Mortality from conveying may not occur immediately. Rodinov (1938) observed that 33% of *S. granarius* died 1 day after turning, and mortality was 92% after 12 days. The percentages of *S. oryzae* surviving when maize was dropped from 0.9, 1.8, 4.3, and 11.0 m were 80.7, 81.1, 63.7, and 46.7, respectively (Joffe and Clark 1963). Impacts of 6.4 m/s and 12.8 m/s, which are equivalent to dropping wheat 3.3 and 6.6 m, resulted in 65 and 90% mortality for *S. granarius* pupae (Bailey 1969). Adults and larvae survived impact better than pupae.

Grain handling also increases the amount of damaged kernels, cracked grains, and grain dust, which are the preferred diet of some species of stored-product insect pests. Chang et al. (1983) showed that for wheat stored in farm bins, the fines decreased from the center to the bin wall and McGregor (1964) demonstrated in the laboratory that in such a gradient *T. castaneum* tends to choose areas with the highest level of fine material.

The amount of fine material present in grain marketing channels differs between wheat and maize. Foster and Holman (1973) found that the level of fine material in wheat did not exceed 1% even after four handlings. In contrast, maize is highly susceptible to breakage in harvesting and handling, particularly at low moisture levels. As a result, fine material in maize may reach very high levels, often increasing as much as 2–2.5% each time it is handled.

The amount of fine material in stored wheat probably has only a minor effect on population growth rates of the major insect pests; however, the amount of fine material in maize may affect the reproductive potential of some insect pests. *C. pusillus* and *C. ferrugineus* and *T. castaneum* are often misrepresented as secondary invaders (i.e., that they only infest grain previously damaged by boring insects). *C. ferrugineus* does require fine material for optimal development on maize (Rilett 1949; Sheppard 1936), but does not require it for optimal development on wheat. Rilett (1949) reported that *C. ferrugineus* increased faster on whole wheat than in coarsely ground wheat. Sinha (1975) also showed that *C. ferrugineus* populations increased faster in wheat without dockage than in wheat containing 5% or 10% dockage. *C. ferrugineus* is able to develop and reproduce on commercial wheat with little or no fine material because a moderate proportion of the kernels are slightly damaged. Tuff and Telford (1964) found that approximately 50% of the wheat grown in the state of Washington had hairline fractures or chipped seed coats, and that this damage was primarily due to threshing. White (1982) found that survival of *T. castaneum* larvae was dependent upon their finding wheat kernels with cracked seed coats, preferably with the germ exposed. The larvae were able to find damaged kernels (germ exposed) equally well whether the percentage was 20% (the amount currently found in commercial wheat in Australia) or 5%.

Internal feeders, such as the *R. dominica*, and *S. granarius* and *S. zeamais*, do not require fine material for optimal growth rates on either wheat or maize. In experiments on grain sorghum, 1.4 times more *S. zeamais* emerged on whole than on halved kernels of sorghum and only a few insects emerged from cracked sorghum (Morrison 1964). *R. dominica* lays its eggs on the exterior of the kernel. After the larvae hatch, they feed on the fine material produced by the boring adults, or bore directly into kernels that have been slightly damaged.

*O. surinamensis* is the only major wheat pest that requires fine material for optimal growth (Fleming 1988), and a lack of damaged or cracked grain could be a limiting factor for young larvae after the population has reached a high density.

The negative effects of fine material on aeration efficiency are probably more important than the direct nutritional effects. Air is blown through grain in the fall to cool it. Fine material in grain restricts air flow and decreases the efficiency of grain cooling (Haque et al. 1978). When fine material is unevenly distributed within the grain mass, the air flow is restricted to localized areas in the grain bulk. Therefore regions with fine material remain essentially unaerated. These zones of moist and warmer grain are more favorable for insect reproduction and allow insects to develop through the cooler months even though the grain has been aerated. It is in these areas that insect-produced "hot spots" frequently develop.

*Cleaning*

Cleaning in-shell peanuts with aspiration and vibration equipment was effective in removing insects under typical infestation levels (Payne et al. 1970). Milling losses due to splitting of the whole kernels increased with insect density. Insect contamination of marketable grade peanuts was increased when loose-shelled kernels were added back after shelling. This may indicate that insects are not as effectively cleaned from loose-shell kernels.

*Milling*

In this section on milling and the section below on residual infestations in flour mills, milling terms are used to refer to the ecology of insects at different points in the milling process. Scott (1951) provides a detailed description of the milling process. Wheat is passed through several break rolls to open the kernels and remove endosperm. After each break roll, sieves are used to segregate the mill stream by particle size and different portions of the mill stream are routed to other grinding and separation equipment according to particle size. Middlings refers to middle-sized particles of endosperm. Sizing rolls and purifiers are used to separate bran from endosperm and reduction rolls reduce particle size of endosperm. Patent flour refers to flour from which some low-grade or clear flour has been removed and it therefore contains less bran than clear flour.

Milling does not kill all *T. confusum* (Cotton and Wagner 1935). Living adults, larvae, and pupae passed uninjured through the first break; living adults and larvae passed uninjured through the second break; and larvae passed uninjured through the third, fourth and fifth breaks and the second sizing rolls. However, eggs of *S. zeamais*, *Cadra cautella* (Walker), and *T. castaneum* were eliminated by milling wheat to flour finer than 210 μm or sieving flour through 210 μm opening mesh (Yamanouchi and Takano 1980).

The susceptibility of a commodity to insects can also vary with stage of milling process. In Canada, Smallman and Loschiavo (1952) found that for *T. confusum* development time is shorter and survival and progeny production are higher when insects are reared on stocks of low-grade flour and tailings from sieves than when reared on stocks from breaks or middlings. In Australia, Miller (1944) obtained similar results with *T. castaneum* and *T. confusum*. The development times of *S. oryzae* and *S. zeamais* were shorter on hulled rice than on rough or polished rice (Takahashi and Mizuno 1982). For both species, longevity and oviposition period were shortest on rough rice and longest on polished rice.

Current commodity harvesting, handing, and processing practices have substantial impact on the survival of stored-product insects. More systematic and quantitative studies are needed to understand and utilize these practices in IPM programs.

## 4. Behavior-Eliciting Chemicals

The behavior of insects in locating and acquiring mates, oviposition sites, and food is an important aspect of their population dynamics. All species of stored-product insects probably use chemical cues to locate these resources.

*Commodity*

Maga (1978) listed 12 volatile compounds released from maize, 23 from oat flakes, 16 from rye, 14 from triticale, 24 from wheat, 64 from barley, and 83 from rice. Seitz and Sauer (1992) have extended these lists considerably, and Maarse (1991) has suggested that careful examination of almost any food would reveal hundreds of volatile chemicals. Many of the volatile chemicals identified in these commodities are the same as or very similar to each other, with short-chain alcohols, aldehydes, fatty acids, ketones, esters, terpenes, and heterocyclic compounds being the most prevalent chemical classes. In addition to the volatile organic chemicals from a stored commodity, water and carbon dioxide are also released continuously. The responses of insects to these gases relative to other released volatiles have yet to be established (Sinha et al. 1986). Water, carbon dioxide, and numerous organic chemicals are also released by bacteria and fungi associated with commodities (Kaminski et al. 1979a,b; Seitz and Sauer 1992; Wilkins and Scholl 1989).

Most of the literature on the response of insects to food odors deals with triglycerides and long-chain fatty acids (Baker and Loschiavo 1987; O'Donnell et al. 1983) that have low volatility. These chemicals are predominantly internal cell components and are usually not available unless the commodity has been damaged. Although much of the available literature would suggest that the commodity volatiles are perceived as food cues (Barrer 1983; Nara et al. 1981; O'Donnell et al. 1983; Pierce et al. 1981; Honda and Ohsawa 1990; Seifelnasr 1991), they may be more important as oviposition cues. Females play a major role in choosing larval food by their choice of an oviposition site. Several workers have demonstrated the presence of ovipositional stimulants in various varieties of maize (Gomez et al. 1983), rice (Arakaki and Takahashi 1982), and wheat (Kanaujia and Levinson 1981a,b), although the actual chemicals have not been identified. Barrer and Jay (1980) and Fletcher and Long (1971) have also studied the roles of volatiles from several commodities in the selection of an oviposition site.

Although most bioassays for food semiochemicals have used adult insects, a few have used larvae. Thus, Baker and Mabie (1973) used *P. interpunctella* larvae to assess the presence of chemicals in extracts of wheat, maize, and peanuts that cause aggregation, and the roles of fatty acids and sucrose as synergistic feeding stimulants. Chippendale and Mann (1972) likewise used larvae of *S. cerealella* to measure the presence of chemicals that cause aggregation and feeding stimulants in wheat germ lipids, and Nara et al. (1981) used *Trogoderma glabrum* (Herbst) larvae to bioassay the presence of volatiles in wheat germ oil that cause aggregation. Commodity odors are clearly important to adults in finding oviposition sites and to larvae in choosing food.

*Pheromones*

Most pheromones of stored-product insects are either sex pheromones or aggregation pheromones (Burkholder 1982). Sex pheromones are released by only one gender, usually the female, and only the other gender responds to the odor; these pheromones are usually produced only by species that have short-lived adults. Aggregation pheromones are produced by one gender, usually the male, but both genders respond to the odor; these pheromones are produced only by species that have long-lived adults. In either case, the significance of the chemical is presumably to bring the two genders together for mating. Studies of the ecological roles of these chemicals in field situations where population densities are low has been extremely limited and much remains unknown. Day–night light cycle may (Barratt 1974; Barrer and Hill 1977; Coffelt et al. 1978; Faustini et al. 1982; Hammack and Burkholder 1976; Qi and Burkholder 1982) or may not (Nammour et al. 1988; Pierce et al. 1987; Walgenbach et al. 1983) influence the emission of pheromones or the response to them, and both occur during the same time period when synchronized by photoperiod (Burkholder and Ma 1985; Cross et al. 1977; Shapas and Burkholder 1978). Production, emission, and utilization of sex or aggregation pheromones by stored-product insects is often influenced by age, nutritional status, and dietary history or previous sexual experience of the insect, and by environmental parameters such as temperature and relative humidity (see reviews by Burkholder and Ma 1985; Chambers 1990; Oehlschlager et al. 1988; and McNeil 1991). Although much of the early literature suggested that only one gender was an emitter of pheromones, more recent literature has shown that in some cases both genders produce chemicals that are required for the successful completion of the mating process (Phelan 1992). Pheromone of one species can also affect another species. With three moth species that have the same major pheromone component, two of these species, *P. interpunctella* and *Ephestia elutella* (Hübner), release an inhibitor that reduces the response of the other *C. cautella*, to the pheromone (Sower et al. 1974; Krasnoff et al. 1984). The inhibitor is thought to reduce interspecies mating.

Stored-product insect pheromones range from relatively simple aliphatic alcohols and aldehydes to highly complex cyclic acetals, esters, ketals, and macrolides (Burkholder and Ma 1985; Mayer and McLaughlin 1991). Many of these chemicals have one or more chiral centers and hence are optically active (Silverstein 1988); others have the complexity of multiple positions of unsaturation (McNeil 1991). Many species produce blends of chemicals, and the role of each chemical in mate location, courtship, and copulation behaviors is often unknown (Burkholder and Ma 1985; Oehlschlager et al. 1988). Individual insects release only picogram to nanogram quantities of these pheromones at any given time (Vick et al. 1973; Phillips et al. 1985; Oehlschlager et al. 1988; Pierce et al. 1989), and for several species their total daily output is no more than a few nanograms (Greenblatt et al. 1977; Tanaka et al. 1986; Coffelt et al. 1978). In addition, many of these compounds are highly volatile and labile, and thus have short half-lives. The active space of pheromone emission of a single insect is no more than a few centimeters to a meter (Mankin et al. 1980), and therefore insects are attracted from only a short distance. Pheromones are important in helping insects to find other insects of the same species. This can ensure mating and may be necessary for mass attack of some foods such as peanuts with hard shells.

*Territorial Markers*

Territorial marking pheromones may be important in reducing the interference among individuals of the same species with one another. These pheromones are used to mark feeding and pupation sites of some moths and oviposition sites of beetles. Pheromones produced by specialized mandibular glands of moth larvae are deposited onto either their silk or food. Working only with last instar *E. kuehniella* larvae, Corbet (1971) suggested that these pheromones regulate moth densities by encouraging emigration from crowded habitats. Mossadegh (1978, 1980) investigated in more detail the silk and mandibular pheromone production by all instars of *P. interpunctella* and found that all stages continuously produced copious quantities of these pheromones while feeding and moving around in their habitat. Unlike the reports of Corbet (1971, 1973), Mossadegh found that intraspecies interaction was not required for the production and release of these chemicals. Mossadegh (1980) found that moth larvae, when given a choice, preferred uncontaminated food over food contaminated with conspecific mandibular gland secretions. Deposition of the mandibular secretions on the substrate may thus serve to mark a territory of each young larva and reduce chances of cannibalism. In addition, Corbet (1973) reported that ovipositing female *E. kuehniella* responded to this larval mandibular secretion. The number of eggs laid was proportional to the amount of the larval secretion. Corbet suggests that levels of secretions proportional to population density allowed females to select oviposition sites that were capable of supporting larval growth but were not overpopulated. These secretions may also have insecticidal, fungicidal, and antibiotic properties (references 7–11 in Mudd 1981). At least 25 different compounds in seven major groupings have been identified for four species of moths (Mudd 1981; Kuwahara et al. 1983; Nemoto et al. 1987a,b). Although each species has some of the same components as other species, the resulting blends are species-specific (Nemoto et al. 1987a). These are highly polar compounds with relatively low volatility and a long-term stability of several years (Mossadegh 1980).

Adult tenebrionid beetles in the genus *Tribolium* have long been known to produce noxious quinones from their paired "defensive gland" reservoirs when they are disturbed or held at high population densities (Roth 1943; Happ 1968; Ladisch et al. 1967). These quinones may also trigger dispersal or inhibit growth of fungi on their food (Sonleitner and Guthrie 1991). In recent years, additional chemicals, including 1-alkenes (Wirtz et al. 1978; Suzuki et al. 1975a), β-hydroxy aromatic ketones (Suzuki et al. 1975b; Howard et al. 1986) and β-hydroxy aromatic esters (Howard and Mueller 1987; Howard 1987) have been found in these glands. They are produced in copious quantities, and are released into the environment, even under uncrowded conditions. These chemicals may serve to mark feeding or ovipositional sites, to repel other insects, or to alter basic physiological processes of other insects (Suzuki et al. 1975a; Howard 1987; Howard and Mueller 1987). Additional studies need to be conducted under conditions similar to those found in the storage environment and with chemicals of known concentration and composition.

*A. obtectus*, *C. maculatus*, and *C. chinensis* have been reported to oviposit in such a manner as to minimize the number of eggs associated with any given legume seed. An extensive literature has developed around the hypothesis that the observed egg distributions are a result of the beetles placing ovipositional deterrent pheromones onto the seeds (Mitchell 1975; Wasserman 1981, 1985; Szentesi 1981; Giga and Smith 1985; Messina et al. 1987; Credland and Wright 1990). Although these authors and many others have built up fairly convincing circumstantial evidence for the function of these compounds, as of yet there are few cases of identified chemicals shown to produce the observed ovipositional patterns. Indeed, the only chemicals identified are cuticular lipids (long-chain hydrocarbons, fatty acids, and triglycerides) found on both genders (Oshima et al. 1973; Yamamoto 1976; Honda et al. 1976; Honda and Ohsawa 1990). The level of response to these chemicals differed between populations (biotypes) (Dick and Credland 1984; Credland 1986). The ecological roles of ovipositional deterring pheromones in bruchids has not been convincingly demonstrated and needs to be investigated further.

Given the large number of behavior-eliciting chemicals that have been identified and the many different aspects of insect ecology that they influence, there should be many opportunities to use these chemicals in monitoring and managing stored-product insects. Studies in storage, transportation, and processing facilities are needed to determine the usefulness of these chemicals.

## II. POPULATION DYNAMICS

By understanding how various factors regulate insect population growth, we can examine the effects of these factors on insect population dynamics in commodities being stored, transported, and processed. Density-regulating factors other than crowding have been emphasized in this chapter because pest populations infesting commodities rarely reach densities at which crowding is important when good pest management programs are used. The dynamics of source populations, populations in raw commodities, and populations in processed commodities will be considered separately. Under the section on source populations, four possible sources of insect infestation are considered. For several commodities, insect infestations occur in the field prior to harvest. Residual insect infestations in storage, transportation, and processing facilities are also important sources of insects that can infest commodities. Stored-product insect pests have been shown to breed on a number of wild hosts and to disperse over considerable distances. Finally, other infested commodities can be an important source of the insects that infest insect-free commodities.

## 1. Sources of Infestation

Stored-product insects are continually leaving infested commodities or food residues, and can disperse over long distances. These insects are well adapted to finding and surviving on small amounts of food widely dispersed in storage, transportation, or processing facilities, and also have been shown to breed on wild hosts. These characteristics make them particularly difficult pests to manage.

### *Field Infestation*

Several species of stored-product insects are known to infest crops in the field prior to harvest (Hagstrum 1985). Because insect species feeding inside kernels are less likely than other species to be killed or cleaned from the commodity by combining and augering, infestations of stored commodities by these internal-feeding species are more likely to be successfully initiated in the field prior to harvest.

The level of infestation of commodities in the field can be low and thus difficult to detect, but even these low-density populations can increase to damaging levels during storage (Hagstrum 1985). Field infestation by *C. maculatus* of cowpeas from 49 fields in the United States (Fig. 10) resulted in an average density of 2.33 adults per bushel emerging every 14 days during the first 42 days of storage. Female *C. maculatus* readily oviposit on cowpea pods, and larvae bore through pods to infest seeds (Messina 1984; Fitzner et al. 1985; Fatunla and Badaru 1983). Field infestation of cowpeas also has been reported for *C. maculatus* in Nigeria (Prevett 1961; Booker 1967; Taylor 1970; Taylor and Aludo 1974). In India, field infestation of pigeonpeas by *C. maculatus* varied with sowing dates from 3.5 to 13% of seed (Patnaik et al. 1986).

A related species, *C. chinensis*, infests cowpeas in the field in South Africa (Oosthuizen and Laubscher 1940) and azuki beans in the field in Japan (Shinoda and Yoshida 1985). In Japan, *C. chinensis* were found in the fields from mid-August through early November, were most abundant in early October, and laid

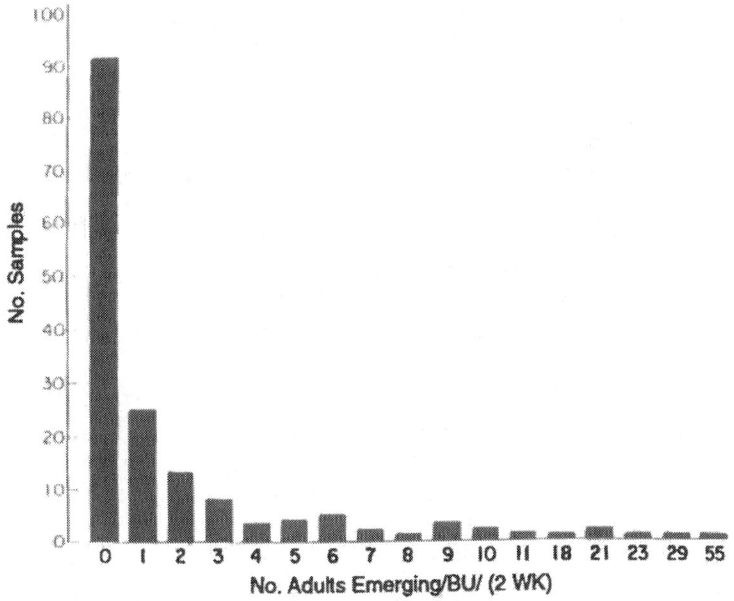


**Figure 10** Frequency distribution of the number of adult *C. maculatus* emerging per 14 day from a bushel of newly harvested cowpeas during first 42 days of storage. (Reprinted from Hagstrum 1985.)

eggs on mature pods from mid-September to mid-October. The percentage of infested beans reached 14.26% with an average of two eggs per pod at harvest. Survival rate from egg to larvae boring into pods was 41%.

Another species. *A. obtectus*, infests beans from July until the end of November in Russia because of the practice of sowing beans during 4–5 consecutive months, and completes two generations per year in the field (Vasil'ev 1935). Adults do not oviposit on green pods but lay eggs through cracks in dry pods on the seeds or inner wall of pods. *Phaseolus vulgaris* and *Phaseolus multiflorus* are the preferred food plants, but several other species of legumes are also attacked. *A. obtectus* rarely survives winter in the field, but completes two generations in storage and is returned to the field when seeds are planted. Larvae and pupae are killed if the infested seeds are sown in damp soil, but in dry soil young adults emerging from seeds can reach the surface from a depth of 5 cm. In California, Larson (1924) found that neglected seeds in warehouses were the primary source of field infestation by *A. obtectus*, and Larson and Fisher (1924) showed that the planting of weevily seed did not contribute to field infestation. Larson and Fisher (1925) suggested that *A. obtectus* breeding on seeds left in bean straw also might contribute to field infestation. Larson (1932) found that early maturing beans were more likely to become infested by *A. obtectus* in the field than late maturing beans and suggested that planting a trap crop of beans that will mature earlier around the edge of a bean field might protect the bean crop from field infestation.

Both wheat and maize are infested in the field by *S. cerealella* in the United States. Field infestation of wheat was low (0.26, 0.02, and 0.04% of kernels in each of 3 years) (Simmons 1927). The infestation level decreased from 0.4% to 0.03% as the distance between field and stored grain increased from 52.7 to 237.7 m. Field infestation of maize in Indiana by *S. cerealella* was studied by taking five ear samples from 164 fields (Russell 1962). When the state was divided into 12 areas, no moth infestations were found in six northern areas and one southern area, but infestations averaged from 6.1 to 12.9% of ears in the other five southern areas. Koone (1952) showed that *S. cerealella* surviving on maize in the field increased from 8% in the roasting-ear stage to 62% as maize dried and was ready to harvest. Weston et al. (1993) found that infestation of maize by *S. cerealella* in central Kentucky decreased with planting date and increased with harvest date. Both of these factors may affect *S. cerealella* through changes in grain moisture content. Maize on which *S. cerealella* could develop to pupa or adult had dried to 31%. This maize would have had a moisture content of 50–55% when eggs were laid. Infestation levels were higher for maize than for wheat.

In other countries, several commodities are infested in the field by *S. cerealella* and there is further evidence that field infestation levels decrease as the distance between field and storage increases. In India, the number of insects infesting maize decreased from 16 per sample of three cobs at 200 m from storage to 4.5 at 1000 m (Singh et al. 1978). In Senegal, field infestations of millet by *S. cerealella* resulted in 10 times more damage to grain from fields close to dwelling houses where grain was stored than to millet from fields away from stored grain (Seck, 1991). In India (Singh et al. 1978) and Bangladesh (Howlader and Matin 1988), the number of *S. cerealella* infesting rice in the field also decreased as the distance between fields and stores increased. In India, the number of *S. cerealella* decreased from 15 per sample of five rice heads at 200 m to 1.5 at 1000 m, and in Bangladesh, the number of insects decreased from 47 per sample of 120 rice heads at 10 m to 16 at 1440 m. Howlader and Matin (1988) also showed that no insects were present 11 days prior to harvest and that the infestation level increased up to harvest. In India, based upon 210 three kg samples of rice collected at threshing yards in 17 districts, field infestations of 29 varieties ranged from 0.33 to 54 *S. cerealella* per kg, and only two samples did not have insects (Sundararaj and Sundararajan 1990). These many studies show that field infestation is a common and widespread source of insect problems in storage.

In Australia, wheat is infested by *S. oryzae* and maize by *S. cerealella* in the field. The level of *S. oryzae* infestating wheat in the field has been shown to be extremely low (Rossiter 1970). Only 42 of 320 200-head wheat samples were infested and usually contained fewer than 15 *S. oryzae* per sample. However, the number of *S. cerealella* infesting maize in the field was higher than the numbers of *S. oryzae* infesting wheat, ranging from 14.5 to 186/0.6 L for 20 early-maturing varieties and from 12.5 to 65.7 for 18 late-maturing varieties (Turner 1976).

Field infestation has been studied most extensively for *S. zeamais*. This insect species has been found to fly to maize fields up to 400 m away from their source population (Chesnut 1972) and to infest most heavily the maize near the edge of the field (Blickenstaff 1960; Kirk 1965; Giles and Ashman 1971). Blickenstaff (1960) found that 30.5% of the ears were infested with an average of 3.6 *S. zeamais* per ear. Tight coverage of the tip of the cob by a thick husk reduced the infestation level, and husk damage increased the infestation level (Bernabe-Adalla and Bernardo 1976). In Louisiana and South Carolina, the number of *S. zeamais* trapped in a maize field prior to tasseling was low, but the number captured increased rapidly through July and then declined in August (Kirk 1965; Williams and Floyd 1970). Both studies used traps to show that weevils entered the field at ear height. The weevils tended to stay in the field if the developing maize was suitable for oviposition (Kirk 1965). In Georgia, Dix and All (1986) found that *S. zeamais* collected from maize ears that had fallen to the ground survived −10°C temperatures better than insects from laboratory cultures, but they could not survive in the northern parts of the state where −15°C winter temperature extremes occurred. Adult insects remained on maize ears and did not burrow into the soil for protection from the cold. In the Philippines, *S. zeamais* began to infest maize fields 4 or 5 weeks before harvest and increased continuously until 93% of cobs were infested at harvest (Schwettmann 1988). An 8-month drought caused *S. zeamais* populations to collapse, but the populations recovered during the next planting season. *S. zeamais* has been shown to develop successfully on maize with moisture contents as high as 60% in Mississippi (Powell and Floyd 1960) and 65% in Kenya (Giles and Ashman 1971). In Kenya, where maize is dried in the field, the offspring of *S. zeamais* infesting the maize crop can develop and reproduce before harvest, thereby increasing the severity of postharvest infestation (Giles and Ashman 1971).

In many countries, several species of stored-product insects infest a number of crops in the field prior to harvest. Studies have shown that infestation levels decreased with distance from storage, indicating that infested commodities are primary sources of field infestation. Infestation was influenced by the crop growth stage. In many cases, field infestations are quite low. However, field infestations are high in cases in which the commodity is left in the field to dry, and insects can complete a generation in the field prior to harvest. Management programs for the species of stored-product insects that infest crops in the field might include the planting of resistant varieties or the use of trap crops.

*Residual Infestations*

Insect populations are known to breed in commodity residues in combines, grain storage bins, flour mills, peanut warehouses, peanut shelling plants, railroad cars, and pallets at port warehouses. Insects emigrating from infested commodities add to the size of residual populations, and insect-free commodities become infested by insects from residual populations. These commodity residues allow insect populations to survive through periods when warehouses are empty (Hagstrum and Stanley 1979). After fumigation, insects outside the gas enclosure can also be important in reinfesting maize (Graham 1970d), rice (Boon and Ho 1988), and probably other commodities.

Several Australian studies have investigated insect populations in grain residues. High densities (70 insects/kg) were found in residual wheat cleaned from combine harvesters in Australia (Sinclair and White 1980). Average grain residues per farm on 57 Australian farms include 270 kg spilled, 65 kg in auger boots, 100 kg in trucks and combine harvester, and 75 kg in other locations (Sinclair 1982). The estimated populations of *S. oryzae*, *R. dominica*, *Cryptolestes* spp, and *T. castaneum* in these residues were 14.6, 10.4, 11.6, and 23.7% of those in stored grain, respectively. These four species represented 85.5% of the insects. The number of insects per kg of grain residue in Australia increased from 44.9 in January to 299.5 in May and then decreased to 66.4 by December. In a small-scale field study that simulated the emigration of insects from spilled grain in Australia, Sinclair and Alder (1984) found that emigration over the course of the study was as high as 36, 53, 54, 61, 80, and 90% of source population for *Cryptolestes pusilloides* (Steele and Howe), *C. ferrugineus*, *R. dominica*, *S. oryzae*, *T. castaneum*, and *C. pusillus*, respectively. Cox and Parish (1991) investigated the refuge-seeking behavior of *C. ferrugineus* in the laboratory and found that the percentage of insects outside the refuge was 15.9% when the refuge contained wheat compared to 76.2% when the refuge contained glass beads. The presence of food in a refuge thus increased the number of insects in the refuge fivefold. The substantial percentage of insects in refuges with or without food explains why many insects are not seen during visual inspections and are not removed by cleaning.

In Canada, *C. ferrugineus* was found in 11.9% and *T. castaneum* was found in 2.3% of 1–1.5 L samples from grain residue in 1752 empty bins on 296 farms (Smith and Barker 1987). In the samples taken between mid-June and mid-August, the average densities of adults and larvae per sample were 0.11 and 0.15 for *C. ferrugineus*, and 0.020 and 0.018 for *T. castaneum* (Barker and Smith 1987). Of the 16 species of stored-grain insects found in these samples, these were the only two species that feed directly on undamaged grain (Smith and Barker 1987). The authors could not explain why some farms were infested and others were not. The proportion of uninfested samples was not consistently correlated with previous infestation, presence of livestock, use of vacuum cleaner to remove residues, insecticide spray, fumigation, spillage, type of grain, or type of granary (Barker and Smith 1990). Using a release-recapture method, the total number of adult *T. castaneum* per bin was estimated to be 33,000±55,000 (Walker 1960). In the United States, the percentage of bins infested and average insect density in stored oats was correlated with the peak insect density in grain previously stored in a bin (Ingemansen et al. 1986).

The residual insect populations in 2367 samples of 0.23 kg of flour from 17 flour mills in the midwestern United States were mainly *T. castaneum*, *Cryptolestes* spp, *R. dominica*, *S. oryzae*, and *L. oryzae* (Good 1937). The average densities of live adults of these five species per sample were 13.25, 1.00, 0.32, 0.17, and 0.17, respectively. Only rarely, and then in small numbers, were *E. kuehniella* and *P. interpunctella* found. Populations of *R. dominica* and *S. oryzae* were found primarily in the wheat sievings and wheat elevator boot (94 and 82%) and were completely absent after the third or fourth break. *T. castaneum* populations averaged between 8.58 and 20.69 insects per sample at these locations and between 14.23 and 74.37 in remaining elevator boots. High populations of 52.17 insects per 0.23 kg sample in the low-grade-flour elevator boot were probably due to this part of the milling system being difficult to reach for cleaning. The patent-flour and clear-flour streams after sieving contained fewer *T. castaneum* (7.92 and 6.47 per sample) because insects were removed. Average densities of *T. castaneum* in purifiers ranged from 17.95 to 33.93. The adults were between 41.4% and 67.6% of the combined adult and immature populations of *T. castaneum*. Because of temperature control in the mills, seasonal changes in *T. castaneum* populations were small, ranging from 12.89 per sample in November–December to 35.66 per sample in April.

In flour mills in England, populations of the two most common species, *Cryptolestes turcicus* (Grouv.) and *E. kuehniella*, increased from March or April until the annual fumigation in June, July, or August (Dyte 1965). The numbers of *C. turcicus* in eight centrifugals were fairly constant during a 5 year period, with population size probably depending upon molds that develop in damp flour residue (Dyte 1966). Insect populations have also been studied in flour mills in Egypt (Hosny et al. 1968) and Greece (Buchelos 1980, 1981), in a barley mill in Japan (Imura 1981), and in food processing facilities in Korea (Sim et al. 1979).

In 16 factories of hand-extended noodles in Japan, pheromone- and food-baited traps caught 6037 adult *L. serricorne* and 22 adult *S. paniceum* (Suezawa et al. 1987). Adult *L. serricorne* populations increased from May through August and then declined. This species was caught at all trap locations, but was most common near wheat flour, products, and trash. Corners of rooms and crevices of floor were most probable sites of adult emergence. The locations and numbers of captures suggest that there were large residual insect populations in factories.

The combined *C. cautella* and *P. interpunctella* population in an empty peanut warehouse increased at a rate of twofold per month from 232 to 2146 adults on residual peanuts (Hagstrum and Stanley 1979). Hagstrum (1984) observed similar growth rates under simulated empty warehouse conditions. Changing the distribution of 48 peanuts from uniform to aggregated did not reduce the number of eggs laid, but did reduce the number of larvae that completed development (Table 4). Population growth rate increased from roughly three- to sevenfold per generation as the number of peanuts per location increased from 2 to 48. The mean number of eggs laid was proportional to the number of peanuts at a location. As the number of locations increased, larval survival declined mainly as a result of females laying eggs at a smaller fraction of the locations, and thus breeding on a smaller number of peanuts. Population growth of residual populations was reduced by egg distribution not conforming to resource distribution. The fact that *C. cautella* did not find all the locations with larval food in a single generation resulted in some of the resource being available for more gradual population growth over several generations.

A study of residual insect populations in 11 peanut shelling plants in the southeastern United States found that 5 of 20 species (*C. cautella*, *P. interpunctella*, *T. castaneum*, *Oryzaephilus mercator* [F.], and *Carpophilus dimidiatus* [F.]) represented 95% of insects collected (Payne and Redlinger 1969). *C. cautella* represented 50% of the insect population. Trends in insect numbers in residual peanuts were more difficult to see than those in peanut-baited traps because insect populations were often destroyed during mill cleanup or when shelling equipment was changed to accommodate another type of peanut. However, during the shelling season, insect densities per kg of peanuts in traps (8–180) were lower than those in residual peanuts from the shelling plant (30–580). In residual peanuts. *O. mercator* and *T. castaneum* were a greater percentage of insects than those in traps. The percentage of traps with insects was reduced 25–50% during the shelling

**Table 4** Oviposition, Survival, and Population Growth of *C. cautella* on 48 Peanuts Distributed among 1–24 Locations.

| Number of peanuts per location | Number of eggs per peanut | Number of larvae per peanut | Population growth rate per generation |
|---|---|---|---|
| 2 | 15.6 | 0.61 | 2.9 |
| 4 | 17.3 | 0.87 | 4.3 |
| 8 | 14.8 | 0.95 | 5.3 |
| 16 | 14.4 | 1.10 | 6.1 |
| 48 | 9.4 | 1.23 | 7.4 |

*Source*: Based on data from Hagstrum (1984).

season by regular sanitation and insect control programs, which included periodic insecticide applications, cleaning of all elevators and equipment, and removal of floor sweeping and waste peanuts from premises. Densities were lower in the winter as a result of cooler weather rather than improved sanitation and insect control practices.

Cogburn (1973b) found that railroad cars delivering processed commodities to ports in the United States generally had enough residual food material to support insect breeding. The levels of residual food were <0.5, 0.5–2, 2–5, and >5 kg in 33, 19, 10, and 14% of the cars, respectively. Many different species of insects were found breeding on these residues, and these insects probably infest commodities during transit. Twelve genera of insects were found in these residues, and five of these genera were found in more than 25% of the cars (i.e., an average per sample of 50 *T. castaneum*, 24 *Trogoderma* spp, 29 *Cryptolestes* spp, 17 *C. cautella*, and 15 *R. dominica* in 62, 49, 33, 25, and 25% of cars, respectively). An average per sample of 6 *L. serricorne* and 10 *Carpophilus piloselius* (Motschulsky) were found in 11 and 13% of cars, respectively. Insects were found crawling on sacks of flour from 85–90% of boxcars unloaded during the summer. Many of the same species breed in food residues on pallets in port warehouses and may also infest commodities during storage. In four port warehouses, the average numbers of insects in 0.9 L of food residues from pallets were 4.83 *T. castaneum*, 6.14 *C. cautella*, 2.18 *L. serricorne*, 1.2 *O. surinamensis*, and 0.96 *C. piloselius* (Cogburn 1973a).

For those species that do not infest commodities in the field and for facilities that do not receive infested commodities, residual populations can be the primary source of infestation of incoming commodities. Industry places heavy emphasis on managing these populations with good sanitation practices.

### *Wild Hosts or Long-Range Dispersal*

Many species of stored-product insects seem to have retained some of their adaptations to wild habitats. Some are capable of long-distance flight, of breeding on many different wild hosts, and of feeding on flowers to increase adult longevity and fecundity. However, these feral populations may be largely emigrants from storage, transportation, and processing facilities, and may not be able to maintain themselves without additional emigration from these facilities.

Stored-product insects have been trapped at locations far from stored commodities (Cogburn and Vick 1981; Sinclair and Haddrell 1985; Strong 1970; Vick et al. 1987; Wohlgemuth et al. 1987). These captures can be explained either by long-distance dispersal or the availability of wild hosts. Chestnut (1972), using the release-recapture method, showed that *S. zeamais* could fly up to 400 m. In a 35 $m^2$ room, male *C. cautella* were observed to fly 300 m during an average 10 min flight (Hagstrum and Davis 1980).

Several studies have reported stored-product insects breeding on wild hosts or have shown that they can develop on wild hosts in the laboratory. In Japan, *C. chinensis* was found to breed on 4 of 11 wild legumes examined (Shinoda et al. 1991). In the laboratory, 85% of the eggs laid on *Vigna angularis* and *Dunbaria vilosa* emerged as adults in about the same amount of time as on azuki beans. For two additional host species, adult emergence was 20% and development time was twice as long as that on azuki beans. Shinoda et al. (1992) found that the population dynamics of *C. chinensis* in wild patches of *V. angularis* and *D. vilosa* were similar to those observed earlier in cultivated azuki beans (Shinoda and Yoshida 1985). However, the densities of adults observed in wild host patches were much lower than in cultivated azuki bean fields. They conclude that this difference may be due to cultivated azuki bean fields being at least seven times larger than wild host patches.

Acorns have been found to be suitable for the development of stored-product insects in several countries. In England, Howe (1965) indicated that *S. oryzae* and *S. granarius* could develop on acorns in the laboratory. In South Africa, Joubert (1966) found that large numbers of *S. oryzae* emerged from acorns collected from rural, pastoral, and urban areas. He also found that these acorns were infested with large numbers of *O. surinamensis* and *C. cautella*, and small numbers of *T. castaneum*. In the United States, Mills (1989) found *S. zeamais* infesting acorns. Joubert (1966) also examined several hundred samples of indigenous grasses and found *S. cerealella* in seven samples representing four grass species. Adult *S. cerealella* reared on red millet were only 5% the size of those reared on maize and they produced significantly fewer eggs.

In Germany, Stein (1990) tested the suitability of six fruits from indigenous trees and shrubs for the development of 10 species of stored-product insects in the laboratory. *Trogoderma granarium* Everts and *S. cerealella* did not develop on any fruit, and *T. confusum* developed only on beechnuts. Undamaged acorns, horse chestnuts, and beechnuts were not suitable for the development of any of the species tested. *Prunus aucuparia* fruits, damaged or dehusked acorns of *Quercus robur*, and chestnuts were suitable for the development of *O. surinamensis*, *R. dominica*, *S. paniceum*, *E. kuehniella*, and *P. interpunctella*. Acorns were also suitable for the development of *S. granarius* and *S. oryzae*. Hipberries (*Rosa carina*) were suitable only for *R. dominica* and *S. paniceum*. Acorns of *Quercus borealis* and horse chestnuts were suitable for *O. surinamensis*, *E. kuehniella*, and *P. interpunctella*, and horse chestnuts were also suitable for *S. paniceum*. The number of adults emerging was generally lower on wild fruits than stored commodities, except for *S. granarius* on the acorns of *Quercus robur*.

Stored-product insects have also been found in rodent burrows (*Rattus rattus*) and ant nests. Khare and Agrawal (1964) found, in 80 rodent burrows in India, an average of 38 *R. dominica*, 45 *S. oryzae*, 27 *O. surinamensis*, and 34 *T. castaneum*. In 40 ant nests, they found an average of 12 *S. oryzae*, 11 *O. surinamensis*, and 6 *T. castaneum*. Wright et al. (1990) did not find stored-product insects in nests of eastern woodrat (*Neotoma floridana*), but they did show that *R. dominica* could develop in the laboratory on some of the plant material taken from the nest, including acorns with damaged husks (*Quercus muehlenbergii*), hackberry (*Celtis occidentalis*), and buckbrush fruit (*Symphoricarpus orbiculatus*). *R. dominica* did not develop on eight other fruits, nuts, and seeds found in the nest.

Flowers are visited by adult stored-product insects and may be a source of food that increases insect longevity. Fourteen species of flowers were visited by *S. oryzae* from mid-April through May in Japan (Yoshida and Takuma 1959). Sunflowers (*Helianthus annus* L.) were attractive to *S. zeamais* for 17 days between opening of inflorescence and dropping of stamens (Williams and Floyd 1971). Weevils were found on 50% of the sunflowers examined and sometimes exceeded 100 per head. In Greece, *L. serricorne* adults were found from March to November on flowers of six of nine genera of thistles examined (Buchelos 1989).

*Other Infested Commodities*

Infested commodities can be more important as a source of stored-product insects than residues because of their larger volume. In Louisiana, Williams and Floyd (1970) caught with sticky traps up to 80 *S. zeamais* per two weeks near 200 bushels of maize stored with husk intact (Fig. 11). Emigration increased from spring into summer and then declined in the fall. Trapping studies in the United States have shown that insect flight activity outside farm storage bins is extensive for wheat (Schwitzgebel and Walkden 1944) and maize (Throne and Cline 1989). These insects can infest crops in the field or newly stored grain, or reinfest grain after fumigation. Throne and Cline (1991) found that maize-filled bait packets in the vicinity of bins quickly became infested, and Schwitzgebel and Walkden (1944) found that small numbers of insects entered grain bins each day from April through October. In Australia, Barrer (1983) showed that grain odor doubled the number of insects finding grain in flat storage. Trapping studies in Canada and the United States also have shown that the flight activity of *R. dominica* is extensive from May to November outside elevators (Fields et al. 1993) and around wheat fields or wheat stored on farms (Fig. 12). This species has also been trapped several km from wheat fields or stored wheat on the Konza prairie in Kansas. These data probably indicate that insects are flying long distances, but insects could also be breeding on wild hosts.

The movement of insects from infested to uninfested lots can also occur with processed commodities. For growing populations of *T. castaneum* in 130 g bags of flour, emigration rate was directly proportional to adult age and to adult population density in bags (Hagstrum and Gilbert 1976). The emigration rate was only 3.3% during the 7 day period between eclosion of adults and first oviposition. However, the average time adults remained in flour was only 11 days, an additional 12% emigrated on day 8, and emigration was 23% per day thereafter. These emigrants tend to spend only a short time in each patch of food residue. During a 16 h period,

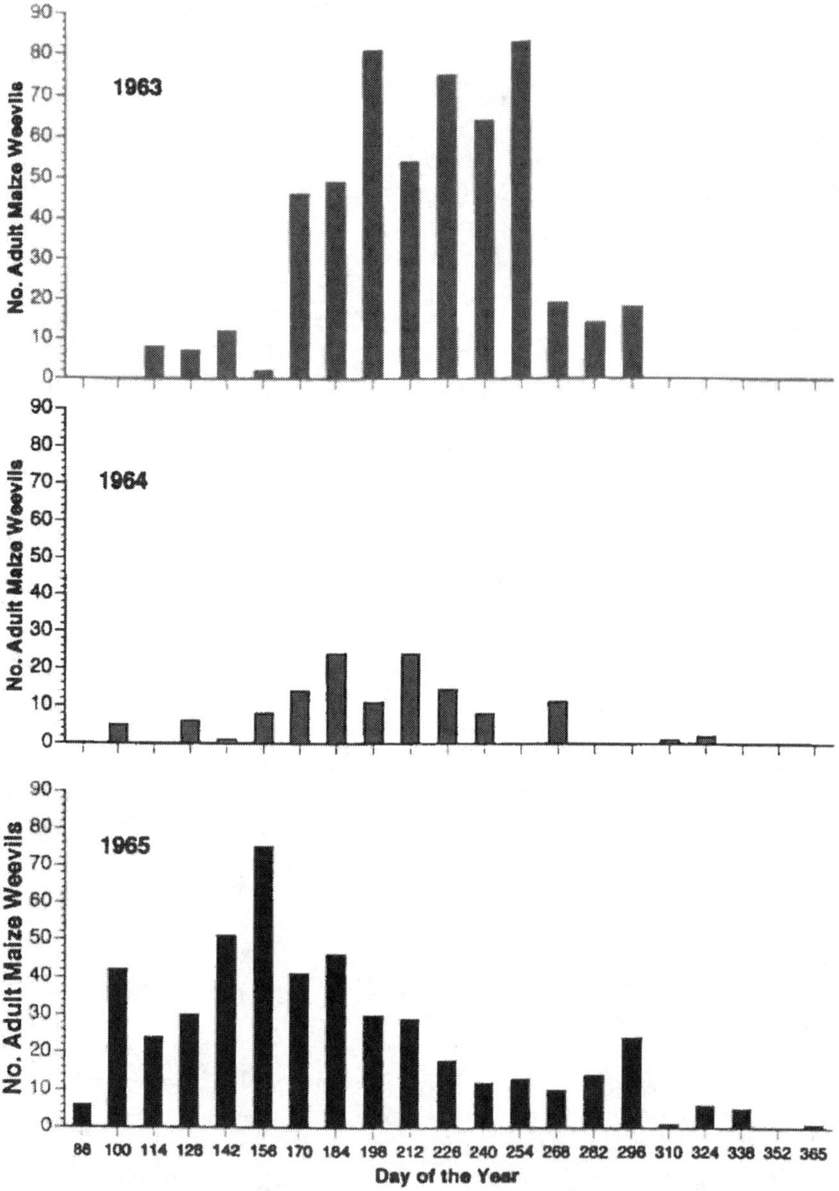


**Figure 11** Seasonal changes in the number of *S. zeamais* emigrating from corn storage bin in Louisiana from March 27 (Julian date 86) to December 30 (Julian date 365) during 1963, 1964, and 1965. (Redrawn with permission from Williams and Floyd 1970.)

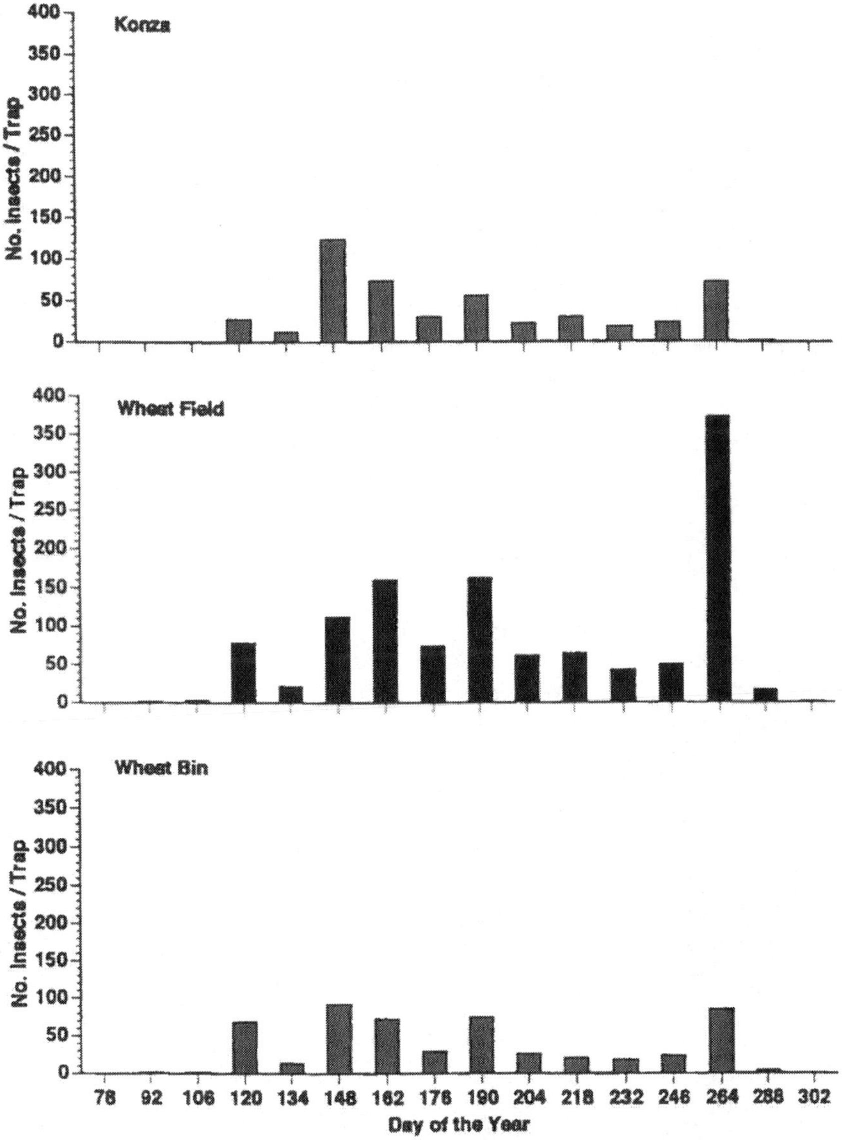


**Figure 12** Number of *R. dominica* adults caught per 2 weeks in pheromone-baited sticky traps from April 2 (Julian date 92) to October 29, 1986 (Julian date 302) on Konza prairie (several km from wheat fields or stored wheat), near wheat fields, and near wheat storage bins on farms in Kansas. (Based on data from Hagstrum, unpublished.)

a single female *T. castaneum* visited from 0 to 5 one g piles of flour (average of 1.8, n = 15) in a 1.2 × 1.2 m arena in which 16 piles were evenly spaced in a 4 × 4 grid (Hagstrum, unpublished data). These females laid from one to eight eggs per pile of flour visited (average of 3.8 eggs per pile). Because of their high mobility, stored-product insects find commodities soon after they are stored. In food distribution warehouses, Vick et al. (1986) found that the numbers of *C. cautella* and *P. interpunctella* trapped in the vicinity of bird seed and chicken feed were significantly higher than those trapped near other commodities, and suggested that these emigrants could be infesting other commodities stored in warehouses.

The extensive emigration of insects from commodities means that without good pest management, uninfested commodities are likely to become infested. Often this occurs when new commodities are brought into a facility containing infested commodities. However, if incoming commodities are not carefully monitored for insects, the reverse can occur. A good stock rotation plan can reduce the chances of having infested commodities in a facility because it limits the time available for pest population growth.

## 2. Population Growth

Insect ecologists first attempted to describe and predict insect population dynamics using simple exponential equations and life table statistics, such as finite rate of increase (Southwood 1980). However, computer simulation models are a more accurate and flexible method of examining population dynamics than finite rate of increase or life table methods. Population growth models have been developed for several species of stored-product insects including *C. ferrugineus* (Hagstrum and Throne 1989; Kawamoto et al. 1989), *T. castaneum* (Hagstrum and Throne 1989; White 1985), *R. dominica* (Hagstrum and Throne 1989), *S. oryzae* (Longstaff and Cuff 1984; Hagstrum and Flinn 1990), and *O. surinamensis* (Hagstrum and Flinn 1990). These models forecast population growth based upon equations that predict the effects of temperature and moisture on insect development time and egg production. Under unfavorable conditions, models will also need to incorporate insect survival. The predictions of these simulation models can explain roughly 90% of the actual changes in insect density in stored grain. The similarity of model predictions to actual densities indicates that the temperature and moisture content of a commodity are the primary factors regulating population growth under favorable conditions. Figure 13 compares the predicted growth rates for five species of beetles at 32°C on grain of 14% moisture content, and Figure 14 shows the predicted growth rate of *R. dominica* populations under different grain temperature and moisture conditions. White (1988b) has reported similar average growth rates of fivefold per month for *R. dominica* and 10-fold per month for *T. castaneum* on wheat in flat storage.

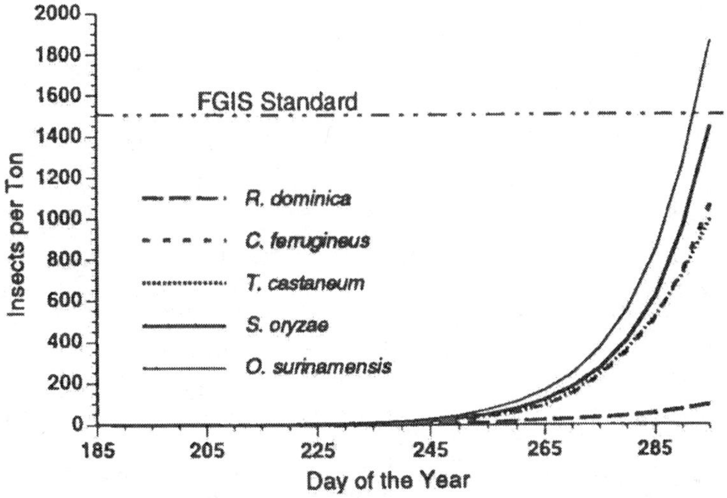


**Figure 13** Predicted population growth of five species of stored-grain insects at 32°C and 14% wheat moisture content. (Redrawn from Hagstrum and Flinn 1990.)

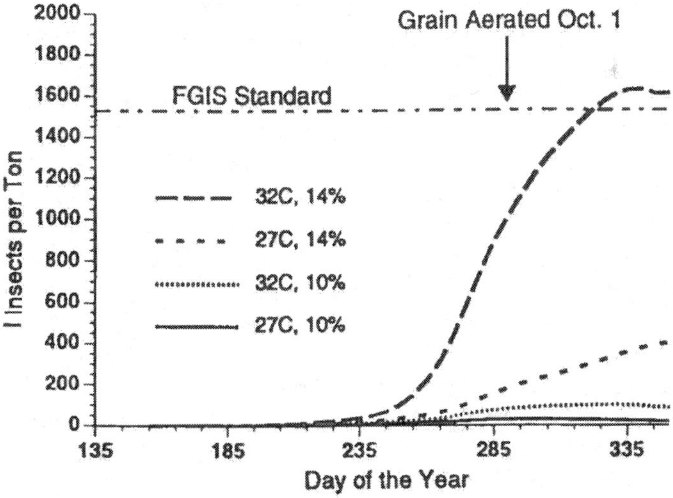


**Figure 14** Predicted effects of temperature and wheat moisture content on the growth of *R. dominica* populations. (Redrawn from Flinn and Hagstrum 1990.)

## 3. Storage of Raw Commodities

The population dynamics of insects in stored raw commodities is discussed using examples of pyralid moths infesting stored wheat, maize, and citrus pulp; other insects infesting stored wheat, maize, and sorghum; and insects infesting stored cowpeas.

*Pyralid Moths*

The population trends of several species of pyralid moths infesting stored raw commodities have been shown to vary with commodity and climate. Most species of stored-product pyralid moths diapause as last instars in response to short photoperiod (Cox and Bell 1991). They eclose, fly, mate, and oviposit at dawn, dusk, or both (Graham 1970a; Hagstrum et al. 1977; Steele 1970; Moriarty 1959). In addition, *C. cautella* oviposit in response to falling temperatures (Hagstrum and Tomblin 1973).

One of the most detailed ecological studies is that by Richards and Waloff (1946) in England. Wheat temperatures tended to remain below 20°C throughout the year, and the majority of adult *E. elutella* emerged in late June in 1943 and 10 days later in 1944. Peak adult densities were estimated to be roughly 40,000 and 200,000 in 1943 and 1944, respectively. Offspring of these adults diapaused as larvae and the autumn generation was much smaller than the summer generation, with only a few additional adult moths emerging in early or late October, respectively. The average egg production was estimated to be 124 eggs per female. The age structure of larvae during the storage period is shown in Figure 15. The first two instars were spent inside the wheat embryo. Larvae feed entirely on wheat embryos and each larva consumed 48 wheat embryos during development. The duration of the wandering period of mature larvae was estimated to be 56 h in 1943 and 24.5 h in 1944. Most larvae did not move far, but some larvae were recovered up to 15.2 m from the release point. During 1943 and 1944, the percentage of damaged wheat embryos decreased from 14 and 15% at the surface to 1% at 25 and 50 cm depth, and some damaged kernels were found as deep as 150 cm. This tendency for moth larvae to feed near the surface is, in part, a result of females ovipositing near the surface. For a related species, *P. interpunctella*, Schmidt (1982) showed that egg deposition was within 4.5–5 cm of the surface in rye and 8 cm in maize. The naturally present pathogen, *Bacillus thuringensis* Berliner, was the main cause of larval mortality for *E. elutella* (Richards and Waloff 1946). Larvae remained in diapause through the winter and diapausing larvae in the empty warehouse were the source of the initial infestation when wheat was first stored. Waloff and Richards (1946) studied the locomotory behavior of *E. elutella* adults and larvae in more detail in the laboratory. They examined the effects of environmental factors such as light intensity and temperature gradients. Populations in granaries in Germany differed from those in granaries in England in that the number of adult *E. elutella* emerging in the autumn generation tended to be

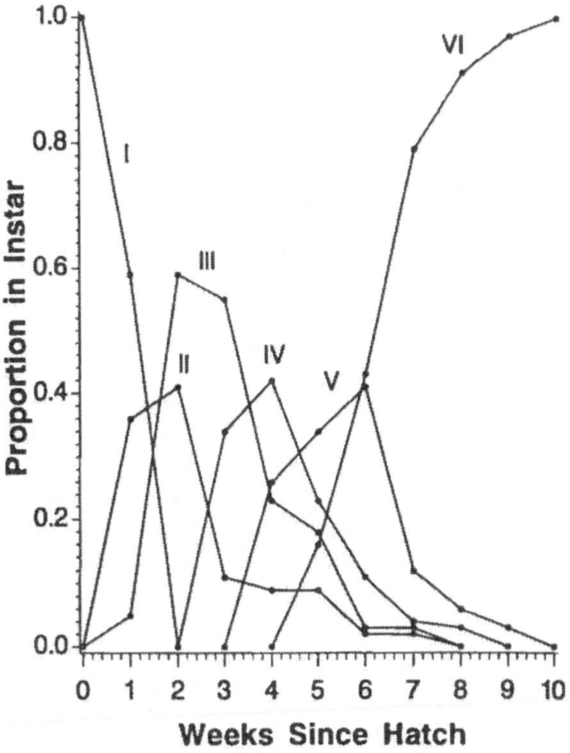


**Figure 15** Age-structure of larval population of *E. elutella* in wheat storage in England. Roman numerals designate curves for 1st through 6th instars. (Data from Richards and Waloff 1946).

larger than the number in the summer generation (Reichmuth et al. 1980). This suggests a lower level of diapause than in England.

In bagged maize stored in Kenya (Graham 1970a,b) and citrus pulp stored in the United States (Hagstrum and Sharp 1975), *C. cautella* had three discrete generations per year. Populations increased roughly 50-fold between the first and second generation and then at a much lower rate. Reduction in population growth was due primarily to a mite predator, *Blattisocius tarsalis* (Berlese), in Kenya and a pathogen, *B. thuringensis*, and a wasp parasite, *Habrobracon hebetor* Say, in the United States. Graham (1970c) found that larvae moved away from areas in which they had fed to pupate. This could be a response to the mandibular gland secretion produced by feeding larvae. Since the parasite *H. hebetor* uses these mandibular gland secretions to find hosts (Strand et al. 1989), this could reduce the chances of larvae being parasitized before they could pupate.

In a citrus pulp warehouse, the source of *C. cautella* infestation was a residual population of diapausing larvae in the warehouse (Hagstrum and Sharp 1975). Twenty-five diapausing last-instars were collected in a citrus pulp warehouse after it had been empty for 2 months. Insects first infested stacks of citrus pulp near the wall, and then spread towards the middle of the room. When a new stack of citrus pulp was stored next to a heavily infested stack, the number of larvae moving to the new stack was equivalent to one-third of those infesting the old stack. Diapausing larvae represented 59% of the larvae in new stack but only 40% of the original population. The numbers of diapausing and nondiapausing larvae increased throughout the storage period (Fig. 16). Peaks in population around the 10th, 18th, 24th, and 31st weeks of storage suggest that generation times were 6–8 weeks. The ratio of diapausing to nondiapausing larvae was highest between larval population peaks and this is probably a result of diapausing larvae remaining active after nondiapausing larvae of the same generation have pupated. After 25 weeks of storage, diapausing larvae were more numerous than nondiapausing larvae. Increasing population density resulted in an increase in the incidence of diapause during the storage period (Hagstrum and Silhacek 1980).

These large numbers of diapausing *C. cautella* larvae increase the effectiveness of *H. hebetor*, because diapausing host larvae are in a susceptible stage longer than nondiapausing host larvae. Also, many of the host larvae stung by *H. hebetor* crawl away before paralysis is complete and the parasite can lay her eggs (Hagstrum 1983). Many of these paralyzed hosts escaping parasite oviposition are unable to complete development and remain suitable for more than a month for host-feeding by adult parasites. Host-feeding increases parasite longevity. These hosts also remain suitable for parasite oviposition and development of their offspring.

*Insects in Grain*

In the United States, newly harvested wheat is generally uninfested when it is stored on the farm (Cotton and Winburn 1941, Hagstrum 1989), or at the elevator (Chao et al. 1953). Insects usually enter stored wheat in small numbers from April through October (Schwitzgebel and Walkden 1944). *C. ferrugineus* disperse from the top of the grain into the grain mass (Hagstrum 1989). This results in a logarithmic decrease in insect density from the top to the bottom of the grain mass (Fig. 17). Unpublished data from this study show that *R. dominica* is less mobile and tends to remain in the center of the top 60 cm of grain but *T. castaneum* behaves like *C. ferrugineus*. The mobility of these and other species of stored-product insects within the grain mass also has been studied in the laboratory (Surtees 1965, and references cited).

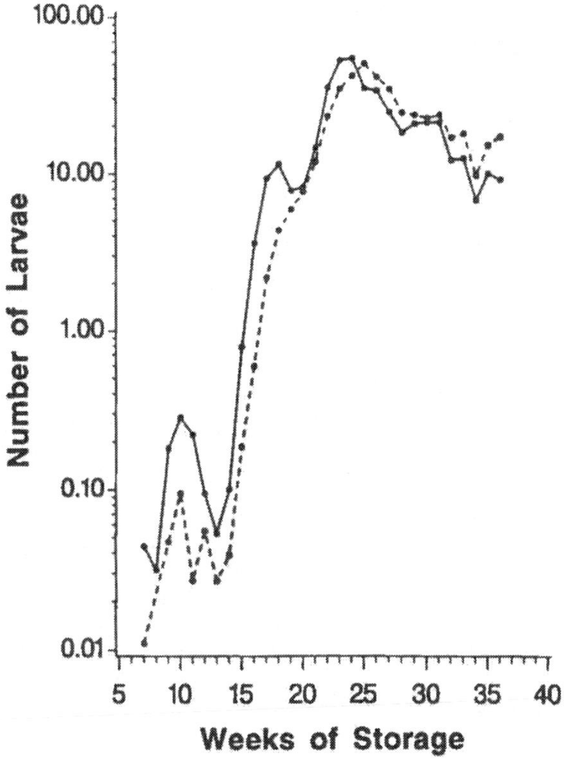


**Figure 16** Seasonal trends of nondiapausing (——) and diapausing (----) *C. cautella* larvae in Florida citrus pulp warehouse. (Redrawn from Hagstrum and Sharp 1975.)

Populations of *C. ferrugineus* increased steadily during the first three months of the storage period (Hagstrum 1987). However, the population growth rate declined as a result of parasitization by *Cephalonomia waterstoni* (Gahan) and then falling temperatures in the fall. The ability of *C. waterstoni* to locate hosts efficiently by following chemical trails left by host larvae as they move through the grain (Howard and Flinn 1990) probably explains why they substantially reduce the host population. Grain stored in farm bins cools in the fall from the outside toward the center, and large temperature gradients occur (Hagstrum 1987). Similar temperature gradients have been observed in flat storage in Australia (White 1988a) and during ocean vessel transport (Paulsen et al. 1991). Moisture can also vary between locations in a grain mass. Laboratory studies have shown that insects exposed to gradients select their preferred temperature (Waterhouse et

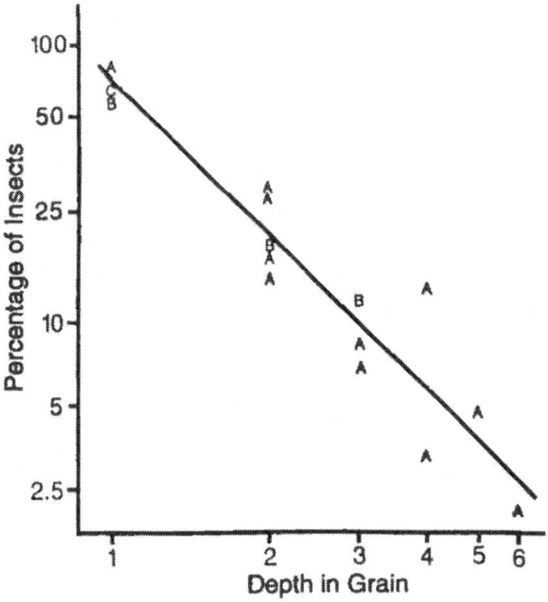


**Figure 17** Logarithmic decrease in the percentage of *C. ferrugineus* population in successive 0.75 m layers of wheat. A, B, and C indicate data point for one, two, or three bins, respectively. (Reprinted from Hagstrum 1989.)

al. 1971) or moisture (Weston and Hoffman 1991). Flinn et al. (1992a) developed a model to predict the effects of these temperature gradients on insect population growth.

In Canada, insect infestations of grain being moved from elevators to railroad cars increased steadily from May to November (Smith 1985). *C. ferrugineus* was found in 38% and 53% of the lots loaded during 2 years. Grain loaded into half of these cars contained only larvae. No more than 6% of the infested carloads were discovered at the elevators receiving shipment. In another study (Loschiavo 1974), 18–20% of railcars were infested by *C. ferrugineus*. Railroad transport did not affect the number of *C. ferrugineus* larvae, but did reduce the number of *C. ferrugineus* adults and *T. castaneum* larvae (Smith and Loschiavo 1978). Grain cleaning at terminal elevators removed a high percentage of adult insects. However, cleaning was ineffective in removing *C. ferrugineus* larvae because they feed within the wheat embryo (Smith 1972).

Hagstrum and Heid (1988) developed a model that simulates population growth of *R. dominica* and the movement of wheat through the marketing system in the United States. The model predicts 96.5% of the seasonal variation in the number of insects found in grain reaching the ports for export. Insect densities increased from two insects per bushel in March to 12 insects per bushel in August and then declined in response to natural cooling. Grain temperatures decline at a rate of 2°C per week from October through January. The time that insect populations have to grow depends upon the rate of grain movement through the marketing system. The marketing system was modeled by classifying it into farmer-owned and elevator wheat stocks. Roughly 20% of the wheat crop was still owned by the farmers when harvest began in June. Grain was stored at 32°C and 12% moisture content. The 20, 40, and 40% of the new crop harvested in June, July, and August, respectively, were added to the farmer-owned stocks in weekly increments of 5, 10, and 10% of the crop. Farmer-owned stocks were sold to elevators and elevator stocks were exported or milled at a rate of 2% per week.

Wheat can also be infested by *T. glabrum* and *T. inclusum*, particularly near the bin walls and grain surface (Hagstrum 1987). *T. glabrum* populations can reach high densities (White and McGregor 1957). In Arizona, *T. granarium* infestations of barley and sorghum have been studied (Nutting and Gerhardt 1964). Using the release–recapture method, they recaptured the majority of insects within 30 cm of the release site, but recovered a few as far away as 150 cm. Burges (1962) reported substantial migration of larvae in malt storage following an abrupt change in temperature or during unloading of malt, with some larvae moving over 12 m. In these studies, large larvae tended to be the predominant stage. *T. glabrum* larvae that did not pupate within 6 weeks (Beck 1971a) and *T. granarium* larvae that did not pupate within 7 weeks (Burges 1959) tended to remain in the larval stage for a long time, their respiratory rate was reduced, and these larvae could survive without feeding for a year or more. The prolonged larval development of *T. granarium* resulted from low temperature, crowding, or reduced quality of diet, whereas increased temperature or a combination of fresh food and reduced crowding stimulated pupation (Nair and Desai 1973). For *T. glabrum*, prolonged larval development was a direct response to the absence of food and pupation was initiated by feeding (Beck 1971a). Pupation also was stimulated by long photoperiod and pupation of females by the presence of male pupae (Beck 1971b). During the prolonged larval development, both species molted periodically and *T. granarium* would feed periodically if food was available. *T. glabrum* became smaller with each molt. A prolonged larval stage has also been reported for other species in the same family including *T. variabile* (= *parabile*) (Burges 1961), *Dermestes maculatus* DeGeer and *T. inclusum* (Beck 1971a), *Attagenus brunneus* Faldermann (= *elongatulus*) (Barak and Burkholder 1977) and *Attagenus unicolor* (Brahm) (= *megatoma*) (Baker 1982).

The population dynamics of insect infestations of bulk stored grain has also been studied in a number of other countries including Egypt (Aboul-Nasr et al. 1973), India (Simwat and Chahal 1980; Singh 1977), Japan (Kiritani 1958; Yoshida and Kawano 1958, 1959), Scotland (Cole and Cox 1981), and Zambia (Hindmarsh and MacDonald 1980).

Several additional studies need to be mentioned because of their unique contributions. Psocid infestations of stored grain are quite common, but have been thoroughly studied only by Sinha (1988) in Canada. Five species, *Lepinotus reticulatus* Enderlein, *Liposcelis bostrychophilus* Badonnel, *Liposcelis liparus* Broadhead, *Liposcelis rugosus* Badonnel, and *Liposcelis subfascus* Broadhead, were found in farm-stored wheat, oats, and barley, and were most abundant during the late summer and fall. Only one serious outbreak was observed involving *L. reticulatus* in wheat. In Nigeria, Ayertey and Ibitoye (1987) used an approach previously used with cowpeas to study the ecology of insects in the grain marketing system. They purchased grain from markets during each month of the year. The numbers of *S. oryzae*, *S. zeamais*, *R. dominica*, and *S. cerealella* infesting maize and sorghum were related to moisture content and region of the country. In the United States, the succession of nine insect species in maize during an 8 year storage period was studied by Arbogast and Mullen (1988). Initially *S. cerealella* was the dominant species. During the next 4 years, *O. surinamensis*, *C. ferrugineus*, and *S. zeamais* were dominant at various times. The wasp parasite, *Anisopteromalus calandrae* (Howard), was also abundant when its host, *S. zeamais*, was dominant. *T. castaneum* was present in significant numbers throughout much of the storage period and predominant for much of the third through fifth years. *L. oryzae*, *Cynaeus angustus* (LeConte), and *T. inclusum* were the dominant species after 5 years when grain was heavily damaged.

*Insects in Cowpeas*

Populations of *C. maculatus* and *Bruchidius atrolineatus* (Pic) that infest cowpeas in the field increase in numbers after storage (Hagstrum 1985; Monge and Huignard 1991). In the United States, cowpeas are harvested in October and the population growth of *C. maculatus* is often delayed 18 weeks by cold winter temperatures (Fig. 18). In Niger, both species reproduced on ripening pods in September, and were abundant when cowpeas were harvested in October and stored in traditional 1 $m^3$ mud-built stores. Both species have a reproductive diapause. Initially, *B. atrolineatus* was the dominant species and most adults were reproductively active. Cool temperatures resulted in high levels of reproductive diapause during the next generation. For both species, insects in reproductive diapause tended to leave the stored commodity in large numbers. The majority of *C. maculatus* were reproductively active during the dry season from December through May. The incidence of reproductive diapause increased from June to August during the rainy season in response to increasing seed moisture content. One egg parasite, *Uscana lariophaga*, and two larval parasites, *Eupelmus vuilleti* (Crawford) and *Dinarmus basalis* (Rondani), attack both bruchids in the field. The two larval parasites increased in numbers during storage.

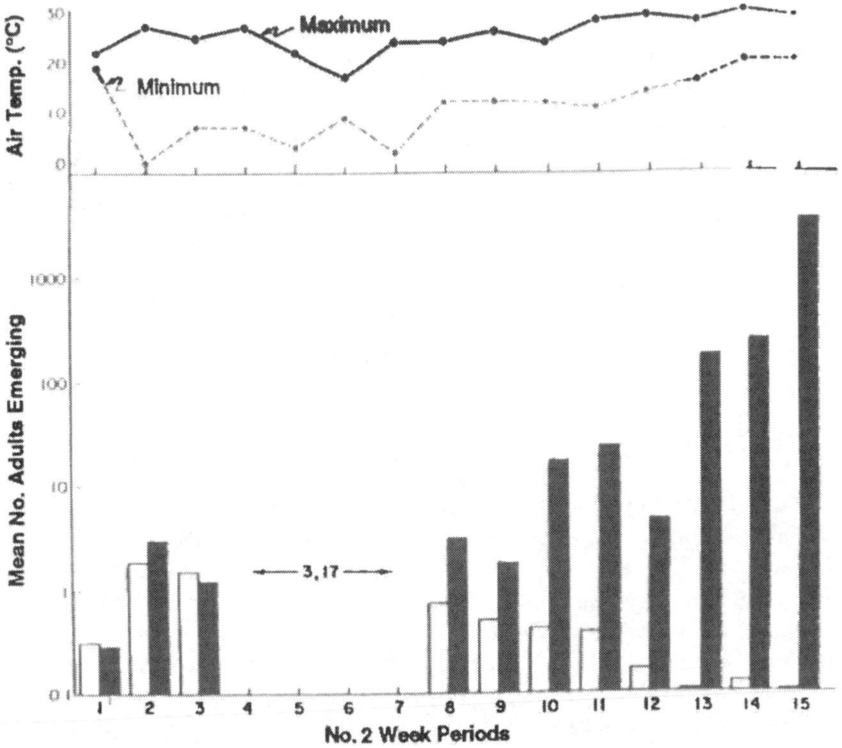


**Figure 18** Seasonal trends in temperature and average *C. maculatus* adult populations during storage of bagged cowpeas in Florida. The solid bars show means for populations that increased and the open bars show means for populations that died out. Only 3 and 17 adults emerged from 4th to 7th weeks of storage from populations that died out and increased, respectively. (Reprinted from Hagstrum 1985.)

Reproductive diapause has been reported for *Callosobruchus analis* (F.) (Tiwary and Verma 1989) and *Zabrotes subfasciatus* (Boh.) (Kapila and Pajni 1987), but reproduction was not reduced as much as that of *C. maculatus*. The average number of *C. analis* adults dispersing to bags 1.5, 3.0, 4.6, 6.1, and 7.6 m away from an infested bag was 2.5, 8.2, 9.8, 5.5, and 1, respectively. The percentage of migrating insects in reproductive diapause was 1.5 times higher than in the source population, and the percentage of reproductively active insects was 0.26 times that of the source population. Laboratory studies have shown that the presence of beans suppressed emigration of *C. chinensis* during the first 5 days of adult life (Shinoda and Yoshida 1984). In Nigeria, Caswell (1961) found that bruchid damage to cowpeas purchased from retail traders increased from 10% of seed in December to 50% of seed between May and September.

The population dynamics of insects in stored commodities are affected by many factors including seasonal changes in temperature, temperature gradients, insect locomotory behavior, semiochemicals, commodity handling and processing practices, larval diapause, reproductive diapause, parasites, predators, and pathogens. There are few comprehensive studies, and therefore population dynamics must be described by piecing together parts of many studies. In considering these descriptions, the reader must keep in mind that each facility is to some extent unique, that industrial practices vary between locations and have changed over time, and that, in some cases, pieces have been taken from a number of studies done under vastly different climates.

## 4. Processed Commodities

The management of insect pests in both raw and processed commodities requires an understanding of the dynamics of residual insect populations in facilities. However, the process of residual insect populations finding commodities is more important for processed than raw commodities because the tolerance for insects in commodities is much lower for processed than raw commodities. The dynamics of residual insect populations was discussed earlier in this chapter. Surveys are available to provide some measure of the likelihood of insects finding processed commodities. However, the relationship between residual infestations and probability of insects finding and entering processed commodities has not been studied. Surveys include records of infested commodities imported into the United States (Olsen 1981; Olsen et al. 1987; Zimmerman 1990), England (Howe and Freeman 1955), and Japan (Kiritani et al. 1959); inspections of packaged-food warehouses in nine countries (Highland 1978); sanitary inspections of bakeries, food stores, warehouses, and restaurants in the United States (Hankin and Welch 1991); inspections of food-handling areas of ships (Evans and Porter 1965); inquiries at Danish Pest Infestation Laboratory (Hallas et al. 1977); and questionnaires sent to households in Canada (Loschiavo and Sabourin 1982) and England (Turner and Maude-Roxby 1989). Much of the data are not reported in such a way that the percentage of facilities or food samples infested by insects can be calculated. Evans and Porter (1965) found *Tribolium* spp in 7% of food-handling areas inspected and Zimmerman (1990) found stored-product Coleoptera in 12.6% of imported food samples. Hallas et al. (1977) found that inquiries about *P. interpunctella* and *O. surinamensis* increased between 1965 and 1975, and concluded that this was due to an increase in indoor temperatures. Loschiavo and Sabourin (1982) found that 13.2 and 15.3% of houses, and 18.0 and 18.8% of apartments were infested with *O. mercator* during a 2 year study. Turner and Maude-Roxby (1989) found that 15.6 and 19.2% of households were infested with Psocoptera during a 2 year study. Both studies showed that 20.4–78% of consumers discard infested food without contacting the store where it was purchased or the manufacturer.

Although surveys do provide some insight into the probability of insects finding processed commodities, studies are badly needed on the relationship between the dynamics of residual pest populations and the probability of insects finding commodities.

## III. SUMMARY

Insect ecology is the study of factors regulating insect distribution and abundance. The main factors regulating insect population growth in the absence of pest management are temperature, moisture, and food. Insect population growth is increased as a result of shorter developmental times, higher egg production, and increased survival. Models have been developed to predict the effects of temperature and moisture on insect population growth. Many of the practices used to harvest, dry, move, and process commodities influence insect survival. The probability of insects infesting stored commodities is determined in part by their behavior in locating mates, oviposition sites, and food. Density-regulating factors other than crowding are most important because managed pest populations rarely reach high densities. Stored-product insects are constantly leaving infested commodities and food residues, and can disperse over long distances. These insects are well adapted to finding and surviving on small amounts of food widely dispersed in storage, transportation, or processing facilities, and have also been shown to breed on wild hosts. These characteristics make them particularly difficult pests to manage. A number of crops are commonly infested in the field prior to harvest by several species of stored-product insects. Infested commodities are the primary sources of stored-product insects infesting crops in the field. Infestation level in the field is influenced by the crop growth stage and can be high when the crop is left in the field to dry. Insect populations breeding on food residues in facilities also can be the primary source of insects that infest commodities. Industry places heavy emphasis on managing residual pest populations with good sanitation. Many species of stored-product insects seem to have retained some of their adaptations to wild habitats. Stored-product insects are capable of long-distance flight, of breeding on many wild hosts, and of feeding on flowers to increase adult longevity and fecundity. However, many of these feral populations may be emigrants from storage, transportation, and processing facilities, and may not be able to maintain themselves without these facilities. Infested commodities tend to be the most important source of insects that infest insect-free commodities because of their large volumes compared with food residues. Stock rotation plans are commonly used to reduce the chances of having infested commodities in a facility. The population dynamics of insects in stored commodities are affected by many factors including seasonal changes in temperature, temperature gradients, insect locomotory behavior, semiochemicals, commodity handling and processing practices, larval diapause, reproductive diapause, parasites, predators, and pathogens. Surveys of insects in processed commodities provide some insight into the probability of their finding commodities, but little information is available on the relationship between the dynamics of residual pest populations and the probability of insects finding commodities.

## REFERENCES

Aboul-Nasr, S., Salama, H. S., Ismail, I. I., and Salem, S. A. (1973). Ecological studies on insects infesting wheat grains in Egypt. *Z. Ang. Entomol.*, *73*:203–212.

Arakaki, N., and Takahashi, F. (1982). Oviposition preference for rice weevil *Sitophilus zeamais* Motschulsky (Coleoptera: Curculionidae), for polished and unpolished rice. *Jpn. J. Appl. Entomol. Zool.*, *26*:166–171.

Arbogast, R. T. (1976). Population parameters for *Oryzaephilus surinamensis* and *O. mercator*: Effect of relative humidity. *Environ. Entomol.*, *5*:738–742.

Arbogast, R. T. (1989). Detritus as a factor influencing population growth rates of three tenebrionid beetles in stored corn. *J. Entomol. Sci.*, *24*:454–459.

Arbogast, R. T., and Mullen, M. A. (1988). Insect succession in a stored-corn ecosystem in southeast Georgia. *Ann. Entomol. Soc. Amer.*, *81*:899–912.

Ayertey, J. N., and Ibitoye, J. O. (1987). Infestation of maize and sorghum seeds by *Sitophilus*, *Rhyzopertha* and *Sitotroga* in three contiguous climatic zones in Nigeria. *Insect Sci. Applic.*, *8*:981–987.

Bahr, I. (1973). Investigations on the reductions of pest populations in grain by pneumatic conveyance. *Nachr. Pflanz. DDR*, *27*:232–237.

Bahr, I. (1975). The incidence of damage caused by the lesser grain borer (*Rhyzopertha dominica* F.) and the action of a suction and pressure blower on infestation of grain. *Nachr. Pflanz. DDR*, *29*:228–231.

Bailey, S. W. (1969). The effects of physical stress in the grain weevil *Sitophilus granarius* *J. Stored Prod. Res.*, *5*:311–324.

Baker, J. E. (1982). Termination of larval diapause-like condition in *Attagenus megatoma* (Coleoptera: Dermestidae) by low temperature. *Environ. Entomol.*, *11*:506–508.

Baker, J. E. (1988). Development of four strains of *Sitophilus oryzae* (L.) (Coleoptera: Curculionidae) on barley, corn (maize), rice and wheat. *J. Stored Prod. Res.*, *24*:193–198.

Baker, J. E., and Loschiavo, S. R. (1987). Nutritional ecology of stored-product insects. In Nutritional Ecology of Insects, Mites, Spiders, and Related Invertebrates (ed. Slansky, F., and Rodriguez, J. G.). John Wiley & Sons, New York, p. 321.

Baker, J. E., and Mabie, J. A. (1973). Feeding behavior of larvae of *Plodia interpunctella* *Environ. Entomol.*, *2*:627–632.

Baker, J. E., Woo, S. M., Throne, J. E., and Finney, P. L. (1991). Correlation of alpha-amylase inhibitor content in Eastern soft wheats with development parameters of the rice weevil (Coleoptera: Curculionidae). *Environ. Entomol.*, *20*:53–60.

Barak, A. V., and Burkholder, W. E. (1977). Studies on the biology of *Attagenus elongatulus*

Casey (Coleoptera: Dermestidae) and the effects of larval crowding on pupation and life cycle. *J. Stored Prod. Res.*, *13*:169–175.

Barker, P. S., and Smith, L.B. (1987). Spatial distribution of insect species in granary residues in the prairie provinces. *Can. Entomol.*, *119*:1123–1130.

Barker, P. S., and Smith, L. B. (1990). Influence of granary type and farm practices on the relative abundance of insects in granary residues. *Can. Entomol.*, *122*:393–400.

Barratt, B. I. P. (1974). Timing of production of a sex pheromone by females of *Stegobium paniceum* (L.) (Coleoptera, Anobiidae) and factors affecting male response. *Bull. Ent. Res.*, *64*:621–628.

Barrer, P. M. (1976). The influence of delayed mating on the reproduction of *Ephestia cautella* (Walker) (Lepidoptera: Phycitidae). *J. Stored Prod. Res.*, *12*:165–169.

Barrer, P. M. (1983). A field demonstration of odour-based, host-food finding behavior in several species of stored grain insects. *J. Stored Prod. Res.*, *19*:105–110.

Barrer, P. M., and Hill, R. J. (1977). Some relationships between the "calling" posture and sexual receptivity in unmated females of the moth, *Ephestia cautella. Physiol. Entomol.*, 2:255–260.

Barrer, P. M., and Jay, E. G. (1980). Laboratory observations on the ability of *Ephestia cautella* (Walker) (Lepidoptera: Phycitidae) to locate, and to oviposit in response to a source of grain odour. *J. Stored Prod. Res.*, *16*:1–7.

Beck, S. D. (1971a). Growth and retrogression in larvae of *Trogoderma glabrum* (Coleoptera: Dermestidae). 1. Characteristics under feeding and starvation conditions. *Ann. Entomol. Soc. Amer.*, *64*:149–155.

Beck, S. D. (1971b). Growth and retrogression in larvae of *Trogoderma glabrum* (Coleoptera: Dermestidae). 2. Factors influencing pupation. *Ann. Entomol. Soc. Amer.*, *64*:946–949.

Bellows, T. S . (1982). Simulation models for laboratory populations of *Callosobruchus chinensis* and *C. maculatus. J. Animal Ecol.*, *51*:597–623.

Bernabe-Adalla, C., and Bernardo, E. N. (1976). Correlation between husk characters and weevil infestation of 51 varieties and lines of maize in the field. *Phil. Agr.*, *60*:121–129.

Birch, L. C. (1945). The mortality of the immature stages of *Calandra oryzae* L. (small strain) and *Rhyzopertha dominica* Fab. in wheat of different moisture contents. *Aust. J. Exp. Biol. Med. Sci.* **23**:141–145.

Blickenstaff, C. C. (1960). Effect of sample location within fields on corn earworm and rice weevil infestation and damage. *J. Econ. Entomol.*, *53*:745–747.

Booker, R. H. (1967). Observations on three bruchids associated with cowpea in northern Nigeria. *J. Stored Prod. Res.*, *3*:1–15.

Boon, K. S., and Ho, S. H. (1988). Factors influencing the post-fumigation reinfestation of *Tribolium castaneum* (Herbst) (Coleoptera: Tenebrionidae) in a rice warehouse. *J. Stored Prod. Res.*, *24*:87–90.

Bryan, J. M., and Elvidge, J. (1977). Mortality of adult grain beetles in sample delivery systems used in terminal grain elevators. *Can. Entomol.*, *109*:209–213.

Buchelos, C. T. (1980). Moth populations at a typical flour mill. *Ann. Instit. Phytopath. Benaki*, *12*:188–197.

Buchelos, C. T. (1981). Coleoptera populations at flour mills and related areas. *Ann. Instit. Phytopath. Benaki*, *13*:6–29.

Buchelos, C. T. (1989). A contribution to *Lasioderma* spp. and other Coleoptera collected from thistles in southern Greece. *Entomologia Hellenica*, 7:7–12.

Burges, H. D. (1959). Studies on the dermestid beetle, *Trogoderma granarium* Everts. II. The occurrence of diapause larvae at a constant temperature and their behaviour. *Bull. Entomol. Res.*, *50*:407–422.

Burges, H. D. (1961). The effect of temperature, humidity and quantity of food on the development and diapause of *Trogoderma parabile* Beal. *Bull. Ent. Res.*, *51*:685–696.

Burges, H. D. (1962). Diapause, pest status and control of the Khapra beetle, *Trogoderma granarium* Everts. *Ann. Appl. Biol.*, *50*:614–617.

Burkholder, W. E. (1982). Reproductive biology and communication among grain storage and warehouse beetles. *J. Georgia Entomol. Soc.*, *17*:1–10.

Burkholder, W. E., and Ma, M. (1985). Pheromones for monitoring and control of stored-product insects. *Annu. Rev. Entomol.*, *30*:257–272.

Caswell, G. H. (1961). The infestation of cowpeas in the western region of Nigeria. *Trop. Sci.*, *3*:154–158.

Chambers, J. (1990). Overview on stored-product insect pheromones and food attractants. *J. Kansas Entomol. Soc.*, *63*:490–499.

Chang, C. S., Converse, H. H., and Martin, C. R. (1983). Bulk properties of grain as affected by self-propelled rotational type grain spreaders. *Trans. Am. Soc. Agric. Eng.*, *16*: 129–133.

Chao, Y., Simkover, H. G., Telford, H. S., and Stallcop, P. (1953). Field infestation of stored grain insects in eastern Washington. *J. Econ. Entomol.*, *46*:905–907.

Chesnut, T. L. (1972). Flight habits of the maize weevil as related to field infestation of corn. *J. Econ. Entomol.*, *65*:434–435.

Chippendale, G. M., and Mann, R. A. (1972). Feeding behavior of Angoumois grain moth larvae. *J. Insect Physiol.*, *18*:87–94.

Cline, L. D. (1991). Progeny production and adult longevity of *Cryptolestes pusillus* (Coleoptera: Cucujidae) on broken and whole corn at selected humidities. *J. Econ. Entomol.*, *84*:120–125.

Coffelt, J. A., Sower, L. L., and Vick, K. W. (1978). Quantitative analysis of identified compounds in pheromone gland rinses of *Plodia interpunctella* and *Ephestia cautella* at different times of day. *Environ. Entomol.*, 7:502–505.

Cogburn, R. R. (1973a). Stored-product insect populations in port warehouses of the Gulf coast. *Environ. Entomol.*, 2:401–407.

Cogburn, R. R. (1973b). Stored-product insect populations in boxcars delivering flour and rice to Gulf coast ports. *Environ. Entomol.*, 2:427–431.

Cogburn, R. R., and Bollich, C. N. (1990). Heritability of resistance to stored-product insects in three hybrid populations of rice. *Environ. Entomol.*, *19*:268–273.

Cogburn, R. R., and Vick, K. W. (1981). Distribution of Angoumois grain moth, almond moth, and Indian meal moth in rice fields and rice storages in Texas as indicated by pheromone-baited adhesive traps. *Environ. Entomol.*, *10*:1003–1007.

Cole, D. B., and Cox, P. D. (1981). Studies on three moth species in a Scottish port silo, with special reference to overwintering *Ephestia kuehniella* (Zeller) (Lepidoptera: Pyralidae). *J. Stored Prod. Res.*, *17*:163–181.

Corbet, S. A. (1971). Mandibular gland secretion of larvae of the flour moth, *Anagasta kuehniella*, contains an epideictic pheromone and elicits oviposition movements in a hymenopteran parasite. *Nature, 232*:481–484.

Corbet, S. A. (1973). Oviposition pheromone in larval mandibular glands of *Ephestia kuehniella. Nature, 243*:537–538.

Cotton, R. T., and Wagner, G. B. (1935). Effect of milling process on insects. *Amer. Miller, 63*:58–60.

Cotton, R. T., and Winburn, T. F. (1941). Field infestation of wheat by insects attacking it in farm storage. *J. Kansas Entomol. Soc., 14*:12–16.

Cox, P. D., and Bell, C. H. (1991). Biology and ecology of moth pests of stored foods. In Ecology and Management of Food-Industry Pests (ed. Gorham, J. R.). Assoc. Off. Anal. Chem., Arlington, VA, pp. 181–193.

Cox, P. D., and Parish, W. E. (1991). Effects of refuge content and food availability on refuge-seeking behavior in *Cryptolestes ferrugineus* (Stephens) (Coleoptera: Cucujidae). *J. Stored Prod. Res., 27*:135–139.

Credland, P. F. (1986). Effect of host availability on reproductive performance in *Callosobruchus maculatus* (F.) (Coleoptera: Bruchidae). *J. Stored Prod. Res., 22*:49–54.

Credland, P. F., and Wright, A. W. (1990). Oviposition deterrents of *Callosobruchus maculatus* (Coleoptera: Bruchidae). *Physiol. Entomol., 15*:285–298.

Cross, J. H., Byler, R. E., Silverstein, R. M., Greenblatt, R. E., Gorman, J. E., and Burkholder, W. E. (1977). Sex pheromone components and calling behavior of the female dermestid beetle, *Trogoderma variable* Ballion. *J. Chem. Ecol., 3*:115–125.

Currie, J. E. (1967). Some effects of temperature and humidity on the rates of development, mortality and oviposition of *Cryptolestes pusillus* (Schonherr) (Coleoptera, Cucujidae). *J. Stored Prod. Res., 3*:97–108.

Dick, K., and Credland, P. F. (1984). Egg production and development of three strains of *Callosobruchus maculatus* (F.) (Coleoptera: Bruchidae). *J. Stored Prod. Res., 20*: 221–227.

Dix, D. E., and All, J. N. (1986). Population density and sex ratio dynamics of overwintering maize weevils (Coleoptera: Curculionidae) infesting field corn. *J. Entomol. Sci., 21*:368–375.

Dobie, P. (1974). The laboratory assessment of the inherent susceptibility of maize varieties to post-harvest infestation by *Sitophilus zeamais* Motsch. (Coleoptera, Curculionidae). *J. Stored Prod. Res., 10*:183–197.

Dyte, C. E. (1965). Studies on insect infestations in the machinery of three English flour mills in relation to seasonal temperature changes. *J. Stored Prod. Res., 1*:129–144.

Dyte, C. E. (1966). Studies on the abundance of *Cryptolestes turcicus* (Grouv.) (Coleoptera, Cucujidae) in different machines of an English flour mill. *J. Stored Prod. Res., 1*: 341–352.

El-Sawaf, S. K. (1956). Some factors affecting the longevity, oviposition, and rate of development in the Southern cowpea weevil, *Callosobruchus maculatus* F. *Bull. Soc. Entomol. Egypte, 40*:29–95.

Evans, B. R., and Porter, J. E. (1965). The incidence, importance, and control of insects found in stored food and food-handling areas of ships. *J. Econ. Entomol., 58*:479–481.

Evans, D. E. (1983). The influence of relative humidity and thermal acclimation on the survival of adult grain beetles in cooled grain. *J. Stored Prod. Res., 19*:173–180.

Evans, D. E. (1987). The survival of immature grain beetles at low temperatures. *J. Stored Prod. Res.*, *23*:79–83.

Farrar, M. D., and Reed, R. H. (1942). Insect survival in drying grain. *J. Econ. Entomol.*, *35*:923–928.

Fatunla, T., and Badaru, K. (1983). Resistance of cowpea pods to *Callosobruchus maculatus*. *J. Agric. Sci., Camb.*, *100*:205–209.

Faustini, D. L., Giese, W. L., Phillips, J. K., and Burkholder, W. E. (1982). Aggregation pheromone of the male granary weevil, *Sitophilus granarius* (L.). *J. Chem. Ecol.*, *8*:679–687.

Fields, P. G. (1992). The control of stored-product insects and mites with extreme temperatures. *J. Stored Prod. Res.*, *28*:89–118.

Fields, P. G., Van Loon, J., Dolinski, M. G., Harris, J. L., and Burkholder, W. E. (1993). The distribution of *Rhyzopertha dominica* (F.) in western Canada. *Can. Entomol.*, *125*: 317–328.

Fitzner, M. S., Hagstrum, D. W., Knauft, D. A., Buhr, K. L., and McLaughlin, J. R. (1985). Genotypic diversity in the suitability of cowpea (Rosales: Leguminosae) pods and seeds for cowpea weevil (Coleoptera: Bruchidae) oviposition and development. *J. Econ. Entomol.*, *78*:806–810.

Fleming, D. A. (1988). The influence of wheat kernel damage upon the development and productivity of *Oryzaephilus surinamensis* (L.) (Coleoptera: Silvanidae). *J. Stored Prod. Res.*, *24*:233–236.

Fletcher, L. W., and Long, J. S. (1971). Influence of food odors on oviposition by the cigarette beetle on nonfood materials. *J. Econ. Entomol.*, *64*:770–771.

Flinn, P. W., and Hagstrum, D. W. (1990). Simulations comparing the effectiveness of various stored-grain management practices used to control *Rhyzopertha dominica* (Coleoptera: Bostrichidae). *Environ. Entomol.*, *19*:725–729.

Flinn, P. W., Hagstrum, D. W., Muir, W. E., and Sudayappa, K. (1992a). Spatial model for simulating changes in temperature and insect population dynamics in stored grain. *Environ. Entomol.*, *21*:1351–1356.

Flinn, P. W., McGaughey, W. H., and Burkholder, W. E. (1992b). Effects of fine material on insect infestation: A review. *N. Cent. Regional Res. Publ. 332*, pp. 24–30.

Foster, G. H., and Holman, L. E. (1973). Grain breakage caused by commercial handling methods. USDA/ARS Marketing Res. Rept. No. 968.

Giga, D. P., and Smith, R. H. (1983). Comparative life history studies of four *Callosobruchus* species infesting cowpeas with special reference to *Callosobruchus rhodesianus* (Pic) (Coleoptera: Bruchidae). *J. Stored Prod. Res.*, *19*:189–198.

Giga, D. P., and Smith, R. H. (1985). Oviposition markers in *Callosobruchus maculatus* F. and *Callosobruchus rhodesiansus* Pic. (Coleoptera: Bruchidae): Asymmetry of interspecific responses. *Agric. Ecosystems and Environ.*, *12*:229–234.

Giles, P. H., and Ashman, F. (1971). A study of pre-harvest infestation of maize by *Sitophilus zeamais* Motsch. (Coleoptera, Curculionidae) in the Kenya highlands. *J. Stored Prod. Res.*, *7*:69–83.

Gomez, L. A., Rodriguez, J. G., Poneleit, C. G., and Blake, D. F. (1983). Relationship between some characteristics of the corn kernal pericarp and resistance to the rice weevil (Coleoptera: Curculionidae). *J. Econ. Entomol.*, *76*:797–800.

Good, N. E. (1937). Insects found in the milling streams of flour mills in the southwestern milling area. *J. Kansas Entomol. Soc.*, *10*:135–148.

Gordon, D. M., and Stewart, R. K. (1988). Demographic characteristics of the stored-products moth *Cadra cautella*. *J. Animal Ecol.*, *57*:627–644.

Graham, W. H. (1970a). Warehouse ecology studies of bagged maize in Kenya I. The distribution of adult *Ephestia (Cadra) cautella* (Walker) (Lepidoptera, Phycitidae). *J. Stored Prod. Res.*, *6*:147–155.

Graham, W. H. (1970b). Warehouse ecology studies of bagged maize in Kenya II. Ecological observations of an infestation by *Ephestia (Cadra) cautella* (Walker) (Lepidoptera, Phycitidae). *J. Stored Prod. Res.*, *6*:157–167.

Graham, W. H. (1970c). Warehouse ecology studies of bagged maize in Kenya III. Distribution of the immature stages of *Ephestia (Cadra) cautella* (Walker) (Lepidoptera, Phycitidae). *J. Stored Prod. Res.*, *6*:169–175.

Graham, W. M. (1970d). Warehouse ecology studies of bagged maize in Kenya. IV. Reinfestation following fumigation with methyl bromide gas. *J. Stored Prod. Res.*, *6*:177–180.

Greenblatt, R. E., Burkholder, W.E., Cross, J. H., Cassidy, R. F. Jr., Silverstein, R. M., Levinson, A. R., and Levinson, H. Z. (1977). Chemical basis for interspecific responses to sex pheromones of *Trogoderma* species. *J. Chem. Ecol.*, *3*:337–347.

Hagstrum, D.W. (1983). Self-provisioning with paralyzed hosts and age, density, and concealment of hosts as factors influencing parasitization of *Ephestia cautella* (Walker) (Lepidoptera: Pyralidae) by *Bracon hebetor* Say (Hymenoptera: Braconidae). *Environ. Entomol.*, *12*:1727–1732.

Hagstrum, D. W. (1984). Growth of *Ephestia cautella* (Walker) population under conditions found in an empty peanut warehouse and response to variations in the distribution of larval food. *Environ. Entomol.*, *13*:171–174.

Hagstrum, D. W. (1985). Preharvest infestation of cowpeas by the cowpea weevil (Coleoptera: Bruchidae) and population trends during storage in Florida. *J. Econ. Entomol.*, *78*:358–361.

Hagstrum, D. W. (1987). Seasonal variation of stored wheat environment and insect populations. *Environ. Entomol.*, *16*:77–83.

Hagstrum, D. W. (1989). Infestation by *Cryptolestes ferrugineus* (Coleoptera: Cucujidae) of newly harvested wheat stored on three Kansas farms. *J. Econ. Entomol.*, *82*:655–659.

Hagstrum, D. W., and Davis, L. R. (1980). Mate-seeking behavior of *Ephestia cautella*. *Environ. Entomol.*, *9*:589–592.

Hagstrum, D. W., and Flinn, P. W. (1990). Simulations comparing insect species differences in response to wheat storage conditions and management practices. *J. Econ. Entomol.*, *83*:2469–2475.

Hagstrum, D. W., and Flinn, P. W. (1992). Integrated pest management of stored-grain insects. In Storage of Cereal Grains and Their Products (ed. Sauer, D. B.). Amer. Assoc. Cereal Chem., St. Paul, MN, pp. 535–562.

Hagstrum, D. W., and Gilbert, E. E. (1976). Emigration rate and age structure dynamics of *Tribolium castaneum* populations during growth phase of a colonizing episode. *Environ. Entomol.*, *5*:445–448.

Hagstrum, D. W., and Heid, W. G. (1988). U.S. wheat-marketing system: An insect ecosystem. *Bull. Entomol. Soc. Amer.*, *34*:33–36.
Hagstrum, D. W., and Milliken, G. A. (1988). Quantitative analysis of temperature, moisture, and diet factors affecting insect development. *Ann. Entomol. Soc. Amer.*, *81*: 539–546.
Hagstrum, D. W., and Milliken, G. A. (1991). Modeling differences in insect developmental times between constant and fluctuating temperatures. *Ann. Entomol. Soc. Amer.*, *84*:369–379.
Hagstrum, D. W., and Sharp, J. E. (1975). Population studies on *Cadra cautella* in a citrus pulp warehouse with particular reference to diapause. *J. Econ. Entomol.*, *68*:11–14.
Hagstrum, D. W., and Silhacek, D. L. (1980). Diapause induction in *Ephestia cautella*: An interaction between genotype and crowding. *Entomol. Exp. Appl.*, *28*:29–37.
Hagstrum, D. W., and Stanley, J. M. (1979). Release-recapture estimates of the population density of *Ephestia cautella* (Walker) in a commercial peanut warehouse. *J. Stored Prod. Res.*, *15*:117–122.
Hagstrum, D. W., and Throne, J. E. (1989). Predictability of stored-wheat insect population trends from life history traits. *Environ. Entomol.*, *18*:660–664.
Hagstrum, D. W., and Tomblin, C. F. (1973). Oviposition by the almond moth, *Cadra cautella* in response to falling temperature and onset of darkness. *Ann. Entomol. Soc. Amer.*, *66*:809–812.
Hagstrum, D. W., and Tomblin, C. F. (1975). Relationship between water consumption and oviposition by *Cadra cautella* (Lepidoptera: Phycitidae). *J. Georgia Entomol. Soc.*, *10*:358–363.
Hagstrum, D. W., Stanley, J. M., and Turner, K. W. (1977). Flight activity of *Ephestia cautella* as influenced by the intensity of ultraviolet or green radiation. *J. Georgia Entomol. Soc.*, *12*:231–236.
Hallas, T., Mourier, H., and Winding, O. (1977). Seasonal variation and trends for some indoor insects in Denmark. *Entomol. Meddr.*, *45*:77–88.
Hammack, L., and Burkholder, W. E. (1976). Circadian rhythm of sex pheromone-releasing behaviour in females of the dermestid beetle, *Trogoderma glabrum*: Regulation by photoperiod. *J. Insect Physiol.*, 22:385–388.
Hankin, L., and Welch, K. (1991). Insects found during sanitary inspections. *Dairy Food Environ. Sanitation*, *11*:575–576.
Happ, G. M. (1968). Quinone and hydrocarbon production in the defensive glands of *Eleodes longicollis* and *Tribolium castaneum* (Coleoptera: Tenebrionidae). *J. Insect Physiol.*, *14*:1821–1837.
Haque, E., Foster, G. H ., Chung, D. S., and Lai, F. S. (1978). Static pressure drop across a bed of corn mixed with fines. *Trans. Amer. Soc. Agric. Eng.*, *21*:997–1000.
Highland, H. A. (1978). Insects infesting foreign warehouses containing packaged foods. *J. Georgia Entomol. Soc.*, *13*:251–256.
Hindmarsh, P. S., and MacDonald, I. A. (1980). Field trials to control insect pests of farm-stored maize in Zambia. *J. Stored Prod. Res.*, *16*:9–18.
Honda, H., and Ohsawa, K. (1990). Chemical ecology for stored product insects. *J. Pesticide Sci.*: *15*:263–270.
Honda, H., Oshima, K., and Yamamoto, I. (1976). Oviposition marker of azuki bean weevil,

*Callosobruchus chinensis* (L.), Proceedings; Joint USA–Japan Seminar on Stored Products Pests, Kansas State University, pp. 116–128.

Hosny, M. M., Hassanein, M. H., and Kamel, A. H. (1968). Ecological studies on *Anagasta kuehniella* and *Corcyra cephalonica* infesting flour mills in Cairo. *Bull. Soc. Entomol. Egypte*, *52*:445–456.

Howard, R. W. (1987). Chemosystematic studies of the Triboliini (Coleoptera: Tenebrionidae): phylogenetic inferences from the defensive chemicals of eight *Tribolium* spp., *Palorus ratzeburgi* (Wissmann), and *Latheticus oryzae* Waterhouse. *Ann. Entomol. Soc. Amer.*, *80*:398–405.

Howard, R. W., and Flinn, P. W. (1990). Larval trails of *Cryptolestes ferrugineus* (Coleoptera: Cucujidae) as kairomonal host-finding cues for the parasitoid *Cephalonomia waterstoni* (Hymenoptera: Bethylidae). *Ann. Entomol. Soc. Amer.*, *83*:239–245.

Howard, R. W., and Mueller, D. D. (1987). Defensive chemistry of the flour beetle *Tribolium brevicornis* (LeC.): Presence of known and potential prostaglandin synthetase inhibitors. *J. Chem. Ecol.*, *13*:1707–1723.

Howard, R. W., Jurenka, R. A., and Blomquist, G. J. (1986). Prostaglandin synthetase inhibitors in the defensive secretion of the red flour beetle *Tribolium castaneum* (Herbst) (Coleoptera: Tenebrionidae). *Insect Biochem.*, *16*:757–760.

Howe, R. W. (1952). The biology of the rice weevil, *Calandra oryzae*. *Ann. Appl. Biol.*, *39*:168–180.

Howe, R. W. (1957). A laboratory study of the cigarette beetle, *Lasioderma serricorne* (F.) (Col., Anobiidae) with a critical review of the literature on its biology. *Bull. Entomol. Res.*, *48*:9–56.

Howe, R. W. (1960). The effects of temperature and humidity on the rate of development and the mortality of *Tribolium confusum* Duval (Coleoptera, Tenebrionidae). *Ann. Appl. Biol.*, *48*:363–376.

Howe, R. W. (1965). *Sitophilus granarius* (L.) (Coleoptera, Curculionidae) breeding in acorns. *J. Stored Prod. Res.*, *1*:99–100.

Howe, R. W., and Freeman, J. A. (1955). Insect infestation of West African produce imported into Britain. *Bull. Entomol. Res.*, *46*:643–668.

Howlader, A. J., and Matin, A. S. M. A. (1988). Observations on the pre-harvest infestation of paddy by stored grain pests in Bangladesh. *J. Stored Prod. Res.*, *24*:229–231.

Ingemansen, J. A., Reeves, D. L., and Walstrom, R. J. (1986). Factors influencing stored-oat insect populations in South Dakota. *J. Econ. Entomol.*, *79*:518–522.

Imura, O. (1981). Stored-product insects in a barley mill. II. Fauna and seasonal changes of species diversity. *Jpn. J. Ecol.*, *31*:139–146.

Jacob, T. A., and Fleming, D. A. (1982). Observations on the influence of free water on the fecundity and longevity of *Dermestes lardarius* L. (Col., Dermestidae). *Entomol. Mon. Mag.*, *118*:127–131.

Joffe, A., and Clarke, B. (1963). The effect of physical disturbance or "turning" of stored maize on the development of insect infestations. II. Laboratory studies with *Sitophilus oryzae* (L.). *S. Afr. J. Agric. Sci.*, *6*:65–84.

Joubert, P. C. (1966). Field infestations of stored-product insects in South Africa. *J. Stored Prod. Res.*, 2:159–161.

Kaminski, E., Stawicki, S., Wasowicz, E., Giebel, H., and Zawirska, R. (1979a). Volatile fatty acids produced by some of the bacteria occurring in cereals. *Bull. Acad. Pol. Sci.*, *27*:7–12.

Kaminski, E., Stawicki, S., Wasowicz, E., Giebel, H., Przybylski, R., Zawirska, R., and Zalewski, R. (1979b). Volatile flavour compounds produced by bacteria cultivated on grain media. *Acta Aliment. Pol.*, *5*:263–274.

Kanaujia, K. E., and Levinson, H. Z. (1981a). Phagostimulatory responses and oviposition behavior of *Sitophilus granarius* L. to newly harvested and stored wheat grains. *Z. Ang. Entomol.*, *91*:417–424.

Kanaujia, K. R., and Levinson, H. Z. (1981b). Oviposition and feeding responses of *Sitophilus granarius* L. to newly harvested and stored wheat as well as to extracts thereof. *Med. Fac. Landbouww. Rijksuniv. Gent.*, *46*:519–534.

Kapila, R., and Pajni, H. R. (1987). Polymorphism in *Zabrotes subfasciatus* (Boh.) (Coleoptera: Bruchidae). *Bull. Entomol.*, *28*:132–137.

Kawamoto, H., Woods, S. M., Sinha, R. N., and Muir, W. E. (1989). A simulation model of population dynamics of the rusty grain beetle, *Cryptolestes ferrugineus* in stored wheat. *Ecol. Model.*, *48*:137–157.

Keever, D. W., Wiseman, B. R., and Widstrom, N. W. (1988). Effects of threshing and drying on maize weevil populations in field-infested corn. *J. Econ. Entomol.*, *81*: 727–730.

Khare, B. P., and Agrawal, N. S. (1964). Rodent and ant burrows as sources of insect innoculum in the threshing floors. *Indian J. Entomol.*, *26*:97–102.

Khare, B. P., Shukla, R. P., and Singh, B. (1979). Development of stored-grain insects in mechanically damaged wheat grain. *Indian J. Agric. Sci.*, *49*:609–612.

Kiritani, K. (1958). On the distribution and seasonal prevalence of stored grain insects in a farm premises. *Botyu-Kagaku*, *23*:164–172.

Kiritani, K., Muramatsu, T., and Yoshimura, S. (1959). Fauna of storage pests of South East Asian produce imported into Japan. *Osaka Shokubutsu Boeki (Plant Prot.)*, *7*: 184–207.

Kirk, V. M. (1965). Some flight habits of the rice weevil. *J. Econ. Entomol.*, *58*: 155–156.

Kokubu, H., and Mills, R. B. (1980). Susceptibility of thirteen stored product beetles to entanglement by *Trogoderma* hastisetae. *J. Stored Prod. Res.*, *16*:87–92.

Komson, A., and Stewart, R. K. (1968). The effect of temperature and diet upon the development of the sawtoothed grain beetle, *Oryzaephilus surinamensis* (Linn.) and of the merchant grain beetle, *Oryzaephilus mercator* (Fauv.), (Coleoptera, Cucujidae). *Thai. J. Agric. Sci.*, *1*:40–51.

Koone, H. D. (1952). Maturity of corn and life history of the Angoumois grain moth. *J. Kansas Entomol. Soc.*, *25*:103–105.

Krasnoff, S. B., Vick, K. W., and Coffelt, J. A. (1984). (Z,E)-9,12-tetradecadien-1-ol: a component of the sex pheromone of *Ephestia elutella* (Hubner) (Lepidoptera: Pyralidae). *Environ. Entomol.*, *13*:765–767.

Kuwahara, Y., Nemoto, T., Shibuya, M., Matsuura, H., and Shiraiwa, Y. (1983). 2-Palmitoyl- and 2-oleoyl-cyclohexane-1,3-dione from feces of the Indian meal moth, *Plodia interpunctella*: Kairomone components against a parasitic wasp, *Venturia canescens*. *Agric. Biol. Chem.*, *47*:1929–1931.

Ladisch, R. K., Ladisch, S. K., and Howe, P. M. (1967). Quinoid secretions in grain and flour beetles. *Nature*, *215*:939–940.

Larson, A. O. (1924). The effect of weevily seed beans upon the bean crop and upon the dissemination of weevils, *Bruchus obtectus* Say and *B. quadrimaculatus* Fab. *J. Econ. Entomol.*, *17*:538–548.

Larson, A. O. (1932). Field control of the common bean weevil. *Calif. Dept. Agric.*, *11*: 400–408.

Larson, A. O., and Fisher, C. K. (1924). The possibilities of weevil development in neglected seeds in warehouses. *J. Econ. Entomol.*, *17*:632–637.

Larson, A. O., and Fisher, C. K. (1925). The role of the bean straw stack in the spread of bean weevils. *J. Econ. Entomol.*, *18*:696–703.

LeCato, G. L. (1975a). Interactions among four species of stored-product insects in corn: A multifactorial study. *Ann. Entomol. Soc. Amer.*, *68*:677–679.

LeCato, G. L. (1975b). Red flour beetle: Population growth on diets of corn, wheat, rice or shelled peanuts supplemented with eggs or adults of the Indian meal moth. *J. Econ. Entomol.*, *68*:763–765.

Lefkovitch, L. P. (1967). A laboratory study of *Stegobium paniceum* (L.) (Coleoptera: Anobiidae). *J. Stored Prod. Res.*, *3*:235–249.

Lefkovitch, L. P. (1968). Interaction between four species of beetles in wheat and wheat-feed. *J. Stored Prod. Res.*, *4*:1–8.

Lhaloui, S., Hagstrum, D. W., Keith, D. L., Holtzer, T. O., and Ball, H. J. (1988). Combined influence of temperature and moisture on red flour beetle (Coleoptera: Tenebrionidae) reproduction on whole grain wheat. *J. Econ. Entomol.*, *81*:488–489.

Longstaff, B. C., and Cuff, W. R. (1984). An ecosystem model of the infestation of stored wheat by *Sitophilus oryzae*: A reappraisal. *Ecol. Model.*, *25*:97–119.

Loschiavo, S. R. (1974). The detection of insects by traps in grain-filled boxcars during transit, Proceedings, 1st International Working Conference on Stored-Product Entomology, Savannah, Georgia, pp. 639–650.

Loschiavo, S. R., and Sabourin, D. (1982). The merchant grain beetle, *Oryzaephilus mercator* (Silvanidae: Coleoptera), as a household pest in Canada. *Can. Entomol.*, *114*:1163–1169.

Lum, P. T. M., and Flaherty, B. R. (1970). Regulating oviposition by *Plodia interpunctella* in the laboratory by light and dark conditions. *J. Econ. Entomol.*, *63*:236–239.

Maarse, H. (1991). Volatile Compounds in Foods and Beverages. Marcel Dekker, New York.

Maga, J. A. (1978). Cereal volatiles, a review. *J. Agric. Food Chem.*, *26*:175–178.

Mankin, R. W., Vick, K. W., Mayer, M. S., Coffelt, J. A., and Callahan, P. S. (1980). Models for dispersal of vapors in open and confined spaces: Applications to sex pheromone trapping in a warehouse. *J. Chem. Ecol.*, *6*:929–950.

Mayer, M. S., and McLaughlin, J. R. (1991). Handbook of Insect Pheromones and Sex Attractants. CRC Press, Boca Raton, FL.

McGaughey, W. H., Speirs, R. D., and Martin, C. R. (1990). Susceptibility of classes of wheat grown in the United States to stored-grain insects. *J. Econ. Entomol.*, *83*:1122–1127.

McGregor, H. E. (1964). Preference of *Tribolium castaneum* for wheat containing various percentages of dockage. *J. Econ. Entomol.*, *57*, 511–513.

McNeil, J. N. (1991). Behavioral ecology of pheromone-mediated communication in moths and its importance in the use of pheromone traps. *Annu. Rev. Entomol., 36*:407–430.

Menusan, H. (1934). Effects of temperature and humidity on the life processes of the bean weevil, *Bruchus obtectus* Say. *Ann. Entomol. Soc. Amer., 27*:515–526.

Messina, F. J. (1984). Influences of cowpea pod maturity on the oviposition choices and larval survival of a bruchid beetle *Callosobruchus maculatus*. *Entomol. Exp. Appl., 35*:241–248.

Messina, F. J., Barmore, J. L., and Renwick, J. A. A. (1987). Oviposition deterrent from eggs of *Callosobruchus maculatus*: Spacing mechanism or artifact? *J. Chem. Ecol., 13*:219–226.

Miller, L. W. (1944). Investigations of the flour beetles of the genus *Tribolium*. II. Effect of different mill fractions on the larval development and survival of *T. castaneum* (Hbst.) and *T. confusum* (Duv.). *J. Agric. Victoria, 42*:365–373, 377.

Mills, R. B. (1989). *Sitophilus zeamais* Motschulsky breeding in acorns. *J. Kansas Entomol. Soc., 62*:416–418.

Mitchell, R. (1975). The evolution of oviposition tactics in the bean weevil *Callosobruchus maculatus* (F.). *Ecology, 56*:696–702.

Momoi, S., and Sadamori, H. (1982). The influences of temperature on the growth and reproduction of *Stegobium paniceum* (Coleoptera: Anobiidae). *Sci. Rept. Fac. Agric. Kobe Univ., 15*:63–70.

Monge, J. P., and Huignard, J. (1991). Population fluctuations of two bruchid species *Callosobruchus maculatus* (F.) and *Bruchidius atrolineatus* (Pic) (Coleoptera Bruchidae) and their parasitoids *Dinarmus basalis* (Rondani) and *Eupelmus vuilleti* (Crawford) (Hymenoptera, Pteromalidae, Eupelmidae) in a storage situation in Niger. *J. Afr. Zool., 105*:187–196.

Mookherjee, P. B., and Chawla, M. L. (1964). Effect of temperature and humidity on the development of *Callosobruchus maculatus* Fab., a serious pest of stored pulses. *Indian J. Entomol., 26*:345–351.

Moriarty, F. (1959). The 24-h rhythm of emergence of *Ephestia kuhniella* Zell. from the pupa. *J. Ins. Physiol., 3*:357–366.

Morrison, E. O. (1964). The effect of particle size of sorghum grain on development of the weevil *Sitophilus zea-mais*. *J. Econ. Entomol., 57*:390–391.

Mossadegh, M. S. (1978). Mechanism of secretion of the contents of the mandibular glands of *Plodia interpunctella* larvae. *Physiol. Entomol., 3*:335–340.

Mossadegh, M. S. (1980). Inter- and intra-specific effects of the mandibular gland secretion of larvae of the Indian-meal moth, *Plodia interpunctella*. *Physiol. Entomol., 5*: 165–173.

Mudd, A. (1981). Novel 2-acylcyclohexane-1,3-diones in the mandibular glands of lepidopteran larvae. Kairomones of *Ephestia kuehniella* Zeller. *J. C. S. Perkin I*, 2357–2362.

Muir, W. E., Yaciuk, G., and Sinha, R. N. (1977). Effects on temperature and insect and mite populations of turning and transferring farm-stored wheat. *Can. Agric. Eng., 19*: 25–28.

Nair, K. S. S., and Desai, A. K. (1973). The termination of diapause in *Trogoderma granarium* Everts (Coleoptera: Dermestidae). *J. Stored Prod. Res., 8*:275–290.

Nammour, D., Huignard, J., and Pouzat, J. (1988). A female sex pheromone in *Bruchidius atrolineatus* (Pic) (Coleoptera, Bruchidae): Analysis of the factors affecting production or emission of this pheromone. *Physiol. Entomol.*, *13*:185–192.

Nara, J. M., Lindsay, R. C., and Burkholder, W. E. (1981). Analysis of volatile components in wheat germ oil responsible for an aggregation response in *Trogoderma glabrum* larvae. *J. Agric. Food Chem.*, *29*:68–72.

Nemoto, T., Kuwahara, Y., and Suzuki, T. (1987a). Interspecific differences in *Venturia* kairomones in larval feces of four stored phycitid moths. *Appl. Entomol. Zool.*, *22*:553–559.

Nemoto, T., Shibuya, M., Kuwahara, Y., and Suzuki, T. (1987b). New 2-acylcyclohexane-1,3-diones: Kairomone components against a parasitic wasp, *Venturia canescens*, from feces of the almond moth, *Cadra cautella*, and the Indian meal moth, *Plodia interpunctella*. *Agric. Biol. Chem.*, *51*:1805–1810.

Nutting, W. L., and Gerhardt, P. D. (1964). A study of the Khapra beetle, *Trogoderma granarium*, in commercial grain storages in southern Arizona. *J. Econ. Entomol.*, *57*:305–314.

O'Donnell, M. J., Chambers, J., and McFarland, S. M. (1983). Attractancy to *Oryzaephilus surinamensis* (L.), saw-toothed grain beetle, of extracts of carobs, some triglycerides, and related compounds. *J. Chem. Ecol.*, *9*:357–374.

Oehlschlager, A. C., Pierce, A. M., Pierce, H. D., Jr., and Borden, J. H. (1988). Chemical communication in cucujid grain beetles. *J. Chem. Ecol.*, *14*:2071–2098.

Olsen, A. R. (1981). List of stored-product insects found in imported foods entering United States at southern California ports. *Bull. Entomol. Soc. Amer.*, *27*:18–20.

Olsen, A. R., Bryce, J. R., Lara, J. R., Madenjian, J. J., Potter, R. W., Reynolds, G. M., and Zimmerman, M. L. (1987). Survey of stored-product and other economic pests in import warehouses in Los Angeles. *J. Econ. Entomol.*, *80*:455–459.

Oosthuizen, M. J., and Laubscher, F. X . (1940). The cowpea weevil. *J. Entomol. Soc. S. Afr.*, *3*:151–158.

Oshima, K., Honda, H., and Yamamoto, I. (1973). Isolation of an oviposition marker from azuki bean weevil, *Callosobruchus chinensis* (L.). *Agric. Biol. Chem.*, *37*:2679–2680.

Pant, N. C., and Dang, K. (1969). Food value of several stored commodities in the development of *Tribolium castaneum* Herbst. *Indian J. Entomol.*, *31*:147–151.

Patnaik, N. C., Panda, N., and Dash, A. N. (1986). Effect of sowing date and cultivar on field infestation of pulse beetle of pigeonpea. *Intern. Pigeonpea Newsl.*, *5*:44–46.

Paulsen, M. R., Hill, L. D., and Shove, G. C. (1991). Temperature of corn during ocean vessel transport. *Trans. Amer. Soc. Agric. Eng.*, *34*:1824–1829.

Payne, J. A., and Redlinger, L. M. (1959). Insect abundance and distribution within peanut shelling plants. *J. Amer. Peanut Res. Educ. Assoc.*, *1*:83–89.

Payne, J. A., Redlinger, L. M., and Davidson, J. I. (1970). Shelling plant studies with insect-infested peanuts. *J. Amer. Peanut Res. Educ. Assoc.*, *2*:103–108.

Phelan, P. L. (1992). Evolution of sex pheromones and the role of asymmetric tracking. In Insect Chemical Ecology: An Evolutionary Approach (ed. Roitberg, B. D., and Isman, M. B.). Chapman & Hall, New York, p. 265.

Phillips, J. K., Walgenbach, C. A., Klein, J. A., Burkholder, W. E., Schmuff, N. R., and Fales, H. M. (1985). (R*,S*)-5-hydroxy-4-methyl-3-heptanone. Male produced aggregation pheromone of *Sitophilus oryzae* (L.) and *S. zeamais* Motsch. *J. Chem. Ecol.*, *11*:1263–1274.

Pierce, A. M., Borden, J. H., and Oehlschlager, A. C. (1981). Olfactory response to beetle-produced volatiles and host-food attractants by *Oryzaephilus surinamensis* and *O. mercator*, *Can. J. Zool.*, *59*:1980–1990.

Pierce, A. M., Pierce, H. D., Jr., Borden, J. H., and Oehlschlager, A. C. (1989). Production dynamics of cucujolide pheromones and identification of 1-octen-3-ol as a new aggregation pheromone for *Oryzaephilus surinamensis* and *O. mercator* (Coleoptera: Cucujidae). *Environ. Entomol.*, *18*:747–755.

Pierce, A. M., Pierce, H. D., Jr., Oehlschlager, A. C., Czyzewska, E., and Borden, J. H. (1987). Influence of pheromone chirality on response by *Oryzaephilus surinamensis* and *Oryzaephilus mercator* (Coleoptera: Cucujidae). *J. Chem. Ecol.*, *13*:1525–1542.

Powell, J. D., and Floyd, E. H. (1960). The effect of grain moisture upon development of the rice weevil in green corn. *J. Econ. Entomol.*, *53*:456–458.

Prevett, P. F. (1961). Field infestation of cowpea (*Vigna unguiculata*) pods by beetles of the families Bruchidae and Curculionidae in northern Nigeria. *Bull. Entomol. Res.*, *52*:635–645.

Qi, Y. T., and Burkholder, W. E. (1982). Sex pheromone biology and behavior of the cowpea weevil, *Callosobruchus maculatus* (Coleoptera: Bruchidae). *J. Chem. Ecol.*, *8*:527–534.

Reichmuth, C., Schmidt, H. U., Levinson, A. R., and Levinson, H. Z. (1980). Seasonal occurrence of the warehouse moth (*Ephestia elutella* Hbn.) in West Berlin granaries as well as the correspondingly timed application of control measures. *Z. Ang. Entomol.*, *89*:104–111.

Richards, O. W., and Waloff, N. (1946). The study of a population of *Ephestia elutella* Hubner (Lep., Phycitidae) living on bulk grain. *Trans. R. Entomol. Soc., Lond.*, *97*:253–298.

Rilett, R. O. (1949). The biology of *Laemophloeus ferrugineus* (Steph.). *Can. J. Res.*, *27*:112–148.

Rodinov, Z. S. (1938). The mechanism of movement of the granary weevil in a heap of grain. *Zool. Zh.*, *17*:610–616.

Rossiter, P. D. (1970). Field infestation of the rice weevil in wheat. *Queensland J. Agric. Animal Sci.*, 27:119–121.

Roth, L. M. (1943). Studies on the gaseous secretion of *Tribolium confusum* Duval. *Ann. Entomol. Soc. Amer.*, *36*:397–424.

Russell, M. P. (1962). Field infestation of corn in India by the Angoumois grain moth and a rice weevil. *J. Econ. Entomol.*, *55*:814–815.

Schmidt, von H. U. (1982). The depth of egg deposition of the Indian meal moth, *Plodia interpunctella* Hbn., in rye and maize. *Anz. Schädlingskde. Pflanzenschutz. Umweltschutz.*, *55*:1–4.

Schwartz, B. E., and Burkholder, W. E. (1991). Development of the granary weevil (Coleoptera: Curculionidae) on barley, corn, oats, rice and wheat. *J. Econ. Entomol.*, *84*:1047–1052.

Schwettmann, K. D. (1988). On the field attack of the corn weevil *Sitophilus zeamais* Motschulsky (Col., Curculionidae) and its associated fauna in the Philippines. *Anz. Schädlingskde. Pflanzenschutz. Umweltschutz.*, *61*:86–95.

Schwitzgebel, R. B., and Walkden, H. H. (1944). Summer infestation of farm-stored grain by migrating insects. *J. Econ. Entomol.*, *37*:21–24.

Scott, J. H. (1951). Flour milling processes. London: Chapman & Hall Ltd.

Seck, D. (1991). Study of initial infestation of *Sitotroga cerealella* Oliv. (Lepidoptera, Gelechiidae) as a function of millet (*Pennisetum typhoides* (L.)) field location. *Insect Sci. Applic.*, *12*:507–509.

Seifelnasr, Y. E. (1991). Influence of olfactory stimulants on resistance/susceptibility of pearl millet, *Pennisetum americanum* to the rice weevil, *Sitophilus oryzae*. *Entomol. Exp. Appl.*, *59*:163–168.

Seitz, L. M., and Sauer, D. B. (1992). Off-odors in grains. In Off-Flavors in Foods and Beverages. (ed. Charalambous, G.). Elsevier, Amsterdam, pp. 17–35.

Shade, R. E., Furgason, E. S., and Murdock, L. L. (1990). Detection of hidden insect infestations by feeding-generated ultrasonic signals. *Amer. Entomol.*, *36*:231–234.

Shapas, T. J., and Burkholder, W. E. (1978). Diel and age-dependent behavioral patterns of exposure-concealment in three species of *Trogoderma*: Simple mechanisms for enhancing reproductive isolation in chemically mediated mating systems. *J. Chem. Ecol.*, *4*:409–423.

Sheppard, E. H. (1936). Notes on *Cryptolestes ferrugineus* Steph., a cucujid occurring in the *Trichogramma minutum* parasite laboratory of Colorado State College. *Colo. Ag. Expt. St. Tech. Bull.*, 17 pp.

Shinoda, K., and Yoshida, T. (1984). Relationship between adult feeding and emigration from beans of Azuki bean weevil, *Callosobruchus chinensis* Linne (Coleoptera: Bruchidae). *Appl. Entomol. Zool.*, *19*:202–211.

Shinoda, W. K., and Yoshida, T. (1985). Field biology of the azuki bean weevil, *Callosobruchus chinensis* (L.) (Coleoptera: Bruchidae). I. Seasonal prevalence and assessment of field infestation of aki-azuki, autumn variety of *Phaseolus angularis*. *Jpn. J. Appl. Entomol. Zool.*, *29*:14–20.

Shinoda, K., Yoshida, T., and Igarashi, H. (1992). Population ecology of the azuki bean beetle, *Callosobruchus chinensis* (L.), (Coleoptera: Bruchidae) on two wild leguminous hosts: *Vigna angularis* var. *nipponensis* and *Dunbaria villosa*. *Appl. Entomol. Zool.*, 27:311–318.

Shinoda, K., Yoshida, T., and Okamoto, T. (1991). Two wild leguminous host plants of the azuki bean weevil, *Callosobruchus chinensis* (L.) (Coleoptera: Bruchidae). *Appl. Entomol. Zool.*, *26*:91–98.

Siddiqui, W. H., and Barlow, C. A. (1973). Population growth of *Anagasta kuehniella* (Lepidoptera: Pyralidae) at constant and alternating temperatures. *Ann. Entomol. Soc. Amer.*, *66*:579–585.

Silverstein, R. M. (1988). Chirality in insect communication. *J. Chem. Ecol.*, *14*:1981–2004.

Sim, K. S., Park, D. S., Moon, C. K., and Lee, J. C. (1979). Studies on injurious insects for processed foods. *Seoul Univ. J. Pharm. Sci.*, *4*:134–139.

Simmons, P. (1927). Dispersion of the Angoumois grain moth to wheat fields. *J. Agric. Res.*, *34*:459–471.

Simwat, G. S., and Chahal, B. S. (1980). Effect of storage period and depth of stored grains on the insect population and the resultant loss of stored wheat with the farmers. *Bull. Grain Techn.*, *18*:35–41

Sinclair, E. R. (1982). Population estimates of insect pests of stored products on farms on the Darling Downs, Queensland. *Aust. J. Exp. Agric. Anim. Husb.*, *22*:127–132.

Sinclair, E. R., and Alder, J. (1984). Migration of stored-grain insect pests from a small wheat bulk. *Aust. J. Exp. Agric. Anim. Husb.*, *24*:260–266.

Sinclair, E. R., and Haddrell, R. L. (1985). Flight of stored products beetles over a grain farming area in southern Queensland. *J. Aust. Entomol. Soc.*, *24*:9–15.

Sinclair, E. R., and White, G. G. (1980). Stored products insect pests in combine harvesters on the Darling Downs. *Queensland J. Agric. Anim. Sci.*, *37*:93–99.

Singh, K. (1977). Seasonal variations in the population of insect pests of stored products in West Bengal. *Indian J. Ecol.*, *4*:212–217.

Singh, K. N., Agrawal, R. K., and Srivastava, P. K. (1978). Infestation of grain moth *Sitotroga cerealella* Oliv. and maize weevil *Sitophilus zeamais* Mots. on standing crops in the field. *Bull. Grain Techn.*, *16*:125–127.

Sinha, R. N. (1975). Effect of dockage in the infestation of wheat by some stored-product insects. *J. Econ. Entomol.*, *68*:699–703.

Sinha, R. N. (1988). Population dynamics of Psocoptera in farm-stored grain and oilseed. *Can. J. Zool.*, *66*:2618–2627.

Sinha, R. N., Waterer, D., and Muir, W. E. (1986). Carbon dioxide concentrations associated with insect infestations of stored grain. 1. Natural infestation of corn, barley and wheat in farm granaries. *Sci. Aliments*, *6*:91–98.

Smallman, B. N., and Loschiavo, S. R. (1952). Mill sanitation studies. I. Relative susceptibilities of mill stocks to infestation by the confused flour beetle. *Cereal Chem.*, *29*:431–440.

Smith, L. B. (1965). The intrinsic rate of natural increase of *Cryptolestes ferrugineus* (Stephens) (Coleoptera, Cucujidae). *J. Stored Prod. Res.*, *1*:35–49.

Smith, L. B. (1972). Wandering of larvae of *Cryptolestes ferrugineus* (Coleoptera: Cucujidae) among wheat kernels. *Can. Entomol.*, *104*:1655–1659.

Smith, L. B. (1985). Insect infestation in grain loaded in railroad cars at primary elevators in southern Manitoba, Canada. *J. Econ. Entomol.*, *78*:531–534.

Smith, L. B., and Barker, P. S. (1987). Distribution of insects found in granary residues in the Canadian prairies. *Can. Entomol.*, *119*:873–880.

Smith, L. B., and Loschiavo, S. R. (1978). History of an insect infestation in durum wheat during transport and storage in an inland terminal elevator in Canada. *J. Stored Prod. Res.*, *14*:169–180.

Sokoloff, A. (1974). The Biology of *Tribolium* with Special Emphasis on Genetic Aspects, Vol. 2. Oxford University Press, London.

Solomon, M.E. (1953). The population dynamics of storage pests. *Trans. IX Intern. Congr. Entomol.*, *2*:235–248.

Sonleitner, F. J., and Guthrie, P. J. (1991). Factors affecting oviposition rate in the flour beetle *Tribolium castaneum* and the origin of the population regulation mechanism. *Res. Popul. Ecol.*, *33*:1–11.

Southwood, T. R. E. (1980). Ecological Methods. Chapman and Hill, London.

Sower, L. L., Vick, K. W., and Tumlinson, J. H. (1974). (Z,E)-9,12-tetradecadien-1-ol: A chemical released by female *Plodia interpunctella* that inhibits the sex pheromone response of male *Cadra cautella*. *Environ. Entomol.*, *3*:120–122.

Steele, R. W. (1970). Copulation and oviposition behavior of *Ephestia cautella* (Walker) (Lepidoptera: Phycitidae). *J. Stored Prod. Res.*, *6*:229–245.

Stein, V. W. (1990). Investigations about the development of stored products insects at fruits of indigenous trees and shrubs. *Anz. Schädlingskde. Pflanzenschutz. Umweltschutz.*, *63*:41–46.

Strand, M. R., Williams, H. J., Vinson, S. B., and Mudd, A. (1989). Kairomonal activities of 2-acylcyclohexane-1,3 diones produced by *Ephestia kuhniella* Zeller in eliciting searching behavior by the parasitoid *Bracon hebetor* (Say). *J. Chem. Ecol.*, *15*:1491–1500.

Strong, R. G. (1970). Distribution and relative abundance of stored-product insects in California: A method of obtaining sample populations. *J. Econ. Entomol.*, *63*: 591–596.

Subramanyam, B., and Hagstrum, D. W. (1993). Predicting developmental times of six stored-product moth species (Lepidoptera: Pyralidae) in relation to temperature, relative humidity, and diet. *Eur. J. Entomol.*, *90*:51–64.

Suezawa, Y., Miyashiro, R., and Tamura, A. (1987). Estimation of population of cigarette beetle, *Lasioderma serricorne* (F.), by a synthetic sex pheromone and food attractant in factories of hand-extended fine noodles. *Nippon Shokuhin Kogyo Gakkaishi*, *34*:635–639.

Sundararaj, R., and Sundararajan, R. (1990). Susceptibility of rice (*Oryza sativa*) to angoumois grain moth (*Sitotroga cerealella*) and its field incidence in Tamil Nadu. *Indian J. Agric. Sci.*, *60*:703–704.

Surtees, G. (1965). Ecological significance and practical implications of behaviour patterns determining the spatial structure of insect populations in stored rain. *Bull. Entomol. Res.*, *56*:201–213.

Suzuki, T., Suzuki, T., Huynh, V. M., and Muto, T. (1975a). Hydrocarbon repellants isolated from *Tribolium castaneum* and *T. confusum* (Coleoptera: Tenebrionidae). *Agri. Biol. Chem.*, *39*:2207–2211.

Suzuki, T., Suzuki, T., Huynh, V. M., and Muto, T. (1975b). Isolation of 2′-hydroxy-4′-methoxy-propiophenone from *Tribolium castaneum* and *T. confusum*. *Agri. Biol. Chem.*, *39*:1687–1688.

Szentesi, A. (1981). Pheromone-like substances affecting host-related behaviour of larvae and adults in the dry bean weevil, *Acanthoscelides obtectus*. *Entomol. Exp. Appl.*, *30*:219–226.

Takahashi, F., and Mizuno, H. (1982). Infestation of rice weevils in rice grain in relation to drying procedures after harvest and the form of the grain at different stages in the milling process. *Environ. Control in Biol.*, *20*:9–16.

Tanaka, Y., Honda, H., Ohsawa, K., and Yamamoto, L. (1986). A sex attractant of the yellow mealworm, *Tenebrio molitor* L. and its role in the mating behavior. *J. Pesticide Sci.*: *11*:49–55.

Taylor, T. A. (1970). On the incidence of the 'active' form of *Callosobruchus maculatus* (F.) on mature cowpea in the field. *Niger. Entomol. Mag.*, 2:66–69.

Taylor, T. A., and Aludo, J. I. S. (1974). A further note on the incidence of 'active' females of *Callosobruchus maculatus* (F.) on mature cowpea in the field in Nigeria. *J. Stored Prod. Res.*, *10*:123–125.

Throne, J. E., and Cline, L. D. (1989). Seasonal flight activity of the maize weevil, *Sitophilus zeamais* Motschulsky (Coleoptera: Curculionidae), and the rice weevil, *S. oryzae* (L.), in South Carolina. *J. Agric. Entomol.*, *6*:183–192.

Throne, J. E., and Cline, L. D. (1991). Seasonal abundance of maize and rice weevils (Coleoptera: Curculionidae) in South Carolina. *J. Agric. Entomol.*, *8*:93–100.

Throne, J. E., and Culik, M. P. (1989). Progeny production and duration of development of rusty grain beetle, *Cryptolestes ferrugineus* (Stephens) (Coleoptera: Cucujidae) on cracked and whole corn. *J. Entomol. Sci.*, *24*:150–155.

Tiwary, P. N., and Verma, K. K. (1989). Studies on polymorphism in *Callosobruchus analis* (Coleoptera: Bruchidae). Part IV—Significance of polymorphism. *Entomography*, *6*: 317–325.

Tuff, D. W., and Telford, H. S. (1964). Wheat fracturing as affecting infestation by *Cryptolestes ferrugineus*. *J. Econ. Entomol.*, *57*:513–516.

Turner, B., and Maude-Roxby, H. (1989). The prevalence of the booklice *Liposcelis bostrychophilus* Badonnel (Liposcelidae, Psocoptera) in British domestic kitchens. *Int. Pest Control*, *31*:93–97.

Turner, J. W. (1976). Resistance of maize to field infestation by *Sitophilus zeamais* Motschulsky and *Sitotroga cerealella* (Oliver). *Queensland J. Agric. Animal Sci.*, *33*: 155–159.

Vasil'ev, I. (1935). *Acanthoscelides obtectus* Say under field conditions in Abkhazua. *Plant Prot.*, *1*:124–130.

Vick, K. W., Coffelt, J. A., and Weaver, W. A. (1987). Presence of four species of stored-product moths in storage and field situations in north-central Florida as determined with sex pheromone-baited traps. *Fla. Entomol.*, *70*:488–492.

Vick, K. W., Drummond, P. C., and Coffelt, J. A. (1973). *Trogoderma inclusum* and *T. glabrum*: Effects of time of day on production of female pheromone, male responsiveness, and mating. *Ann. Entomol. Soc. Amer.*, *66*:1001–1004.

Vick, K. W., Koehler, P. G., and Neal, J. J. (1986). Incidence of stored-product phycitinae moths in food distribution warehouses as determined by sex pheromone-baited traps. *J. Econ. Entomol.*, *79*:936–939.

Walgenbach, C. A., Phillips, J. K., Faustini, D. L., and Burkholder, W. E. (1983). Male-produced aggregation pheromone of the maize weevil, *Sitophilus zeamais*, and interspecific attraction between three *Sitophilus* species. *J. Chem. Ecol.*, *9*:831–841.

Walker, D. W. (1960). Population fluctuations and control of stored grain insects. *Wash. Agric. Exp. Sta. Techn. Bull. 31.*

Waloff, N., and Richards, O. W. (1946). Observations on the behavior of *Ephestia elutella* Hubner (Lep., Phycitidae) breeding on bulk grain. *Trans. R. Entomol. Soc., Lond.*, *97*:299–335.

Wasserman, S. S. (1981). Host induced oviposition preferences and oviposition markers in the cowpea weevil, *Callosobruchus maculatus*. *Ann. Entomol. Soc. Am.*, *74*:242–245.

Wasserman, S. S. (1985). Oviposition behaviour and its disruption in the southern cowpea weevil, *Callosobruchus maculatus* F. (Coleoptera: Bruchidae). *J. Econ. Entomol.*, *78*:89–92.

Waterhouse, F. L., Onyearu, A. K., and Amos, T. G. (1971). Oviposition of *Tribolium* in static environments incorporating controlled temperature and humidity gradients. *Oikos*, *22*:131–135.

Watters, F. L., and Bickis, M. (1978). Comparison of mechanical handling and mechanical handling supplemented with malathion admixture to control rusty grain beetle infestations in stored wheat. *J. Econ. Entomol.*, *71*:667–669.

Weston, P. A., and Hoffman, S. A. (1991). Humidity and tactile responses of *Sitophilus zeamais* (Coleoptera: Curculionidae). *Environ. Entomol.*, *20*:1433–1437.

Weston, P. A., Barney, R. J., and Sedlacek, J. D. 1993. Planting date influences preharvest infestation of dent corn by Angoumois grain moth (Lepidoptera: Gelechiidae). *J. Econ. Entomol.*, *86*:174–180.

White, G. D., and McGregor, H. E. (1957). Epidemic infestations of wheat by a dermestid, *Trogoderma glabrum* (Herbst). *J. Econ. Entomol.*, *50*:382–385.

White, G. G. (1982). The effect of grain damage on development in wheat of *Tribolium castaneum* (Herbst) (Coleoptera: Tenebrionidae). *J. Stored Prod. Res.*, *18*:115–119.

White, G. G. (1985). Population dynamics of *Tribolium castaneum* (Herbst) with implications for control strategies in store wheat. PhD thesis. University of Queensland.

White, G. G. (1988a). Temperature changes in bulk stored wheat in sub-tropical Australia. *J. Stored Prod. Res.*, *24*:5–11.

White, G. G. (1988b). Field estimates of population growth rates of *Tribolium castaneum* (Herbst) and *Rhyzopertha dominica* (F.) (Coleoptera: Tenebrionidae and Bostrychidae) in bulk wheat. *J. Stored Prod. Res.*, *24*:13–22.

White, N. D. G., and Bell, R. J. (1990). Relative fitness of a malathion-resistant strain of *Cryptolestes ferrugineus* (Coleoptera: Cucujidae) when development and oviposition occur in malathion-treated and untreated wheat kernels. *J. Stored Prod. Res.*, *26*:23–37.

Wilkins, C. K., and Scholl, S. (1989). Volatile metabolites of some barley storage molds. *Int. J. Food Microbiol.*, *8*:11–17.

Williams, J. O., and Mills, R. B. (1980). Influence of mechanical damage and repeated infestation of sorghum on its resistance to *Sitophilus oryzae* (L.) (Coleoptera: Curculionidae). *J. Stored Prod. Res.*, *16*:51–53.

Williams, R. N., and Floyd, E. H. (1970). Flight habits of the maize weevil, *Sitophilus zeamais*. *J. Econ. Entomol.*, *63*:1585–1588.

Williams, R. N., and Floyd, E. H. (1971). The maize weevil in relation to sunflower. *J. Econ. Entomol.*, *64*:186–187.

Wirtz, R. A., Taylor, S. I., and Simey, H. G. (1978). Concentrations of substituted p-benzoquinones and 1-pentadecene in the flour beetles *Tribolium madens* (charp.) and *Tribolium brevicornis* (LeC.) (Coleoptera: Tenebrionidae). *Comp. Biochem. Physiol.*, *61C*:287–290.

Wohlgemuth, R., Reichmuth, C., Rothert, H., and Bode, E. (1987). The appearance of moths of the genus *Ephestia* and *Plodia* which are harmful to stored products, outside of warehouses and food processing factories in Germany. *Anz. Schädingskde. Pflanzenschutz. Umweltschutz.*, *60*:44–51.

Wright, V. F., Fleming, E. E., and Post, D. (1990). Survival of *Rhyzopertha dominica* (Coleoptera, Bostrichidae) on fruits and seeds collected from woodrat nests in Kansas. *J. Kansas Entomol. Soc.*, *63*:344–347.

Yamamoto, I. (1976). Approaches to insect control based on chemical ecology—case studies. In Environ. Quality and Safety, Vol. 5, Global Aspects of Chemistry, Toxicology, and Technology Applied to the Environment (ed. Coulson, F., and Korte, F.). G. Thieme, Stuttgart, pp. 73–77.

Yamanouchi, M., and Takano, T. (1980). The effectiveness of milling and sifting of wheat in destruction and elimination of eggs of some stored-product insects. *Res. Bull. Plant Prot. Japan*, *16*:99–100.

Yoshida, T. (1974). Effect of sun-drying on the mortality of immature rice weevils, *Sitophilus zeamais* Motsch. (Coleoptera, Curculionidae). *Sci. Rep. Faculty Agric. Okayama Univ.*, *43*:11–17.

Yoshida, T., and Kawano, K. (1958). Seasonal fluctuation of the number of insects in the grain stored at farm-house. 1. The ecological studies of the pests infesting stored grains Part 2. *Mem. Faculty Liberal Arts Ed. Miyazaki Univ. Nat. Sci.*, *5*:11–23.

Yoshida, T., and Kawano, K. (1959). Fauna and community structure of the insects in the grain stored at farm-houses. The ecological studies of pests infesting stored grains. Part 3. *Mem. Faculty Liberal Arts Ed. Miyazaki Univ. Nat. Sci.*, *7*:33–61.

Yoshida, T., and Takuma, T. (1959). Seasonal fluctuation of the number of the flower-visiting rice weevil, *Sitophilus oryzae* Linne. *Jpn. J. Appl. Entomol. Zool.*, *3*:281–285.

Zimmerman, M. L. (1990). Coleoptera found in imported stored-food products entering southern California and Arizona between December 1984 through December 1987. *Coleopterists Bull.*, *44*:235–240.

# 4

# Sampling

**Bhadriraju Subramanyam**
*University of Minnesota, St. Paul, Minnesota*

**David W. Hagstrum**
*U.S. Department of Agriculture, Manhattan, Kansas*

Many insect species are associated with raw and processed stored commodities. The number of insects of a species in grain stored in a bin or in a gunny sack or in a packet of flour is called a population. A population is a group of living individuals (insects) of a species set in a frame that is limited and defined with respect to both time and space (Pearl 1937). A researcher may be interested in determining the kinds and numbers of insect species infesting a commodity, or in accurately determining changes in insect numbers with an increase or decrease in grain temperature and moisture. A grain or pest manager may be interested in determining if insect populations have exceeded a threshold to warrant the use of fumigation or other suitable control option. Grain or pest managers might also be interested in knowing the effectiveness of a control measure on insect populations. Therefore, estimating insect populations is essential for research purposes and for making pest management decisions. It is easy to count the total number of insects in a kilogram of grain sample, a box of cereal, or a packet of flour. However, time and money constraints do not permit us to count all the insects in an entire grain mass stored in a bin/silo or all insects in flour packets stored in a warehouse. In such situations, only a portion of the insect population is counted to allow us to make inferences about the population. For example, to determine insects present in a grain bin, grain samples are removed with a grain trier and the number of insects in the samples is counted. Sampling is the process and art of taking samples to make inferences about the population. Insects are sampled to determine various characteristics of the population, such as density or number of a species occupying a given area, dispersion or the arrangement of individuals in space, changes in birth and death rates, relative numbers of various insect stages, and changes in insect numbers over time (Pedigo 1994).

Practical sampling programs have been developed for insects associated with field and orchard crops (Pedigo and Buntin 1994). The quantitative techniques used to develop sampling programs for field and orchard crop insect pests can also be applied to stored-product insect pests. Only recently has there been interest in analyzing stored-product insect sampling data using quantitative techniques (Hagstrum et al. 1985; Lippert and Hagstrum 1987; Hagstrum et al. 1988a,b; Subramanyam and Harein 1990; Subramanyam et al. 1993). To date, we are aware of only a few published sampling programs for stored-product insects that can be used for making pest management decisions (Hodges et al. 1985; Hagstrum 1994). In this chapter, we provide basic concepts of sampling, discuss types of population density estimates, describe characteristics of samples and associated sampling statistics, and show methods for developing sampling plans or programs. Throughout this chapter, sampling techniques are discussed with reference to insects. However, the same techniques can be applied to other pests, such as stored-product mites, diseases, or weeds. Most of the quantitative procedures described are illustrated using stored-product insect sampling data. It is our hope that researchers and pest managers working with stored-product pests will use the quantitative procedures to design and implement an appropriate sampling program specific to their situation.

## I. SAMPLING CONCEPTS

Before sampling is initiated, it is important to define the purpose or objectives of sampling. These objectives help us to know what area needs to be sampled, what type of sampling device to use, how to proceed in collecting samples, and how to analyze the collected sampling data. As explained above, samples are removed to make inferences about the population under study. The entire population is generally referred to as the sampling universe. In other words, the sampling universe represents the habitat in which the population occurs. Samples are removed from the sampling universe to describe the population. For example, if a grain trier is used to sample insects in the top meter of the grain in a bin, the sampling universe is the entire top meter of the grain bulk. Defining sampling universe will help in determining where samples need to be taken and avoid areas where insect populations do not occur.

### 1. Sample Unit

A sample or sampling unit is the fraction of the habitable space from which insect counts are made. The sampling universe can be assumed to contain a known number of sampling units. Conversely, the sampling units make up a population. For example, when probing grain with a grain trier, the amount of grain in the trier is the sampling unit. When using traps, the sampling unit is the trap's effective capture area and duration of trapping. However, it is difficult to determine the trap's effective area. Therefore, for practical purposes, when traps are used for sampling insects, the sampling unit is a trap.

## 2. Size of Sample Unit

The size of the sampling unit, to some extent, will be determined by the choice of the sampling device. In general, the ratio of area or volume of the insect being sampled to the area or volume of the sample unit should be negligible. Taking small-size samples is often more efficient than taking large-size samples (Downing 1979). Even though increasing the size of the sample reduces the number of samples needed, the time (and cost) required to process each sample generally increases with size of the sample. With a large number of small samples one can increase the chances of obtaining a representative sample by sampling more locations. However, samples must be large enough to recover some insects at the lowest density of interest, because insect density cannot be estimated if none of the samples contain insects.

## 3. Sampling Techniques

The method employed to sample insects using a sampling unit constitutes a sampling technique. The method includes the sampling unit selection and the process of collecting sampling data. The selection of a sampling unit is based on the insect species being sampled. Bauwin and Ryan (1974) reviewed devices and protocols used for sampling grain; White et al. (1990), Barak et al. (1990), and Pinniger (1990) reviewed various devices used for sampling stored-product insects.

Insects can be sampled in a random or nonrandom fashion. Random sampling ensures that every sampling unit has an equal chance of being selected or that the selection of one sampling unit does not influence the choice of another sampling unit. Random selection is achieved by assigning all sampling units in the sampling universe a number and drawing sampling units of a known size at random using a random number table, which is provided in many statistical texts (Snedecor and Cochran 1980; Zar 1984). Using a random number table is feasible if the number of sampling units is small. In most sampling situations in agriculture, the sampling units in the sampling universe are too numerous to be counted reliably, and a number assigned to each for random selection. In these cases, random coordinate selection is a viable approach. For example, grain trier in the top 1 m of stored wheat can be taken by randomly selecting an angle and distance from the bin center. Restricted or stratified random sampling involves dividing the sampling universe into quarters or subdivisions and allocating sampling units to each quarter or subdivision at random. Randomness ensures that sampling estimates (means) are unbiased and provides a valid estimate of sampling error (variance). Bias in sampling estimates is a tendency to err in one direction, whereas variation (variance) in sampling estimates tends to be positive or negative (see Legg and Moon [1994] for a complete discussion on the importance of variance and bias in pest management).

In nonrandom sampling, the selection of the sampling unit influences the sampling of another unit. In nonrandom sampling, the first sampling unit is selected at random, and the remaining sample units are selected in a systematic fashion (Cochran 1977; Legg and Moon 1994). Legg and Moon (1994) discuss several variations of systematic sampling commonly used when sampling field crop insect pests. Nonrandom sampling in stored-grain situations may include taking grain samples or placing traps at equal distances or at regular intervals along a north–south or east–west transect of a bin (Hagstrum et al. 1985; Lippert and Hagstrum 1987; Subramanyam et al. 1989, 1993). Other examples include taking samples from fixed locations or at fixed intervals from a moving grain stream (Bauwin and Ryan 1974). For bagged commodities, taking samples from every 20th bag at three heights above the floor along all four sides of a stack is an example of nonrandom sampling. For nonrandom sampling, the sample-to-sample variation in the number of insects per sample depends on how the sampling locations were chosen and on the distribution of insects in the area sampled. The variance or error estimate based on nonrandom sampling may not be very accurate. It is important to bear in mind that taking more samples from as many parts of the sampling universe is essential for obtaining a representative sample of the population being studied. When densities are high and sampling statistics suggest that only a small number of samples are needed to estimate insect populations, more than the recommended number of samples may need to be taken to ensure that samples are representative of the actual insect population density. If repeated sampling provides similar results, the samples are believed to be representative of the population under study. Binns and Nyrop (1992) suggest that having a representative sample may be more important than the benefits of reducing bias by random sampling.

## 4. Normal Distribution

The heights of people, lengths of insect wings, number of insects in a sample, or percentage of samples with insects follow a normal distribution or a bell-shaped curve. In other words, the frequency distribution (number of observations or outcomes belonging to a classes; for example, there may be five people who are 1.65 m [5.5 ft] tall, 25 people who are 1.83 m [6.0 ft] tall, and so on) of heights or wing lengths is said to be approximately normal. Even if the individual sample observations are not normal, the distribution of sample means (based on large sample sizes [≥ 30 samples]) from the population follows a normal distribution (Snedecor and Cochran 1980). Most statistical hypothesis tests assume that the sample data are normally distributed.

### 5. Sampling Program

A sampling program uses sampling techniques to obtain a sample and make an estimate (Pedigo 1994). Sampling programs indicate which sampling unit to use, how many sampling units to use, and when to sample for insects, and the spatial distribution or dispersion of insects based on sampling data. To date, sampling programs for stored-product insects have been developed to estimate mean number of insects per grain or trap samples, to determine spatial distribution of insects (Hagstrum et al. 1985; Subramanyam and Harein 1990), and to detect the presence or absence of insects in a sampling unit (Lippert and Hagstrum 1987; Subramanyam et al. 1993).

## II. POPULATION DENSITY ESTIMATES AND ESTIMATION METHODS

### 1. Absolute Estimates

Absolute density estimates provide information on number of insects per unit of a commodity such as insects/kg of grain or number of moths/m$^2$ area. Devices that give absolute estimates include a grain trier (Fig. 1), which is used to remove samples of raw commodities stored in bulk or bags (Bauwin and Ryan 1974; Golob 1976; Hagstrum et al. 1985). Pelican samplers (Fig. 1) are inserted into a moving grain stream to collect samples periodically. Absolute estimates have been made for insect populations in farm bins (Hagstrum et al. 1985; Meagher et al. 1986; Hagstrum 1989; Reed et al. 1991), flat storage units (White 1988), elevators (Smith 1985), and bagged grain (Golob 1976; Hodges et al. 1985). Absolute densities have been estimated by sampling insects in grain residues left over in combines (Sinclair and White 1980), empty farm bins (Barker and Smith 1987), peanut shelling equipment (Payne and Redlinger 1969), and flour milling equipment (Dyte 1965). Suction traps (Taylor and Agbaji 1974; Leos-Martinez et al 1986), rotary traps (Barnes et al. 1939), and vehicle traps (Sinclair and Haddrell 1985) are some examples where insect numbers have been estimated per unit volume of air. For pest management decision-making, it would be useful to establish a relationship between the number of insects per unit volume of air and number of insects per unit of commodity.

### 2. Indirectly Derived Absolute Estimates

Absolute estimates can be obtained with sampling modes that give only relative estimates using the mark–release–recapture, removal trapping, and nearest

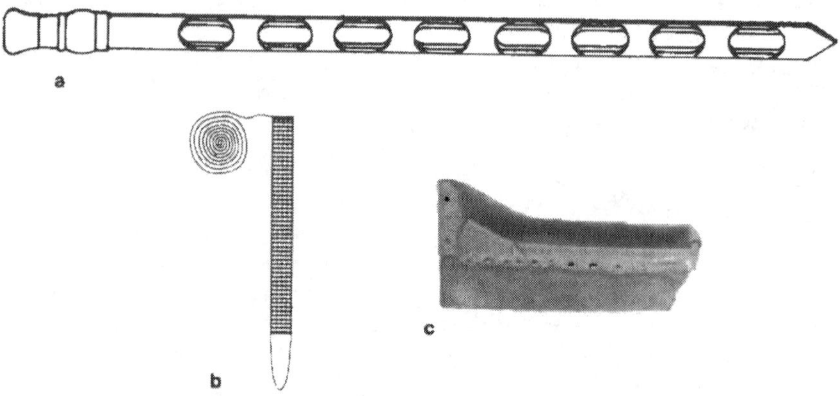


**Figure 1** Devices used for sampling stored-product insects. (a) The grain trier is probed into the grain mass, and live insects from the grain sample in each trier are recovered by sifting. (b) The perforated probe trap, or Storgard WB Probe II, is made of polyethylene plastic and is used for sampling several species of beetles in stored commodities. (c) The Pelican sampler is used to collect samples from a moving grain stream.

neighbor methods (Southwood 1978). By releasing a known numbers of marked (with radioactive or fluorescent dyes) insects, the natural population density can be estimated based on the ratio of marked to unmarked insects in samples using Equation 1,

$$Q = \frac{mn}{r} \tag{1}$$

where $Q$ = estimated density of natural population, $m$ = number of marked insects that are released, $n$ = total number of individuals in samples, and $r$ = number of marked individuals recaptured. Although the method is simple, satisfying the assumptions upon which the method is based is complicated. In using the method, the released insects are assumed to mix thoroughly with the natural population, and the density of the natural population is assumed not to change as a result of births, deaths, immigration, or emigration between release and recapture. Many variations of the release–recapture method have been developed to deal with these assumptions (Seber 1982). Using the release–recapture method, Walker (1960) estimated the total number of red flour beetles, *Tribolium castaneum* (Herbst), in empty grain bins to be 33,000. Hagstrum and Stanley (1979) used the release–recapture method to estimate, with suction traps, the absolute densities of almond moth, *Cadra cautella* (Walker), in an empty and a stocked peanut warehouse.

Recently, a self-marking recapture method developed by Wileyto et al. (1994) should make the use of the release–recapture method more practical by eliminating the need to release marked insects into food storage or processing facilities.

Removal trapping has not been used to estimate stored-product insect populations. This technique of absolute estimation may be useful when a large number of traps are used in facilities. This method estimates insect densities from the reduction in trap catch due to previous trapping (Pruess and Saxena 1977; Southwood 1978).

The distance to the nearest neighbor may be useful in cases where visual counts are being made of insects on walls or stacks of bags. The mobility of insects and the risk that the sampler will fail to find the nearest neighbor have limited its application in entomology (Ruesink and Kogan 1982; Southwood 1978). For less active or immobile insects, this technique would be useful in mapping the position of individuals in space and time to determine the underlying population spatial and temporal structure.

## 3. Extraction Methods

Absolute estimates require efficient methods of extracting insects from the commodity. For grain, many methods have been developed for detecting and counting kernels infested with immature stages of weevils (*Sitophilus* spp.), lesser grain borer (*Rhyzopertha dominica* [F.]), and Angoumois grain moth (*Sitotroga cerealella* [Olivier]) (Pedersen 1992). Russell (1988) compared four of these methods for detection of granary weevil, *Sitophilus granarius* (L.). Schatzki and Fine (1988) have shown that with the x-ray method false positives decreased exponentially with insect maturity. Other insects can generally be separated by sieving commodities. Insects can be removed from small samples with a hand sieve (Hagstrum 1987), and from larger samples with an inclined sieve (Hagstrum 1989; Fig. 2). Recoveries of red flour beetle, rice weevil, and lesser grain borer with three passes of wheat over the inclined sieve were 92.4, 96.4, and 100%, respectively (White 1983). Sieve tables have been used to separate several species of insects from peanuts (Prevett 1964), and cowpea weevils from cowpea samples (Hagstrum 1985). Berlese funnels have been used to separate insect larvae and mites from grain or flour (Golob et al. 1975; Smith 1977; Imura 1979). These studies have shown that recovery rates varied with size of the sample, stage and species of the insect, and type of commodity. With 200 or 300 g samples, the recovery rates ranged from 47 to 98% for 13 insect species.

## 4. Relative Estimates

Unlike absolute estimates, which are based on insect counts in a known amount of grain sample, volume of air, or in a given land area, relative estimates are based on the type of sampling device used. For example, the number of insects captured in a

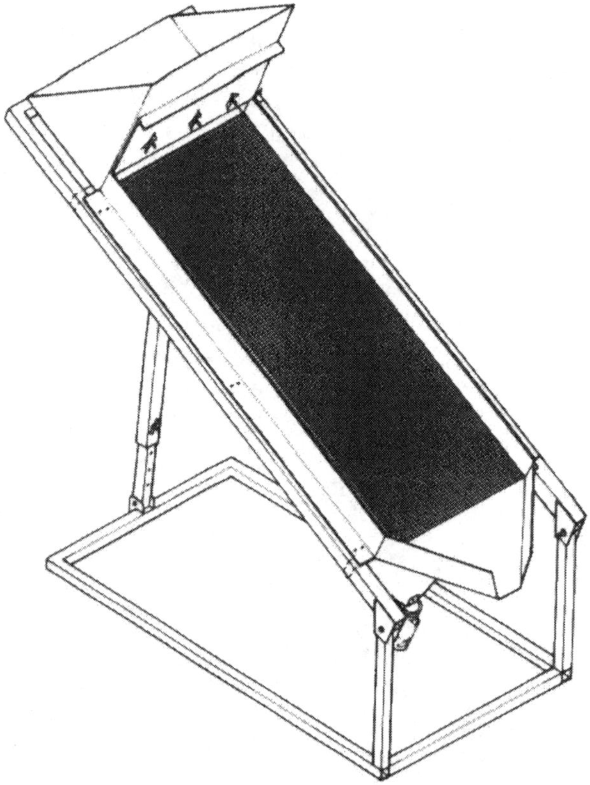

**Figure 2** An inclined sieve for separating insects from large volumes of grain (also see Hagstrum 1989).

trap is a relative estimate. Relative estimates of stored-product insects have been obtained with sticky traps (Vick et al. 1990), perforated probe traps (White et al. 1990; Figs. 1 and 6), food-baited traps (Pinniger 1990; Subramanyam et al. 1992; Subramanyam 1992), light traps (Rees 1985), corrugated cardboard pupation sites (Hagstrum and Sharp 1975; McGaughey 1978), or by visual search (Gentry 1984). Traps are an effective insect detection and monitoring tool because most stored-product insects are very active. Hagstrum and Davis (1980) showed that the male almond moth flew an average of 300 m (0.2 miles) during a 10 min flight. Traps have been used to locate infestations in grocery distribution warehouses (Vick et al. 1986). The effectiveness of traps, especially probe traps, can be increased by using pheromone and food lures (Chambers 1990).

## 5. Converting Relative to Absolute Estimates

Sampling programs can be developed based on relative density estimates. However, in pest management the relative estimates are valuable only if they can be related to absolute density estimates or some measure of insect damage or monetary loss. The relationship between relative and absolute estimates and between relative estimates and commodity damage can be determined by regression techniques. For stored-product insects, equations have been developed by comparing the results of sampling methods that provide paired absolute and relative estimates (Lippert and Hagstrum 1987; Hodges et al. 1985), or by using sampling data from populations of several known densities (Hagstrum et al. 1988b, 1990, 1991). Probe trap catch can be converted to an insect density per kilogram of grain by dividing by the trap efficiencies shown in Figure 3 (Lippert and Hagstrum 1987). Trap efficiency is the number of insects caught in traps divided by the number of insects in a grain sample. The regression line in Figure 3 indicates that the trap efficiency increased with insect density in traps. The number of insects in traps was roughly six times the number in grain samples when traps caught two insects, 15 times when traps caught eight insects, and 25 times when traps caught 14 insects. Figure 3 also illustrates the difficulty of converting trap catch to insect density per grain sample. The wide scattering of points indicates that the relationship between trap catch and insect density in grain samples is quite variable. The conversion will work well only for those observations that fall close to the regression line. Also, over 500 pairs of trap catches and grain samples were taken to obtain the 45 points shown on this graph. For the other pairs, trap catch was zero, no insects were found in grain samples, or no insects were found in both the trap and grain samples.

Traps depend on insect mobility, which varies with insect species, environmental conditions, and the trapping period. Equations that adjust trap catch for the insect species caught, the duration of trapping (Fig. 4), or the effects of temperature on insect activity are examples of conversions that are needed to estimate insect population densities accurately (Cuperus et al. 1990). The capture rate of insects in probe traps can be increased by leaving them in the grain for longer time periods (Fig. 4). The increase in capture rate was greatest for rusty grain beetle, *Cryptolestes ferrugineus* (Stephens) and lowest for the lesser grain borer. Grain samples only recover insects present in a volume of grain when it is taken. Probe traps can recover insects from a much larger volume, but they tend to recover only a percentage of the insects (Wright and Mills 1984). This often results in probe traps detecting insects (beetles) when grain samples fail to detect them (Subramanyam and Harein 1989). If capture rates are not adjusted for trapping period and converted to densities per kilogram of grain, we are likely to overestimate insect populations in grain samples.

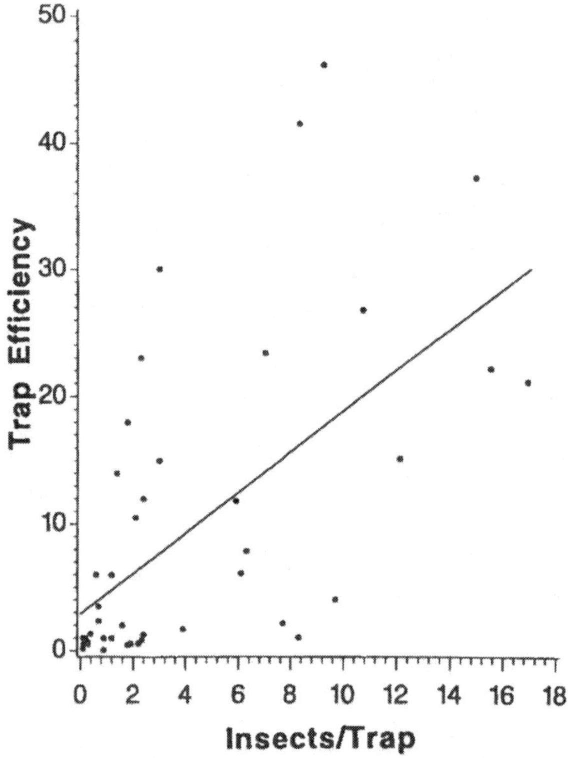


**Figure 3** Relationship between probe trap efficiency (number of insects in a trap ÷ number of insects in a grain sample) and trap catch. (Redrawn from Lippert and Hagstrum 1987.)

## 6. Population Indices

Population indices estimate insect numbers indirectly by estimating levels of commodity damage or products of insect activity (Pedigo 1994). Examples are the use of insect-damaged kernels, insect trails in flour residues on the floor, or the build-up of silk produced by moth larvae, as an indicator of insect infestation level. Visual inspection is a common method of obtaining population indices. These indices have the advantage of being more readily observed than the insects, and may accumulate over time. Loss assessment methods provide indices of insect population levels because they involve directly measuring insect-induced weight loss of a commodity. Boxall (1991) summarizes the literature on postharvest losses and loss assessment methods. Reed (1987) provides a good analysis of the problem of relating commodity weight loss to insect numbers.

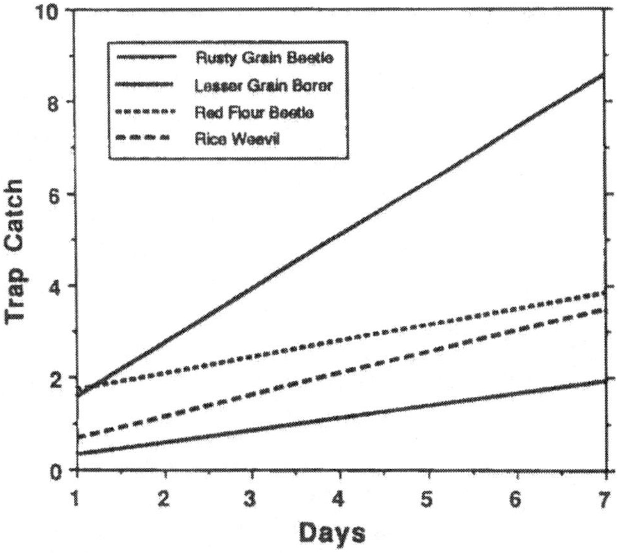


**Figure 4** Catch of adults of four stored-product insect species in probe traps as a function of trapping period. (From Hagstrum and Flinn 1992.)

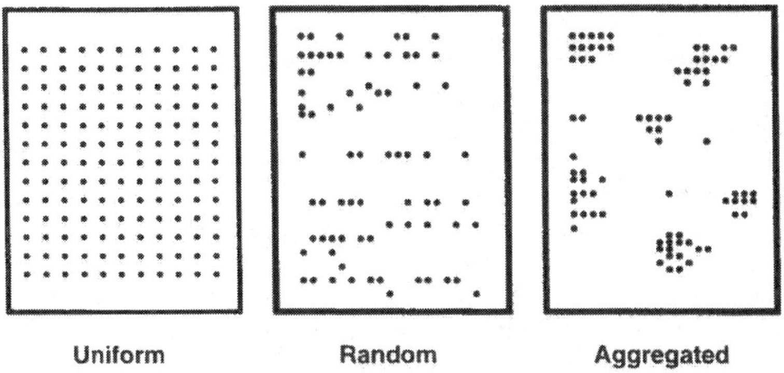


**Figure 5** A two-dimensional representation of individuals distributed in uniform, random, and aggregated fashions in space.

Automation of insect monitoring with acoustical sensors has been used to detect and estimate insect populations without taking grain samples (Hagstrum et al. 1991, 1994). The method is based on a strong correlation between insect numbers and number of sounds insects produce while feeding or moving in the grain. Estimates of insect numbers based on insect sounds are as accurate as those made by taking grain samples (Hagstrum 1994). Automation should allow grain to be monitored intensively so that insects can be detected at lower densities. The automated system would allow insect populations in all of the bins at an elevator to be monitored from a computer in the office. It should be possible to develop similar automated insect monitoring systems for processed commodities.

## III. BASIC SAMPLING STATISTICS

Several mathematical formulas describe the mean and variance of the samples, spatial pattern or dispersion of insects; relationship between number of sampling units with insects and insect density; precision in estimating densities; and number of samples required to estimate densities with a specific degree of precision. These formulas constitute sampling statistics, and will be discussed below in some detail. In the following discussion, mean density is used both in an absolute and relative sense, and refers to mean number of insects in a grain sample or mean number of insects in a trap.

When several samples are collected, the number of insects of a species in each sample is counted. If $n$ is the number of samples collected and $x_1$, $x_2$, $x_3$ $\ldots\ldots\ldots + x_n$ represent number of insects in each sample, then the average or mean number of insects per sample ($\bar{x}$) is

$$\bar{x} = \frac{(x_1 + x_2 + x_3 \ldots\ldots\ldots\ldots + x_n)}{n} \tag{2}$$

Variance is the sum of squared deviations of each sample observation from the mean. The simplest formula to calculate the variance for $n$ samples is

$$s^2 = \frac{(x_1 - \bar{x})^2 + (x_2 - \bar{x})^2 + (x_3 - \bar{x})^2 \ldots\ldots\ldots + (x_n - \bar{x})^2}{n - 1} \tag{3}$$

The square-root of the variance ($\sqrt{s^2} = s$) is the sample standard deviation. The standard deviation measures the amount of deviation of each sample observation from the mean, $x_i - \bar{x}$ (where, $i = 1, 2, 3, \ldots\ldots n$), and it plays an important role in determining how closely or accurately we can estimate the population mean from the sample mean (Snedecor and Cochran 1980). Like sample variance, variance for the mean can be calculated as

$$s_{\bar{x}}^2 = \frac{s^2}{n} \tag{4}$$

From Equation 4, the standard error of mean is calculated as

$$s_{\bar{x}} = \frac{s}{\sqrt{n}} \tag{5}$$

Table 1 shows the results of five sampling efforts, in which 10 samples are taken in each sampling effort and the number of insects in each sample counted. Table 1 clearly shows that the number of insects vary from one sample to the next. Means among sampling efforts are also not similar. These differences can be attributed to variance.

Variance comes from biological and sampling sources, and is termed biological error and sampling error, respectively. Biological error in estimates is attributed to differences in microclimate, such as grain temperature, moisture, humidity, nutritional quality of grain, behavior, and genetics of insects. The microclimate within a grain bulk is not uniform and, as a result, insects tend to prefer certain areas of the bulk as opposed to less conducive areas. Thus, many sample units fail to capture insects, while a few sample units tend to capture high numbers of insects. It is difficult or impossible to know the exact contribution of each of the factors to biological error. Sampling error is associated with sampling units and can be estimated (Legg and Moon 1994). The errors associated with sampling

**Table 1** Number of Insects in Each Sample in Five Hypothetical Sampling Efforts

| | Sampling effort | | | | |
|---|---|---|---|---|---|
| Sample number | 1 | 2 | 3 | 4 | 5 |
| 1 | 0 | 3 | 0 | 1 | 2 |
| 2 | 2 | 3 | 0 | 1 | 0 |
| 3 | 5 | 0 | 9 | 1 | 0 |
| 4 | 8 | 1 | 13 | 1 | 2 |
| 5 | 0 | 5 | 0 | 0 | 1 |
| 6 | 0 | 8 | 4 | 1 | 4 |
| 7 | 2 | 4 | 1 | 0 | 9 |
| 8 | 1 | 0 | 1 | 0 | 8 |
| 9 | 1 | 2 | 0 | 2 | 7 |
| 10 | 4 | 1 | 0 | 1 | 8 |
| Number of samples ($n$) | 10 | 10 | 10 | 10 | 10 |
| Mean ($\bar{x}$) | 2.3 | 2.7 | 2.8 | 0.8 | 4.1 |
| Variance ($s^2$) | 6.90 | 6.23 | 21.07 | 0.40 | 12.77 |
| Standard deviation ($s$) | 2.63 | 2.50 | 4.59 | 0.63 | 3.57 |
| Standard error ($s_{\bar{x}}$) | 0.83 | 0.79 | 1.45 | 0.20 | 1.13 |

error vary depending on whether sampling units are selected in a random or nonrandom fashion.

In general, low sample variance indicates that the insect counts do not differ greatly from one sample to the next. Conversely, a large variance about the mean indicates large differences in insect counts among samples. As variance increases, the estimate (such as mean density) tends to become less reliable or uncertainty about the estimated mean increases. Taking more samples, and taking samples where insects are present, may reduce variance provided the distribution of insects among sample units is approximately close to one another. The calculation of mean and variance is the first step in developing a sampling program, and is important for describing the spatial distribution of insects.

## IV. SPATIAL DISTRIBUTION OR DISPERSION OF INSECTS

The sample-to-sample variation in number of insects can be used to characterize the spatial distribution or dispersion patterns of insect populations. The dispersion patterns, which are largely determined by insect behavior in relation to environmental changes, can be classified as uniform, random, or aggregated (Fig. 5). In a uniformly distributed population, there is some degree of repulsion among individuals, and the sample variance is less than the sample mean ($s^2 < \bar{x}$). In a randomly distributed population, there is an equal chance of an insect occupying any point in space, and the presence of one individual does not influence the distribution of another individual (Davis 1994). Randomly distributed populations follow a Poisson distribution (Zar 1984, pp. 406–408), and the sample variance is equal to the mean ($s^2 = \bar{x}$). In an aggregated population, most of the samples contain a few or no insects, while a few samples contain high insect numbers. For an aggregated population, the variance is greater than the mean ($s^2 > \bar{x}$). Insects are mostly distributed in an aggregated fashion, especially at intermediate to high densities.

The frequency distribution of insects based on sampling data can be described by complex mathematical formulas, called the probability distribution functions. Randomly distributed data are best described by a Poisson distribution (Zar 1984), whereas the aggregated distribution is best described by a negative binomial distribution (Southwood 1978). In addition, when the count data are based on whether insects are present or absent in a sample unit, the data can be fit to a binomial distribution (Zar 1984; Jones 1994). Hagstrum et al. (1985) described insect populations sampled with a grain trier in four bins of stored wheat using a negative binomial, Poisson, logarithmic, Neyman type A, and Thomas probability distributions. The insect count data from all four bins was described by the negative binomial distribution, indicating that insects in the top 1 m of the grain mass were aggregated. For practical purposes, it may not be necessary to describe sampling data by these probability distribution functions. Several indices have been developed to classify dispersion patterns of insects. Sampling data characterized by these indices have been used for developing and implementing many sampling plans (see Pedigo and Buntin 1994).

### 1. Dispersion Pattern Indices

If the number of samples taken is less than 30, it may be practical to classify spatial dispersion patterns using several indices. The dispersion patterns of insects can be characterized using the following information: $\bar{x}$, $s^2$, and $n$.

*Variance-to-Mean Ratio*

This ratio is useful for determining whether the sampling data are randomly distributed or whether the data departs from randomness. As discussed earlier, for insects distributed in an aggregated fashion (i.e., $s^2 > \bar{x}$), the variance-to-mean ratio ($s^2/\bar{x}$) is greater than 1. An $s^2/\bar{x}$ ratio of 1 indicates a randomly distributed population, while an $s^2/\bar{x}$ ratio less than 1 indicates a uniformly distributed population. A simple test for departure from randomness is

$$I_D = (n - 1)s^2/\bar{x} \tag{6}$$

where $I_D$ = index of dispersion (Southwood 1978). $I_D$ is distributed as a chi-square ($\chi^2$) with $n - 1$ degrees of freedom (df). Calculated $I_D$ values that fall within a confidence interval bound by $\chi^2$ with $n - 1$ df and selected probability levels ($P$) of 0.95 and 0.5, indicate a random distribution. Table 2 shows the $I_D$ for adults of seven insect species captured in perforated probe traps (Grain Guard traps; Fig. 6) in farm-stored, shelled maize (Subramanyam, unpublished data). Across all species, a majority of observations indicated an aggregated distribution of the insects, with a substantial number of observations showing a random distribution.

*Negative Binomial Distribution Parameter k*

The negative binomial $k$ can be estimated by the method of moments (Southwood 1978) as

$$k = \frac{\bar{x}^2}{s^2 - \bar{x}} \tag{7}$$

The inverse of $k$ for uniform, random, and aggregated distributions is less than 0, equal to 0, and greater than 0, respectively (Davis 1994). The value of $k$ estimated by the maximum likelihood method (see Southwood 1978) is more accurate than $k$ estimated by the method of moments. The value of $k$ is typically around 2. As the distribution approaches a Poisson, the value of $k$ tends to increase. Values of $k < 1$ indicate a distribution that follows a logarithmic series, especially when $k = 0$.

The value of $k$ changes with insect density. Hagstrum et al. (1985) have shown that the value of $k$ for stored-wheat insects was correlated with the mean

**Table 2** Index of Dispersion ($I_D$) for Adults of Seven Insect Species Associated with Stored Shelled Maize Sampled with Perforated Probe Traps

| Insect species | $n$[a] | Mean trap catch range[b] | $P > I_D$, Number of values: $P > 0.95$ $s^2 < \bar{x}$[c] | $0.05 \leq P \leq 0.95$ $s^2 = \bar{x}$[d] | $P < 0.05$ $s^2 > \bar{x}$[e] |
|---|---|---|---|---|---|
| Foreign grain beetle | 44 | 0.11–5.44 | 0 | 27 | 17 |
| Larger black flour beetle | 38 | 0.11–5.89 | 0 | 21 | 17 |
| Rusty grain beetle | 46 | 0.11–17.22 | 1 | 20 | 25 |
| Flat grain beetle | 41 | 0.11–112.00 | 0 | 12 | 29 |
| Squarenosed fungus beetle | 53 | 0.11–104.80 | 1 | 21 | 31 |
| Red flour beetle | 23 | 0.11–422.00 | 0 | 12 | 11 |
| Hairy fungus beetle | 30 | 0.11–31.89 | 0 | 15 | 15 |
| All species | 275 | 0.11–112.00 | 2 | 128 | 145 |

[a]Number of mean-variance pairs.
[b]Each mean is based on nine perforated probe traps (Fig. 6) per bin. Traps were inserted just below the grain surface, and the trapping duration was 14 days. A total of four bins located on farms in Minnesota was sampled between May and November, 1990.
[c]Uniform distribution.
[d]Random distribution.
[e]Aggregated distribution.
*Source*: Subramanyam (unpublished data).

number of insects per sample, and the $k$ values ranged from 0.075 to 5.49. Therefore, when using $k$ in sampling programs (e.g., sequential sampling programs, which are discussed later), it is important to determine if the value of $k$ is consistent over a range of densities of interest to the sampler. If $k$ is stable across the densities, then a common $k$ or $k_c$ can be calculated and used in sampling plans based on the negative bionomial distribution or the Taylor's power law (which is discussed later). A common $k$ can be calculated as follows (Southwood 1978). First, calculate $x'$ and $y'$ as $x' = \bar{x}^2 - s^2/n$ and $y' = s^2 - \bar{x}$. Second, regress each $y'/x'$ ratio on corresponding $\bar{x}$ values. If there is no trend in the data or clustering, then the use of $k_c$ is justified. An approximate estimate of $k_c$ is

$$k_c = \frac{1}{(\Sigma y'/\Sigma x')^*} \tag{8}$$

Lloyd's mean crowding, Morisita's index, and Green's index are a few other indices for describing spatial patterns (see Southwood 1978; Davis 1994).

*$\Sigma$ is a summation symbol. For example, if $y' = 5, 8, 2$ then $\Sigma y' = 5 + 8 + 2 = 15$.

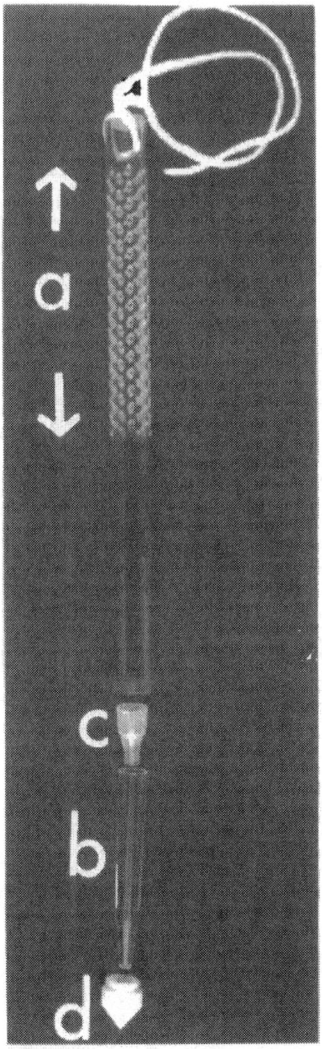


**Figure 6** Perforated probe trap, or Grain Guard trap, made of polycarbonate, that was used to sample adults of seven insect species associated with farm-stored shelled maize. (Subramanyam, unpublished data.) a, Perforated region of the trap; b, insect-collecting vial; c, teflon cap that fits on top of the insect-collecting vial; d, snap cap that fits at the bottom of the trap. The rope on top of the trap facilitates easy retrieval from the grain. See Subramanyam et al. (1993) for a description of Grain Guard and Storgard WB Probe II (shown in Fig. 1).

## 2. Regression Techniques for Describing Dispersion Patterns

The two most commonly used regression techniques for characterizing insect dispersion are the Iwao's patchiness regression and Taylor's power law.

*Iwao's Patchiness Regression*

Lloyd (1967) coined the term "mean crowding"($\overset{*}{x}$) to describe the mean number of other individuals per individual in an average sample unit, and it is expressed as

$$\overset{*}{x} = \bar{x} + \left(\frac{s^2}{\bar{x}} - 1\right) \tag{9}$$

For randomly distributed populations, $s^2/\bar{x} = 1$, and hence $\overset{*}{x} = \bar{x}$. The $\overset{*}{x}/\bar{x}$ ratio is called the "index of patchiness" (Lloyd 1967), and $\overset{*}{x}/\bar{x}$ ratios less than 1, equal to 1, and greater than 1 indicate a uniform, random, and an aggregated distribution, respectively (Davis 1994). Iwao's patchiness regression (Iwao 1968) involves expression $\overset{*}{x}$ and $\bar{x}$ for a species over a range of densities by a linear regression. Mean and variance necessary to calculate $\overset{*}{x}$ can be based on different sample sizes (number of samples) and on several sets of samples obtained from different locations and sampling times. The regression equation is of the following form:

$$\overset{*}{x} = \alpha + \beta\bar{x} \tag{10}$$

where, $\alpha$ is the y-intercept and $\beta$ is the regression slope. $\alpha$ denotes the average number of individuals living in the same sample unit or quadrat per individual, and is termed the "index of basic contagion." The basic component is a single individual if $\alpha = 0$; $\alpha < 0$ indicates repulsion among individuals and $\alpha > 0$ indicates that the basic component is a colony. $\alpha + 1$ gives a measure of clump size. The slope value, $\beta$, indicates whether the clumps are distributed uniformly ($\beta < 1$), randomly ($\beta = 0$), or in an aggregated fashion ($\beta > 1$) (Iwao 1968; Southwood 1978). Both the $\alpha$ and $\beta$ can be tested for departure from 0 and 1, respectively using a two-tailed $t$-test (Snedecor and Cochran 1980) at $n - 2$ df as

$$t = \alpha/SE_\alpha \tag{11}$$

$$t = \beta/SE_\beta \tag{12}$$

where $SE_\alpha$ and $SE_\beta$ are the standard errors of the estimates $\alpha$ and $\beta$, respectively. A random (Poisson) distribution is indicated if $\alpha = 0$ and $\beta = 1$.

Table 3 shows Iwao's patchiness regression estimates for seven beetle species captured in perforated probe traps in farm-stored, shelled corn (Subramanyam, unpublished data). The linear regression model fit the foreign grain beetle (*Ahasverus advena* [Waltl]), larger black flour beetle *Cynaeus angustus* [LeConte]), hairy fungus beetle (*Typhaea stercorea* [L.]), and flat grain beetle (*Cryptolestes pusillus* [Schöenherr]) data well as indicated by the high $r^2$ values ($\geqslant 0.89$). For the remaining three species, and for the combined species' data,

**Table 3** Iwao's Patchiness Regression Estimates for Adults of Seven Insect Species Associated with Stored Shelled Maize Sampled with Perforated Probe Traps[a]

| Insect species | $n^b$ | $\alpha \pm SE_\alpha$ | $\beta \pm SE_\beta$ | $r^2$ |
|---|---|---|---|---|
| Foreign grain beetle | 44 | −1.173 ± 0.256* | 4.112 ± 0.171 | 0.932 |
| Larger black flour beetle | 38 | 0.206 ± 0.465 | 2.844 ± 0.165 | 0.892 |
| Rusty grain beetle | 46 | 15.496 ± 15.646 | 0.988 ± 3.404** | 0.002 |
| Flat grain beetle | 41 | −2.152 ± 2.781 | 3.146 ± 0.087 | 0.971 |
| Squarenosed fungus beetle | 53 | 21.018 ± 24.925 | 3.213 ± 1.101 | 0.143 |
| Red flour beetle | 23 | −1.293 ± 1.176 | 5.986 ± 0.089 | 0.701 |
| Hairy fungus beetle | 30 | 1.504 ± 0.757 | 1.446 ± 0.089 | 0.905 |
| All species | 275 | 4.776 ± 5.119 | 3.144 ± 0.316 | 0.266 |

[a]Additional information on data used for this analysis is given in Table 2.
[b]Number of $\overset{*}{x}$-$\bar{x}$ pairs used in the regression.
*$\alpha$ significantly less than 0 ($P < 0.05$; two-tailed $t$ test); all other $\alpha$ values are not significantly different from 0 ($P > 0.05$).
**$\beta$ not significantly different from 1 ($P = 0.773$; two-tailed $t$ test); all other $\beta$ values are significantly greater than 1 ($P < 0.01$).
*Source*: Subramanyam (unpublished data).

Iwao's regression fit the data poorly (Table 3). Figure 7 shows the $\overset{*}{x}$-$\bar{x}$ regression based on the combined species' data.

*Taylor's Power Law*

Taylor (1961) described the relationship between sample variance ($s^2$) and sample mean ($\bar{x}$) by a power function as

$$s^2 = A\bar{x}^b \tag{13}$$

where, $A$ is a scaling factor dependent on sample unit and $b$ is an index of dispersion with values of $b$ less than 1, equal to 1, and greater than 1 indicating uniform, random, and an aggregated distribution of populations, respectively. $A$ and $b$ are estimated by regressing $s^2$ on $\bar{x}$ after transformation of both $s^2$ and $\bar{x}$ to logarithmic ($\log_{10}$ or ln) scale. The regression equation is expressed as

$$\log_{10} s^2 = a + b \log_{10} \bar{x} \tag{14}$$

where $a$ is the y-intercept and $b$ is the slope of the regression line. Antilogarithm of $a$ ($10^a$) gives $A$. $a$ and $b$ can be tested for departure from 0 and 1, respectively, using a two-tailed $t$ test at $n - 2$ df as

$$t = a/SE_a \tag{15}$$

$$t = b/SE_b \tag{16}$$

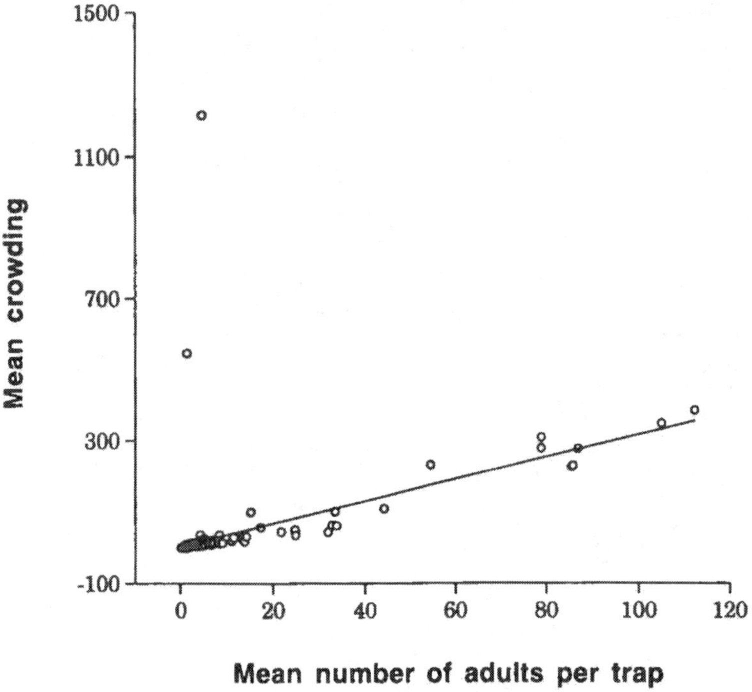


**Figure 7** Iwao's patchiness regression showing the relationship between mean crowding and mean density (mean trap catch). The patchiness regression was fit to the combined species' data. (Subramanyam, unpublished data), and the regression estimates are shown in Table 3.

where $SE_a$ and $SE_b$ are the standard errors associated with $a$ and $b$, respectively. Mean-variance pairs based on samples taken from different locations, sampling times, or those based on different sample sizes can be described by the Taylor's power law for estimating $A$ and $b$. Additional information on interpreting Taylor's estimates $A$ and $b$ is given in Taylor (1984). Kuno (1991) discussed the merits and drawbacks of Taylor's power law. In general, Taylor's power law is an empirical model, and the model fits most sampling data well because of the logarithmic transformation of both the $s^2$ and $\bar{x}$ scales. In addition, the model can be used to describe data over a broad range of insect densities.

Table 4 shows $A$ and $b$ values estimated from mean–variance pairs generated using nine and 20 trap samples per bin. The overlapping 95% confidence intervals for $A$ or $b$ across the studies and trap types indicated that these estimates were similar when nine or 20 traps were used to estimate Taylor's $A$ and $b$. Therefore, sampling plans based on nine traps would be as reliable as those based on 20 traps.

**Table 4** Influence of Number of Samples Used for Estimating Mean and Variance on Taylor's Power Law (TPL) Estimates

| Study | Probe trap | Number of trap samples[a] | $n^b$ | TPL estimate (95% confidence interval)[c] $A$ | $b$ | $r^2$ |
|---|---|---|---|---|---|---|
| 1[d] | Grain Guard | 9 | 275 | 2.69 (2.46–2.97) | 1.70 (1.64–1.78) | 0.927 |
| 2[e] | Grain Guard | 20 | 14 | 3.33 (2.15–5.16) | 1.67 (1.46–1.88) | 0.952 |
| 3[e] | Storgard WB Probe II | 20 | 14 | 3.11 (2.27–4.27) | 1.54 (1.38–1.69) | 0.968 |

[a]The trapping duration was 14 days for all three studies.
[b]Number of mean-variance pairs used in the regression.
[c]The 95% confidence interval was obtained by multiplying standard error of $A$ or $b$ with $z_{\alpha/2}$ at $\alpha = 0.05$ [1.96]. $A$ or $b$ among the three studies was similar ($P > 0.05$) because of overlapping 95% confidence intervals.
[d]Data from stored shelled maize (Table 2).
[e]Taylor's power law analysis of data presented in Subramanyam et al. (1993).

Hagstrum et al. (1985) used Taylor's power law to describe the dispersion of insect species in grain samples removed with a grain trier from different strata (regions) of four bins holding wheat. The regression models were compared using the model comparison procedure (Draper and Smith 1981), which indicated that the dispersion of insects among regions within a bin, insect species, and bins were statistically significant. Subramanyam and Harein (1990) used Taylor's power law to describe the dispersion of six insect species captured in perforated probe traps inserted just below the grain surface in four strata (regions) of a bin holding barley. The trapping duration was 1 week. For the six species, $A$ varied from 0.173 to 2.924 and $b$ varied from 0.858 to 3.066. For five of the six insect species, the value of $b$ was between 0.858 and 1.901. For the rusty grain beetle, the high $b$ value of 3.066 is probably related to the abundance of this species in grain (Subramanyam and Harein 1989), and increased activity and consequently greater capture of these adults in traps (Cuperus et al. 1990; Subramanyam and Harein 1990). The increased capture was a result of placing traps in the grain for 1 week. When the same traps were placed in stored barley for 2 days instead of a week, the $b$ value was 1.574 (Subramanyam and Harein 1990). When sampling stored-product insects with grain triers (Hagstrum et al. 1985) or perforated probe traps (Subramanyam and Harein 1990), a majority of the total variance comes from samples taken within a bin region. Consequently, within-region variances influence Taylor's estimates and variance (standard errors) associated with these estimates. Therefore, sampling designs should try to minimize the variance within bin regions.

The Taylor's estimates, $A$ and $b$ for seven insect species captured in perfo-

**Table 5** Taylor's Power Law Estimates for Adults of Seven Insect Species Associated with Stored Shelled Maize Sampled with Perforated Probe Traps

| Insect species | $n^a$ | $a \pm SE_a$* | $A^b$ | $b \pm SE_b$** | $r^2$ |
|---|---|---|---|---|---|
| Foreign grain beetle | 44 | 0.381 ± 0.049 | 2.404 | 1.623 ± 0.090 | 0.886 |
| Larger black flour beetle | 38 | 0.503 ± 0.037 | 3.184 | 1.661 ± 0.056 | 0.960 |
| Rusty grain beetle | 46 | 0.384 ± 0.069 | 2.421 | 1.572 ± 0.104 | 0.840 |
| Flat grain beetle | 41 | 0.368 ± 0.050 | 2.333 | 1.872 ± 0.054 | 0.969 |
| Squarenosed fungus beetle | 53 | 0.345 ± 0.060 | 2.213 | 1.867 ± 0.078 | 0.918 |
| Red flour beetle | 23 | 0.592 ± 0.057 | 3.908 | 1.692 ± 0.086 | 0.949 |
| Hairy fungus beetle | 30 | 0.388 ± 0.041 | 2.443 | 1.508 ± 0.057 | 0.960 |
| All species | 275 | 0.430 ± 0.210 | 2.692 | 1.700 ± 0.029 | 0.927 |

[a]Number of mean-variance pairs used in the regression.
[b]Antilogarithm of $a$ ($A = 10^a$).
*All $a$ values are significantly greater than 0 ($P < 0.01$; two-tailed $t$ test).
**All $b$ values are significantly greater than 1 ($P < 0.01$; two-tailed $t$ test).
*Source*: Subramanyam (unpublished data).

rated probe traps placed just below the grain surface in stored shelled maize (Subramanyam, unpublished data) are presented in Table 5, and Figure 8 shows the Taylor's power law fit to the combined species' data. Unlike Taylor's estimates from Hagstrum et al. (1985) and Subramanyam and Harein (1990), the values of $A$ and $b$ across the species were essentially similar (Table 5). Therefore, it is acceptable to pool data across the species to estimate $A$ and $b$ values that are representative of all the seven species. Such a pooling may be important when sampling plans are to be developed simultaneously for two or more species.

Data from several studies (Hagstrum and Sharp 1975; Hagstrum 1983, 1985, 1987; Hodges et al. 1985; Smith 1985; White 1985; Meagher et al. 1986; Barker and Smith 1987; Lippert and Hagstrum 1987; Hagstrum et al. 1988a; Subramanyam et al. 1993) were described by Taylor's power law. The resulting values of $A$ and $b$ along with the number of variance-mean pairs used in the regression are provided in Table 6. These studies cover insects or insect stages sampled from a wide range of situations, and include data from the field and laboratory. As expected, the use of different sample units produced values of $A$ that were variable (range among studies, 1.15–6.31), because $A$ is a scaling factor related to the size of the sample unit (Southwood 1978; Sawyer 1989). However, the value of $b$ was stable across the studies (range, 1.08–1.69) in spite of differences in the size and type of sample units used for sampling insects. Sawyer (1989), through computer simulation, showed that the value of $b$ increased with an increase in sample unit size (quadrat size). Therefore, comparisons of $b$ across species (e.g., Downing 1986; Subramanyam and Harein 1990) may be misleading because different population and behavioral processes could produce similar $b$ values. However, $b$

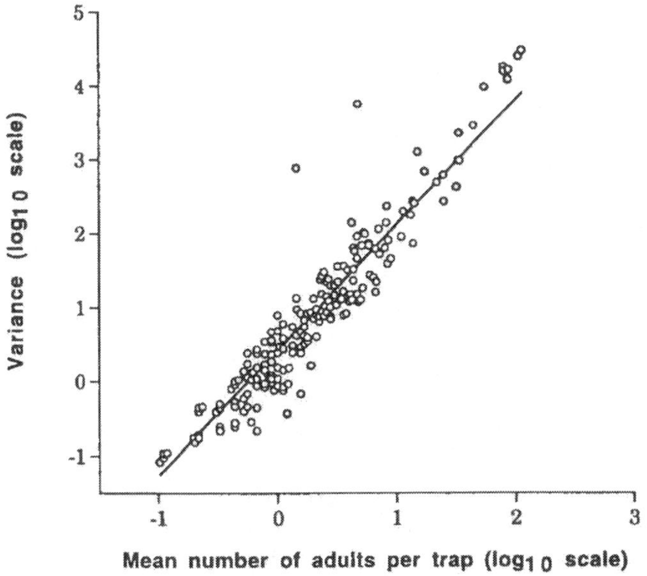


**Figure 8** Taylor's power law fit to the combined species' data (Subramanyam, unpublished data), and regression estimates are shown in Table 5.

values shown in Table 6, and those presented in Hagstrum et al. (1985) and Subramanyam and Harein (1990), clearly showed that the spatial aggregation of stored-product insects in diverse situations and on different spatial scales is somewhat constant.

In the Iwao's patchiness regression and Taylor's power law, both variables (especially $\bar{x}$) used in the regression are measured with error, and the errors associated with $\bar{x}$ or $s^2$ values are not equal to one another. In simple linear regression, the independent variable must be error-free, and if measurement errors are present, they tend to be small or negligible (Snedecor and Cochran 1980). Zar (1984, p. 268) gives five basic assumptions of linear regression analysis. Unless violations of the assumptions are severe, they should not greatly affect the y-intercept and slope estimates of Iwao's patchiness regression or Taylor's power law. A nonparametric regression technique, when both variables are subject to error, was developed by Bartlett (1949). The use of Bartlett's (1949) regression may provide unbiased *A* and *b* estimates. Legg (1986) developed a computer program to facilitate computations using the Bartlett's method. A comparison of *A* and *b* estimates obtained by parametric and nonparametric techniques may be useful.

**Table 6** Taylor's Power Law Estimates, *A* and *b*, for Stored-Product Insects ... ent Sampling Units in Diverse Situations

| Type of study | Sampling unit | $n^a$ | *A* | Reference |
|---|---|---|---|---|
| Empty farm bins | 1500 ml debris[c] | 27 | 6.31 | er and Smith (1987) |
| Farm bins, sorghum | 1 kg[d] | 73 | 3.24 | gher et al. (1986) |
| Farm bins, wheat | 0.5 kg[e] | 357 | 2.19 | strum (1987) |
| Farm bins, wheat | 0.5 kg[e] | 539 | 1.15 | strum et al. (1985) |
| Farm bins, wheat | 0.265 kg[f] | 190 | 3.39 | ert and Hagstrum (1987) |
| Bins and flat units, shelled maize | Probe trap[g] | 79 | 3.56 | amanyam et al. (1993) |
| Grain elevator, wheat | 40 kg[h] | 7 | 3.39 | te (1985) |
| Grain elevator, wheat | 0.5 kg[i] | 165 | 3.02 | h (1985) |
| Bagged rice | 0.12–0.16 kg[j] | 27 | 4.07 | ges et al. (1985) |
| Red flour beetle, all stages | 0.12 kg wheat[k] | 204 | 2.45 | strum et al. (1988a) |
| Moth eggs | No. eggs/3 × 3 cm grid | 136 | 4.79 | strum et al. (1988a) |
| Almond moth larvae (nondiapausing) | 5-cm spool[l] | 39 | 1.78 | strum and Sharp (1975) |
| Almond moth larvae (diapausing) | 5-cm spool[l] | 34 | 1.99 | strum and Sharp (1975) |
| Parasitization | 35 g peanuts/cell[m] | 39 | 1.23 | strum (1983) |
| Almond moth larvae | 35 g peanuts/cell[m] | 42 | 1.51 | strum (1983) |
| Average | | | 2.94 | |

[a]Number of mean-variance pairs used in the regression.
[b]Required sample size to estimate one insect per sample with 100% precision (±1 insect)
[c]Debris (sweepings) collected from empty bins using a brush and spatula.
[d]Insect density expressed on a 1 kg sample basis. Pneumatic probe was used to sample g
[e]Sample collected using a 1.27 m long (compartmented) grain trier.
[f]Sample collected using a deep-bin cup probe.
[g]Insects were captured in Grain Guard and Storgard WB II probe traps.
[h]Sample collected from a grain stream using a Pelican sampler.
[i]Sample scooped from a weigh bucket at the elevator.
[j]Bags were sampled with a 1.5 cm diameter and 45 cm long grain probe.
[k]Samples collected using a small (compartmented) grain trier.
[l]Corrugated paper, 20 mm wide and 60 mm long, rolled into a spool. The corrugations a
[m]Each cell is part of a larger grid consisting of 36 cells. Number of larvae in each cell v

### 3. Utility of Taylor's Power Law Estimates

Taylor's *A* and *b* are useful for predicting variance for independent sets of means. However, the new means substituted for $\bar{x}$ in Equation 13 must be within the range of data originally used to estimate *A* and *b*. In sampling plans, it is generally assumed that the estimated *A* and *b* values of Taylor's approximate the actual sample variance for an independent mean density. To verify this assumption, Taylor's estimates for the combined species' data ($A = 2.692$ and $b = 1.70$; Table 5) were used in Equation 13 to predict the variance for an independent set of means for each of the seven insect species. The combined species' estimates *A* and *b* are typical of all seven species (see Table 5). Independent mean-variance pairs were obtained by sampling four additional bins of shelled maize between May and November 1990 using nine perforated probe traps per bin. The trapping duration was 14 days. Therefore, mean-variance pairs originally used to estimate *A* and *b* (Table 5), and the independent mean-variance pairs were collected using identical sampling techniques and procedures. The variance predicted by Taylor's power law was regressed on actual sample variance, and the results are presented in Table 7. Taylor's power law overestimated the variance for the flat grain beetle and squarenosed fungus beetle (*Lathridius minutus* [L.]), and underestimated the variance for foreign grain beetle, larger black flour beetle, rusty grain beetle, and hairy fungus beetle. This error in variance estimation by Taylor's power law, for

**Table 7** Relationship between Variance Predicted by Taylor's Power Law (TPL) for Independent Sets of Means and the Variance Measured from Sample Data for the Same Independent Means[a]

| Insect species | $n$[a] | Slope ($b$) | $r^2$ |
|---|---|---|---|
| Foreign grain beetle | 12 | 0.348 | 0.801 |
| Larger black flour beetle | 13 | 0.352 | 0.846 |
| Rusty grain beetle | 9 | 0.477 | 0.964 |
| Flat grain beetle | 13 | 2.021 | 0.996 |
| Squarenosed fungus beetle | 12 | 1.589 | 0.579 |
| Red flour beetle | 4 | 0.950 | 0.999 |
| Hairy fungus beetle | 15 | 0.492 | 0.947 |
| All species | 78 | 0.940 | 0.810 |

[a]For each species, the y-intercept value was not significantly different from 0 ($P > 0.05$). Therefore, the regression was forced through the origin. The modified regression equation is: TPL predicted $s^2 = b \times$ actual $s^2$.

[b]Number of observations used in the regression.

the example shown in Table 7, can be corrected by dividing estimated Taylor's predicted variance by the slope of the predicted versus observed variance regression.

Taylor's $b$ can be used to determine an appropriate transformation for the raw counts (Southwood 1978). Variance-stabilizing transformations may be necessary for the analysis of variance. The transformation is achieved using the following expression:

$$z = rh \tag{17}$$

where $z$ is the transformed value, $r$ is the raw insect count or datum, and $h = 1 - 0.5b$. Generally, if $h = 0$, a logarithmic transformation is appropriate; for $h = 0.5$ and $h = -0.5$, a square root and a reciprocal square root transformation, respectively, should be used.

The most important and practical utility of $A$ and $b$ is in developing sampling plans, and in formulating decision rules for pest management. These aspects are discussed in the following sections.

### 4. Relationship between Taylor's Power Law and Negative Binomial *k*

The negative binomial parameter $k$ can be estimated using Taylor's $A$ and $b$ as

$$k = \frac{\bar{x}}{A\bar{x}^{b-1} - 1} \tag{18}$$

This equation allows one to compute $k$ without fitting a negative binomial distribution to data. Figure 9 shows the relationship between $k$ calculated by the method of moments (Equation 7) using actual sampling data presented in Table 2, and $k$ calculated using Equation 18. A majority of $k$ values fall within the 95% confidence intervals for $k$ calculated by Taylor's power law. However, $k$ was highly variable, especially at the lower insect densities, and $k$ estimated by Taylor's power law failed to capture the variability in $k$ that is independent of the mean density. In situations where the change in $k$ with density is not too drastic, $k$ estimated by Equation 18 could be used in sampling programs, especially in binomial sequential sampling plans.

## V. BINOMIAL ESTIMATION METHODS

Unlike previous methods discussed above, the binomial method estimates a mean and variance by categorizing sample units based on the presence or absence of insects. Therefore, these techniques are also called presence/absence or binomial sampling methods. For example, if a grain or trap sample has no insects, then the sample is classified as 0 (uninfested), and a sample containing one or more insects

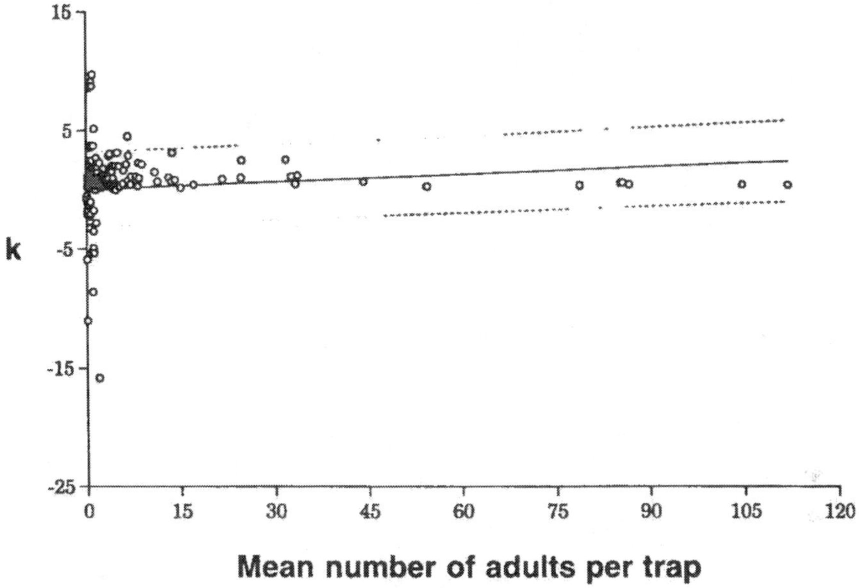


**Figure 9** Estimates of the negative binomial $k$ shown in relation to mean density (mean trap catch). The open circles represent $k$ estimated by the method of moments ($k = \bar{x}^2/s^2 - \bar{x}$). The solid line is the predicted $k$ based on regression of $k$ calculated using Taylor power law ($k = \bar{x}/A\bar{x}^{(b-1)} - 1$) on mean density. The regression equation was $k = 0.044 + 0.021\bar{x}$; $n = 275$; $r^2 = 0.04$; $P < 0.001$. The dashed lines are the 95% confidence intervals for the predicted $k$. Figure based on combined species' data. (Subramanyam, unpublished data.)

is classified as 1 (infested). The insect count data in Table 1 are used to illustrate the binomial estimation methods (see Table 8). Let us consider that $n$ samples were taken from the sampling universe. Let us assume that in $y$ samples there were one or more insects of a given species. This would mean that in $n - y$ samples there were no insects. From this information, the proportion or percentage (proportion × 100) of samples with one or more insects ($p$) is calculated as

$$p = y/n \qquad (19)$$

The proportion of samples with no insects ($q$) is calculated as

$$q = 1 - p \qquad (20)$$

The variance of $p$ or $q$ is

$$\text{Var}(p) = \text{Var}(q) = pq/n \qquad (21)$$

**Table 8** Classifying Insect Count Data Based on the Presence or Absence of Insects in the Sample in Each of Five Hypothetical Sampling Efforts

| | Sampling effort | | | | |
|---|---|---|---|---|---|
| Sample number | 1 | 2 | 3 | 4 | 5 |
| 1 | 0 | 1 | 0 | 1 | 1 |
| 2 | 1 | 1 | 0 | 1 | 0 |
| 3 | 1 | 0 | 1 | 1 | 0 |
| 4 | 1 | 1 | 1 | 1 | 1 |
| 5 | 0 | 1 | 0 | 0 | 1 |
| 6 | 0 | 1 | 1 | 1 | 1 |
| 7 | 1 | 1 | 1 | 0 | 1 |
| 8 | 1 | 0 | 1 | 0 | 1 |
| 9 | 1 | 1 | 0 | 1 | 1 |
| 10 | 1 | 1 | 0 | 1 | 1 |
| Number of samples ($n$) | 10 | 10 | 10 | 10 | 10 |
| Proportion infested ($p$) | 0.7 | 0.8 | 0.5 | 0.7 | 0.8 |
| Proportion uninfested ($q$) | 0.3 | 0.2 | 0.5 | 0.3 | 0.2 |
| Variance = $pq/n$ | 0.021 | 0.016 | 0.025 | 0.021 | 0.016 |

The actual number of insects in each sample is given in Table 1.

Because $p$ is based on presence and absence of insects in samples and not on complete counts, different sampling efforts may have a similar $p$ (compare Table 1 with Table 8). The differences in $p$ between certain sampling efforts are primarily caused by variance (Legg and Moon 1994) or, in other words, variation in the spatial distribution of insects may result in a majority of uninfested samples, with a few samples containing high numbers of insects. However, unlike enumerative methods that involve counting all individuals in a sample, the variance among samples for binomial methods is smaller. This is because each sample is scored as 0 or 1. Samples can also be scored using any threshold, typically called the tally threshold (Jones 1994). For example, if 30 samples are collected from a sampling universe, each sample can be scored as containing two or more insects, or five or more insects, and so on. Such a scoring may reduce sample variance, and has practical advantages in pest management over classifying samples as just infested or uninfested (see Jones 1994).

Just as count data can be described by Poisson and negative binomial probability functions, the binomial outcomes from a population of samples can be described by a binomial probability function. Once $p$, $q$, and $n$ are known, fitting a binomial probability function (Eq. 22) to presence/absence data can yield information on the probability that none of the samples contain insects ($P[X = 0]$).

$$P(X = 0) = \frac{n!}{X!\,(n - X)!} p^X q^{n-X} \tag{22}$$

$1 - P(X = 0)$ gives the probability that the samples contain one or more insects. The notation "!" is called a factorial. For example, 5! is $5 \times 4 \times 3 \times 2 \times 1 = 120$, $0! = 1$. If $X = 0$, then $p^0 = 1$. Similarly, when $n = X$, then $q^0 = 1$. As $n$ gets larger and $p$ gets smaller, the binomial distribution tends to be similar to a Poisson distribution. Interested readers should consult Zar (1984, pp. 370–380) and Jones (1994) for additional information on estimating binomial probabilities.

### 1. Relationship between Incidence and Mean Density

Binomial sampling is a viable alternative when counting insects in samples is cumbersome or expensive. Subramanyam et al. (1989) have shown that the time taken to count insects in perforated probe traps (Grain Guard traps; Fig. 6) is related to the number of insects, grain kernels, and grain debris in traps. To use binomial methods for developing sampling plans, it is important to establish the relationship between proportion of sample units with one or more insects and mean density (or mean trap catch). The purpose of such a relationship is twofold: for a given mean density the proportion of infested sample units can be determined, or for a given proportion the mean density can be determined. Numerous equations have been used to relate the nonlinear relationship between incidence and mean density for a variety of insect species and mites infesting field crops (Gerrard and Chiang 1970; Wilson and Room 1983; Nachman 1984; Ward et al 1986; Nyrop et al. 1989; Feng and Nowierski 1992; Jones 1994).

For stored-product insects, Hagstrum et al. (1988) and Subramanyam et al. (1993) used a double-logarithmic model to relate proportion of infested samples and mean density. Fitting the double-logarithmic model to data allowed for quantitative comparisons among species, commodities, and sampling devices.

We present here four equations that describe the relationship between proportion of samples with insects and mean density. The four equations were fit to the combined species' data, in which insects were sampled with perforated probe traps in stored, shelled maize (Subramanyam, unpublished data).

*Model 1*

Wilson and Room (1983) used Taylor's estimates to predict the proportion of infested sample units ($p$) as

$$p = 1 - e^{-\bar{x}[(\ln(A\bar{x}^{b-1}))(1/(A\bar{x}^{b-1})-1)]} \tag{23}$$

where $e = 2.71828$, $A$ and $b$ are Taylor's estimates, and $\bar{x}$ = mean number of insects per sample.

*Model 2*

The Poisson distribution model is initially used to estimate the probability of samples having no insects ($p_0$) as

$$p_0 = e^{-\bar{x}} \tag{24}$$

where e and $\bar{x}$ are described above. The probability (or proportion) of samples with insects is estimated as

$$p = 1 - p_0 \tag{25}$$

*Model 3*

Gerrard and Chiang (1970) and Nachman (1984) related the proportion of infested sample units and mean density as

$$p = 1 - e^{-\delta\bar{x}^{\gamma}} \tag{26}$$

where $e^{-\delta\bar{x}^{\gamma}}$ is the probability of samples without insects ($p_0$), and $\delta$ and $\gamma$ are model parameters (constants) that must be estimated through nonlinear or linear regression. Parameters $\delta$ and $\gamma$ can be estimated iteratively using any nonlinear regression technique. We recommend the technique that does not use any derivatives (DUD method; Ralston and Jennrich [1978]) available under the PROC NLIN procedure of the Statistical Analysis System (SAS Institute 1988). Estimated $\delta$ and $\gamma$ using the DUD method for the combined species' data (see Table 2) were 0.576 and 0.514, respectively.

In sampling literature (e.g., Feng and Nowierski 1992, Nyrop et al. 1989), the Nachman model is linearized to estimate the model parameters. The relationship between mean density and proportion of empty sample units $p_0$ [(or $1 - p) = e^{-\delta\bar{x}^{\gamma}}$] can be linearized by regression of $\ln(\bar{x})$ on $\ln(-\ln(p_0))$, and is expressed as

$$\ln(\bar{x}) = a' + b' \ln(-\ln(p_0)) \tag{27}$$

where $p_0 = (1 - p)$, $a'$ and $b'$ are the y-intercept and slope values of the linearized model, respectively, and these values are not equivalent to $\delta$ and $\gamma$ of the nonlinear model. With the linear model, mean density can be predicted from Equation 27 as

$$\bar{x} = 2.71828^{\ln(\bar{x})} \tag{28}$$

For predicting the proportion of empty sample units from a given mean, $\ln(-\ln(p_0))$ is regressed on $\ln(\bar{x})$, and the resulting regression equation is

$$\ln(-\ln(p_0)) = c' + d' \ln(\bar{x}) \tag{29}$$

where $c'$ and $d'$ are y-intercept and slope, respectively. The variance estimate of $\ln(-\ln(p_0)$ is

$$\text{Var}(\ln \ln p_0) = \text{mse}/N + [\ln(\bar{x}) - \text{avg}\,\bar{x}]^2 s_d^2 + \text{mse} \tag{30}$$

where mse = mean square error from regression, $N$ = number of data points in the regression, avg $\bar{x}$ is the average of the independent variable values, and $s_d^2$ is the variance of the slope $d'$ (or square of the slope standard error).

The linear model is preferable, because equations have been developed to calculate the variance associated with the predicted mean density and the sample sizes necessary to estimate mean density with a desired level of precision (see Feng and Nowierski 1992; Nyrop et al. 1989; Jones 1994). The variance of the predicted mean in Equation 27 is the sum of predicted variance ($\text{Var}_p(\bar{x})$) and sampling variance ($\text{Var}_s(\bar{x})$) (Binns and Bostanian 1990; Nyrop and Binns 1991). The predicted variance is calculated as

$$\text{Var}_p(\bar{x}) = \{\text{mse}/N + [\ln(-\ln(p_0)) - \text{avg}\ln\ln p_0]^2 \text{SE}_b^2 + \text{mse}\} \tag{31}$$

where mse = mean square error from the regression, $N$ = number of data points in the regression, avg $\ln \ln p_0$ is the average of the independent variable $\ln(-\ln(p_0))$, and $\text{SE}_b^2$ is the variance of the slope $b'$ (or square of slope standard error). The sampling variance is calculated as

$$\text{Var}_s(\bar{x}) = \frac{p_0(1 - p_0)\,b'^2}{n} \tag{32}$$

where $n$ is the number of samples used to determine $p_0$ for predicting $\bar{x}$. The total variance ($\text{Var}(\ln \bar{x})$) associated with $\bar{x}$ is

$$\text{Var}(\ln(\bar{x})) = \text{Var}_p(\bar{x}) + \text{Var}_s(\bar{x}) \tag{33}$$

The confidence interval for the predicted mean is calculated as

$$\ln(\bar{x}) \pm z_{\alpha/2}(\text{Var}(\ln(\bar{x})))^{0.5*} \tag{34}$$

where $z_{\alpha/2}$ is the upper $\alpha/2$ of the standard normal distribution (at $\alpha = 0.05$, $z_{\alpha/2}$ = 1.96). The variance can be expressed on arithmetic scale as

$$\text{Var}(\bar{x}) = \bar{x}^2(\text{Var}(\ln(\bar{x}))) \tag{35}$$

Figure 10 shows the linear regressions of $\ln(\bar{x})$ on $\ln(-\ln(p_0))$ and $\ln(-\ln(p_0))$ on $\ln(\bar{x})$ based on the combined species' data obtained by capturing insects in stored maize with probe traps (also see Table 2). The relationship shown in Figure 10A was used to predict means and variance components $\text{Var}_p(\bar{x})$ and $\text{Var}_s(\bar{x})$, from proportion of samples without insects (independent data) using Equations 27, 28, 31–33, and 35. The results of this exercise are shown in Table 9.

---

*For example, $(\text{Var}(\ln(\bar{x})))^{0.5}$ is the same as $\sqrt{\text{Var}(\ln(\bar{x}))}$.

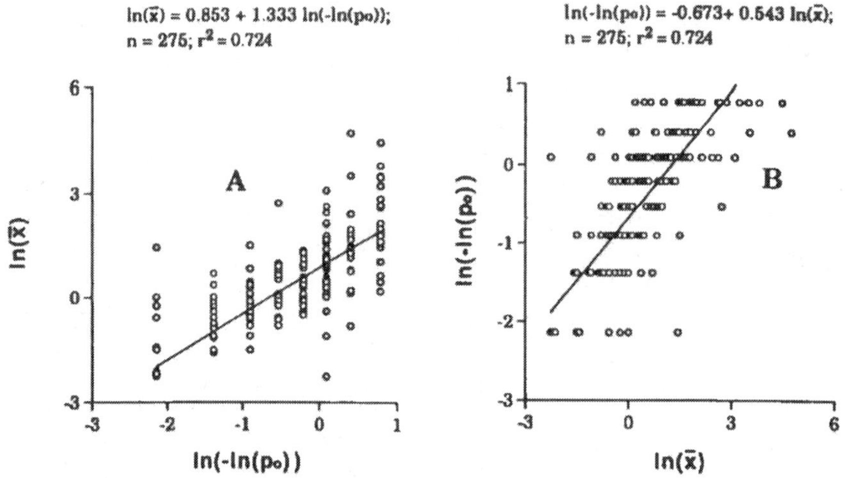


**Figure 10** Regressions showing the relationship between the proportion of sample units without adults and mean density (mean trap catch). These relationships are valuable for predicting mean densities from proportion of empty sample units, and vice versa. Figure based on combined species' data. (Subramanyam, unpublished data.)

**Table 9** Estimating Mean Density and Associated Prediction and Sampling Variance from Proportion of Samples without Insects[a]

| $p_0$[b] | $p$[c] | $\bar{x}$[d] | $\ln(\bar{x})$ | $\mathrm{Var}_p(\bar{x})$[e] | $\mathrm{Var}_s(\bar{x})$[f] | Total variance[g] |
|---|---|---|---|---|---|---|
| 0.1 | 0.9 | 7.133 | 1.965 | 0.635 | 0.018 | 33.206 |
| 0.3 | 0.7 | 3.305 | 1.100 | 0.631 | 0.041 | 2.953 |
| 0.5 | 0.5 | 1.440 | 0.364 | 0.629 | 0.049 | 1.406 |
| 0.9 | 0.1 | 0.117 | −2.147 | 0.634 | 0.018 | 0.009 |

[a]Linear regression of $\ln(\bar{x})$ on $\ln(-\ln(p_0))$ (Fig. 10A). The y-intercept, $a' \pm SE_{a'} = 0.853 \pm 0.061$; slope, $b' \pm SE_{b'} = 1.333 \pm 0.05$; mean square error, mse = 0.626. Each proportion, $p_0$ or $p$, is based on $n = 9$ samples.
[b]Proportion of samples without adults.
[c]Proportion of samples with one or more adults.
[d]Predicted mean number of adults per sample expressed on arithmetic scale, $[\bar{x} = e^{(\ln(\bar{x}))}]$.
[e]Prediction variance is based on a logarithmic (ln) scale.
[f]Sampling variance is based on a logarithmic (ln) scale.
[g]Total variance $[\mathrm{Var}_p(\bar{x}) + \mathrm{Var}_s(\bar{x})]$, is expressed on arithmetic scale $[\bar{x}^2[\mathrm{Var}_p(\bar{x}) + \mathrm{Var}_s(\bar{x})]$.

*Model 4*

The double-logarithmic model has been used to relate proportion of samples with insects and mean density for different species and stages of stored-product insects sampled with various devices, commodities, and marketing systems (Table 10). The model is expressed as

$$p = 1 - (Ae^{(-B\bar{x})} + (1 - A)e^{(-C\bar{x})}) \quad (36)$$

where parameters *A*, *B*, and *C* can be estimated by the DUD method of PROC NLIN procedure (SAS Institute 1988). Estimated *A*, *B*, and *C* for the combined species' data (see Table 2) were 0.473, 2.208, and 0.124, respectively. The model explains mechanistically the sample units occupied by insects in two steps. The first step is the logarithmic increase in sample units occupied by more than one insect with an increase in mean density. The second step is a logarithmic increase in the number of insects occupying the infested sample units. Subramanyam et al. (1993), using independent data sets, showed that this model accurately predicted mean densities of insects based on proportion of sample units (traps) with insects. The model predictions explained 84–90% of the variation in the actual mean densities (trap catches) (see Subramanyam et al. 1993). Computer programs for fitting the model to data and for estimating the variance associated with the proportions are available from Bh. Subramanyam or D. W. Hagstrum. The equations for calculating variances for the predicted means and for estimating sample sizes with a given degree of precision have not been developed. However, the prediction error statistic for mean density (Efron and Gong 1983) can be calculated for these binomial regression models (models 1–4). Models that give smaller prediction errors must be chosen.

Because the double logarithmic model satisfactorily described the $p$-$\bar{x}$ relationship for stored-product insect sampling data (Table 10), equations for estimating variance associated with means and for calculating sample sizes will facilitate the use of this model in developing sampling plans.

Figure 11 shows the nonlinear relationship between proportion of samples with insects and mean density for the combined species' data based on the four models described above. For all four models, the proportion of infested sample units was similar at 0.01–0.5 insects per trap (Fig. 11B). With the Poisson model, the proportion of infested sample units increased rapidly for a unit increase in trap catch, especially at >0.5 insects per trap. For example, the model predicted that 90% of the traps contained insects at a density of 2.5 insects per trap. At the same density, the other models predicted that about 60–62% of the traps contained insects (Fig. 11B). The aggregated distribution of insects explains the poor fit of the Poisson model to data. In general, the predicted proportion of infested sample units across mean densities was similar with the Wilson and Room (1983), Nachman (1984), and double-logarithmic models.

**Table 10** Parameters *A*, *B*, and *C* of Double-Logarithmic Model Describing the Relationship between Proportion of Sample Units with Insects and Mean Density for Stored-Product Insects Sampled with Different Sampling Units in Diverse Situations

| Type of study | *A* | *B* | *C* | $p^a$ | Reference[b] |
|---|---|---|---|---|---|
| Empty farm bins | 0.10 | 5.16 | 0.19 | 0.25 | Barker and Smith (1987) |
| Farm bins, sorghum | 0.26 | 1.25 | 0.17 | 0.30 | Meagher et al. (1986) |
| Farm bins, wheat | 0.32 | 2.60 | 0.21 | 0.43 | Hagstrum (1987) |
| Farm bins, wheat | 0.29 | 2.77 | 0.41 | 0.51 | Hagstrum et al. (1985) |
| Farm bins, wheat | 0.52 | 1.41 | 0.11 | 0.44 | Lippert and Hagstrum (1987) |
| Bins and flat units, shelled corn | 0.41 | 1.88 | 0.11 | 0.41 | Subramanyam et al. (1993) |
| Grain elevator, wheat | 0.34 | 4.29 | 0.17 | 0.44 | White (1985) |
| Grain elevator, wheat | 0.59 | 1.42 | 0.11 | 0.49 | Smith (1985) |
| Bagged rice | 0.34 | 1.32 | 0.23 | 0.39 | Hodges et al. (1985) |
| Red flour beetle, all stages | 0.19 | 1.18 | 0.32 | 0.35 | Hagstrum et al. (1988a) |
| Moth eggs | 0.09 | 1.74 | 0.36 | 0.35 | Hagstrum et al. (1988a) |
| Almond moth larvae (nondiapausing) | 0.39 | 1.88 | 0.34 | 0.51 | Hagstrum and Sharp (1975) |
| Almond moth larvae (diapausing) | 0.23 | 3.31 | 0.47 | 0.51 | Hagstrum and Sharp (1975) |
| Parasitization | 0.23 | 3.15 | 0.48 | 0.51 | Hagstrum (1983) |
| Almond moth larvae | 0.24 | 2.49 | 0.65 | 0.58 | Hagstrum (1983) |
| Average | 0.30 | 2.39 | 0.29 | 0.43 | |

[a]Proportion of samples with insects at a mean density of one insect per sample.
[b]See Table 6 for information on sampling unit, sampling device, and sample size used for fitting the double-logarithmic model to various data sets.

Binomial methods are not appealing when estimating mean density with a fixed precision, because more samples are needed to estimate density as opposed to methods based on complete insect counts in sample units (Nachman 1984; Kuno 1986). The primary reason for this is that limited information (e.g., presence or absence) is obtained from each sample. In addition, sample-to-sample variation associated with insect dispersion adds to the uncertainty in the incidence/density relationship. Therefore, binomial methods are more suited for developing sequential sampling plans in which the presence/absence information is used to classify the pest status for making management decisions (Nyrop et al. 1989; Binns and Nyrop 1992; Binns 1994). The use of binomial methods for classifying pest status is discussed later.

## VI. PRECISION AND NUMBER OF SAMPLES

Precision refers to the closeness of estimates in repeated measurements or sampling, and should not be confused with accuracy. Accuracy is the nearness of a measurement to the actual value of a variable being measured (Zar 1984). Therefore, accurate estimates or measurements are free of bias (Legg and Moon 1994).

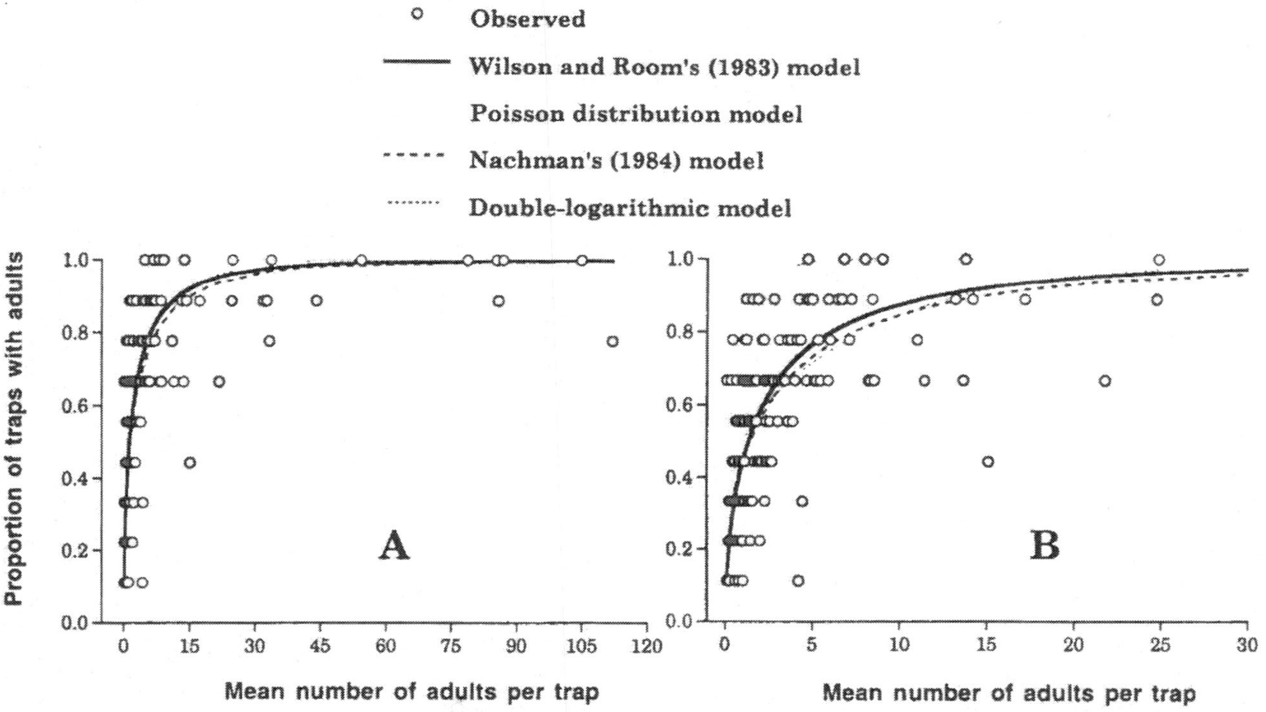


**Figure 11** Nonlinear relationship between the proportion of sample units with one or more insects and mean density (mean trap catch). This relationship is shown over a broad range of densities (A) and over a narrow range of densities (B). The open circles ($n$ = 275) represent the combined species' data (Subramanyam, unpublished data), and the lines represent predictions of the four models fit to the data (see text for model equations).

In real-world situations, obtaining accurate estimates is impossible because estimates are biased in spite of probability (random) sampling, and the bias associated with biased estimates is unknown (Fowler and Witter 1982). Therefore, in routine sampling, although accuracy is what we desire, it is precision that we are able to measure (Fowler and Witter 1982).

Precision is the difference between each mean ($\bar{x}$) and the average group of means (avg$\bar{x}$) (Legg and Moon 1994), and is expressed as

$$\text{Precision} = \bar{x} - \text{avg}\,\bar{x} \tag{37}$$

Precision can be expressed in the original units of measurement, and these measures of scale of precision are the standard error of a mean ($s_{\bar{x}}$) and the standard error of a proportion ($s_p$)

$$s_{\bar{x}} = (s^2/n)^{0.5} \tag{38}$$

$$s_p = (pq/n)^{0.5} \tag{39}$$

The coefficient of variation for means ($CV_{\bar{x}}$) and binomial proportions ($CV_p$) are scale-free measures of precision:

$$CV_{\bar{x}} = s_{\bar{x}}/\bar{x} \tag{40}$$

$$CV_p = s_p/p \tag{41}$$

In sampling literature, $CV_{\bar{x}}$ or $CV_p$ is often referred to as relative variation (RV), and is expressed as a percentage (100 RV). It is also common to calculate precision as a predetermined proportion ($d$) of the mean. This predetermined proportion can be expressed in terms of the half-width of the confidence interval (${}_{1/2}CI_{\bar{x}}$) as

$${}_{1/2}CI_{\bar{x}} = d\bar{x} \tag{42}$$

where $d = (CV_{\bar{x}})z_{\alpha/2}$ or $d = s_{\bar{x}}z_{\alpha/2}$. $z_{\alpha/2}$ is the upper $\alpha/2$ of the standard normal distribution (at $\alpha = 0.05$, $z_{\alpha/2} = 1.96$). A $t_{\alpha/2}$ may also be used instead of $z_{\alpha/2}$, if estimates are based on $< 30$ samples. Most statistical texts give the $z_{\alpha/2}$ and $t_{\alpha/2}$ values (e.g., Zar 1984, Snedecor and Cochran 1980).

The lower and upper confidence limits for the estimate ($\bar{x}$) are calculated as

$$\text{Lower confidence limit} = \bar{x} - {}_{1/2}CI_{\bar{x}} \tag{43}$$

$$\text{Upper confidence limit} = \bar{x} + {}_{1/2}CI_{\bar{x}} \tag{44}$$

It is impossible to know how close our estimated sample mean ($\bar{x}$) or proportion ($p$) is to the (unknown) population mean or proportion. Therefore, we construct a confidence interval from sample data to make a confident statement about the true population mean or proportion. For example, at a specified confidence level $(1 - \alpha)$ 100 (95% in our case, because we chose $\alpha = 0.05$), we could state that 95% of the confidence intervals would capture the true population mean or proportion (Snedecor and Cochran 1980). When the population distribution is not normal or is unknown, or when small number of samples are used for calculating means and variances, it is appropriate to use the 75% error bound approach for interval estimation (Fowler and Hauke 1979).

For data described by Taylor's power law, Poisson, binomial, and negative binomial distributions, equations for calculating precision ($d$) at various mean densities are provided in Table 11. The resulting $d$ values for these distributions can be used in the above equations to construct intervals about the mean or proportion at any confidence level.

In pest management sampling or decision-making programs, a high degree of precision is desirable. The precision in the estimates of means or proportions is said to be high if the $\bar{x}$ or $p$ values in repeated sampling are close to one another, and low if they are farther away from one another (Legg and Moon 1994). Karandinos (1976) and Ives and Moon (1984) gave equations for calculating the number of samples required to achieve a desired precision in estimations for various probability distributions and Taylor's power law (Table 11). These equations were rearranged to calculate precision of estimates for a given sample size. In Table 11, $t_{\alpha/2}$ may be used instead of $z_{\alpha/2}$ for density estimates based on small sample sizes (<30 samples). In addition, the $z_{\alpha/2}$ or $t_{\alpha/2}$ values at any $\alpha$ can be included in the calculations; however, it is customary to use an $\alpha$ of 0.05.

In developing sampling plans for estimation or use in pest management, the precision must be usually set at 10% ($d = 0.10$) or 20% ($d = 0.20$). In general, sampling costs (in terms of time and money) will determine the maximum number of samples that can be used. The achievable precision in estimates must be balanced with the sampling costs.

**Table 11** Formulas for Estimating Precision, and for Determining Sample Sizes Required to Achieve a Desired Precision, as a Function of the Underlying Distribution

| | Formula for calculating | |
|---|---|---|
| Sampling distribution | Precision ($d$)[a] | Sample size ($n$) |
| Taylor's power law | $z_{\alpha/2}(A\bar{x}^{(b-2)}/n)^{0.5}$[b] | $(z_{\alpha/2}/d)^2 A\bar{x}^{(b-2)}$ |
| Poisson | $z_{\alpha/2}(1/n\bar{x})^{0.5}$ | $(z_{\alpha/2}/d)^2(1/\bar{x})$ |
| Binomial | $z_{\alpha/2}(q/np)^{0.5}$ | $(z_{\alpha/2}/d)^2(q/p)$ |
| Negative binomial | $z_{\alpha/2}(1/\bar{x} + 1/k)^{0.5}(1/n)^{0.5}$ | $(z_{\alpha/2}/d)^2(1/\bar{x} + 1/k)$ |
| Normal | $z_{\alpha/2}(s/\bar{x})(1/n)^{0.5}$ | $(z_{\alpha/2}d)^2(s^2/\bar{x}^2)$ |

[a]Expressed as half-width of the confidence interval of the parameter (mean or proportion).
[b]For example, $(A\bar{x}^{(b-2)}/n)^{0.5}$ is the same as $\sqrt{A\bar{x}^{(b-2)}/n}$.

For stored-wheat insects sampled with a grain trier (Hagstrum et al. 1985) or deep-bin cup probe and perforated probe traps (Lippert and Hagstrum 1987), and for stored-barley insects sampled with perforated probe traps (Subramanyam and Harein 1990), the precision achievable in estimating densities and sample sizes required for estimating densities at various precision levels were calculated. In these papers, the precision was expressed as *d*, and the precision equation was based on Taylor's power law. Although the calculations were correct, in these papers, precision was referred to as accuracy, following the terminology of Ruesink and Kogan (1982). Subramanyam and Harein (1990) examined the effects of increasing the number of trap samples on the precision in estimating densities of six insect species. They also determined the sample sizes needed to estimate densities at two fixed precision levels (0.25 and 0.50). For all six species, the precision was low when one was estimating low insect densities, and the precision improved with an increase in insect density and number of traps used. For example, using two traps instead of one gave 1.41 times more precise estimates. Irrespective of the trap catch, the incremental improvement in the average precision per trap decreased with an increase in the number of traps used. Increasing the number of traps from 2 to 6, 6 to 10, and 10 to 20 improved the precision per trap by 0.43-, 0.32-, and 0.14-fold, respectively (Subramanyam and Harein 1990). Figure 12 shows how precision (*d*), shown as a 95% confidence interval for estimating a fixed mean density, changes with the number of samples used (Hagstrum and Flinn 1992). For estimating a given mean density, increasing the number of samples gives improved estimates of precision, as indicated by the narrow confidence intervals. Figure 13A shows the precision achievable (using Taylor's *A* and *b* values for the combined species' data [Table 5] and Equation in Table 11) when 9 and 20 traps are used to estimate densities of 0.1–15 adults per trap. The estimates are more precise with 20 traps than with 9 traps.

Hagstrum et al. (1985) and Subramanyam and Harein (1990) showed that the number of samples required to estimate insect densities with fixed precision levels was inversely related to insect density. In other words, more samples are required to estimate insects at low densities than at high densities. Also, more samples would be required to estimate densities with a high precision. For example, estimating densities at a fixed precision level of 0.25 required four times more traps than estimating densities at a precision of 0.50 (Subramanyam and Harein 1990). Figure 13B shows that more traps are required to estimate insect densities at a precision of 0.2 than at a precision of 0.5. Lippert and Hagstrum (1987), using presence/absence sampling, showed that for stored wheat insects the probability of insect detection was greater when five traps were used instead of one trap. They also reported that the chance of detecting infestations was approximately equal for 10 or 20 traps.

Figure 14 shows the precision of binomial estimates as a function of mean density. First, the proportion of sample units with insects was predicted using

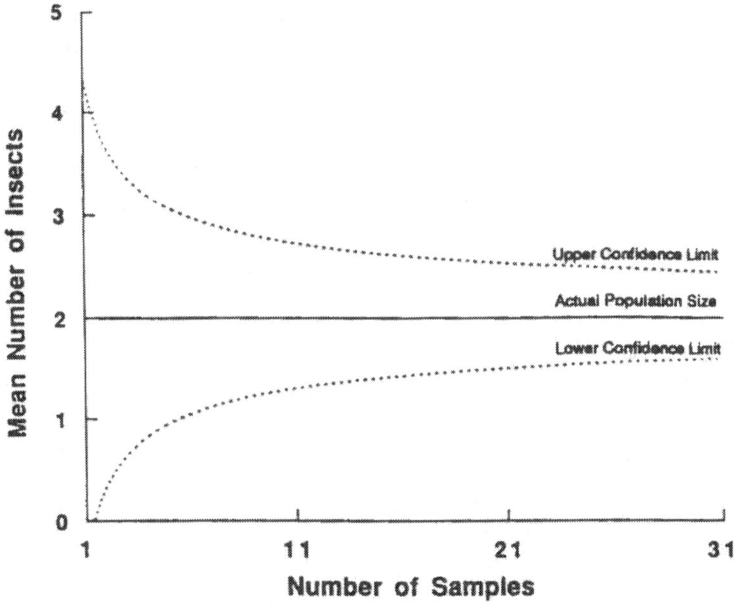


**Figure 12** Variation in population estimates in relation to number of samples. The dashed lines represent precision equal to the confidence interval at $\alpha = 0.05$. (From Hagstrum and Flinn 1992.)

Wilson and Room's (1983) model (Equation 23) for mean densities ranging from 0.01 to 5 adults per trap. Taylor's *A* and *b* values were based on the combined species' data (Table 5). The precision and number of samples required to estimate these proportions were calculated using the appropriate equations given in Table 11 for the binomial distribution. The predicted proportion of samples with insects, precision, and sample sizes was plotted as a function of the mean density. The precision improved with an increase in mean density or an increase in proportion of the sample units with insects (Fig. 14A). However, the number of samples decreased with mean density, but increased as the proportion of infested sample units and mean density increased (Fig. 14B). This sample size curve is typical for binomial sampling data.

The precision can be improved by taking more samples. Characterizing insect distribution and using the published formulas (Karandinos 1976; Ives and Moon 1984) may improve the precision of estimates, provided the variation among samples is not influenced by the sampling plan. Changing the sample unit or the sampling plan may improve precision, by reducing sample-to-sample variation. Sampling data may inherently show a high degree of variability (conse-

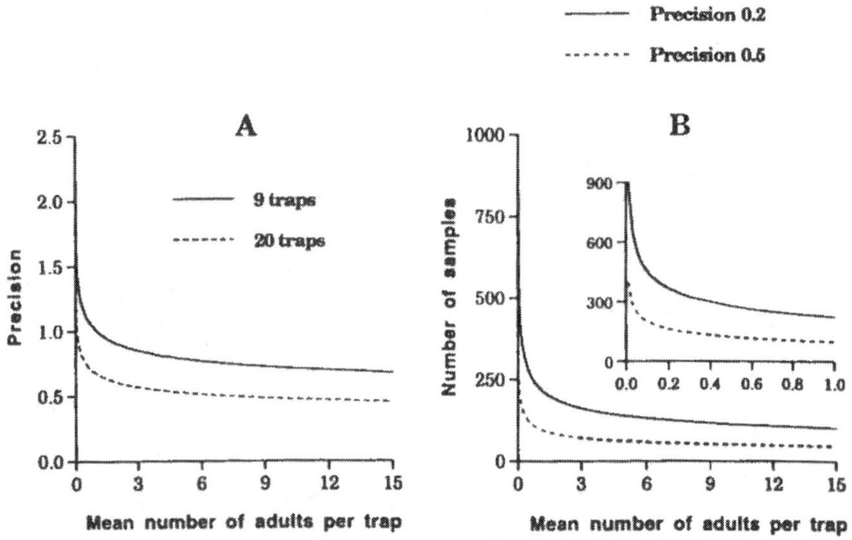


**Figure 13** Precision achievable in estimating mean densities when 9 and 20 samples are used (A), and number of samples required to estimate densities at fixed levels of precision (B). Taylor's estimates *A* and *b* used to determine precision and sample sizes (Table 11) are based on the combined species' data (Table 5).

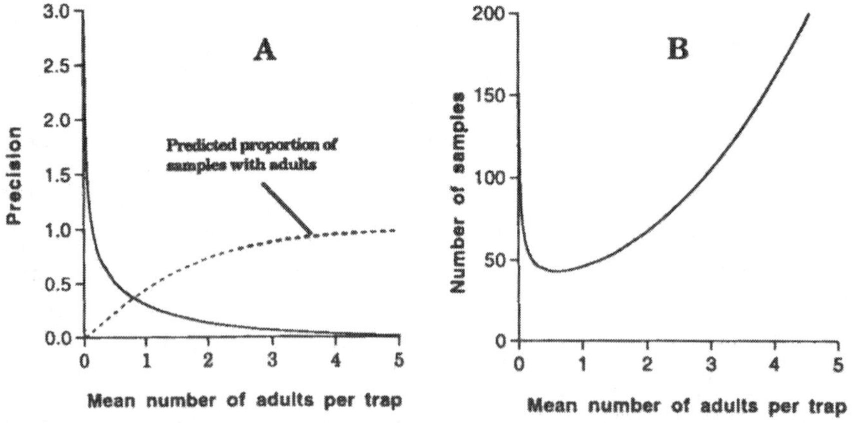


**Figure 14** Precision and sample sizes for binomially distributed data, plotted as a function of mean density. The proportion of samples with insects at mean densities of 0.01–5 insects per sample was estimated with Wilson and Room's (1983) model (Equation 23) using *A* and *b* estimates of the combined species' data (Table 5). The proportions were then used in precision and sample size equations for binomial distribution (Table 11).

quently a low degree of precision) that is not within the control of the sampler. In such situations, the use of presence/absence estimation methods may improve precision.

## VII. SEQUENTIAL SAMPLING

In pest management, the manager is faced with two types of risk when making a management decision: the risk of deciding to control pests when it is not needed and the risk of not controlling pests when a control is in fact warranted. Sequential sampling is designed to minimize these risks. The information from sample units is used to determine if the density of pests or proportion of sample units with insects has exceeded a threshold for intervention. More samples are taken when the density or presence of pests in sample units is close to the threshold, and vice versa. In essence, this sampling procedure balances the economy of sampling with the expected quality of control decisions (Binns 1994). In sequential sampling the cumulative number of sample units collected and the cumulative number of insects or presence of insects found in them is counted and plotted on a graph. Sampling stops when the predetermined stop lines are crossed (Fig. 15), and sampling continues if the information from sample units falls within the stop lines. The upper stop line is usually the economic or action threshold, at which economic damage occurs and pest control is essential. The lower stop line is determined arbitrarily, and is usually set as a certain percentage of the threshold. Below the lower stop line, pest damage is considered to be noneconomic, and usually no control action is taken. For example, let us assume that for lesser grain borer adults in wheat the economic threshold is 2 insects per kilogram of grain, and the lower noneconomic threshold is 0.5 insects per kilogram of grain. The pest manager is interested in determining if lesser grain borers in kilogram lots of samples are at or above the upper threshold [$d_1$ or $p_1$] (hypothesis, $H_1$) or are at or below the lower threshold [$d_0$ or $p_0$] (hypothesis, $H_0$). The pest manager also has to make another decision that concerns the probability of incorrectly classifying the population with respect to the upper and lower thresholds. These probabilities are expressed as type I ($\alpha$) or type II ($\beta$) error rates: $\alpha$ = the probability of accepting $H_1$ when $H_0$ is true; $\beta$ = the probability of accepting $H_0$ when $H_1$ is true. The $\alpha$ and $\beta$ presented here should not be confused with the y-intercept and slope of Iwao's patchiness regression. In sequential sampling, the managers can specify error rates to minimize the risk of incorrect assessment. It is important to understand that failing to treat when pest outbreak occurs may be more costly than treating unnecessarily.

Sequential sampling plans have not been developed for stored-product insects. However, these techniques have been developed and implemented for field crop insects (see Binns 1994; Hutchison 1994). The techniques presented here should enable researchers, pest managers, and others to develop sequential sampling plans for stored-product insects.

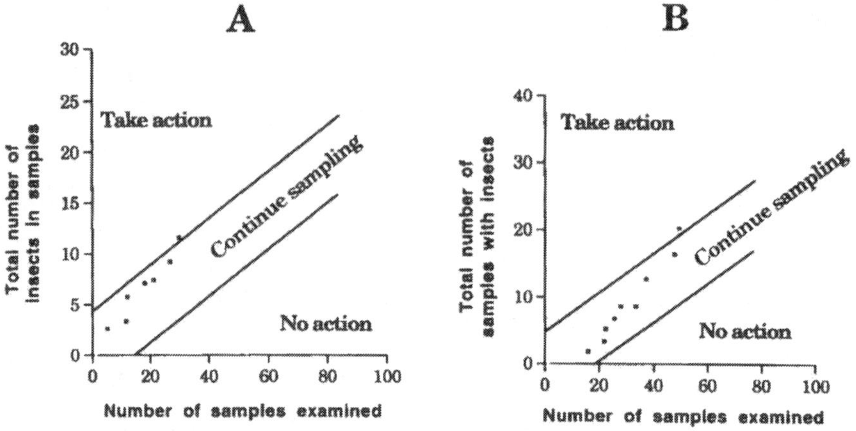


**Figure 15** Diagrammatic representation of sequential sampling decision or stop lines based on complete counts of insects in samples (A) or on presence/absence of insects in samples (B). The solid squares are sampling data, and the sampling continues until plotted sample data cross the upper decision line. These figures are not based on any data.

The sequential probability ratio test (SPRT) and Iwao's confidence interval method are two sequential sampling techniques for classifying pest status. To use SPRT, the distribution of the population must be characterized. The SPRT stop lines can be calculated for populations that follow a Poisson, negative binomial, binomial, and normal distributions. The upper and lower stop lines have a common slope but different intercepts (also see Fig. 15). Formulas for the intercept and slope for each of the underlying distributions are given in Table 12.

Given a particular population distribution, the upper and lower decision lines are calculated as follows:

Upper decision or stop line =

$$T_u = (\text{intercept}) \ln((1 - \beta)/\alpha) + n\,(\text{slope}) \quad (45)$$

Lower decision or stop line =

$$T_l = (\text{intercept}) \ln(\beta/(1 - \alpha)) + n\,(\text{slope}) \quad (46)$$

where $\alpha$ and $\beta$ are the specified error rates, and the intercept and slope can be calculated using formulas given in Table 12 for different number of sample units ($n$) examined. The upper and lower decision lines for the Iwao's confidence interval method (Binns 1994), which is based on Iwao's patchiness regression, are calculated as

Upper decision or stop line =

$$T_u = nd_c + t\{n[(\alpha + 1)d_c + (\beta - 1)d_c^{2}]\}^{0.5} \quad (47)$$

**Table 12** Intercept and Slope Formulas for Developing Sequential Probability Ratio Test (SPRT) Plans

| Sampling distribution | Intercept | Slope |
|---|---|---|
| Poisson[a] | $\frac{1}{\ln(d_1/d_0)}$ | $\frac{(d_1 - d_0)}{\ln(d_1/d_0)}$ |
| Negative binomial[a,b] | $\frac{1}{\ln\{[d_1(k + d_0)]/[d_0(k + d_1)]\}}$ | $\frac{k \ln\{(k + d_1)/(k + d_0)\}}{\ln\{[d_1(k + d_0)]/[d_0(k + d_1)]\}}$ |
| Binomial[c] | $\frac{1}{\ln\{(p_1q_0)/(p_0q_1)\}}$ | $\frac{\ln(q_0/q_1)}{\ln\{(p_1q_0)/(p_0q_1)\}}$ |
| Normal[d] | $\frac{\sigma^{2\,e}}{d_1 - d_0}$ | $\frac{(d_1 + d_0)}{2}$ |

[a]$d_1$ = upper critical density; $d_0$ = lower critical density.
[b]$k$ = negative binomial parameter $k$ (common $k$ or $k_c$).
[c]$p_1$ and $p_0$ are upper and lower critical proportions, respectively. $q_0 = 1 - p_0$; $q_1 = 1 - p_1$.
[d]Distribution of data based on large samples (> 30) has a "bell-shaped" curve, where the frequency ($f_i$) of an observation of size ($X_i$) is expressed as $f_i = (1/\sigma\sqrt{2\pi})e^{-(X_i-\mu)^2/2\sigma^2}$. $\pi$, a constant, is the ratio between the circumference and the diameter of a circle, and has a value of 3.142; e = 2.71828. See Zar (1984, pp. 79–96) for additional information on normal distribution.
[e]Population variance. The sample variance ($s^2$) can be substituted for $\sigma^2$.

Lower decision or stop line =

$$T_1 = nd_c - t\{n[(\alpha + 1)d_c + (\beta - 1)d_c^{\,2}]\}^{0.5} \tag{48}$$

where $n$ = number of samples taken, $d_c$ = critical pest density, $\alpha$ and $\beta$ are y-intercept and slope estimates, respectively, from Iwao's patchiness regression (not to be confused with type I and type II error rates discussed above), and $t$ is $t$ distribution statistic (at any probability level; usually 0.01, 0.05, or 0.10). SPRT has two class limits and hence two error rates. Iwao's method has one class limit (critical density), and tests the hypothesis that the density estimate from a set of samples is equal to the critical density against the alternative hypothesis that the density estimate is not equal to the critical density. Therefore, this method uses only the type I error rate. In sequential sampling the cumulative number of insects divided by the cumulative number of samples gives the estimated mean density or proportion.

The two most important properties of SPRT and Iwao's confidence interval method are the operating characteristic (OC) function and the average sample number (ASN). Fowler and Lynch (1987) give formulas for calculating OC and ASN curves for SPRT plans based on Poisson, negative binomial, binomial, and normal distributions. Binns (1994) provided a computer program for calculating OC and ASN functions for SPRT and Iwao's methods. We recommend interested readers to these publications for specific information on computing OC and ASN functions. The OC curve shows the probability of accepting the hypothesis ($H_0$) that the pest density is at or below a threshold, when the pest density is actually above the threshold. In other words, OC is the probability of not to use a control measure as a function of insect density or it is the probability of a correct decision. ASN curve shows the average number of samples (observations) required for each value of insect density. In general, ASN increases if the lower and upper decisions lines are closer together, and if the type I ($\alpha$) and type II ($\beta$) error rates are set low. Figure 16 is a diagrammatic representation of OC and ASN curves as a function of mean insect density. The shapes of the curves vary with the underlying distribution, decision limits, and the error rates (Fowler and Lynch 1987). Binns (1994) discusses sequential decision lines based on more than two class limits (three-decision SPRT), especially where pest density needs to be categorized as low, medium, or high. Interested readers should consult Watters (1955) and Fowler and Lynch (1987). Binns (1994) outlined some practical difficulties in using three-decision SPRT.

In the sequential plans discussed above, there is no limit on the maximum number of samples to be examined before a decision is made. If the density of pest is midway between the upper and lower thresholds, a decision to intervene or not to intervene is difficult to reach (Binns 1994). As a result, time is wasted by taking more samples. Establishing a maximum sample size reduces unnecessary sampling effort. The maximum sample size limit must not be low (twice or greater

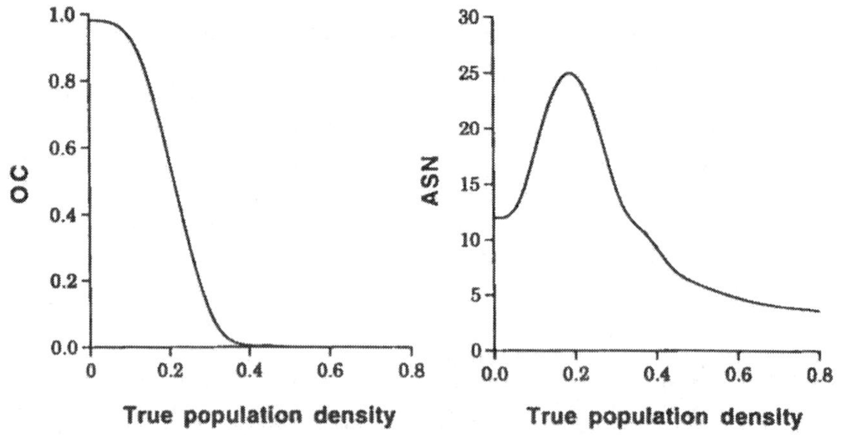


**Figure 16** Typical operating characteristic (OC) and average sample number (ASN) curves essential for determining the performance of sequential sampling plans. The curves are not based on any data.

than ASN). The effect of setting the maximum number of samples on OC and ASN functions must be verified through simulation (Binns 1994).

When the relationship between the proportion of infested sample units ($p$) and mean density ($\bar{x}$) is established, binomial sequential classification procedures can be developed by using a confidence interval around the incidence corresponding to the critical threshold density (Bechinski and Stoltz 1985; Nyrop et al. 1989). If $P_t$ is the incidence that corresponds to an insect density ($\bar{x}$) considered to be an economic or action threshold, then the upper and lower confidence limits around $P_t$ are expressed as

$$\text{Upper confidence limit} = T_u = nP_t + nz_{\alpha/2}\,[P_t(1 - P_t)/n]^{0.5} \tag{49}$$

$$\text{Lower confidence limit} = T_l = nP_t - nz_{\alpha/2}\,[P_t(1 - P_t)/n]^{0.5} \tag{50}$$

where $n$ = number of samples; $z_{\alpha/2} = 1.96$, is the value of upper $\alpha/2$ of the standard normal distribution at $\alpha = 0.05$. $t_{\alpha/2}$ can be used instead of $z_{\alpha/2}$, for $n < 30$. These equations can be solved for different $n$ (for example, $n = 1$ through $n = 30$). This method is somewhat similar to the Iwao's confidence interval method. If the cumulative number of infested sample units exceeds $T_u$, the population is classified as exceeding the critical threshold. If the cumulative number of infested units is below $T_l$, then the population is classified as being below the critical threshold. Sampling continues if cumulative infested units fall within $T_u$ and $T_l$. Nyrop et al. (1989) gave methods for evaluating the performance criteria using OC and ASN functions. These sequential plans can be refined through simulation (Overholt et al. 1990; Jones 1994; Binns 1994; Nyrop et al. 1989). Maximum efficiency is attained if 50% of the sample units are infested at the critical density (Ives and Moon 1984). Wilson et al. (1983) modified the sample size equations of Taylor's power law and the binomial distribution (Table 11) to classify pest status based on insect counts or presence/absence of insects in samples.

Sequential classification procedures based on pest density generally require fewer samples, on average, than sequential methods for estimating density with a fixed level of precision (see section below). Binomial sequential estimation methods likewise require more samples than binomial sampling to classify a density sequentially (Nyrop et al. 1989).

In situations where the proportion of sample units with insects increases rapidly for a unit increase in mean density, estimating insect density using binomial sampling methods may not be feasible. When density must be estimated with a high degree of precision (for example, in research programs, or for high value crops where low insect densities can cause considerable economic losses), sampling methods based on counting actual numbers of insects in sample units is necessary. Sequential sampling can be used to estimate insect density with a fixed level of precision (Hutchison 1994). Green (1970) developed an equation that incorporates Taylor's $A$ and $b$ for estimating mean density sequentially with a fixed level of precision as

$$\ln(T_n) = \frac{\ln(\mathrm{CV}_{\bar{x}}^2/A)}{\{[(b-2)+(b-1)]/[(b-2)\ln(n)]\}} \tag{51}$$

$T_n$ = cumulative number of insects in samples, $\mathrm{CV}_{\bar{x}}$ = scale-free measure of precision, and $n$ = number of samples examined. If the precision is expressed as a fixed proportion of the mean ($d$) set to half the width of a confidence interval with $1-\alpha$ probability, then $T_n$ can be estimated by

$$\log_{10}(T_n) = \frac{\ln[d^2/(z_{\alpha/2})^2 A]}{(b-2)+[(b-1)/(b-2)]\ln(n)} \tag{52}$$

Kuno (1969) developed sequential stop lines for estimating densities at a fixed precision level as

$$T_n = \frac{(1+\alpha)}{[\mathrm{CV}_{\bar{x}}^2 - ((\beta-1)/n)]} \tag{53}$$

where $\alpha$ and $\beta$ are the basic contagion and clumping parameter, respectively, from Iwao's patchiness regression. For $\alpha = 0$ and $\beta = 1$, which is indicative of a Poisson distribution, Equation 53 becomes

$$T_n = \frac{(1+\alpha)}{\mathrm{CV}_{\bar{x}}^2} \tag{54}$$

The common $k$ of the negative binomial distribution can be incorporated into Equation 53 by replacing $1/(\beta - 1)$ with $k$ and where $\alpha = 0$ (Kuno 1991).

$$T_n = \frac{1}{[\mathrm{CV}_{\bar{x}}^2 - (1/nk)]} \tag{55}$$

A fixed precision sequential plan for estimating mean trap catches of stored-product insects in shelled maize based on Equation 51 is shown in Figure 17. The data were based on combined catches of seven insect species in shelled maize, sampled with probe traps (Subramanyam, unpublished data), and Taylor's estimates for these data are shown in Table 5. The stop lines were generated for precision levels of 0.20, 0.25, and 0.30, by calculating $T_n$ for different $n$. $T_n/n$ gives the mean trap catch or density, along the line at a fixed precision level. As expected, for estimating trap catches or density at high precision (0.20) more samples must be examined than when one is estimating density at a low precision (0.30).

The performance of the sampling plans with respect to the fixed precision can be studied through Monte Carlo or bootstrap simulation. In Monte Carlo method, the population is assumed to fit a probability distribution (also see Snedecor and

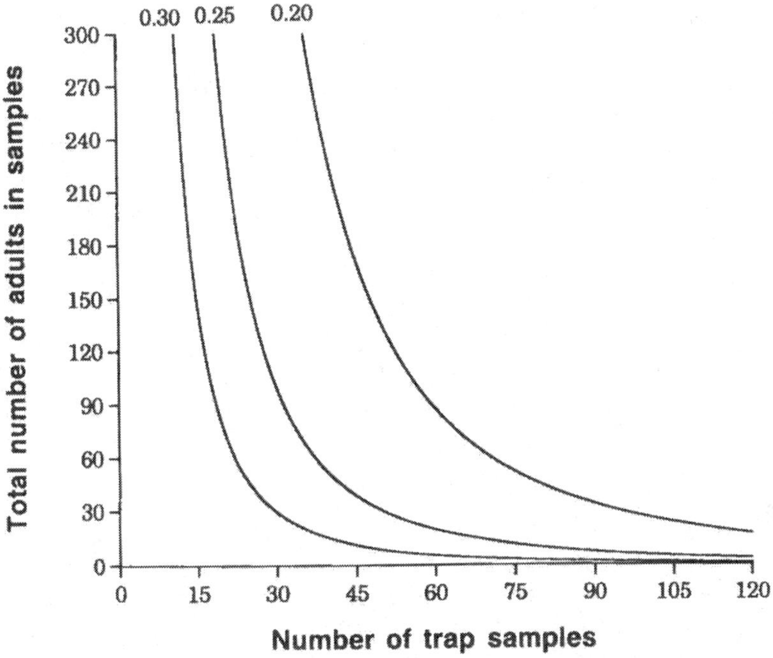


**Figure 17** Green's (1970) fixed precision stop lines used for sequential estimation of insect density. The stop lines are based on Taylor's *A* and *b* for the combined species' data (Table 5), and each line can be used to predict insect density at a fixed level of precision (0.20, 0.25, or 0.30). The precision is expressed as $CV_{\bar{x}}$ (Equation 40). All along the line, density can be estimated by dividing the cumulative number of adults by the cumulative number of samples.

Cochran 1980, pp. 9, 15). In bootstrap method, no underlying distribution is assumed of the data, but actual insect counts are resampled from files containing the data (see Hutchison [1994] for the bootstrap simulation program). Hutchison et al. (1988), using Monte Carlo and bootstrap simulations on data of pea aphid on alfalfa, showed that the precision levels obtained in estimating mean densities were not fixed but stochastic (variable). The bootstrap simulation showed that the actual precision levels obtained were lower than the fixed precision level. For example, the use of a precision of 0.35 in calculating stop lines gave the required fixed precision level of 0.25, at low aphid densities (Hutchison et al. 1988).

To use sequential sampling as an alternative to fixed size sampling, the following criteria must be satisfied (Binns 1994):

1. An estimate of a critical threshold that justifies taking a control action must be

established. This threshold could be mean density, damage, or proportion of infested units that corresponds to economic damage or density.

2. If SPRT is used, the spatial structure of the population (based on sample data) must fit an underlying distribution. If the data fails to fit any distribution, SPRT can be used assuming a normal distribution, provided the data are based on large number of samples.
3. The value of the negative binomial $k$ must be stable over the range of mean densities that are of interest to the pest manager. The stability of $A$ and $b$ of Taylor's for a species must also be verified.
4. Care must be taken to prevent early stopping from giving biased results.
5. OC, ASN, costs, and critical thresholds must be continually checked through field sampling or simulations, and the sampling plans must be adjusted accordingly.

## VIII. OTHER SAMPLING TECHNIQUES

The use of random or systematic sampling requires all sampling units in the universe to be listed before the units are selected randomly or systematically. If the sampling universe is large, it may be impractical to list all sample units and select units at random. Several alternative sampling techniques exist that balance the cost involved in locating sample units and the relative net precision (Cochran 1977), which for a sampling unit is inversely proportional to the variance obtained for a fixed cost. These techniques and calculations are discussed in Cochran (1977), and will be mentioned here briefly.

### 1. Cluster Sampling

The sampling unit consists of a group of smaller units called elements or subunits. A sampling unit (or the number of elements per unit) that gives the smaller variance for a given cost or the smaller cost for a given variance must be chosen. For example, when sampling grain with a trier, a sampling unit could be two trier samples, five trier samples, 10 trier samples, and so on. Usually, but not always, the variance increases with an increase in the sampling unit size, but the sampling costs will be lower for a larger sample unit than a smaller sample unit. However, it would be more costly to process a larger sample than a smaller sample. Mean-variance pairs calculated from sample units that consist of several subunits can be described by Taylor's power law, and the Taylor's equation is slightly modified as

$$s^2 = u^{(1-b)} A \bar{x}_u^{\,b} \tag{56}$$

where $u$ = number of subunits per sample unit. $A$ and $b$ are Taylor's coefficients, and $\bar{x}_u$ = mean density per sample unit (density per subunit × number of subunits)

### 2. Hierarchial, Nested, or Multistage Sampling

This sampling is similar to cluster sampling, except that the subunits are further subsampled. The premise behind this technique is that if subunits within a unit give similar results, it is not economical to measure all subunits (Cochran 1977). First a sample of units, called primary units, is selected. Subunits or secondary units are sampled from each primary unit. Each secondary unit can be further subsampled, resulting in tertiary units, and so on. For instance, if four pallets from 20 pallets in a warehouse are selected at random, then a pallet is the primary unit. Let us assume that each pallet has 100 bags of stored seed. From each pallet, five bags are selected at random; these constitute secondary units. From each bag, two samples are taken with a suitable device, and number of insects in each sample is counted. These constitute tertiary units. Here the two samples are nested within bags, and bags are nested within pallets. The objective with nested sampling is to estimate the components of variance due to samples, bags, and pallets, and the variance contributed at each level of classification to the total variance (Hagstrum et al. 1985). Depending on the variance, and the costs involved in sampling, the sampler can decide whether to increase the number of primary, secondary, or tertiary units to be sampled. Cochran (1977), Snedecor and Cochran (1980), and Zar (1984) describe nested designs and statistical analysis of nested data. This technique can also be used for estimating proportions (Cochran 1977). The optimal number of sampling units to be examined depend on whether the budget or time for sampling is limited or unlimited. Cochran (1977) gives formulas for calculating optimal number of samples at each level that incorporate cost considerations.

### 3. Double Sampling

Double sampling involves taking a large preliminary sample. The information in the preliminary sample is used to decide on the size of the second sample or whether to take a second sample at all. Nyrop and Wright (1985) developed double sampling plans for the Colorado potato beetle. This sampling is better suited for binomial sequential classification than for estimating insect densities, because insect density estimation procedures are complex (Binns and Nyrop 1992). Cochran (1977) provides additional details, including computational procedures, on double sampling.

### 4. Variable-Intensity Sampling

Samples are taken from as many locations as possible from the sampling universe to obtain a representative sample. Sampling is done more intensively when the information from samples (e.g., mean) is close to the critical threshold (e.g., economic threshold), and less intensively when it is far away from the threshold (Hoy et al. 1983; Hoy 1991). Variable-intensity sampling procedures can be developed using Taylor's power law, nested analysis of variance, and negative binomial distribution (Binns and Nyrop 1992).

Sampling plans have also been developed that incorporate a time component (Bechniski and Hescock 1991). Sampling plans have been developed for two or more pest species (Binns and Nyrop 1992), and for a pest and its natural enemy (Nyrop 1988).

## IX. SAMPLING SCHEDULE

Samples must be taken often enough to detect insect populations before they reach unacceptable levels. Because stored-product insects multiply rapidly, samples must be taken frequently to detect infestations early. The time interval between sampling should be based on the population dynamics of the pest. Population dynamics or changes in pest density over time due to environmental and genetic influences must be known. Monthly sampling is often recommended because many insects can develop from eggs to adults during this time period. Deciding when to take the first sample must be based on prior knowledge of insect incidence and abundance. For example, grain that is harvested and stored during the warmer parts of the year (temperatures > 20°C) must be sampled immediately after harvest, and at monthly intervals thereafter. However, if grain temperatures for extended periods are <20°C, which are too cold for insect activity and reproduction, the sampling interval could be longer than a month. Frequent sampling helps in detecting infestations early. Applying control measures early, especially when insects are present at low densities, helps in maintaining populations below economically damaging levels. As a consequence, it takes longer for insect populations to build up and reach economically damaging levels.

If a large number of samples is taken, managers can sample less frequently and still be confident that insect populations will not grow to unacceptable levels before they sample again. Taking large number of samples also ensures that the samples are representative, and the estimates are closer to the unknown population parameters. With accurate estimates of current population densities, insect population growth models (Hagstrum and Throne 1989; Hagstrum and Flinn 1990) can be used to predict when to sample again. These models also allow researchers and managers to adjust the sampling interval for changes in the egg-to-adult developmental time, which are influenced by temperature and moisture of the stored commodity.

The type of sampling device used for stored-product insects must be based on insect activity, which is largely dependent on the temperature. Because traps depend on insect movement, they should not be used for sampling insects that are inherently less mobile or those that are inactive due to cooler temperatures.

## X. UTILIZATION OF SAMPLING DATA

The business of storing, processing, and distributing raw and processed food products can be more profitable when a sampling program provides accurate estimates of insect densities so that insect pest management decisions are consistently correct. Incorrect decisions are made when a sampling program underestimates or overestimates insect densities. Money is wasted when pest control measures are used unnecessarily because of overestimates, and profit is reduced when insect densities are underestimated resulting in damage to the commodity because insect infestations were not controlled.

The importance of the precision of estimates to the utilization of sampling data can be illustrated using a graph showing the ranges of insect density estimates likely to be obtained with different sample sizes (Fig. 12). When only one sample is taken from a population with an actual density of two insects per kilogram, the number of insects per sample will range from zero to four, because insects are not uniformly distributed within the grain mass. If the sample you take has no insects, you would incorrectly conclude that the grain is uninfested and that no insect control is needed. The confidence intervals indicate that you are likely to find some insects if two or three samples are taken. However, you could still find less than two insects and underestimate population density. The precision of estimates as indicated by the distance between upper and lower confidence limits increases rapidly up to five samples but much more slowly with each additional sample. This means that for insect populations at this density there is little advantage in taking more than five samples, and you would accomplish more by sampling at a later date. However, for insect populations with lower densities, more samples would be beneficial. Because you cannot be sure that the insect density is below two even when 30 samples are taken, you would have to control insects when density estimates are between 1 and 1.5 insects if you want to be sure that populations have not reached two insects per kilogram. Therefore, for this type of sampling situation, the development of sequential sampling appears to be a viable approach. The utilization of sampling information in making pest management decisions is discussed in more detail in the chapter on integrated pest management.

Sampling should also be used to determine whether control measures were successful, by comparison of pre- and posttreatment counts or presence/absence of insects in the samples. However, it is important to bear in mind that after applying control measures a majority of the insects are killed, and the density and spatial structure of insects in the sampling universe may be altered. Therefore, the number of samples needed to estimate the posttreatment density reliably may be different from the number of samples needed to estimate the pretreatment density.

## XI. SUMMARY

In most agricultural situations, sampling the entire sampling universe to determine a population attribute (mean density or proportion) is impossible or cost-prohibitive. Therefore, these population parameters are unknown. Thus, samples of a known size are removed from the sampling universe to make inferences about these population parameters. Designing a sampling program involves choosing an appropriate sample unit, determining the number of samples to use, selecting locations to be sampled, and the time of sampling. Small sample units are often more efficient than large sample units, but the costs associated with obtaining the samples are higher for the former than the latter. Sample locations should be selected to provide a representative sample. Samples must be taken frequently to detect insect populations before they reach unacceptable levels. Sampling methods that provide absolute estimates are generally needed for studying insect ecology, but for pest management less accurate estimates of insect density may be considered to reduce sampling costs. However, it is important to establish a relationship between relative density estimates and absolute pest density estimates or damage.

Mapping the location of insects within the sampling universe is difficult, especially if the insects are mobile or active. Therefore, information from the sample units is used to describe the spatial distribution or sampling distribution of the insect populations. Fitting probability distribution functions or dispersion indices to sampling data permits development of sampling plans for estimation or decision-making in pest management (Binns and Nyrop 1992). A few sampling plans, for estimating insect density, have been developed for stored-product insects (Hagstrum et al. 1985; Hodges et al. 1985; Subramanyam and Harein 1990; Subramanyam et al. 1993). We hope researchers and pest managers working with stored-product insects will use the sampling concepts and formulas presented in this chapter to develop sampling plans for cost-effective management of insect pests. The four basic pieces of information needed to use the numerous equations presented in this chapter are the mean, variance, number of samples used to estimate the sample mean and variance, and proportion of sample units with insects. Most of the computations are easy and can be done on a hand calculator, while fitting a few models or equations to sample data may require the use of a computer. Most of the equations have been illustrated using examples based on published and unpublished stored-product insect sampling data. Numerous references have been provided for those interested in additional information not covered in this chapter. Sampling is an integral component of integrated pest management. The business of storing, processing, and distributing food products can be more profitable if pest management decisions are based on practical sampling programs that provide precise estimates of population density or proportion.

## ACKNOWLEDGMENTS

The unpublished data of Bh. Subramanyam used in this chapter are based on 1990 research supported by grants from the United States Department of Agriculture to the Minnesota Pesticide Impact Assessment Program. Bh. Subramanyam thanks Tim Schenk, Mike Kubly, Matt Buehling, and Doreen Dekker for assistance in field sampling, and Mike Tufte for spending countless hours entering sampling data into the computer and for verifying that data were entered correctly.

## REFERENCES

Barak, A. V., Burkholder, W. E., and Faustini, D. L. (1990). Factors affecting the design of traps for stored-product insects. *J. Kansas Entomol. Soc.*, *63*:466–485.

Barker, P. S., and Smith, L. B. (1987). Spatial distribution of insect species in granary residues in the prairie provinces. *Can. Entomol.*, *119*:1123–1130.

Barnes, D. F., Fisher, C. K., and Kaloostian, G. H. (1939). Flight habits of the raisin moth and other insects as indicated by the use of a rotary net. *J. Econ. Entomol.*, *32*: 859–863.

Bartlett, M. S. (1949). Fitting a straight line when both variables are subject to error. *Biometrics*, *5*:207–212.

Bauwin, G. R., and Ryan, H. L. (1974). Sampling, inspection, and grading of grain. In Storage of Cereal Grains and Their Products (ed. Christensen, C. M.). American Association of Cereal Chemists, St. Paul, MN, pp. 115–157.

Bechinski, E. J., and Hescock, R. (1991). Fixed-time classification and detection plans for alfalfa snout beetle (Coleoptera: Curculionidae). *J. Econ. Entomol.*, *84*:1388–1395.

Bechinski, E. J., and Stoltz, R. L. (1985). Presence-absence sequential decision plans for *Tetranychus urticae* (Acari: Tetranychidae) in garden-seed bean, *Phaseolus vulgaris*. *J. Econ. Entomol.*, *78*:1475–1480.

Binns, M. R. (1994). Sequential sampling for classifying pest status. In CRC Handbook of Sampling Methods for Arthropods in Agriculture (ed. Pedigo, L. P., and Buntin, G. D.). CRC Press, Boca Raton, FL, pp. 137–174.

Binns, M. R., and Bostanian, N. J. (1990). Robustness in empirically based binomial decision rules for integrated pest management. *J. Econ. Entomol.*, *83*:420–427.

Binns, M. R., and Nyrop, J. P. (1992). Sampling insect populations for the purpose of IPM decision making. *Annu. Rev. Entomol.*, *37*:427–453.

Boxall, R. A. (1991). Post-harvest losses to insects—a world overview. In Biodeterioration and Biodegradation 8 (ed. Rossmoore, H. W.). Elsevier, New York, pp. 160–175.

Chambers, J. (1990). Overview of stored-product insect pheromones and food attractants. *J. Kansas Entomol. Soc.*, *63*:490–499.

Cochran, W. G. (1977). Sampling Techniques. 3rd ed. Wiley, New York.

Cuperus, G. W., Fargo, W. S., Flinn, P. W., and Hagstrum, D. W. (1990). Variables affecting capture rate of stored-grain insects in probe traps. *J. Kansas Entomol. Soc.*, *63*:486–489.

Davis, P. M. (1994). Statistics for describing populations. In CRC Handbook of Sampling Methods for Arthropods in Agriculture (ed. Pedigo, L. P., and Buntin, G. D.). CRC Press, Boca Raton, FL, pp. 33–54.

Downing, J. A. (1979). Aggregation, transformation, and the design of benthos sampling programs. *J. Fisheries Res. Board Canada*, *36*:1454–1463.

Downing, J. A. (1986). Spatial heterogeneity: evolved behavior or mathematical artefact? *Nature*, *6085*:255–257.

Draper, N. R., and Smith, N. (1981). Applied Regression Analysis. 2nd ed. Wiley, New York.

Dyte, C. E. (1965). Studies on insect infestations in the machinery of three English flour mills in relation to seasonal temperature changes. *J. Stored Prod. Res.*, *1*:129–144.

Efron, B., and Gong, G. (1983). A leisurely look at the bootstrap, the jackknife, and cross-validation. *Am. Stat.*, *37*:36–48.

Feng, M. G., and Nowierski, R. M. (1992). Spatial distribution and sampling plans for four species of cereal aphids (Homoptera: Aphidadae) infesting spring wheat in south-western Idaho. *J. Econ. Entomol.*, *85*:830–837.

Fowler, G. W., and Hauke, D. (1979). A distribution-free method for interval estimation and sample size determination. *Res. Inv. Notes BLM*, *19*:1–8.

Fowler, G. W., and Lynch, A. M. (1987). Sampling plans in insect management based on Wald's sequential probability ratio test. *Environ. Entomol.*, *16*:345–354.

Fowler, G. W., and Witter, J. A. (1982). Accuracy and precision of insect density and impact estimates. *Great Lakes Entomol.*, *15*:103–117.

Gentry, J. W. (1984). Inspection techniques. In Insect Management for Food Storage and Processing (ed. F. J. Baur). American Association of Cereal Chemists, St. Paul, MN, pp. 33–42.

Gerrard, D. J., and Chiang, H. C. (1970). Density estimation of corn root-worm egg populations based upon frequency of occurrence. *Ecology*, *51*:237–245.

Golob, P. (1976). Techniques for sampling bagged produce. *Trop. Stored Prod. Inf. Bull.*, *31*:37–48.

Golob, P., Ashman, F., and Evans, N. (1975). The separation of live stored product insect larvae from flour or sievings using a modified Tullgren funnel. *J. Stored Prod. Res.*, *11*:17–23.

Green, R. H. (1970). On fixed precision level sequential sampling. *Res. Popul. Ecol.*, *12*:249–251.

Hagstrum, D. W. (1983). Self-provisioning with paralyzed hosts and age, density, and concealment of hosts as factors influencing parasitization of *Ephestia cautella* (Walker) (Lepidoptera: Pyralidae) by *Bracon hebetor* Say (Hymenoptera: Braconidae). *Environ. Entomol.*, *12*:1727–1732.

Hagstrum, D. W. (1985). Preharvest infestation of cowpeas by the cowpea weevil, *Callosobruchus maculatus* (Coleoptera: Bruchidae) and population trends during storage in Florida. *J. Econ. Entomol.*, *78*:358–361.

Hagstrum, D. W. (1987). Seasonal variation of stored wheat environment and insect populations. *Environ. Entomol.*, *16*:77–83.

Hagstrum, D. W. (1989). Infestation by *Cryptolestes ferrugineus* of newly-harvested wheat stored on three Kansas farms. *J. Econ. Entomol.*, *82*:655–659.

Hagstrum, D. W. (1994). Field monitoring and prediction of stored-grain insect populations. *Postharvest News Info.*, *5*:39–45.

Hagstrum, D. W., and Davis. L. R. (1980). Mate-seeking behavior of *Ephestia cautella*. *Environ. Entomol.*, *9*:589–592.

Hagstrum, D. W., and Flinn, P. W. (1990). Simulations comparing insect species differences in response to wheat storage conditions and management practices. *J. Econ. Entomol.*, *83*:2469–2475.

Hagstrum, D. W., and Flinn, P. W. (1992). Integrated pest management of stored-grain insects. In Storage of Cereal Grains and Their Products (ed. Sauer, D. B.). American Association of Cereal Chemists, St. Paul, MN, pp. 535–562.

Hagstrum, D. W., and Sharp, J. E. (1975). Population studies on *Cadra cautella* in a citrus pulp warehouse with particular reference to diapause. *J. Econ. Entomol.*, *68*:11–14.

Hagstrum, D. W., and Stanley, J. M. (1979). Release-recapture estimates of the population density of *Ephestia cautella* (Walker) in a commercial peanut warehouse. *J. Stored Prod. Res.*, *15*:117–122.

Hagstrum, D. W., and Throne, J. E. (1989). Predictability of stored-wheat insect population trends from life history traits. *Environ. Entomol.*, *18*:660–664.

Hagstrum, D. W., Flinn, P. W., and Fargo, W. S. (1991). How to sample grain for insects. In FGIS Handbook on Management of Grain, Bulk Commodities, and Bagged Products, Oklahoma State Univ. Coop. Ext. Serv. Circ. E-912, pp. 65–69.

Hagstrum, D. W., Flinn, P. W., and Shuman, D. (1994). Acoustical monitoring of stored-grain insects: an automated system. In Proceedings of the Sixth International Working Conference on Stored-Product Protection, 17–21 April 1994, Canberra, Australia, pp. 403–405.

Hagstrum, D. W., Meagher, R. L., and Smith, L. B. (1988a). Sampling statistics and detection or estimation of diverse population of stored-product insects. *Environ. Entomol.*, *17*:377–380.

Hagstrum, D. W., Milliken, G. A., and Waddell, M. S. (1985). Insect distribution in bulk-stored wheat in relation to detection or estimation of abundance. *Environ. Entomol.*, *14*:655–661.

Hagstrum, D. W., Vick, K. W., and Webb, J. C. (1990). Acoustical monitoring of *Rhyzopertha dominica* (Coleoptera: Bostrichidae) populations in stored wheat. *J. Econ. Entomol.*, *83*:625–628.

Hagstrum, D. W., Vick, K. W., and Flinn, P. W. (1991). Automated acoustical monitoring of *Tribolium castaneum* (Coleoptera: Tenebrionidae) populations in stored wheat. *J. Econ. Entomol.*, *84*:1064–1068.

Hagstrum, D. W., Webb, J. C., and Vick, K. W. (1988b). Acoustical detection and estimation of *Rhyzopertha dominica* (F.) larval populations in stored wheat. *Fla. Entomol.*, *71*: 441–447.

Hodges, R. J., Halid, H., Rees, D. P., Meik, J., and Sarjono, J. (1985). Insect traps tested as an aid to pest management in milled rice stores. *J. Stored Prod. Res.*, *21*:215–219.

Hoy, C. W. (1991). Variable-intensity sampling for proportion of plants infested with pests. *J. Econ. Entomol.*, *84*:148–157.

Hoy, C. W., Jennison, C., Shelton, A. M., and Andaloro, J. T. (1983). Variable-intensity sampling: a new technique for decision making in cabbage pest management. *J. Econ. Entomol.*, *76*:139–143.

Hutchison, W. D. (1994). Sequential sampling to determine population density. In CRC Handbook of Sampling Methods for Arthropods in Agriculture (ed. Pedigo, L. P., and Buntin, G. D.). CRC Press, Boca Raton, FL, pp. 207–244.

Hutchison, W. D., Hogg, D. B., Poswal, M., Berberet, R. C., and Cuperus, G. W. (1988). Implications of the stochastic nature of Kuno's and Green's fixed precision stop lines: sampling plans for the pea aphid (Homoptera: Aphididae) in alfalfa as an example, *J. Econ. Entomol.*, *81*:749–758.

Imura, O. (1979). The extraction of insects of stored products from samples using a modified Tullgren funnel. *Jpn. J. Appl. Entomol. Zool.*, *23*:134–140.

Ives, P. M., and Moon, R. D. (1987). Sampling theory and protocol for insects. In Crop Loss Assessment and Pest Management (ed. Teng, P. S.). American Phytopathological Society Press, St. Paul, MN, pp. 49–75.

Iwao, S. (1968). A new regression method for analyzing the aggregation pattern of animal populations. *Res. Popul. Ecol.*, *10*:1–20.

Iwao, S. (1975). A new method of sequential sampling to classify populations relative to a critical density. *Res. Popul. Ecol.*, *16*:281–288.

Jones, V. P. (1994). Sequential estimation and classification procedures for binomial counts. In CRC Handbook of Sampling Methods for Arthropods in Agriculture (ed. Pedigo, L. P., and Buntin, G. D.). CRC Press, Boca Raton, FL, pp. 175–206.

Karandinos, M. G. (1976). Optimal sample size and comments on some published formulae. *Bull. Entomol. Soc. Am.*, 22:417–421.

Kuno, E. (1969). A new method of sequential sampling to obtain the population estimates with a fixed level of precision. *Res. Popul. Ecol.*, *11*:127–136.

Kuno, E. (1986). Evaluation of statistical precision and design of efficient sampling for the population estimation based on frequency of occurrence. *Res. Popul. Ecol.*, *28*: 305–319.

Kuno, E. (1991). Sampling and analysis of insect populations. *Annu. Rev. Entomol.*, *36*: 285–304.

Legg, D. E. (1986). An interactive computer program for calculating Bartlett's regression method. *N. Cent. Comput. Inst. Software J.*, 2:1–23.

Legg, D. E., and Moon, R. D. (1994). Bias and variability in statistical estimates. In CRC Handbook of Sampling Methods for Arthropods in Agriculture (ed. Pedigo, L. P., and Buntin, G. D.). CRC Press, Boca Raton, FL, pp. 55–69.

Leos-Martinez, J., Granovsky, T. A., Williams, H. J., Vinson, S. B., and Burkholder, W. E. (1986). Estimation of aerial density of the lesser grain borer (Coleoptera: Bostrichidae) in a warehouse using dominicalure traps. *J. Econ. Entomol.*, *79*:1134–1138.

Lippert, G. E., and Hagstrum, D. W. (1987). Detection or estimation of insect populations in bulk-stored wheat with probe traps. *J. Econ. Entomol.*, *80*:601–604.

Lloyd, M. (1967). Mean crowding. *J. Anim. Ecol.*, *36*:1–30.

McGaughey, W. H. (1978). Moth control in stored grain: efficacy of *Bacillus thuringiensis* on corn and method of evaluation using small bins. *J. Econ. Entomol.*, *71*:835–839.

Meagher, R. L., Mills, R. B., and Rubison, R. M. (1986). Comparison of pneumatic and manual probe sampling of Kansas farm-stored grain sorghum. *J. Econ. Entomol.*, *79*: 284–288.

Nachman, G. (1984). Estimates of mean population density and spatial distribution of *Tetranychus urticae* (Acarina: Tetranychidae) and *Phytoseiulus persimilis* (Acarina: Phytoseidae) based upon the proportion of empty sampling units. *J. Appl. Ecol.*, *21*: 903–913.

Nyrop, J. P. (1988). Sequential classification of prey–predator ratios with application to

European red mite (Acari: Tetranychidae) and *Typlodromus pyri* (Acari: Phytoseidae) in New York apple orchards. *J. Econ. Entomol.*, *80*:14–21.

Nyrop, J. P., and Binns, M. R. (1991). Quantitative methods for designing and analyzing sampling programs for use in pest management. In CRC Handbook of Pest Management in Agriculture (ed. Pimentel, D. R.). CRC Press, Boca Raton, FL, pp. 67–132.

Nyrop, J. P., and Wright, R. J. (1985). Use of double sample plans in insect sampling with reference to the Colorado potato beetle, *Leptinotarsa decemlineata* (Coleoptera: Chrysomelidae). *Environ. Entomol.*, *16*:644–649.

Nyrop, J. P., Agnello, A. M., Kovach, J., and Reissig, W. H. (1989). Binomial sequential classification plans for European red mite (Acari: Tetranychidae) with special reference to performance criteria. *J. Econ. Entomol.*, *82*:482–490.

Overholt, W. A., Knutson, A. E., Smith, J. W., Jr., and Gilstrap, F. E. (1990). Distribution and sampling of southwestern corn borer (Lepidoptera: Pyralidae) in preharvest corn. *J. Econ. Entomol.*, *83*:1370–1375.

Payne, J. A., and Redlinger, L. M. (1969). Insect abundance and distribution within peanut shelling plants. *J. Am. Peanut Res. Educ. Assoc.*, *1*:83–89.

Pearl, R. (1937). On biological principles affecting populations: human and other. *Am. Nat.*, *71*:50–68.

Pedersen, J. R. (1992). Insects: identification, damage and detection. In Storage of Cereal Grains and Their Products (ed. Sauer, D. B.). American Association of Cereal Chemists, St. Paul, MN, pp. 435–489.

Pedigo, L. P. (1994). Introduction to sampling arthropod populations. In CRC Handbook of Sampling Methods for Arthropods in Agriculture (ed. Pedigo, L. P., and Buntin, G. D.). CRC Press, Boca Raton, FL, pp. 1–11.

Pedigo, L. P., and Buntin, G. D. (eds.) (1994). CRC Handbook of Sampling Methods for Arthropods in Agriculture. CRC Press, Boca Raton, FL.

Pinniger, D. B. (1990). Food-baited traps: past, present and future. *J. Kansas Entomol. Soc.*, *63*:533–538.

Prevett, P. F. (1964). The distribution of insects in stacks of bagged groundnuts in northern Nigeria. *Bull. Entomol. Res.*, *54*:689–713.

Pruess, K. P., and Saxena, K. M. (1977). Estimation of insect populations by removal sampling. *Univ. Nebr. Dept. Entomol. Rept.*, *4*:1–30.

Ralston, M. L., and Jennrich, R. I. (1978). DUD, a derivative-free algorithm for nonlinear least squares. *Technometrics*, *20*:7–14.

Reed, C. (1987). The precision and accuracy of the standard volume weight method of estimating dry weight losses in wheat, grain sorghum and maize, and a comparison with the thousand grain mass method in wheat containing fine material. *J. Stored Prod. Res.*, *23*:223–231.

Reed, C., Wright, V. F., Mize, T. W., Pedersen, J., and Evans, J. B. (1991). Pitfall traps and grain samples as indicators of insects in farm-stored wheat. *J. Econ. Entomol.*, *84*: 1381–1387.

Rees, D. P. (1985). Review of the response of stored product insects to light of various wavelengths, with particular reference to the design and use of light traps for population monitoring. *Trop. Sci.*, *25*:197–213.

Ruesink, W. G., and Kogan, M. (1982). The quantitative basis of pest management:

sampling and measuring. In Introduction to Insect Pest Management (ed. Metcalf, R. L., and Luckmann, W. H.). Wiley, New York, pp. 315–352.

Russell, G. E. (1988). Evaluation of four analytical methods to detect weevils in wheat: granary weevil, *Sitophilus granarius* (L.), in soft white wheat. *J. Food Prot.*, *51*: 547–553.

SAS Institute. (1988). SAS/STAT User's Guide, release 6.03 ed. SAS Institute, Cary, NC.

Sawyer, A. J. (1989). Inconstancy of Taylor's b: simulated sampling with different quadrat sizes and spatial distribution, *Res. Popul. Ecol.*, *31*:11–24.

Schatzki, T. F., and Fine, T. A. (1988). Analysis of radiograms of wheat kernels for quality control. *Cereal Chem.*, *65*:233–239.

Seber, G. A. F. (1982). The Estimation of Animal Abundance and Related Parameters. Macmillan, New York.

Sinclair, E. R., and Haddrell, R. L. (1985). Flight of stored products beetles over a grain farming area in southern Queensland. *J. Aust. Entomol. Soc.*, *24*:9–15.

Sinclair, E. R., and White, G. G. (1980). Stored products insect pests in combine harvesters on the Darling Downs. *Queensland. J. Agric. Anim. Sci.*, *37*:93–99.

Smith, L. B. (1977). Efficiency of Berlese-Tullgren funnels for removal of the rusty grain beetle, *Cryptolestes ferrugineus*, from wheat samples, *Can. Entomol.*, *109*:503–509.

Smith, L. B. (1985). Insect infestation in grain loaded in railroad cars at primary elevators in southern Manitoba, Canada. *J. Econ. Entomol.*, *78*:831–834.

Snedecor, G. W., and Cochran, W. G. (1980). Statistical Methods, 7th ed. Iowa State University Press, Ames, IA.

Southwood, T. R. E. (1978). Ecological Methods. Chapman and Hall, London.

Subramanyam, Bh. (1992). Relative retention of adult sawtoothed grain beetles in potential food baits. *Postharvest Biol. Technol.*, 2:73–82.

Subramanyam, Bh., and Harein, P. K. (1989). Insects infesting barley stored on farms in Minnesota. *J. Econ. Entomol.*, *82*:1817–1824.

Subramanyam, Bh., and Harein, P. K. (1990). Accuracies and sample sizes associated with estimating densities of adult beetles (Coleoptera) caught in probe traps in stored barley. *J. Econ. Entomol.*, *83*:1102–1109.

Subramanyam, Bh., Harein, P. K., and Cutkomp, L. K. (1989). Field tests with probe traps for sampling adult insects infesting farm-stored grain. *J. Agric. Entomol.*, *6*:9–21.

Subramanyam, Bh., Hagstrum, D. W., and Schenk, T. C. (1993). Sampling adult beetles (Coleoptera) associated with stored grain: comparing detection and mean trap catch efficiency of two types of probe traps. *Environ. Entomol.*, *22*:33–42.

Subramanyam, Bh., Wright, V. F., and Fleming E. E. (1992). Laboratory evaluation of food baits for their relative ability to retain three species of stored-product beetles (Coleoptera). *J. Agric. Entomol.*, *9*:117–127.

Taylor, L. R. (1961). Aggregation, variance and the mean. *Nature* (London), *189*:732–735.

Taylor, L. R. (1984). Assessing and interpreting spatial distribution of insect populations. *Annu. Rev. Entomol.*, *29*:321–357.

Taylor, T. A., and Agbaje, L. A. (1974). Flight activity in normal and active forms of *Callosobruchus maculatus* (F.) in a store in Nigeria. *J. Stored Prod. Res.*, *10*:9–16.

Vick, K. W., Koehler, P. G., and Neal, J. J. (1986). Incidence of stored-product phycitinae moths in food distribution warehouses as determined by sex pheromone-baited traps. *J. Econ. Entomol.*, *79*:936–939.

Vick, K. W., Mankin, R. W., Cogburn, R. R., Mullen, M., Throne, J. E., Wright, V. F., and Cline, L. D. (1990). Review of pheromone-baited sticky traps for detection of stored-product insects. *J. Kansas Entomol. Soc.*, *63*:526–532.

Walker, D. W. (1960). Population fluctuations and control of stored grain insects. *Wash. Agric. Exp. Stn. Tech. Bull.*, *31*.

Ward, S. A., Chambers, R. J., Sunderland, K., and Dixon, A. F. G. (1986). Cereal aphid populations and the relation between mean density and spatial variance. *Neth. J. Pl. Path.*, *92*:127–132.

Watters, W. E. (1955). Sequential sampling in forest insect surveys. *For. Sci.*, *1*:68–79.

White, G. G. (1983). A modified inclined sieve for separation of insects from wheat. *J. Stored Prod. Res.*, *19*:89–91.

White, G. G. (1985). Population dynamics of *Tribolium castaneum* (Herbst) with implications for control strategies in stored wheat. PhD thesis, University of Queensland.

White, G. G. (1988). Field estimates of population growth rates of *Tribolium castaneum* (Herbst) and *Rhyzopertha dominica* (F.) (Coleoptera: Tenebrionidae and Bostrychidae) in bulk wheat. *J. Stored Prod. Res.*, *24*:13–22.

White, N. D. G., Arbogast, R. T., Fields, P. G., Hillmann, R. C., Loschiavo, S. R., Subramanyam, Bh., Throne, J. E., and Wright, V. F. (1990). The development and use of pitfall and probe traps for capturing insects in stored grain. *J. Kansas Entomol. Soc.*, *63*:506–525.

Wileyto, E. P., Ewens, W. J., and Mullen, M. A. (1994). Markov-recapture population estimates: a tool for improving interpretation of trapping experiments. *Ecology*, *75*: 1109–1117.

Wilson, L. T., and Room, P. M. (1983). Clumping patterns of fruit and arthropods in cotton, with implications for binomial sampling. *Environ. Entomol.*, *12*:50–54.

Wilson, L. T., Pickel, C., Mount, R. C., and Zalom, F. G. (1983). Presence–absence sequential sampling for cabbage aphid (Homoptera: Aphididae) on Brussels sprouts. *J. Econ. Entomol.*, *76*:476–479.

Wright, V. F., and Mills, R. B. (1984). Estimation of stored-product insect populations in small bins using two sampling techniques. Proceedings of the Third International Working Conference on Stored-Product Entomology, 23–28 October 1983, Kansas State University, Manhattan, KS, pp. 672–679.

Zar, J. H. 1984. Biostatistical Analysis, 2nd ed. Prentice Hall, Englewood Cliffs, NJ.

# 5

# Physical Control

**Paul G. Fields**
*Agriculture and Agri-Food Canada, Winnipeg, Manitoba, Canada*

**William E. Muir**
*The University of Manitoba, Winnipeg, Manitoba, Canada*

Physical control is the manipulation of the physical environment so that insect populations do not increase or are reduced and eliminated. The physical attributes refer to temperature, relative humidity, moisture content, structures containing the commodity (silos, elevators, bags, packaging), forces on commodity (compression, impaction), and irradiation. Gases are also a component of the physical environment, but they will be considered in the chapter on chemical control under fumigants.

Stored-product insects have been controlled using physical means for thousands of years. The basic rule of good seed storage—keep seeds cool and dry—was employed in neolithic times in the Nile delta by placing seeds in clay jars in the ground (Levinson and Levinson 1989). High temperature was used to control the moth *Sitotroga cerealella* (Oliver) in 16th century France (Oosthuizen 1935). In the last 50 years the organochlorines, organophosphates, and fumigants have been the method of choice for insect control for most stored-product managers. Recently there has been a movement away from chemical control methods for two basic reasons. The first is that insecticides are toxic to species other than those they are intended to control. Predators and parasites that keep the pest species in check are also killed by insecticide treatments. Other animals such as fish, birds, and mammals can be adversely effected by insecticide treatment. In 1984, liquid fumigants were deregistered in Canada and the United States because of concerns over toxic residues, and concerns for worker safety have placed more restrictions on the procedures for pesticide application. Also, the use of methyl bromide will be restricted or discontinued because of the damage it causes to stratospheric ozone. The second reason is that the continued use of a single insecticide leads to resistance to this insecticide within the pest population. An example of this is the ineffectiveness of malathion in the United States (Subramanyan and Harein 1990) and Australia (Collins and Wilson 1986).

There have been several reviews of physical control of stored-product insects (Banks 1976, 1986; Evans 1987; Fleurat-Lessard 1987; Watters 1991; Fields 1992; Banks and Fields 1994). The purpose of this review is to highlight recent advances and suggest future developments.

## I. LOW TEMPERATURE

Lowering the temperature of commodities to prevent spoilage is a technique as common as the everyday refrigerator. It is also used extensively in the fall and winter with bulk storages in temperate climates in conjunction with ambient air aeration. Refrigerated air aeration is used in tropical climates or in the summer in temperate areas. Details will be given in the section on temperature control below. There are two basic effects of low temperature: reducing the development rate, feeding, and fecundity; and decreasing survival.

The optimal temperature for fecundity and development of stored-product insects is between 25 and 33°C (Birch 1945; Howe 1965; Lhaloui et al. 1988) (Table 1). At low temperatures fecundity is reduced and insects develop more slowly. This lengthens the time before populations increase to a point where they cause significant damage (Longstaff and Evans 1983; Flinn and Hagstrum 1990). Temperatures between 13 and 25°C will slow development. Howe (1965) reviewed the temperatures that slow insects' development. It should be noted that the temperatures listed are not the developmental thresholds, but temperatures that

**Table 1** Response of Stored-Product Insects to Temperature[a]

| Zone | Temperature (°C) | Effect |
|---|---|---|
| Lethal | 50–60 | Death in minutes |
| | 45–50 | Death in hours |
| Suboptimal | 35 | Development stops |
| | 33–35 | Develoment slows |
| Optimal | 25–33 | Maximum rate of development |
| Suboptimal | 13–25 | Development slows |
| | 13–20 | Development stops |
| Lethal | 5 | Death in days (unacclimated), movement stops |
| | −10–5 | Death in weeks to months (acclimated) |
| | −25–−15 | Death in minutes, insects freeze |

[a]Species, stage of development, and moisture content of food will influence the response to temperature.

would allow only two generations per year. For most stored-product insects 20°C will stop development; the major exception is *Sitophilus granarius* that can develop at 15°C. Mites in damp grain only stop developing at 2°C. Although no development occurs at these temperatures, the insects and mites remain alive for long periods and will cause damage if the temperature of the commodity rises.

To disinfest commodities within reasonable durations, which may be 2 months for bulk grain in long-term storage or 24 h for chocolate bars that are to be shipped in a few days, cooler temperatures are necessary. Many factors, such as temperature, species, stage, acclimation, and relative humidity, determine the length of time needed to kill all individuals (Fields 1992).

The relationship between duration of exposure (y-axis) and temperature (x-axis) for a given mortality level is usually a concave (j-shaped) curve (Evans 1987) (Fig. 1). Several models have been developed to describe cold-related mortality (Kawamoto et al. 1989; Turnock et al. 1983; Brokerhof et al. 1992), although not all species are described by these models (Strail and Reichmuth 1984).

Some species are more cold-hardy than others. In general *Tribolium castaneum* (Herbst), *Tribolium confusum* Jacquelin du Val, and *Oryzaephilus merca-*

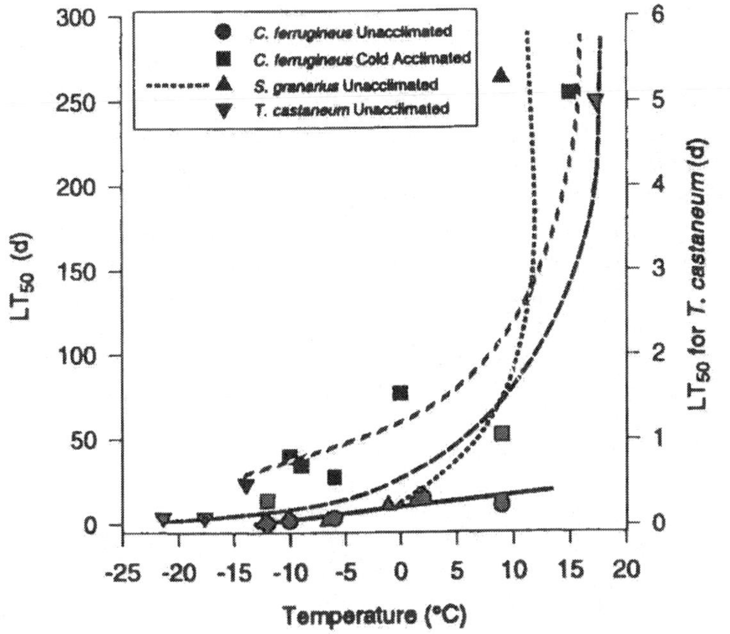


**Figure 1** The lethal time for 50% of the population for some stored-product insects at temperatures below their developmental thresholds. (From Fields 1992.)

*tor* (Fauvel) are the most cold-susceptible, whereas *Trogoderma granarium* Everts, *Sitophilus granarius* (L.), *Ephestia elutella* (Hübner), *Ephestia kuehniella* (Zeller), and *Plodia interpunctella* (Hübner) are the most cold-tolerant species (Table 2). Actual durations to control species at a given temperature have been summarized (Fields 1992).

The stage of development—egg, larva, pupa, or adult—will affect an insect's ability to withstand low temperatures. For many species the egg is the most susceptible stage (Fields 1992). For a few species there has been detailed work showing that the age of the egg can affect susceptibility (Watters 1966; Cline 1970; Daumal et al. 1974; Johnson and Wofford 1991). Larvae are the most cold-tolerant stage for *Rhyzopertha dominica* (F.), and *Sitophilus oryzae* (L.) (David et al. 1977). This is true for the diapausing larvae of *E. kuehniella* (Cox 1987) and may hold true for the other Pyralidae: *P. interpunctella*, *Cadra cautella* (Walker), and *E. elutella* (Fields 1992). The adult is the most cold-hardy stage for *Cryptolestes ferrugineus* (Stephens) (Smith 1970) and *T. confusum* (Nagel and Sheppard 1934). For *S. granarius*, the larvae are as cold-hardy as the adults (Howe and Hole 1968; David et al. 1977).

In most field applications when one is using cold temperature to control stored-product pests, insects are exposed to temperatures between 20 and 10°C before being exposed to lethal cold temperatures. The exception to this rule is when commodities are rapidly cooled in freezers to control insect populations. When insects are exposed to cool temperatures (20–10°C) they become acclimated and their cold-hardiness increases by 2–10-fold (Smith 1970; Evans 1983; Fields 1990, 1992). There are only two studies, both with *S. granarius*, demonstrating that insects do not increase in cold-hardiness if exposed to cool temperatures before being exposed to cold temperatures (Robinson 1926; Evans 1983). The implications for low-temperature control are that laboratory studies often underestimate the cold-hardiness of insects because of inadequate acclimation before cold treatment. Also rapid cooling of commodities from 25°C should prevent cold-acclimation and may be a useful tool to allow the rapid control of insects at higher temperatures than insects that have been cold-acclimated. The risk of this approach is that insects can cause damage at 25°C, but most insects stop feeding at 20°C.

## II. HIGH TEMPERATURE

Temperatures for the maximum rate of multiplication are only 5°C lower than the temperatures that stop development. *S. oryzae* has a maximum rate of development at 29°C and stops developing at 35°C, and the temperatures for *R. dominica* are 34°C and 39°C (Birch 1945). Other nonlethal effects are reduced fecundity and fertility (Kirkpatrick and Tilton 1973; Vardel and Tilton 1980; Arbogast 1981).

High-temperature mortality is dependent on temperature, duration, species, stage of development, acclimation, and relative humidity. Unlike low-temperature studies, high-temperature studies use temperatures that change rapidly, do not remain constant for the majority of the test, and most exposures last only a few minutes. The temperature regimens are largely determined by the method used to heat the commodity.

Kirkpatrick and Tilton (1972) tested the heat-tolerance of insects at 49°C. The most tolerant to the least tolerant were *Lasioderma serricorne* (F.) ≥ *Cryptolestes pusillus* (Schönherr) = *R. dominica* > *S. oryzae* = *T. castaneum* = *Trogoderma variabile* (Ballion) > *S. granarius* = *Gibbium psylloides* (Czenpinski) > *Cathartus quadricollis* (Guérin-Méneville) = *O. mercator* > *T. confusum* = *Oryzaephilus surinamensis* (L.). Evans and Dermott (1981) used *R. dominica* and *S. oryzae* to test their fluidized bed because these species are the most heat-tolerant pests in Australia. There has been little extensive work on the susceptibility of different developmental stages. Oosthuizen (1935) ranked *T. confusum* pupae > eggs > larvae > adults tolerant of temperatures at 44°C, but found no differences at 50°C. Larvae of boring insects are more difficult to control with heat treatment because the seed buffers the larvae and pupae from short periods of extreme heat (Dermott and Evans 1978). Acclimation to high temperatures can occur, as shown by the threefold increase in heat tolerance by acclimating *S. granarius* (Gonen 1977). However, Evans (1981) used *S. oryzae* and *R. dominica* to show that prior exposure to moderately high temperatures had no effect on survival at 55°C.

The lower the relative humidity or moisture content of the grain the more susceptible insects are to heat treatments (Mellanby 1932; Kirkpatrick and Tilton 1973; Evans 1981; Vardel and Tilton 1981). These effects are greatest between 40 and 45°C (Fields 1992). The differences at the higher temperatures reached in fluidized beds are due to moist grain not heating as much as dry grain does (Evans 1981).

## III. TEMPERATURE CONTROL

### 1. Unventilated

Temperatures of stored grain are affected by several factors that can be controlled or modified to improve the physical control of stored-grain insects. The temperature of the grain as it is moved into storage can have a longlasting effect on temperatures in unventilated storages. The initial storage temperature affects the temperature of the grain at the center of the bin for up to 5 months in a 6 m diameter bin and up to 20 months in a 10 m diameter bin (Yaciuk 1973). High initial storage temperatures occur when the grain is harvested on hot, sunny days (e.g., 5–8°C above the ambient air temperature) (Prasad et al. 1978; Williamson 1964); or when grain is not cooled sufficiently after drying with heated air.

Grain temperatures in unventilated storages follow the outside ambient-air temperatures. Because bulks of stored grains or oilseeds have low thermal diffusivities (even lower than fiberglass) the temperatures of the stored grain change

**Table 2** Relative Cold-Hardiness of Some Stored-Product Insects

| Reference | Site | Stage[b] | Most susceptible[c] | | | Most tolerant |
|---|---|---|---|---|---|---|
| Mansbridge[a] 1936 | Field | l or a | *T. castaneum*<br>*T. confusum* | *C. turcicus*<br>*C. cautella*<br>*L. serricorne*<br>*R. dominica*<br>*S. oryzae* | *O. surinamensis*<br>*S. granarius* | *E. elutella*<br>*E. kuehniella*<br>*P. interpunctella*<br>*T. granarium*<br>*T. molitor* |
| Solomon and Adamson[a] 1955 | Field | e, l, p, a | *O. mercator* | *C. cautella*<br>*S. oryzae*<br>*S. granarius*<br>*T. castaneum*<br>*T. confusum* | *S. cerealella* | *C. ferrugineus*<br>*C. turcicus*<br>*E. elutella*<br>*E. kuehniella*<br>*O. surinamensis*<br>*P. interpunctella*<br>*T. granarium*<br>*T. molitor* |
| Mathlein 1961 | Field and lab | l or a | *C. cautella*<br>*O. surinamensis*<br>*R. dominica*<br>*S. oryzae* | *C. ferrugineus*<br>*E. elutella*<br>*E. kuehniella*<br>*T. granarium*<br>*S. granarius* | | |

| | | | | | | |
|---|---|---|---|---|---|---|
| David et al. 1977 | Lab | e, l, p, a | *R. dominica* | *S. oryzae* | *S. granarius* | |
| Bahr 1978 | Field | l,a | *C. pusillus*<br>*O. mercator*<br>*R. dominica*<br>*T. castaneum*<br>*T. confusum* | *S. cerealella*<br>*S. oryzae*<br>*S. zeamais* | *C. turcicus*<br>*C. ferrugineus*<br>*O. surinamensis*<br>*S. granarius* | |
| Mullen and Arbogast 1979 | Lab | e | *C. cautella*<br>*O. surinamensis*<br>*T. castaneum* | *C. maculatus* | *L. serricorne* | |
| Evans 1983 | Lab | a | *T. castaneum* | *O. surinamensis*<br>*C. ferrugineus*<br>*S. oryzae*<br>*R. dominica* | *S. granarius* | |
| Evans 1987 | Lab | l,p | *O. surinamensis*<br>*T. castaneum* | *S. oryzae* | *S. granarius* | *R. dominica* |
| Wohlgemuth 1989 | Field | l,a | *T. confusum* | *C. turcicus*<br>*O. surinamensis* | *T. granarium* | *S. granarius* |

[a]Only a partial list of species studied.

[b]e, egg; l, larva; p, pupa; a, adult.

[c]Within a given study, species in the same column had similar cold-hardiness.

slowly. The temperature at the center of a bulk can lag behind the ambient air temperature by 4 months in a 4 m diameter bin and 7 months in an 8 m diameter bin (Yaciuk 1973). Storing grain in small-diameter bins, rather than in larger, more economical bins, can improve the physical control of insects during fall and winter in a temperate climate by reducing the center grain temperatures and decreasing the time during which the grain at the center is above the low-temperature thresholds of the insects (Yaciuk et al. 1975). Shading the bins from solar radiation and painting or constructing the outside surfaces of the bins with materials that have high thermal radiation emissivities at ambient temperatures can reduce storage temperatures, particularly in tropical climates (Ferreira and Muir 1981).

### 2. Ventilated

Aeration is the forced movement of air at near-ambient conditions through stored grain to bring the grain to a uniform temperature near the temperature of the ambient air. Aeration normally has little effect on the moisture content of the stored grain because the amount of air required to change the grain temperature is much less than that needed to change the grain moisture content. The temperature of a grain bulk can be brought to the air temperature in about 10 days at an airflow rate of 1 (L/s)/m$^3$ of grain, while to dry wheat from 19% to 14% may require over 20 days at 12 (L/s)/m$^3$ (Sanderson et al. 1989).

During aeration, a front, in which the grain comes into thermal equilibrium with the ventilating air, initially develops in the grain where the forced air enters the grain bulk (Sanderson et al. 1988a; Sutherland et al. 1983). The grain downstream from this temperature front stays near its initial temperature until the leading edge of the front passes through (Fig. 2).

Under most climatic conditions, ambient air temperatures are lowest at night and coincide with the highest relative humidities. Aerating with this cool, moist air can cause an increase in moisture content of the grain near the entrance of the air. The amount of rewetting increases with airflow rate, relative humidity, and temperature of the air (Sokhansanj et al. 1983). Rewetting tends to occur slowly and the wetting front spreads out and disappears with continued aeration with drier air (Sanderson et al. 1988b). Although this rewetting is a concern, control of an insect infestation may be a much greater concern. The risk of rewetting does indicate, however, that aeration should be strictly limited to the time duration needed to cool the grain.

Ambient air cooling is used mainly for control of insects in bulk grain storage. However, it has been also used in controlling insects in flour mills as an alternative to fumigation. The facilities to be frozen must be carefully prepared before the freeze-out by removing temperature-sensitive equipment, draining water from plumbing and equipment, and reducing product stocks to reduce thermal mass and eliminate refuges for insects. Outside air temperatures should be below −17°C for 3 days for effective control (Worden 1987).

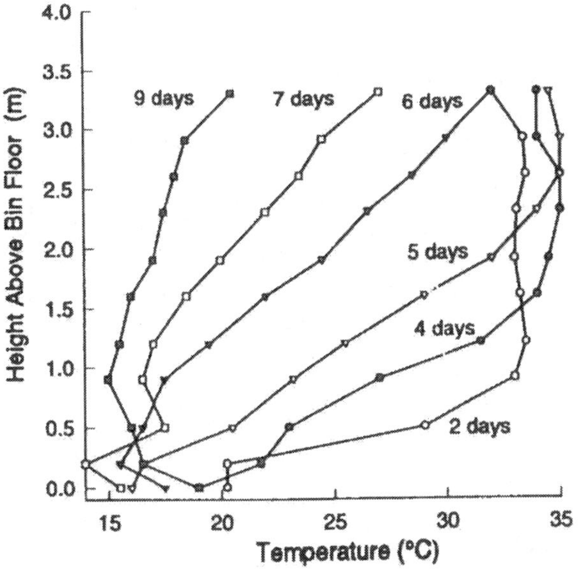


**Figure 2** Grain temperatures during the initial cooling of a bin of wheat aerated continuously with upward airflow of 0.8 (L/s)/$m^3$ of grain at various times after aeration was begun. (Redrawn from Sanderson 1986.)

## 3. Airflow Rate

The rate of airflow through the grain bulk along any given streamline determines the rate of cooling. The fan moves the air through the grain against the airflow resistance provided by the air ducts, perforated metal flooring, grain, and exhaust ducts. These resistances increase with increased air velocity (or flow rate) and, therefore, the airflow resistance can be reduced by decreasing grain depth or increasing cross-sectional areas of the ducts, perforated metal, and grain bulk. The main resistance is provided by the grain and is affected by the type of grain (e.g., for the same air velocity the resistance to airflow of wheat can be three times that of maize) (ASAE 1990). Filling a bin with a spreader can result in airflow resistances 1.5 times greater than spout- or stream-filled (Friesen and Huminicki 1986). Moving the air horizontally through barley, for example, can reduce airflow resistance by one-half compared with vertical air flow. Changing the shape of the grain bulk also has a great effect because as the cross-sectional area decreases, both the air velocity and the length of the flow path through the grain increase. For the same grain volume and airflow rate per unit volume of grain, doubling the grain depth can increase the airflow resistance and power input to the fan by five times.

Aeration systems are frequently designed and operated so that the grain is not ventilated and cooled uniformly (Burrell et al. 1982). To obtain uniform cooling, and thus insect control, the floor should be completely perforated, and the bin should be filled using a well-designed spreader.

## 4. Mathematical Models

Mathematical models have been used to study and predict the rate and uniformity of cooling. They are also useful to compare fan management policies (Metzger and Muir 1983; Smith et al. 1992; Wilson 1990).

A simplified method (Eq. 1) for calculating the approximate time to cool a grain bulk by uniform aeration was presented by Navarro and Calderon (1982):

$$\theta = (M \times \Delta T \times c)/(Q \times \Delta H \times E) \quad (1)$$

where:
$\theta$ = cooling time (h)
$M$ = mass of grain to be cooled (kg)
$\Delta T$ = difference between initial and final grain temperatures (°C)
$c$ = specific heat of grain, kJ/(kg·°C)
$Q$ = mass flow rate of air (kg/h)
$\Delta H$ = change in enthalpy of air between entering and leaving the grain bulk (kJ/kg)
$E$ = correction factor (dimensionless)

The correction factor is required because the change in enthalpy of the ventilating air, $\Delta H$, decreases as the cooling front passes through the grain. Navarro and Calderon (1982) recommended a correction factor of 0.5 while Sanderson et al. (1988a) recommended 0.4. The accuracy of the predicted cooling time is also affected by changes in ambient air conditions and uniformity of airflow through the grain bulk.

A simpler but less accurate prediction of cooling time can be based on the assumptions that at an airflow rate of 1 (L/s)/$m^3$ requires a cooling time of about 10 days and that the rate of cooling is directly proportional to airflow rate. More accurate predictions can be obtained by using computer simulation techniques (Sanderson et al. 1989; Wilson 1990).

## 5. Fan Management

The simplest fan management policy is to operate aeration fans whenever the average ambient air temperatures are more than 5–10°C below the maximum grain temperature. The fans are run continuously until the cooling front has passed through the grain bulk. Depending on the relative costs of grain spoilage, automatic or manual control, fan-energy costs, and local weather conditions, more complex fan management policies can be instituted.

Energy costs and lower grain temperatures can be obtained by operating the fans only at night (Metzger and Muir 1983). This control can be carried out manually or automatically by an inexpensive and easily checked time-clock.

Automatic control based on the temperature differential between the grain and the outside air is the most rational system because the objective is to cool the grain to the ambient air temperature. Such a control is more complex and requires the installation of at least one temperature sensor in the grain bulk near the air exit and one in the ambient air (Metzger and Muir 1983; Armitage 1986).

When grain is stored at temperatures above the optimum range for the insect pests, it may be advantageous to delay aeration until late summer when ambient air temperatures have dropped below the optimum range for the insects. If such a policy is used to control insects, the grain must be dry and its condition must be monitored regularly to reduce the risk of rapid deterioration at these high temperatures.

## 6. Refrigerated Aeration

The physical control of insects by aeration in grain stored in hot climates frequently requires refrigerated air. Aeration with chilled air provides satisfactory insect control in Texas (Maier and Bakker-Arkema 1992; Maier et al. 1992), Australia (Sutherland 1986), and Israel (Navarro and Calderon 1982). In all locations the grain was cooled to 15°C, with energy costs similar to those for insecticide treatments. To keep energy costs down and to maintain low temperatures in grain near the bin periphery, the silos must be insulated. Reheating the chilled air to prevent moisture adsorption from the high-humidity air leaving the refrigerator coils is beneficial (Maier et al. 1992).

## 7. High Temperature

High-temperature disinfestation has been studied and commercially developed in Australia, where ambient air temperatures are not low enough to control insects by aeration (Thorpe 1983). Most heated-air grain-driers cannot heat the grain uniformly and efficiently to the desired temperature. Fluidized beds (Evans 1981; Thorpe 1983), spouted beds (Claflin et al. 1986), pneumatic conveyors (Sutherland et al. 1989), a counterflow heat exchanger (Lapp et al. 1986), high-frequency waves (Nelson and Kantack 1966), microwaves (Locatelli and Traversa 1989), infrared waves (Kirkpatrick and Tilton 1972), and solar radiation (McFarlane 1989; Murdock and Shade 1991; Kitch et al. 1992) have been used satisfactorily to disinfest seeds. Maximum kernel temperatures and kernel residence times must be carefully controlled to obtain disinfestation without thermal damage to the kernels (Evans et al. 1983).

Heat sterilization or superheating has been used occasionally to disinfest flour mills and food plants since the 1990s. The area to be disinfested is closed off, and heaters raise the temperature between 50 and 60°C for at least 24 h. Using fans and running the machinery empty aid in the distribution of the heat. Heat-sensitive equipment must be removed, insulated, or air conditioned. Heat-sensitive sprinkler heads must be able to withstand the heat without activating. After the treatment, bearings should be checked for lubrication loss, slack belts tightened, and other materials such as plastics and rubber checked for damage (Imholte 1984; Sheppard 1984). There is also a concern that heating can cause damage to concrete buildings and roofing materials.

## IV. IRRADIATION

Since there are several recent reviews of irradiation of foods (Urbain 1986; Anderson 1989; Watters 1991) the focus here will be on a simple overview of the topic. Irradiation of 40 different foods is permitted in over 30 countries. In the 20 countries that have commercial operations, most are used to prevent sprouting of potatoes and onions or the microbial infestations of spices and meats (Anderson 1989). The first operation to disinfest grain at a commercial facility was at the port of Odessa, Ukraine, where effective control was obtained at 0.2 kGy, using 40 kW at a flow rate of 200 t/h (Zakladnoi et al. 1982).

There are two types of ionizing radiation: gamma rays and electron beam irradiation. Nonionizing irradiation refers to electromagnetic radiation (radio waves, infrared waves, visible light, and microwaves) that does not contain enough energy to dislodge electrons from molecules. Ionizing radiation damages organisms by causing the production of ions or free radicals: charged molecules that are highly reactive. In addition to ionization, chemical bonds can also be broken. Gamma radiation, with cobalt 60 as the radioactive source, is the most common commercial method of food irradiation. It can penetrate 20–50 cm into solid foods, requires shielding to protect workers, and must be replaced periodically since it has a half life of 5.3 years. Electron beam irradiation is used to treat grain in Odessa, Ukraine, using an accelerator powered by electricity to accelerate electrons to speeds high enough to cause ionization. Electrons penetrate only 1 mm into solid food, and grain treated must be only 1 kernel deep at the point of application (Zakladnoi et al. 1982).

Most of the major stored product insects have been tested to determine their sensitivity to irradiation. To disinfest grain or flour, doses of 0.2–1.0 kGy are required (Golumbic and Davis 1966; Watters and MacQueen 1967; Watters 1968; Tilton et al. 1974). Unlike after chemical treatments, irradiated insects can survive for several weeks, although they may feed less and may be infertile.

Irradiation at 10 kGy does not produce toxins in treated foods and the World Health Organization recommends that foods treated at this or lower doses do not require testing to prove that irradiation does not produce toxins (Urbain 1986). Irradiation can reduce vitamins A, C, E, B1 (thiamin), and K. The amount of reduction is dependent on the food irradiated, dose, and other factors (Urbain 1986). The doses of irradiation needed to kill insects also kill the seed, making this type of control unsuitable for malting barley (Watters and MacQueen 1967) or seed stocks. Bread quality is affected only after wheat is irradiated at doses above 2.5 kGy (Lee 1959), but no effects could be seen at the lower doses needed to control stored-product insects (Watters and MacQueen 1967; Rao et al. 1975; Urbain 1986).

## V. MECHANICAL ACTION

There are two basic approaches to control stored-product insect populations using mechanical means: indirect (manipulation of environment) or direct (manipulation of insect). Additional details are covered in the chapters of this volume on ecology and on handling and processing. One indirect method is the reduction of dockage (broken seeds, dust, and weed seeds) and seeds with cracks in the endosperm. Many insects, such as *C. ferrugineus*, *Oryzaephilus* spp., and *T. castaneum*, are classified as secondary pests because they require a break in the seed coat to infest cereal grains. The presence of dockage increases the populations of these insects (Sinha 1975). Also these species have higher populations when the seed coat is damaged (Tuff and Telford 1964; Li and Arbogast 1991). Mechanized harvesting can cause 50% of wheat kernels to be cracked or broken (Tuff and Telford 1964; Kline 1973; Bourgeois 1993). Movement of grain and effectiveness of cleaners will also affect the amount of dockage present and the integrity of the seed coat. Maize is especially sensitive to breaking during movement (Baker et al. 1986).

Another indirect method is simply good sanitation (Foulk 1990a,b; Chowaniec 1986) such as removing residual food sources: grain in granaries, spilt grain in handling facilities, flour in flour mills (Mills and Pederson 1990). Handling and storage equipment can be designed so that it is easier to keep clean and does not leave food residues.

A direct method of controlling insect infestations is by removing insects from the infected commodity. Equipment used to remove dockage should also remove insects that are outside of the seed. This would not work for the immature stages of *Sitophilus* spp. or *R. dominica*. Flour mills often pass flour through fine sieves that remove foreign matter, including insects, just before it is packaged.

By far the most extensively used direct means of mechanical control are the Entoleters (Fig. 3), which use centrifugal force to impact insects or seeds containing insects (Stratil and Wohlgemuth 1989; Watters 1991; Wanzenreid 1992). Entoleters are used primarily in flour mills, where they are placed in the production line before the wheat kernels are milled. Kernels infested with primary feeders such as *Sitophilus* spp. or *R. dominica* break apart and are separated from the intact kernels. The speed at which the rotors turn and hence the velocity of impact can be

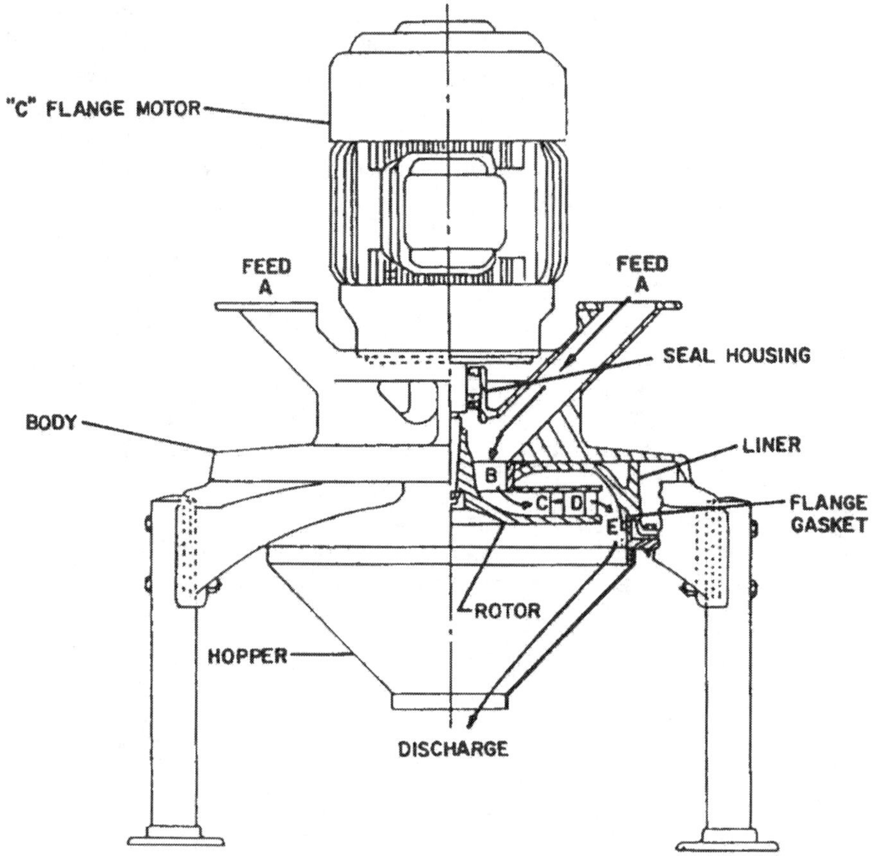


**Figure 3** An impact device used to kill insects. (Courtesy of Entoleter Inc., Hamden, CT.)

adjusted so that insects are killed, but sound kernels are not broken (Watters 1991). Entoleters are also used for the finished product, flour, to kill any insects (Stratil and Wohlgemuth 1989; Wanzenreid 1992). However not all insects run through an entoleter die immediately (Stratil and Wohlgemuth 1989).

Simply turning the grain can control many stored-product insects (Banks 1986). Several studies have investigated the effect of disturbing or impacting infested grain or insects (Joffe and Clarke 1963; Bailey 1962, 1969; Loschiavo 1978; Ungsunantwiwat and Mills 1979). However, there has been only one extensive field study (Joffe 1963), which showed that moving maize every 2 weeks reduced *Sitophilus* spp by 87%, *Tribolium* spp by 75%, and *Cryptolestes* spp by 89%.

Modern field studies investigating the effect of shock on insect pests directly are restricted to the work of Bahr (summarized in Bahr 1990). He showed that some species were particularly susceptible to damage during pneumatic conveying of grain. Thus, conveying of severely infested rye at 38 t/h produced 99.8% and 96% mortality of all stages of *O. surinamensis* and *C. ferrugineus*, respectively. Conveying of wheat through an industrial vacuum conveying system gave high mortality (>90%) of several species of stored grain pests (Bahr 1990). It was particularly effective against *T. granarium* and *O. surinamensis* where 99.6, and 100% kill was obtained after four passes through the system.

Turning the grain is thought to kill insects by crushing them; however, there are other reasons why moving grain can cause mortality. Neonate *Acanthoscelides obtectus*, the common bean weevil, need over 24 h to enter a kidney bean. To do this they need to wedge themselves against neighboring beans to bore into the seed. Rolling beans every 8 h reduces *A. obtectus* populations by 97% because neonate larvae are never able to finish an entry hole (Quentin et al. 1991).

## VI. PACKAGING

There has been extensive work on the capacity of packaging to prevent insects from infesting food products, and as there are several reviews (Newton 1988; Highland 1991) of this topic our discussion here will be brief. Packaging provides a physical barrier that prevents or impedes the infestation by insects. This is important since companies are held responsible for insects in their products, even if the infestation occurs in stores outside the company's control. Three factors determine if a commodity will be infested: insect species, packaging material, and commodity packaged.

Insects vary in their capacity to penetrate packaging: *L. serricorne*, *Stegobium paniceum* (Linnaeus), *P. interpunctella*, *C. cautella*, *Corcyca cephalonica* (Stainton), and *Trogoderma variable* can all penetrate intact packages. *Rhyzopertha dominica* can penetrate packages but are rarely found in packaged foods. The other stored-product insects (*T. castaneum*, *T. confusum*, *C. ferrugineus*, *C. pusillus*, *O. mercator*, and *O. surinamensis*) must have a small opening to enter most packaged goods (Highland 1991). However, *T. castaneum*, *O. mercator*, and *C. pusillus* can penetrate the less-resistant packaging materials (Cline 1978). Holes larger than 2 $mm^2$ will allow most stored-product adult insects to enter packages, whereas holes smaller than 0.3 $mm^2$ will prevent the entry of most stored-product insects (Cline and Highland 1981). Neonate larvae can enter much smaller orifices since their head capsule width is often less than 0.3 mm (Kahn 1983). A convenient measure of package integrity is the pressure packages can sustain, which is related to the susceptibility of the packaging to insect infestation (Yerington 1978).

Packaging materials differ in their capacity to prevent penetration. Listed in order of ease of penetration they are cellophane, polyethylene, paper polyvinyl chloride, aluminum foil, polyester, polypropylene, and polycarbonate (Cline 1978). However, most packages have small holes that allow entry. Packaging materials also vary in slipperiness, which affects the capacity of insects to move from package to package. In general, packaging materials in order of increasing difficulty for adult beetles to climb are paper, ethylenetetrafluoroethylene (ETFE), polyester (PE), aluminum foil, cellophane, fluorinated ethylenepropylene (FEP), glass, polyvinylchloride (PVC), and polypropylene (PP). Packaging materials are ranked differently for larvae than for adults, with the easiest to most difficult to climb being paper, glass, PET, ETFE, PVC, cellophane, FEP, foil, PE, and PP (Cline 1978).

Insecticides can be inserted into the packaging materials to render them more insect resistant. Currently in the United States the only insecticide registered for this use is synergized pyrethrum (Newton 1988). Multiwall packages are necessary to prevent the migration of the pesticide into the commodity.

Packaging is also a factor when one is considering heat sterilization of packaged product. With the use of long-wave irradiation, commodities can be heated at high enough temperatures to kill any insects in the package. Packaging must allow the transmission of long-wave radiation and be resistant enough to the heat and pressure created during heating (Fleurat-Lessard 1989). Airtight packaging with modified atmosphere is used to maintain freshness of high-value products and these atmospheres will also prevent the development of insects also (Cline and Highland 1987).

Combining the parasite *Habrobracon hebetor* Say with insect-resistant packaging reduced infestations by *C. cautella*. The proposed mechanisms are that the larvae were exposed to the parasites for longer periods of times in the treatments with insect-resistant packaging (Cline et al. 1984; Cline and Press 1990).

## VII. INERT DUSTS

Clays were used as an insecticide by aboriginal peoples in North America and Africa thousands of years ago (Ebeling 1971; Golob and Webley 1980). The work on inert dusts and stored product insects began in the 1920s (Headlee 1924) and a few reviews deal with inert dusts for insect control (Ebeling 1971; Golob and Webley 1980; Ross 1981; Quarles 1992a,b). The main advantage of inert dusts is that they are nontoxic. Diatomaceous earth is registered as a food additive in the United States and silica aerogels have a rat oral $LD_{50}$ of 3160 mg/kg rat (Ebeling 1971). They also provide continued protection and do not affect baking quality (Desmarchelier and Dines 1987; Alydryhim 1990). There are four basic types of inert dusts; earth, diatomaceous earth, silica aerogels, and nonsilica dusts. Clays, sands, and earth have been traditional insecticides that are used as a layer on top of stored seed (Golob and Webley 1980). Diatomaceous earth is the fossilized siliceous remains of diatoms. Diatoms are microscopic unicellular aquatic plants that have a fine shell made of opaline silica ($SiO_2 + n\ H_2O$). The shells of diatoms have built up over million of years, and in some places the deposits can be hundreds of meters thick (Ross 1981). The main constituent of these deposits is silica ($SiO_2$) although there are small amounts of other minerals (aluminum, iron oxide, lime, magnesium, and sodium). The principal current use of diatomaceous earth is for filters. Silica aerogels are produced by drying an aqueous solution of sodium silicate (Quarles 1992b). Silica aerogels are very light powders that are not hydroscopic. The fourth type is nonsilica dusts, such as rock phosphate, which have been used in Egypt (Fam et al. 1974), and tests have showed that lime (calcium oxide) has some activity (Golob and Webley 1980).

The principal mode of action for inert dusts is that they cause insects to desiccate. Insects die when they lose about 60% of their water or about 30% of their total body weight (Ebeling 1971). Silica aerogels can absorb as much as three times their own weight in oils. It is believed that as insects move through the grains the silica aerogels absorb waxes from the insects' cuticle (Le Patourel et al. 1989). In addition to the absorption of cuticular waxes, diatomeous earth abrades the cuticle. Because the stored-product insects live in very dry environments with limited access to free water, water retention is crucial to their survival. Also, since insects are very small they have a large surface area with respect to their body weight and therefore have greater problems retaining water than larger animals. Insects protect against desiccation in many ways but the waxy cuticle, which inert dusts destroy, is one of the main mechanisms in maintaining water balance (Noble-Nesbitt 1991).

There are two points that support the desiccation hypothesis. First, inert dusts are more toxic in drier grain (Carlson and Ball 1962; Fam et al. 1974; Le Patourel 1986). Second, insects treated with inert dust have a greater rate of water loss than untreated insects (Carlson and Ball 1962), although this was not the case for *S. oryzae* or *S. granarius*.

A summary of susceptibilities is given in Table 3. In general, *Tribolium* species are the most resistant, whereas *C. pusillus* tends to be the least resistant to inert dusts. In general, the capacity to survive dry conditions is correlated with resistance to inert dusts (Nair 1957; Carlson and Ball 1962), although this is not demonstrated in all studies (La Patourel 1986). Several possible factors would determine the effectiveness of inert dusts: greater capacity of insects to gain water from their food, greater water reabsorption during excretion, less water loss through the cuticle, type of cuticular wax, or amount of movement through grain. Not all the mortality seen in inert dust treated grain may be due to desiccation (Carlson and Ball 1962).

Inert dusts are used to a limited extent commercially. In Australia, Canada, and the United States, diatomaceous earth is registered to be used as a crack and crevice treatment and as a residual grain protectant. In Australia it is used mainly as a crack and crevice treatment to disinfest structures before new grain is stored in them. It can be used to protect feed grains, but not grains destined to be delivered

**Table 3** Relative Susceptibility of Stored-Product Insects to Inert Dusts Ranked from Most to Least Susceptible

| Inert dust | Insect species | Reference |
|---|---|---|
| Magnesite | *Callosobruchus chinenus* (L.) | Nair 1957 |
| | *R. dominica* | |
| | *S. oryzae* | |
| | *T. castaneum* | |
| Diatomaceous earth (Perma-guard) | *C. pusillus* | Carlson and Ball 1962 |
| | *S. oryzae* | |
| | *S. granarius* | |
| | *R. dominica* | |
| | *O. surinamensis* | |
| | *T. parabile* | |
| | *T. castaneum* | |
| | *T. confusum* | |
| Silica aerogel (Dri-Die) | *T. castaneum* | Kamel et al. 1964 |
| | *R. dominica* | |
| | *S. oryzae* | |
| | *S. granarius* | |
| Silica aerogel (Sipernat 22s) | *C. pusillus* | Le Patourel 1986 |
| | *O. surinamensis* | |
| | *T. castaneum* | |
| | *S. granarius* | |
| Diatomaceous earth with silica aerogel (Dryacide) | *T. castaneum* | Desmarchelier and Dines 1987 |
| | *R. dominica* | |
| | *S. oryzae* | |
| | *S. granarius* | |
| Silica aerogel (SG-67) | *O. mercator* | White and Loschiavo 1989 |
| | *T. confusum* | |
| Diatomaceous earth with silica aerogel (Dryacide) | *S. granarius* | Aldryhim 1990 |
| | *T. castaneum* | |

Comparisons of susceptibility are only valid within a study, not between studies.

to the grain-handling authorities. In Canada, diatomaceous earth can be used to protect grain destined for farm or local use, but cannot be delivered to a commercial elevator. In the United States, diatomaceous earth can be used through out the grain-handling system. In India during the 1960s, 70% of the grain was treated with activated kaolin clay (Ebeling 1971). Egypt used rock phosphate and sulfur (Ebling 1971).

The main problems with the use of inert dusts are that they decrease the bulk density and flowability of grain, are dusty to apply, and are ineffective in some cases. Because inert dusts adhere to the surface of the kernels and increase the friction between the grains, grain does not flow as easily, increasing angles of repose, and decreasing bulk density. Diatomaceous earth, at 2 kg/t caused a 4.4 kg/hl decrease in bulk density in corn and 6.2 kg/hl decrease in wheat (LaHue 1966). Since desiccation is the mode of action, in moist grain diatomaceous earth does not control insects as well as in dry grain (LaHue 1966). Application of inert dusts can be undesirable because of the dust generated. To alleviate this, aqueous applications for surface treatments are used in Australia, although this reduces the effectiveness of the inert dusts (Maceljski and Korunic 1971). Silica aerogel glued onto cards was able to kill insects that walked across the cards (Loschiavo 1988a). However mortality was reduced when insects were given access to food (Loschiavo 1988b). Application of diatomaceous earth to empty storage structures via the aeration system is another method to reduce worker exposure to dust. Finally, there has been concern that diatomaceous earth will increase wear on machinery, but there are no data to support or refute this hypothesis.

## VIII. SUMMARY

Physical control methods were some of the first methods used to control stored-product insects. Their use was phased out with the introduction of the modern synthetic insecticides. In the future we expect physical control to be used to a greater extent and synthetic chemicals to be used to a lesser degree because of the concerns about chemical residues on grain, worker safety, the environment, and insecticide-resistant populations.

Extreme temperatures are currently the most widely used physical control method. Insects cannot grow and reproduce below 13°C or above 35°C. The duration required to control insects depends upon temperature, species, stage of development, acclimation, and relative humidity. The temperature of commodities depends upon ambient air temperature, thermal mass, and ventilation equipment. Irradiation has been registered for use in many countries but has only been used at a commercial scale in the former Soviet Union. Mechanical action, turning grain, vacuum conveyance, and Entoleters can reduce insect populations, but there is concern over insects found within the kernels and damage to the grain. Packaging acts as a physical barrier that prevents the spread of insects within processed food. Inert dusts control insects by damaging the cuticle and causing the insects to desiccate. There are a variety of inert dusts but the most common use in use in grain today are derived from the shells of diatoms. The main problem with inert dusts is that they decrease bulk density and flowability. The main advantages are that they are not toxic to mammals and provide continued protection from insects.

## REFERENCES


Aldryhim, Y. N. (1990). Efficacy of the amorphous silica dust, Dryacide, against *Tribolium confusum* Duv. and *Sitophilus granarius* (L.) (Coleoptera; Tenebrionidae and Curculionidae), *J. Stored Prod. Res.*, *26*:207–210.

Anderson, J. (1989). Food Irradiation: An alternative food processing technology, Food Development Division, Agriculture Canada, p. 17.

Arbogast, R. L. (1981). Mortality and reproduction of *Ephestia cautella* and *Plodia interpunctella* exposed as pupae to high temperatures, *Environ. Entomol.*, *10*:708–710.

Armitage, D. M. (1986). Pest control by cooling and ambient air drying, *Int. Biodeterior. Supp.*, *22*:13–20.

ASAE (1990). Resistance to airflow of grains, seeds, other agricultural products, and perforated metal sheets, *Standards 1990*, Am. Soc. Agric. Engl., St. Joseph, MI, p. 371.

Bahr, I. (1978). Uberwinterungsversuche mit Schadinsekten der Getreidevorráte in ungeheizten, Ráumen. *Nachr.-Bl. Pflanzenschutz Berl.*, *32*:224–230.

Bahr, I. (1990). Reduction of stored product insects during pneumatic unloading of ship cargoes, Proceedings, Fifth International Working Conference on Stored-Product Protection, Bordeaux, France, pp. 1135–1144.

Bailey, S. W. (1962). The effects of percussion on insect pests of grain, *J. Econ. Ent.*, *55*:301–304.

Bailey, S. W. (1969). The effects of physical stress in the grain weevil *Sitophilus granarius*, *J. Stored Prod. Res.*, *5*:311–324.

Baker, K. D., Stroshine, R. L., Magee, K. J., Foster, G. H., and Jacko, R. B. (1986). Grain damage and dust generation in a pressure pneumatic conveying system, *Trans. A. S. A. E. Am. Soc. Agric. Eng.*, *29*:840–847.

Banks, H. J. (1976). Physical control of insects—recent developments, *J. Aust. Entomol. Soc.*, *15*:89–100.

Banks, H. J. (1986). Impact, physical removal and exclusion for insect control in stored products, Proceedings 4th International Working Conference on Stored-Product Protection, Tel Aviv, Israel, pp. 165–184.

Banks, H. J., and Fields, P.G. (1995). Physical methods for insect control, In Stored-grain Ecosystems (ed. Jayas, D. S., White, N. D. G., and Muir, M. E.). Marcel Dekker, New York, pp. 353–409.

Birch, L. C. (1945). The influence of temperature, humidity and density on the oviposition of the small strain of *Calandra oryzae* L. and *Rhizopertha dominica* FAB. (Coleoptera), *Aust. J. Exp. Biol. Med. Sci.*, *23*:197–203.

Bourgeois, L. (1993). Vigour loss of what caused by threshing. M.Sc. thesis, University of Manitoba, Winnipeg, Canada.

Brokerhof, A. W., Banks, H. J., and Morton, R. (1992). A model for time–temperature–mortality relationships for eggs of the webbing clothes moth, *Tineloa bisselliella* (Lepidoptera: Tineidae), exposed to cold, *J. Stored Prod. Res.*, *28*:269–277.

Burrell, N. J., Smith, E. A., and Armitage, D. M. (1982). Air distribution from ventilation ducts under grain, *J. Agric. Eng. Res.*, *27*:337–354.

Carlson, S. D., and Ball, H. J. (1962). Mode of action and insecticidal value of a diatomaceous earth as a grain protectant, *J. Econ. Entomol.*, *55*:964–969.

Chowaniec, T. (1986). Sanitation costs—the bottom line, *Assoc. Oper. Millers Bull.* Dec. 4867–4871.

Claflin, J. K., Evans, D. E., Fane, A. G., and Hill, R. J. (1986). The thermal disinfestation of wheat in a spouted bed, *J. Stored Prod. Res.*, *22*:153–161.

Cline, D. L. (1970). Indian-meal moth egg hatch and subsequent larval survival after short exposures to low temperature, *J. Econ. Entomol.*, *63*:1081–1083.

Cline, L. D. (1978). Clinging and climbing ability of larvae of eleven species of stored-product insects on nine flexible packaging materials and glass, *J. Econ. Entomol.*, *71*:689–691.

Cline, L. D., and Highland, H. A. (1981). Minimum sizes of holes allowing passage of adults of stored-product coleoptera, *J. Georgia Ent. Soc.*, *16*:525–531.

Cline, L. D., and Highland, H. A. (1987). Survival of four species of stored-product insects confined with food in vacuumized and unvacuumized film pouches, *J. Econ. Entomol.*, *80*:73–76.

Cline, L. D., and Press, J. W. (1990). Reduction in almond moth (Lepidoptera: Pyralidae) infestations using commercial packaging of foods in combination with the parasitic wasp, *Bracon hebetor* (Hymenoptera: Braconidae), *J. Econ. Entomol.*, *83*:1110–1113.

Cline, L. D., Press, J. W., and Flaherty, B. R. (1984). Preventing the spread of the almond moth (Lepidoptera: Pyralidae) from infested food debris to adjacent uninfested packages, using the parasite *Bracon hebetor* (Hymenoptera: Braconidae), *J. Econ. Entomol.*, *77*:331–335.

Collins, P. J., and Wilson, D. (1986). Insecticide resistance in the major coleopterous pests of stored grain in southern Queensland, *Queensland J. Agric. Anim. Sci.*, *43*: 107–114.

Cox, P. D. (1987). Cold tolerance and factors affecting the duration of diapause in *Ephestia kuehniella* Zeller (Lepidoptera: Pyralidae), *J. Stored Prod. Res.*, *23*:163–168.

Daumal, J., Jourdheuil, P., and Tomassone, R. (1974). Variabilité des effects létaux des basses temperatures en fonction du stage de développement embryonnaire chez la pyrale de la farine (Anagasta kuhniella Zell., Lepid., Pyralidae), *Ann. Zool.-Ecol. Anim.*, *6*:229–243.

David, M. H., Mills, R. B., and White, G. D. (1977). Effects of low temperature acclimation on developmental stages of stored-product insects, *Envir. Entomol.*, *6*:181–184.

Dermott, T., and Evans, D. E. (1978). An evaluation of fluidized-bed heating as a means of disinfesting wheat, *J. Stored Prod. Res.*, *14*:1–12.

Desmarchelier, J. M., and Dines, J. C. (1987). Dryacide treatment of stored wheat: Its efficacy against insects, and after processing, *Aust. J. Exp. Agric.*, *27*:309–312.

Ebeling, W. (1971). Sorptive dusts for pest control, *Annu. Rev. Entomol.*, *16*:123–158.

Evans, D. E. (1981). The influence of some biological and physical factors on the heat tolerance relationships for *Rhyzopertha dominica* (F.) and *Sitophilus oryzae* (L.) (Coleoptera: Bostrychidae and Curculionidae), *J. Stored Prod. Res.*, *17*:65–72.

Evans, D. E. (1983). The influence of relative humidity and thermal acclimation on the survival of adult grain beetles in cooled grain, *J. Stored Prod. Res.*, *19*:173–180.

Evans, D. E. (1987). The survival of immature grain beetles at low temperatures, *J. Stored Prod. Res.*, *23*:79–83.

Evans, D. E., and Dermott, T. (1981). Dosage-mortality relationships for *Rhyzopertha dominica* (F.) (Coleptera: Bostrychidae) exposed to heat in a fluidized bed, *J. Stored Prod. Res.*, *17*:53–64.

Evans, D. E., Thorpe, G. R., and Dermott, T. (1983). The disinfestation of wheat in a continuous-flow fluidized bed, *J. Stored Prod. Res.*, *19*:125–137.

Fam, E. Z., El-Nahal, A. K. M., and Fahmy, H. S. M. (1974). Influence of grain moisture on the efficiency of silica aerogel and katelsous used as grain protectants, *Bull. Ent. Soc. Egypt Econ. Ser.*, *8*:105–114.

Ferreira, W. A., and Muir, W. E. (1981). Efeito do diametro e dos materiais da Parede do silo metalico na temperatura de graos armazenados. (Effect of bin diameter and wall materials on the temperature of grain stored in metal bins), *Rev. Bras. Armaz.*, *6*:11–18.

Fields, P. G. (1990). The cold hardiness of *Cryptolestes ferrugineus* and the use of ice-nucleating active bacteria, Proceedings Fifth International Working Conference Stored-Product Protection, Bordeaux, France, pp. 1183–1191.

Fields, P. G. (1992). The control of stored-product insects and mites with extreme temperatures, *J. Stored Prod. Res.*, *28*:89–118.

Fleurat-Lessard, F. (1987). Control of storage insects by physical means and modified environmental conditions. Feasibility and application. In British Crop Protection Council Monograph (ed. Lawson, T. J.). British Crop Protection Council, London, p. 209–218.

Fleurat-Lessard, F. (1989). La désinsectisation des stocks de farine, *Ind. Céréales*, *60*:17–22.

Flinn, P. W., and Hagstrum, D. W. (1990). Simulations comparing the effectiveness of various stored-grain management practices used to control *Rhyzopertha dominica* (Coleoptera: Bostrichidae), *Environ. Entomol.*, *19*:725–729.

Foulk, J. D. (1990a). Pest-related aspects of sanitation audits, *Pest Control Tech.*, *18*:32.

Foulk, J. D. (1990b). Pest occurrence and prevention in the foodstuffs container manufacturing industry, *Dairy Food Envir. Sanit.*, *10*:725–730.

Friesen, O. H., and Huminicki, D. N. (1986). Evaluation of grain airflow resistance characteristics and air delivery systems, *Can. Agric. Eng.*, *28*:107–115.

Golob, P., and Webley, D. J. (1980). The use of plants and minerals as traditional protectants of stored products, *Rep. Trop. Prod. Instit.*, G138, p. 32.

Golumbic, C., and Davis, D. F. (1966). Radiation disinfestation of grain and seeds, *Int. Atom. Energy Agency*, Vienna, Sm *73/69*:473.

Gonen, M. (1977). Susceptibility of *Sitophilus granarius* and *S. oryzae* (Coleoptera: Curculionidae) to high temperature after exposure to supra-optimal temperature, *Ent. Exp. Appl.*, *21*:243–248.

Headlee, T. J. (1924). Certain dusts as agents for the protection of stored seeds from insect infestation, *J. Econ. Entomol.*, *17*:298–307.

Highland, H. A. (1991). Protecting packages against insects. In Ecology and Management of Food-Industry Pests (ed. Gorham, J. R.). Association of Official Analytical Chemists, Arlington, p. 345–350.

Howe, R. W. (1965). A summary of estimates of optimal and minimal conditions for population increase of some stored products insects, *J. Stored Prod. Res.*, *1*:177–184.

Howe, R. W., and Hole, B. D. (1968). The susceptibility of developmental stages of *Sitophilus granarius* (L.) (Coleoptera, Curculionidae) to moderately low temperatures, *J. Stored Prod. Res.*, *4*:147–156.

Imholte, T. J. (1984). A guide to the sanitary design of food plants and food plant equipment, Engin. Food Safety Sanit. Technical Institute of Food Safety, Crystal, p. 31.

Joffe, A. (1963). The effect of physical disturbance or "turning" of stored maize on the development of insect infestations I. Grain elevator studies, *S. Afr. J. Agric. Sci.*, *6*:55–68.

Joffe, A., and Clarke, B. (1963). The effect of physical disturbance or turning of stored maize on the development of insect infestations. II. Laboratory studies with *Sitophilus oryzae* (L.), *S. Afr. J. Agric. Sci.*, *6*:65–84.

Johnson, J. A., and Wofford, P. L. (1991). Effects of age on response of eggs of indianmeal moth and navel orangeworm (Lepidoptera: Pyralidae) to subfreezing temperatures, *J. Econ. Entomol.*, *84*:202–205.

Kahn, M. A. (1983). Untersuchungen über die Invasion von Eilaven von vorratsschädlichen Insekten durch verschieden grosse poren des Verpackungsmaterials, *Anzeriger für Schäclingskunde, Pflanzenshutz, Umweltschatz*, *56*:65–67.

Kamel, A. H., Fam, E. Z., and Ezzat, T. M. (1964). Studies on Drione dust as a grain protectant, *Agr. Res. Rev.* (UAR, Ministry Agr.), *42*:52–69.

Kawamoto, H., Woods, S. M., Sinha, R. N., and Muir, W. E. (1989). A simulation model of population dynamics of the rusty grain beetle, *Cryptolestes ferrugineus* in stored wheat, *Ecol. Modelling*, *48*:137–157.

Kirkpatrick, R. L., and Tilton, E. W. (1972). Infrared radiation to control adult stored-product Coleoptera, *J. Georgia Entomol. Soc.*, *7*:73–75.

Kirkpatrick, R. L., and Tilton, E. W. (1973). Elevated temperatures to control insect infestations in wheat, *J. Georgia Entomol. Soc.*, *8*:264–267.

Kitch, L. W., Ntoukam, G., Shade, R. E., Wolfson, J. L., and Murdock, L. L. (1992). A solar heater for disinfesting stored cowpeas on subsistence farms, *J. Stored Prod. Res.*, *28*:261–267.

Kline, G. L. (1973). Mechanical damage levels in shelled corn from farms, *ASAE*, 73-331:1–7.

Lapp, H. M., Madrid, F. J., and Smith, L. B. (1986). A continuous thermal treatment to eradicate insects from stored wheat, Paper 86-3008, *Am. Soc. Agric. Eng.*, St. Joseph, MI, p. 14.

LaHue, D. W. (1966). Evaluation of malathion, synergized pyrethrum, and diatomaceous earth on shelled corn as protectants against insects in small bins, U.S.D.A., *A.R.S. Marketing Res. Rept.*, *768*:10.

Lee, C. (1959). Baking quality and maltose value of flour irradiated with $Co^{60}$, *Cereal Chem.*, *36*:70–77.

Le Patourel, G. N. J. (1986). The effect of grain moisture content on the toxicity of a sorptive silica dust to four species of grain beetle, *J. Stored Prod. Res.*, *22*:63–69.

Le Patourel, G. N. J., Shawir, M., and Moustafa, F. I. (1989). Accumulation of mineral dusts from wheat by *Sitophilus oryzae* (L.) (Coleoptera: Curculionidae), *J. Stored Prod. Res.*, *25*:65–72.

Levinson, H. Z., and Levinson, A. R. (1989). Food storage and storage protection in ancient Egypt, *Bol. San. Veg.*, *17*:475–482.

Lhaloui, S., Hagstrum, D. W., Keith, D. L., Holtzer, T. O., and Ball, H. J. (1988). Combined influence of temperature and moisture on red flour beetle (Coleoptera: Tenebrionidae) reproduction on whole grain wheat, *J. Econ. Entomol.*, *81*:488–489.

Li, L., and Arbogast, R. T. (1991). The effect of grain breakage on fecundity, development, survival, and population increase in maize of *Tribolium castaneum* (Herbst) (Coleoptera, Tenebrionidae), *J. Stored Prod. Res.*, *27*:87–94.

Locatelli, D. P., and Traversa, S. (1989). Microwaves in the control of rice infestations, *Ital. J. Food Sci.*, *2*:53–62.

Longstaff, B. C., and Evans, D. E. (1983). The demography of the rice weevil, *Sitophilus oryzae* (L.) (Coleoptera: Curculionidae), submodels of age-specific survivorship and fecundity, *Bull Ent. Res.*, *73*:333.

Loschiavo, S. R. (1978). Effect of disturbance of wheat on four species of stored-product insects, *J. Econ. Entomol.*, *71*:888–893.

Loschiavo, S. R. (1988a). Safe method of using silica aerogel to control stored-product beetles in dwellings, *J. Econ. Entomol.*, *81*:1231–1236.

Loschiavo, S. R. (1988b). Availability of food as a factor in effectiveness of a silica aerogel against the merchant grain beetle (Coleoptera: Cucujidae), *J. Econ. Entomol.*, *81*: 1237–1240.

Maceljski, M., and Korunic, Z. (1971). The results of investigations of the use of inert dusts in water suspensions against stored-product insects, *Pl. Prot. Belgrade*, *22*:119–129.

Maier, D. E., and Bakker-Arkema, F. W. (1992). Storage of cereal grains using chilled aeration, International Symposium on Stored-grain Ecosystems, Winnipeg, Canada, p. 32.

Maier, D. E., Moreira, R. G., and Bakker-Arkema, F. W. (1992). Comparison of conventional and chilled aeration of grains under Texas conditions, *Appl. Eng. Agric.*, *8*: 661–667.

Mansbridge, G. H. (1936). A note on the resistance to prolonged cold of some insect pests of stored products, *Proc. R. Entomol. Soc. Lond.*, *11*:83–86.

Mathlein, R. (1961). Studies on some major storage pests in Sweden, with special reference to their cold resistance, *Meddn. St. Vaxtsk Anst.*, *12:83*:1–49.

McFarlane, J. (1989). Preliminary experiments on the use of solar cabinets for thermal disinfestation of maize cobs and some observations on heat tolerance in *Prostephanus truncatus* (Horn) (Coleoptera: Bostrichidae), *Trop. Sci.*, *29*:75–89.

Mellanby, K. (1932). The influence of atmospheric humidity on the thermal death point of a number of insects, *J. Exp. Biol.*, *9*:222–231.

Metzger, J. F., and Muir, W. E. (1983). Aeration of stored wheat in the Canadian Prairies, *Can. Agric. Eng.*, *25*:127–137.

Mills, R., and Pedersen, J. (1990). A Flour Mill Sanitation Manual. Eagan Press, St. Paul, pp. 1–164.

Mullen, M. A. and Arbogast, R. T. (1979). Time–temperature–mortality relationship for various stored-product insect eggs and chilling times for selected commodities, *J. Econ. Entomol.* *72*:476–478.

Murdock, L. L., and Shade, R. E. (1991). Eradication of cowpea weevil (Coleoptera: Bruchidae) in cowpeas by solar heating, *Am. Entomol.*, *37*:228–231.

Nagel, R. H., and Shepard, H. H. (1934). The lethal effect of low temperatures on the various stages of the confused flour beetle, *J. Agric. Res.*, *48*:1009–1016.

Nair, M. R. G. K. (1957). Structure of waterproofing epicuticular layers in insects in relation to inert dust action, *Ind. J. Entomol.*, *10*:37–49.

Navarro, S., and Calderon, M. (1982). Aeration of grain in subtropical climates. FAO Agriculture Services Bulletin 52, Rome.

Nelson, S. O., and Kantack, B. H. (1966). Stored-grain insect control studies with radiofrequency energy, *J. Econ. Entomol.*, *59*:588–594.

Newton, J. (1988). Insects and packaging—a review, *Int. Biodet.*, *24*:175–187.

Noble-Nesbitt, J. (1991). Cuticular permeability and its control. In Physiology of the Insect Epidermis (ed. Binningham, K., and Retnakaran, A.). CSIRO Publications, East Melbourne, pp. 252–273.

Oosthuizen, M. J. (1935). The effect of high temperature on the confused flour beetle, *Minn. Tech. Bull.*, *107*:1–45.

Prasad, D. C., Muir, W. E., and Wallace, H. A. H. (1978). Characteristics of freshly harvested wheat and rapeseed, *Trans. ASAE*, *21*:782–784.

Quarles, W. (1992a). Diatomaceous earth for pest control, *IPM Pract.*, *14*(5/6):1–11.

Quarles, W. (1992b). Silica gel for pest control, *IPM Pract.*, *14*(7):1–11.

Quentin, M. E., Spencer, J. L., and Miller, J. R. (1991). Bean tumbling as a control measure for the common bean weevil, *Acanthoscelides obtectus*, *Entomol. Exp. Appl.*, *60*: 105–109.

Rao, S. R., Hoseney, R. C., Finney, K. F., and Shogren, M. D. (1975). Effect of gamma irradiation of wheat on breadmaking properties, *Cereal Chem.*, *52*:506–509.

Robinson, W. (1926). Low temperature and moisture as factor in the ecology of rice weevil, *Sitophilus oryza* L. and the granary weevil, *Sitophilus granarius* L., *Univ. Minn. Agric. Exp. Stat.*, *41*:1–40.

Ross, T. E. (1981). Diatomaceous earth as a possible alternative to chemical insecticides, *Agric. Environ.*, *6*:43–51.

Sanderson, D. B. (1986). Evaluation of stored-grain ecosystems ventilated with near-ambient air. M.Sc. thesis, University of Manitoba, Winnipeg, Canada.

Sanderson, D. B., Muir, W. E., and Sinha, R. N. (1988a). Intergranular air temperatures of ventilated bulks of wheat, *J. Agric. Eng. Res.*, *40*:33–48.

Sanderson, D. B., Muir, W. E., and Sinha, R. N. (1988b). Moisture contents within bulks of wheat ventilated with near-ambient air: Experimental results, *J. Agric. Eng. Res.*, *40*:45–55.

Sanderson, D. B., Muir, W. E., Sinha, R. N., Tuma, D., and Kitson, C. I. (1989). Evaluation of a model of drying and deterioration of stored wheat at near-ambient conditions, *J. Agric. Eng. Res.*, *42*:219–233.

Sheppard, K. O. (1984). Heat sterilization (superheating) as a control for stored-grain pests in a food plant. In Insect Management for Food Storage and Processing (ed. Baur, F. J.). American Association of Cereal Chemists, St. Paul, pp. 193–200.

Sinha, R. S. (1975). Effect of dockage in the infestation of wheat by some stored-product insects, *J. Econ. Entomol.*, *68*:699–703.

Smith, E. A., Jayas, D. S., Muir, W. E., Alugusundaram, K., and Kalbande, V. H. (1992). Simulation of grain drying in bins with partially perforated floors, Part I: Isotraverse lines, *Trans. ASAE*, *35*:909–915.

Smith, L. B. (1970). Effects of cold-acclimation on supercooling and survival of the rusty grain beetle, *Cryptolestes ferrugineus* (Stephens) (Coleoptera: Cucujidae) at subzero temperatures, *Can. J. Zool.*, *48*:853–858.

Sokhansanj, S., Lampman, W. P., and MacAulay, J. D. (1983). Investigation of grain tempering on drying tests, *Trans. ASAE, 26*:293–296.

Solomon, M. E., and Adamson, B. E. (1955). The powers of survival of storage and domestic pests under winter conditions in Britain, *Bull. Ent. Res., 46*:311–355.

Stratil, V. H., and Wohlgemuth, R. (1989). Untersuchungen zum Wirkungsmechanismus von Prallmaschinen auf vorratsschadliche Insekten, *Anz. Schadlingskde, Pflanzenschutz, Umweltschutz, 62*:41–47.

Stratil, V. H. H., and Reichmuth, C. H. (1984). Uberlebensdauer von Eiern der vorratsschadlichen Motten Ephestia cautella (Wlk.) und Ephestia elutella (Hbn.) (Lepidoptera, Pyralidae) bei Temperaturen unterhalb ihres Entwicklungsminimums, *Z. Ang. Ent., 97*:63–70.

Subramanyam, B., and Harein, P. K. (1990). Status of malathion and pirimiphosmethyl resistance in adults of red flour beetle and sawtoothed grain beetle infesting farm-stored corn in Minnesota, *J. Agric. Entomol.*, 7:127–136.

Sutherland, J. W. (1986). Grain aeration in Australia. In Preserving Grain Quality by Aeration and In-Store Drying (ed. Champ, B. R., and Highley, E.). Australian Centre for International Agricultural Research, Canberra, Australia, pp. 206–218.

Sutherland, J. W., Banks, P. J., and Elder, W. B. (1983). Interaction between successive temperature or moisture fronts during aeration of deep grain beds, *J. Agric. Eng. Res., 28*:1–19.

Sutherland, J. W., Fricke, P. W., and Hill, R. J. (1989). The entomological and thermodynamic performance of pneumatic conveyor wheat disinfestor using heated air, *J. Agric. Eng. Res., 44*:113–124.

Tilton, E. W., Brower, J. H., and Cogburn, R. R. (1974). Insect control in wheat flour with gamma irradiation, *Int. J. Appl. Radiat. Isotopes, 25*:301–305.

Thorpe, G. R. (1983). High temperature disinfestation of grain. Division of Chem. and Wood Technol. Res. Rev., CSIRO, Australia, pp. 41–47.

Tuff, D. W., and Telford, H. S. (1964). Wheat fracturing as affecting infestation by *Cryptolestes ferrugineus, J. Econ. Ent., 57*:513–516.

Turnock, W. J., Lamb, R. J., and Bodnaryk, R. P. (1983). Effects of cold stress during pupal diapause on the survival and development of *Mamestra configurata* (Lepidoptera: Noctuidae), *Oecologia, 56*:185–192.

Ungsunantwiwat, A., and Mills, R. B. (1979). Influence of medium and physical disturbances during rearing on the development and numbers of *Sitophilus* progeny, *J. Stored Prod. Res., 15*:37–42.

Urbain, W. M. (1986). Food Irradiation. Academic Press, New York, p. 351.

Vardell, H. H., and Tilton, E. W. (1980). Heat sensitivity of the lesser grain borer, *Rhyzopertha dominica* (F.), *J. Georgia Entomol. Soc., 16*:117–121.

Vardell, H. H., and Tilton, E. W. (1981). Control of the lesser grain borer, *Rhyzopertha dominica* (F.), and the rice weevil, *Sitophilus oryzae* (L.), in wheat with a heated fluidized bed, *J. Kansas Entomol. Soc., 54*:481–485.

Wanzenried, H. (1992). Infestation control without use of chemicals, *Assoc. Operative Millers Bull. July*, pp. 6069–6073.

Watters, F. L. (1966). The effects of short exposures to subthreshold temperatures on subsequent hatching and development of eggs of *Tribolium confusum* Duval (Coleoptera: Tenebrionidae), *J. Stored Prod. Res.*, 2:81–90.

Watters, F. L. (1968). An appraisal of gamma irradiation for insect control in cereal foods, *Manitoba Entomologist*, 2:37–45.

Watters, F. L. (1991). Physical methods to manage stored-food pests. In Ecology and Management of Food-Industry Pests (ed. J. R. Gorham). Association of Official Analytical Chemists, Arlington, pp. 399–414.

Watters, F. L., and MacQueen, K. F. (1967). Effectiveness of gamma irradiation for control of five species of stored-product insects, *J. Stored Prod. Res.*, *3*:223–234.

White, N. D. G., and Loschiavo, S. R. (1989). Factors affecting survival of the merchant grain beetle (Coleoptera: Cucujidae) and the confused flour beetle (Coleoptera: Tenebrionidae) exposed to silica aerogel, *J. Econ. Entomol.*, *82*:960–969.

Williamson, W. F. (1964). Temperature changes in grain dried and stored on farms, *J. Agric. Eng. Res.*, *9*:32–47.

Wilson, S. (1990). Aerate: A PC Program for Predicting Aeration System Performance, CSIRO Division of Entomology, Highett, Australia, p. 74.

Wohlgemuth, V. R. (1989). Überlebensdauer vorratsschädlicher Insekten in Getreidekühllägern, *Anz. Schädlingskde Pflanzenschutz Umweltschutz*, *62*:114–119.

Worden, G. C. (1987). Freeze-outs for insect control, *Assoc. Operative Millers Bull. Jan*:4903–4904.

Yaciuk, G. (1973). Temperatures in grain storage systems, Ph.D. thesis, University of Manitoba, Winnipeg, Canada.

Yaciuk, G., Muir, W. E., and Sinha, R. N. (1975). A simulation model of temperatures in stored grain, *J. Agric. Eng. Res.*, *20*:245–250.

Yerington, A. P. (1978). Insects and package seal quality, *Mod. Pack, June*:41–42.

Zakladnoi, G. A., Menishenin, A. I., Pertsovskii, E. S., Salimov, R. A., Cherepkov, V. G., and Krsheminskii, V. S. (1982). Industrial application of radiation deinsectification of grain, *Soviet Atomic Energy*, *52*:74–78.

# 6

# Biological Control

**John H. Brower**

*U.S. Department of Agriculture, Manhattan, Kansas*

**Lincoln Smith**

*CIAT, Cali, Colombia*

**Patrick V. Vail**

*U.S. Department of Agriculture, Fresno, California*

**Paul W. Flinn**

*U.S. Department of Agriculture, Manhattan, Kansas*

All insect pest populations tend to increase exponentially as long as there is adequate food, suitable environment, and no predators or parasites. Historically, pest control has focused on the use of pesticides, exclusion (packaging), and adverse environmental conditions (desiccation, modified atmosphere, or temperature extremes) to suppress stored-product insects. Biological control employs parasites, predators, or pathogens (microorganisms that cause disease) to suppress pest populations. Although biological control may seem new to the field of stored-product insects, it was first used in 1911 against the Mediterranean flour moth (Froggatt 1912) and has a long history in other areas of agriculture (Simmonds et al. 1976). Interest in biological control is increasing as consumers become more intolerant of pesticide residues and as availability of conventional pesticides decreases due to pest resistance, developmental costs, and government registration and safety requirements (Waage 1991). The practicality of biological control is exemplified by the dominant role it plays in integrated pest management in greenhouses in northern Europe (van Lenteren and Woets 1988). Previous reviews of biological control of stored-product pests include those by Arbogast (1984b,c), Haines (1984), Brower (1990), Nilakhe and Parker (1990), Brower et al. (1991a), Burkholder and Faustini (1991), and Gordh and Hartman (1991).

# I. GENERAL THEORY

## 1. Natural Control

The theory of biological control is based on "natural control," which can be observed in the balance of predator–prey and parasite–host populations (Huffaker et al. 1976). We are not knee-deep in rabbits, despite the abundance of grass, because of mortality caused by many "natural enemies" such as foxes, hawks, and pathogens. Nonlethal natural enemies, such as mild diseases or intestinal worms, also help slow population growth by delaying or reducing reproduction, and by increasing risk of mortality due to other causes. Biological control employs beneficial organisms as pest control operators. Rather than spraying an insecticide, parasites or predators are reared and released to search out the pests and attack them. If the natural enemies are sufficiently effective to maintain the pest population below the economic injury level (Fig. 1; see also Hagstrum and Flinn, Chap. 10), no other control measures are needed. Insect-specific pathogens may likewise be manipulated or mass-produced and applied to control insect populations.

Natural enemies can be classified into several types based on their life history, ecology, and population dynamics. Predators generally feed on many prey during their lifetime, usually attacking individuals smaller than themselves. Parasites can be classified into parasitoids, macroparasites, and microparasites. Parasitoids are insects whose immature stage develops as a parasite (either internal or

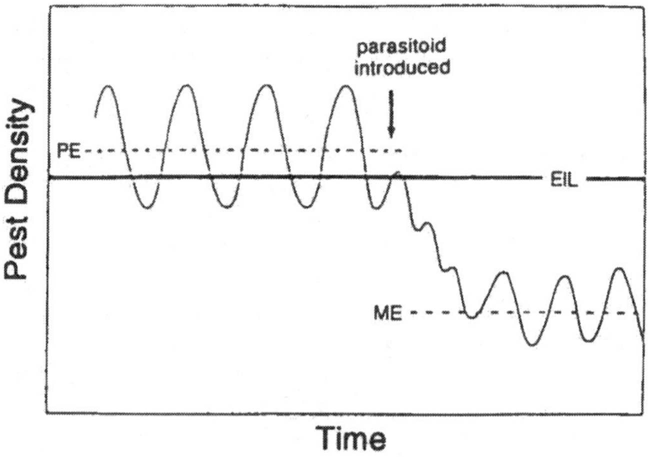


**Figure 1** Reduction of a pest population equilibrium (PE) from a damaging level above the economic injury level (EIL) to a modified equilibrium (ME) below the EIL by release of a biological control agent. (Modified from Smith and van den Bosch [1967], with permission.)

external) on another insect. The adult female searches for the host, upon finding a host she stings the host, often paralyzing it, and lays an egg (dipteran parasitoids do not sting their host and exhibit diverse oviposition strategies; see discussion of *Clausicella* below). The developing parasitoid feeds on the host and kills it before emerging as an adult. Solitary parasitoids produce one offspring per host, whereas gregarious parasitoids produce many. Macroparasites are comprised of metazoans such as insect parasitoids, mites, and nematodes, and are commonly referred to as "parasites." Microparasites are represented by microbial pathogens, such as viruses, bacteria, fungi, protozoa, and rickettsia, which cause contagious diseases. Microparasites are generally more conducive to formulation as conventional insecticides, genetic engineering, and patenting than are the other types of natural enemies, and are discussed in a separate section below.

Although parasites have often been considered to be host-specialists and predators to be generalists, there is actually much variation among species of both groups in the degree of specialization. Host specificity is important in understanding how many pest species can be attacked by a particular biological control agent and also what nontarget species could be affected. Habitat specificity is also an important component. Most natural enemies currently being studied are found naturally occurring in stored commodities in the regions where they would be used, they appear to be well-adapted to storage environments, and probably pose little risk to nontarget species. However, the straw itch mite, *Pyemotes tritici* (Lagreze-Fossat and Montague), which parasitizes a wide variety of insects (Bruce 1983), can cause dermatitis in people (Harwood and James 1979). This significantly reduces its attractiveness as a biological control agent in situations where people may come in contact with the commodity. Bethylids are the only stored-product parasitoids known to sting people, and such stings are only likely to occur when the organisms are handled (Nagy 1968).

It is important to realize the effect of pesticides on natural control. Predators and parasites are usually more susceptible to pesticides than are the target pests (Croft 1990). After a pesticide application eliminates pest populations and their natural enemies, if any pests survive or invade the commodity, the pest population will rapidly increase because of the absence of their natural enemies (Metcalf 1982). This phenomenon has been observed in many agricultural situations and is known as "pest resurgence" (Fig. 2). This situation is aggravated when the pest populations develop pesticide resistance, and treatment actually helps increase pest populations by killing their natural enemies but not the pests. For example, repeated application of malathion failed to control a nuisance population of almond moth, *Cadra cautella* (Walker), in bagged citrus pulp (used for animal feed), but when insecticides were discontinued the pest population quickly declined to innocuous levels due to the combined effects of an indigenous pathogen (*Bacillus thuringiensis*), predaceous mite (*Blattisocius tarsalis* [Berlese]), facultative insect predator (*Tribolium castaneum* [Herbst]), and parasitoid (*Habrobracon*

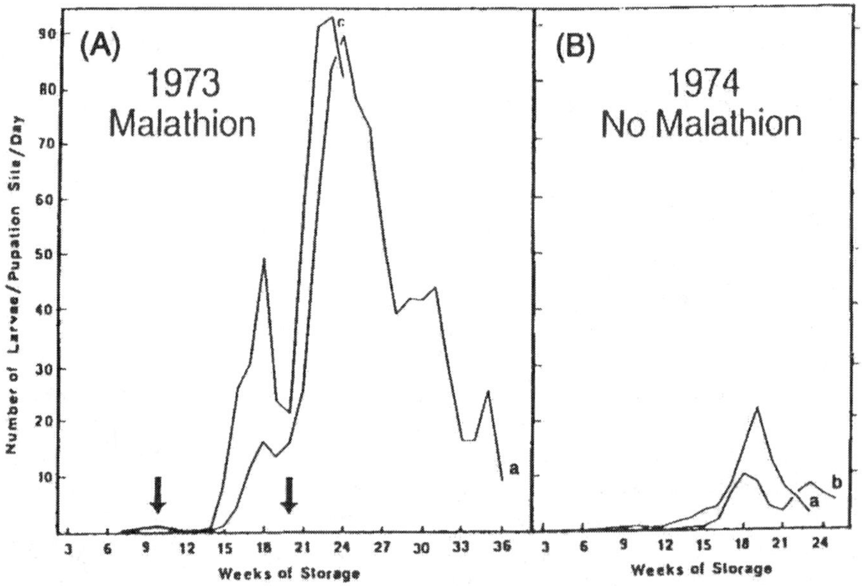


**Figure 2** (A) Resurgence of *Cadra cautella* population in dried citrus pulp warehouse after two applications of malathion. (B) *C. cautella* population in dried citrus pulp warehouse was held at innocuous levels by naturally occurring biological control agents when malathion treatments were discontinued (a, b, and c represent different locations within the warehouse). (Modified after Hagstrum and Sharp 1975.)

[Bracon] *hebetor* [Say]) (Hagstrum and Sharp 1975). A related problem, "secondary pest outbreak," occurs when a broad-spectrum pesticide eliminates the target pest and all natural enemies, and the ecological vacuum is filled by another, formerly innocuous, pest.

## 2. Biological Control

The "classic" approach to biological control has been to search for natural enemies in foreign countries to import and release against a pest. Although this approach has achieved many successes (Hall et al. 1980), it is appropriate only for cases in which effective natural enemies are not already present. This applies particularly to alien pests that accidentally enter a new geographic region. Thus, when the larger grain borer, *Prostephanus truncatus* (Horn), appeared in Africa (Dick 1988), the approach was to determine the origin of the infestation (Central America), go there and identify its important natural enemies (e.g., *Teretriosoma nigrescens* Lewis, a predatory beetle), determine that they were host-specific and would be safe to import (Rees 1991), and release them in numbers sufficient to establish self-sustaining populations (Markham and Herren 1990). This approach is extremely cost-effective because the only expenses are those of the initial research and development, whereas the benefits are self-perpetuated year after year (Huffaker et al. 1976). One shortcoming of classic biological control, however, is that efficient natural enemies capable of adapting to the new region are not always easily found, a situation that has beleaguered the gypsy moth biological control program in North America (Waters et al. 1976). Another important aspect is that classic biological control depends primarily on the investment of public, rather than private, institutions. This is because of the inability to collect royalty payments for naturally occurring biological control agents, which generally cannot be patented. Furthermore, agents that spread naturally provide pest control regardless of whether a producer paid for the initial importation. This circumstance doomed the first commercial attempt at biological control in stored products (Froggatt 1912), as explained below.

Most stored-product insect pests are probably not suitable for classic biological control because their natural enemies have been distributed as widely as the pests. Furthermore, stored products generally can tolerate only very low pest populations (low economic injury level; see Hagstrum and Flinn, Chap. 10), particularly if they have been processed. In this case, "augmentative" biological control, which involves more manipulation than classic biological control, is appropriate (Parrella et al. 1992). "Inundation" of the commodity or its environment by repeated releases of mass-reared natural enemies can reduce the pest population below the economic injury level by artificially increasing the density of natural enemies (Fig. 1). "Inoculation" involves a single release of natural enemies to establish a population whose successive generations would be expected to provide sustained control. "Conservation" includes a variety of techniques to attract, retain, and favor natural enemies. Techniques include applying artificial host odors that attract natural enemies (e.g., kairomones: odors from pests that attract parasitoids or predators), food supplements, factitious hosts, inactivated natural hosts, oviposition sites, shelter, and favorable environmental conditions (Ridgway and Vinson 1977; Powell 1986; Hagstrum 1983).

Augmentative biological control requires the development of economical techniques for mass-rearing, storage, transportation, and application of live natural enemies. The quality of natural enemies is also a serious concern because it is sensitive to rearing conditions (contamination of cultures by other species or strains, inbreeding, inadvertent selection for inferior "laboratory strains," poor nutrition, behavioral effects of crowding), storage period, and release conditions (Hoy et al. 1991; Bigler 1991). These problems are relatively minor for microbial insecticides, such as Dipel, which contains *Bacillus thuringiensis* (*Bt*) spores, but can be serious for insect parasitoids and predators. Although many companies in the United States are now selling insect parasitoids and predators (Hunter 1992), there is currently no system for ensuring quality or for establishing efficacy or application methods and rates. However, some of these issues are being addressed by the recently formed Association of Natural Bio-control Producers.

## 3. Biological Control in Stored Products

### *Insect Pathogens*

The application of microbial agents for biological control of stored-product pests is similar in technique and philosophy of use to the application of chemical protectants. However, there are several ways to use these agents for population regulation and reduction of damage. The most common use has been similar to that of insecticides (protectants) where immediate and possibly long-term control is provided. *Bt* and viruses have been used in this way to date. Inoculative type introductions may also be used either by direct inoculation of infestations or the use of diseased or contaminated individuals of the same species. Space treatments could also be used to reduce sources of the pests within warehouses and other sources of infestation. However, the value of this application of microbials has not yet been verified.

There are a number of advantages to the use of microbial control agents for stored-product insect control (Table 1). Generally there are no adverse effects of these pathogens on parasitoids and predators. However, indirect effects may occur because all of the organisms are competing for the same host(s). It is likely that a number of the bacteria and viruses, when used as protectants, could provide long-term control in bulk storage and also through marketing channels. Such treatments could also reduce the number of fumigations required for some commodities. Until recently resistance to microbial control agents had not been demonstrated. However, McGaughey (1985) demonstrated high levels of resistance of Indianmeal moth *Plodia interpunctella* (Hübner), to *Bacillus thuringiensis* and Vail and

**Table 1** Advantages and Disadvantages of the Use of Microbial Agents for Stored-Product Insect Control

| Advantages | Disadvantages |
|---|---|
| Safety for nontarget organisms (beneficials, vertebrates, etc.) | High specificity |
| May provide long-term protection | Low heat, UV stability |
| No demonstrated resistance in many cases | Requires strict quality controls in production |
| Can be stored for extended periods | Control not immediate |
| Registration procedures streamlined | Only in vivo production systems available for some organisms |
| | Patent problems |

Tebbets (1990) demonstrated a difference in response of geographic strains of Indianmeal moth to a granulosis virus. Because of the general opinion that many of these organisms are safe, registration procedures have been streamlined by the United States Environmental Protection Agency (EPA).

Although there are more distinct advantages to using microbials, there are also some distinct concerns or limitations to their use. They are generally slower acting than chemical pesticides; however, infected individuals often consume little food and thus damage is restricted. Development may be retarded or arrested completely. The high specificity of some of these organisms, particularly viruses and bacteria, may not provide control of all species in a complex. To solve this problem a pathogen would have to be developed for each species in the complex or a pathogen with a broad host range capable of infecting a number of species would be required. Pathogens have low heat and ultraviolet (UV) light tolerance. However neither of these factors should be critical in storage, where UV is minimal and sustained high temperatures rarely occur.

The production, formulation, and application of microbials demand strict adherence to quality control and application protocols. To date all large-scale production of baculoviruses has been done in vivo utilizing mass-reared host insects. Although in vitro systems (cell cultures) have been developed, none are as yet cost effective. Bacteria, fungi, and nematodes have been shown to be readily produced utilizing fermentation or artificial media. Because most of these organisms are naturally occurring, they may not be readily patentable, which causes some impediments to commercialization.

Compared to production systems, postharvest systems have attributes complementary to the use of microbial control agents. Ultraviolet inactivation is a major problem encountered in microbial use in production systems. Ultraviolet irradiation is not of concern in storage. Also of concern in production agriculture is the diluting effect of plant growth. The location of pests and the environment in postharvest storage also can be more readily defined. All of these characteristics tend to simplify the utilization of microbials and allow them to persist for long periods when used as protectants in storage.

### *Parasites and Predators*

Biological control in commodity storage has some unique advantages. Release of the biological agents within storage structures where they are protected from vagaries of the weather has a big advantage. Biological agents leave no harmful chemical residues on the commodities. Most are harmless to humans and can be applied by relatively unskilled workers. An additional long-term advantage is that stored-product pests (hosts) are not known to develop resistance to their parasites or predators. Biological control agents for storage pests usually are small to very small and therefore inconspicuous, have short life cycles, and have high reproductive potentials. They usually respond in a density-dependent fashion to host abundance, and populations can be self-perpetuating.

Parasites of stored-product pests also have limitations for general use. For example, they may be too host-specific because a range of pest species often is present in infested commodities. Predators tend to attack a wider range of prey and probably should be used in conjunction with parasite releases. Compared to most chemicals, biological agents are relatively slow acting. At present, biological control may be more expensive than traditional chemical controls, but this appears to be changing with the advent of improved mass-rearing techniques, development of artificial diets, and availability of commercial suppliers. Effective use of biological agents may require frequent releases or introductions. One potential disadvantage that must be considered is the possible contamination of the commodity by insect fragments from large numbers of released parasites or predators. This limitation probably precludes release of beneficial insects in or around manufactured food products unless they are already well packaged. The situation is completely different in farm-stored or bulk-stored agricultural commodities that will be cleaned and processed before human use. Studies have shown that insect fragments in processed commodities such as flour come almost exclusively from the primary grain pests that live within the grain kernels (Harris et al. 1952, 1953). External grain feeders, beneficial or incidental insects, and other types of dockage are separated and removed from the grain before it is processed further. Thus, even large numbers of small, lightweight bodies of beneficial insects should not add to the number of insect fragments in processed food commodities.

## II. HISTORY OF BIOLOGICAL CONTROL IN STORED PRODUCTS

### 1. Use of Insect Pathogens

Insect pathogens of interest for microbial control of storage pests have been found in all orders of insects of concern. Pathogens isolated from these insects include viruses, bacteria, protozoa, nematodes, rickettsia, and fungi. The well-known and widely used bacterium *Bacillus thuringiensis* Berliner was first isolated from the Mediterranean flour moth, *Ephestia kuehniella* (Zeller) and described by Berliner in 1915. Based on the works of Berliner and others the potential of this organism was recognized in Europe and its use as a control agent for stored product pests was investigated. From 1927 until 1939, commercial production and field trials of *B. thuringiensis* were conducted in France (Métalnikov and Métalnikov 1935) for the control of Mediterranean flour moth, but World War II interfered with the completion of this work. This was the first project to develop specifically a microbial control agent for stored-product pests. Tests after World War II by Jacobs (1951) with some of the above product, called "Sporeine," also demonstrated the potential of using *Bt* for control of the flour moth. However *Bt* was not to be used commercially for stored product insects until the 1970s.

In the early 1950s interest in microbial control increased as resistance problems began to appear to chlorinated hydrocarbons and other chemical insecticides and the potential of microorganisms for insect control was beginning to develop. Steinhaus and Bell (1953) conducted tests using microorganisms and antibiotics to determine their potential as insect control agents. Several of the bacteria, including *Bt*, caused high levels of mortality in stored product insects. The purpose of the use of antibiotic treatments was to demonstrate their effect on the intracellular organisms (symbiotes?) occurring in certain species and the resultant effects such administration might have on the insects. Generally the response of the Coleoptera to the pathogens/antibiotics were not very encouraging. *Bt* was found to be quite pathogenic to some lepidopteran pests. The lack of viral infections in the Coleoptera was also noted. Steinhaus and Bell (1953) also recognized the potential of microbial agents for the control of stored-product pests, noting that long-term protection of up to a year would be ideal. They also point out that those species inhabiting the interior of the grain or product would be more difficult to control than those feeding on the surface. The potential for resistance to *Bt* as indicated by their tests with the rice weevil was also discussed. These authors recommended further testing for some of these agents and particularly *Bt*. Concerted efforts on microbial control did not begin until the 1960s, even though a number of pathogens had been isolated and studied from stored product pests. Until the 1960s most studies were descriptive but provided background information necessary for the selection and development of microbial control agents. In the 1970s more intensive studies began on the use of some of these organisms as microbial control agents, either as population management tools or as protectants of various stored commodities. It is recognized that storages provide an ideal environment for the use of pathogens: locations of infestations are known and the environment is generally permissive for the use of pathogens. The lack of ultraviolet inactivation, a serious concern in field applications of pathogens, is virtually nonexistent in storage. Extended periods of high temperature are likewise seldom of great concern. Thus, several major impediments to the use of microbials in production agriculture are not present, which affords more predictability to the use of these agents in the storage situation. The present review specifically addresses the use of pathogens as microbial control agents, while recognizing the large component of basic research required to develop these agents. In 1974, Djerassi et al. discussed future control methodologies for insects and factors affecting their ultimate use including economic factors. Among those were insect pathogens. They stated that it is not likely that many pathogens will replace chemical pesticides but rather will be utilized in integrated pest management systems. Such integration will also be discussed in this chapter. Aizawa et al. (1976) reviewed reports on the potential use of pathogens for control of stored product pests and mention many organisms; they also discuss problems associated with their use.

Kellen (1978) likewise discussed different methods of use as related to their inherent virulence and the mechanisms of transmission (e.g., transovarial) of protozoans. He also mentioned that to provide predictable control, epizootics must be artificially induced by artificial introduction of the pathogen(s). Henry (1981) noted that the protozoa are probably significant regulators of various stored-product insects under natural conditions. He mentioned that protozoa could not compete with the more rapid action of bacteria, viruses, and certain fungi. He also describes strategies that could be used to incorporate protozoans into management systems. Arbogast (1984c) discusses the potential utilization of protozoa, *B. thuringiensis*, and baculoviruses. He also mentioned the compatibility of *Bt* with various fumigants, an important factor if *Bt* was to be applied as a protectant prior to fumigation and storage. Hodges (1984) discusses application and introduction techniques for various pathogens in tropical areas. He offers that more research needs to be conducted on the development of procedures integrating pathogens with other control methods. Subramanyam and Cutkomp (1985) reviewed the use of *Bt* for stored-product moth control.

## 2. Early Use of Parasites and Predators

The concept of using beneficial insects for control of stored product pests is certainly not new. In 1887, Geikie reported the complete destruction of a population of *Ephestia kuehniella* (= *Anagasta kuehniella*) in a large flour warehouse in London by the parasitoid *Habrobracon* (Bracon) *hebetor* Say (= *B. brevicornis* Wesmael). Buchwald and Berliner (1910) speculated that this parasite might be a valuable aid in the control of the Mediterranean flour moth in flour mills in Germany. In 1911 a larger parasitoid, *Venturia canescens* (Gravenhorst) (= *Amorphota ephestia* Cameron) was discovered in a flour mill in London, and it was reported that this ichneumon wasp greatly suppressed the numbers of *A. kuehniella* in this mill (Froggatt 1912). A sales circular was prepared that offered this parasitoid for sale to flour mills, and it was proposed to export the species to Australia for sale. However, this first commercial venture failed when it was discovered that *V. canescens* already occurred in a number of flour mills in England and also in several Australian locations (Froggatt 1912). In 1926, Ryabov published a rather detailed study of the parasitoid *Lariophagus distinguendus* Förster and its effects on its hosts, *Sitophilus granarius* (L.) (= *Calandra granaria* L.) and *Sitophilus oryzae* (L.) (= *Calandra oryzae* L.). He was so impressed with the parasitoid's ability to suppress populations of these pests that he advocated the culture and release of the parasitoid as a control technique in stored grain. Probably the most successful use of a beneficial insect to control a storage pest in a commercial situation was reported by Flanders (1930). Cultures of *Trichogramma minutum* Riley on eggs of the Angoumois grain moth, *Sitotroga cerealella* (Olivier), in a commercial insectary became infested with the Indianmeal moth, *Plodia interpunctella*. The braconid *Habrobracon hebetor* (= *Habrobracon juglandis*) was introduced into the moth rooms, where it practically eliminated the

Indianmeal moth while not affecting the nonhost Angoumois grain moth. From about 1910 to the mid 1940s many authors commented on the possible utility of parasitoids, particularly *H. hebetor*, for the control of stored-product pests; but with the widespread use of synthetic pesticides after World War II, little more work was devoted to this area of research. Only in the last few years has any real emphasis been placed on the development of a biologically based strategy for stored-product pest control.

### 3. Regulatory Status

In the 1970s *Bt* was registered for use on stored grains (McGaughey 1982) after definitive tests of the organism as a protectant were conducted. In the 1970s and 1980s the granulosis virus infectious to the Indianmeal moth was shown to have a high level of efficacy as a protectant on grains (McGaughey 1975) and dried fruits and nuts (Hunter et al. 1973). A production and formulation process was patented (Vail 1991) but the virus was not registered for use. All genera of parasitoids and predators known commonly to attack stored-food insect pests were exempted by the EPA from tolerance in stored raw whole grains and packaged food in warehouses so long as the insects do not become a component of the food (Anonymous 1992). These include the following genera of parasitic Hymenoptera, *Trichogramma*, *Bracon* (*Habrobracon*), *Venturia*, *Mesostenus*, *Anisopteromalus*, *Choetospila*, *Lariophagus*, *Dibrachys*, *Habrocytus*, *Pteromalus*, *Cephalonomia*, *Holepyris*, and *Laelius*, and predatory Hemiptera, *Xylocoris*, *Lyctocoris*, and *Dufouriellus*.

## III. BIOLOGICAL CONTROL AGENTS

### 1. Insect Pathogens

*Bacteria*

Many bacteria have been isolated from stored product insects, primarily the Lepidoptera. The most studied of these organisms is *Bacillus thuringiensis* (*Bt*). Bacteria such as *Bt* are facultative pathogens that generally kill by means of insecticidal proteins (toxins) or by invasion of the hemocoelom, thus causing a septicemia. Infections are often acute and cause rapid mortality. Both modes of action may be present in the same organism. In general they can be easily produced in large quantities using fermentation procedures. Since they are primarily active by ingestion, coverage is extremely important.

Early in the 1960s Burges (1964) stated some of the practical considerations of the use of bacteria to control insects in storage conditions including "hardy, long lived, dormant stage, capable of infecting insects in dry conditions and of being mass produced, e.g., some spore forming-bacteria, particularly the *B. thuringiensis* group that form crystals of toxic protein." Fletcher and Long (1971) conducted research on *Bacillus cereus* isolated from cigarette beetle, *Lasioderma serricorne* (Fabricius), larvae, which demonstrated good control of this insect in laboratory tests. This work is important because it is one of the few demonstrations of microbial control of a coleopteran stored-product pest.

*Bacillus thuringiensis* is currently registered for use on grains in the United States and is exempt from tolerance requirements. Several early papers (Kinsinger and McGaughey 1976; Nwanze et al. 1975) demonstrated high levels of susceptibility in *Plodia interpunctella* and *Cadra cautella* larvae. These studies also revealed the relative heat stability of *Bt* when applied to grain, and the authors believed *Bt* and a granulosis virus could be used as protectants for long-term storage of grains (and possibly other commodities). Schesser (1976) evaluated four commercial formulations of *Bt* and found differences in potency between them. Burges and Hurst (1977) showed that cadaver feeding by stored-product lepidopterous larvae may be a potent means of natural spread of the organism and probably is a contributing factor to the efficacy of *Bt* applications. They also noted various factors that may affect efficacy of *Bt* applied to stored products. McGaughey and Kinsinger (1978) showed that the feeding habits of *Sitotroga cerealella* may reduce exposure to the pathogen and thus allow greater survival. McGaughey (1978) also demonstrated decreased susceptibility with increasing age of *P. interpunctella* and *C. cautella* larvae. Kinsinger and McGaughey (1979) showed that Indianmeal moth and almond moth populations differed as much as 6–10-fold, respectively, in responses to *Bt* and suggested that wide range of application rates may be needed for control. In comparing formulations for treating in-shell peanuts, McGaughey (1982) showed that a dust formulation was superior to a wettable powder. The dust provided essentially complete control at 500 mg/kg and provided better protection against damage. Coverage was identified as the principal reason for the increased efficacy of the dust. Mardan and Harein (1984) tested the susceptibility of a malathion-resistant strain of Indianmeal moth to *Bt*. They proposed that tolerance of older larvae to *Bt* could be due to immunity or behavioral characteristics.

Although *Bt* has been found to be very effective against a number of lepidopterous pests, there is a "fly (moth) in the ointment." McGaughey (1985) reported the first documented resistance to *Bt*. By selection, tolerance of *P. interpunctella* was shown to increase resistance by 30-fold in two generations and it reached a plateau of 100-fold after 15 generations. The recessive trait was stable after selection was discontinued. *Plodia interpunctella* strains resistant to *Bt* were also collected from treated bins and indicated resistance can develop quickly in the field. McGaughey points out that the storage environment is ideal because *Bt* is stable on stored grains and the environment remains undisturbed for long periods. McGaughey and Beeman (1988) reported a similar phenomenon with *Cadra cautella*. Considerable research is now being conducted on the resistance mechanism(s) and ways to manage resistance in both stored-product and production pests (Johnson et al. 1990; Van Rie et al. 1990). McGaughey (1986) provided a critical review of the use of *Bt* in stored products, including the previously described work. Later McGaughey and Whalon (1992) reported on managing resistance to *Bt* toxins. They also indicated that none of the approaches offer clear advantages in all situations.

Microbial control of coleopterous pests by bacteria under storage conditions has not been reported. Few references are available even for laboratory studies of bacterial pathogens, although Kumari and Neelgund (1985) did report studies of the activity of several isolates toward the red flour beetle (*Tribolium castaneum*).

*Viruses*

Many viruses have been isolated from stored product pests and include nuclear polyhedrosis viruses (NPV), granulosis viruses (GV), cytoplasmic polyhedrosis viruses, and a number of nonoccluded viruses. Most isolates have been obtained from the order Lepidoptera, although several have been isolated from the Coleoptera. Of the viruses, NPVs and GVs have been studied most intensely and are generally considered the best candidates for microbial control agents because of their high virulence, acute type of infections, and stability. Viruses are strict intracellular parasites and depend on the host cells for reproduction (Table 2). The normal route of infection is by ingestion, although some have been reported to be transmitted on or in the egg (transovarian). Because of their demanding requirements, viruses can only be produced in living hosts or in vitro in insect cell lines.

Many reviews of basic research and microbial control of production pests utilizing entomopathogenic viruses are available; however, none are available for stored-product pests (Cantwell 1974a,b; Summers et al. 1975; Kurstak 1982; Granados and Federici 1986a,b; Fuxa and Tanada 1987).

The above references do provide a wealth of information on insect virology, the potential for development of viruses as microbial control agents, and the limitations or concerns about the use of these organisms in general. Until the 1970s most research conducted on viruses was oriented to insect pests in production agriculture. In the 1970s research intensified on the potential use of viruses as microbial control agents for stored-product insects. With a few exceptions, these investigations were primarily on baculoviruses (NPV and GV) and a few nonoccluded viruses. However, to this date a baculovirus has not been registered for use for control of a stored-product pest. Chapman and Glaser (1915) provided a list of insects dying of wilt (now generally considered to be caused by NPV) and emphasized that these diseases are probably more common and widely distributed than previously believed. Their premise was correct: there are now hundreds of reports of infections by not only the NPVs and GVs but also other types of viruses.

**Table 2** Mode of Entry, Mode of Action, Production Processes, and Stage(s) of M ntrol Potential for Stored-Product Pests

| Pathogen | Normal mode of entry | Mode of action | Production process | Most comr used stag microbial c | nt use and status |
|---|---|---|---|---|---|
| Bacteria | Ingestion | Bacteremia: invasion of hemocoelom. Toxins: *Bt* endotoxin or parasporal body | Fermentation | Spores | *giensis* registered for use t. Demonstrated efficacy table. Commonly used crops. Documented an be engineered, many different levels of y to different hosts. tive against Lepidoptera. effective against |
| Viruses | Ingestion | Invasion of cells | Obligate parasites: intracellular; can be produced with live insects; cell culture methods | Inclusion b | particularly from . Variable specificity. able. Efficacy d in several cases. Can -term protection. Few tudied from Coleoptera. eleases have been |

| | | | | | |
|---|---|---|---|---|---|
| Protozoa | Ingestion | Invasion of cells | Live insects obligate parasites. In vitro on small scale | Spores | ı and basic studies Some large-scale tests ı conjunction with May be best suited to type release for ıuppression. Some with ange have not been |
| Fungi | Contact or ingestion | Mycelial growth; metabolites | Fermentation | Spores | : broad host range. production achieved. relatively short. Few lies and of limited scale. *assiana* produced and :hoslovakia under the :rosil" |
| Nematodes | Contact or ingestion | Penetration of cuticle—often depend on symbiotic bacteria | Artificial media or live insects | Invasive st: | ıone tested for potential ;ents. Require high |

A number of these have been isolated from stored-product pests, primarily the Lepidoptera.

In 1968, Arnott and Smith (1968a,b) described a GV infectious to *P. interpunctella* isolated from a laboratory colony at Cambridge, England. Adams and Wilcox (1968), Thompson and Redlinger (1968), and Hunter and Hoffmann (1970) described a NPV isolated from *C. cautella* that infected many of the major tissues of this insect. Hunter and Dexel (1970) later described a GV from *C. cautella*. These basic studies soon led to more intensive studies of these and other baculoviruses of stored product pests.

The first experiments to develop baculoviruses as microbial control agents were conducted by Hunter (1970), in which the pathogenicity and host range of the *P. interpunctella* (IMM) GV were investigated. Although highly virulent to *P. interpunctella*, the virus failed to infect neonate *Ephestia elutella* (Hübner), *Cadra cautella*, and *Cadra figulilella* (Gregson) first instar larvae. However, Hunter recognized the importance of the ability to control a group (complex) of pests since many commodities are infested by more than one pest. With this in mind, Hunter and Hoffmann (1972) showed that the GV of *P. interpunctella* would not infect neonate *C. cautella*; however, a GV from *Cadra* would infect *P. interpunctella* although the capsules were of abnormal form. Hunter et al. (1973) reported on differences (7-fold) in susceptibility of two strains of *P. interpunctella* to the GV and inferred that they were either naturally present or a result of laboratory selection. Hunter et al. (1973) demonstrated high levels of control (95–99% reduction of moth emergence) when almonds, peanuts, and walnuts were treated with an aqueous suspension of the IMMGV. They considered coverage an important consideration for the use of the virus. McGaughey (1975) and Kinsinger and McGaughey (1976), using a similar preparation, also showed good control of *P. interpunctella* infesting stored wheat and corn and believed that long-term protection could be afforded to these commodities.

Previous studies with the GV of *P. interpunctella* were conducted with freshly prepared aqueous homogenates of diseased larvae. Cowan et al. (1986) believed that a dry formulation would provide several advantages over such preparations by being more easily standardized, having a reduced risk of growth of contaminants, and increased long-term storage. A simple production and formulation scheme was developed and later patented (Vail 1991). The formulation provided high levels of protection from *P. interpunctella* damage to almond meats, unhulled almonds, and processed or unprocessed raisins. When stored, the formulation maintained activity for 12 months at −12°C but some activity was lost at 22°C and drastically reduced at 45°C over the same period. However, the loss in activity probably would not be of great concern for long periods of storage of these commodities, since average storage temperatures are much lower than 45°C. Using a similar formulation, Vail et al. (1991) studied the efficacy and persistence of GV as related to new or incipient infestations on raisins and storage temperatures. As before, sustained storage control was related to temperature; sustained temperatures above 32° and 38°C caused reduction in efficacy and damage increased over a 12 month period. Control of incipient populations was dependent upon dosage and larval age at times of treatment. Control of progeny was at least as good as that of the original treated insects. Differences in GV susceptibility and biology of six *P. interpunctella* populations were noted by Vail and Tebbets (1990). Significantly longer development times were described in the less susceptible populations.

Studies on the autodissemination of the GV by *P. interpunctella* adults were conducted by Kellen and Hoffmann (1987). Virgin females were used to bring male moths to the contamination source. Several methods of transmission were noted, but it occurred predominantly when young larvae fed on contaminated moth cadavers or when small quantities of larval food were available for contamination and ingestion. Infection rates varied from 12 to 52% depending on the mode of transmission. In similar but larger tests (Vail et al. 1993) a pheromone lure and the freeze-dried powder described by Vail and Tebbets (1990) were used. A pheromone was used to attract *P. interpunctella* males to the contamination source. Fifty to 60% mortality was observed when males were introduced 24 h prior to introduction of females. However, when both genders were introduced simultaneously only 16% of the progeny were infected. The combined studies indicated that autodissemination may be useful for population regulation but probably would not offer adequate protection to commodities when used alone.

Several viruses or viruslike structures have been observed in the midgut of the beetle *Tenebrio molitor* L. by several investigators (Zeikus and Steinhaus 1969; Devauchelle 1970). Infection was noted only in about 25% of the insects and only about 25% of the cells. To our knowledge these are the only observations of presumptive viruses in stored-product Coleoptera.

### *Fungi*

Fungi are generally facultative pathogens of insects. However, some are very fastidious in their culture requirements and pose difficulties for mass production. Fungi enter the insect by being ingested, by penetration of the cuticle, and some have been reported to enter through the spiracles in some insects. Often fungi are very dependent upon the environment, particularly as related to germination of the infective stage(s). Like the bacteria, fungal infections are often acute once invasion has occurred. Although most are easy to culture, attenuation has been reported for some fungi after long-term production on artificial media. Once infection occurs, the fungus enters into the hemocoelom and begins to reproduce, eventually filling the insect with various life stages, at which time the insect dies. The most studied of the fungi is *Beauveria bassiana*, a pathogen of many agricultural pests. Because of their dependency upon high ambient humidities and our lack of knowledge of the factors affecting virulence of these pathogens, they have not been developed as microbial control agents of stored-product pests. Fungi can produce toxins harmful to humans and animals and thus some may be difficult to register. Toxin production may vary between isolates as well as with the production procedures used.

A few preliminary investigations have been conducted recently on the entomopathogenic fungi. Ferron and Robert (1975) demonstrated the susceptibility of *Acanthoscelides obtectus* (Say) to several fungi including *Beauveria bassiana*, *Beauveria tenella*, *Metarhizium anisopliae*, and *Paecilomyces fumosa-roseus*. However, the potential of these organisms as microbial control agents was not addressed. Davis and Smith (1977) suggested that conditions for cultivation of fungi may be contributing factors in the production of metabolites toxic to the yellow mealworm, *Tenebrio molitor*, and provided evidence that different fungal metabolites are involved in regulating the growth and mortality of *T. molitor*. They found that gains in fresh weight of larvae are influenced by the species of fungus, by incubation temperature of the fungus, and interactions of the two. A soil isolate of *B. bassiana* was evaluated as a potential microbial control agent of the saw-toothed grain beetle *Oryzaephilus surinamensis* (Fauvel) (Searle and Doberski 1984). It was found to be an effective pathogen under certain conditions. The beetles would not become infected at 90% relative humidity (RH). They concluded that although humidity requirements are high, it is feasible that humidities of 90% RH or more could be attained particularly if the product is infested. "Boverosil" is a commercial formulation of *Beauveria bassiana* registered for use in Czechoslovakia for the control of residual infestations of stored-product pests in grain stores and silos. Hluchý and Samšiňáková (1989) tested this formulation against greater wax moth larvae, *Galleria mellonella* (L.) and adult *Sitophilus granarius*. They also noted high levels of infection when humidity was close to 100% but a sharp drop-off of activity below 90% RH. Comparable mortality of *S. granarius* required two to five times the dosage effective for *G. mellonella*, if conditions are ideal.

*Protozoa*

Protozoa are one of the most extensively studied pathogen groups of stored-product pests. Many insect species are regulated to a greater or lesser extent by these organisms under natural conditions. The protozoa also depend on living insects or cells therefrom to reproduce. Protozoa are transmitted orally but in many cases may be transmitted also in or on the egg or between male and female under natural conditions. Although many of these organisms have been studied, few have been investigated as potential microbial control agents. This is possibly due to their generally slow mode of action and chronic type of infection. Being strict parasites, protozoans also require living hosts to reproduce. A few will reproduce in insect cell lines (Table 2).

More protozoa have been isolated from stored-products pests than any other group. Brooks (1971) discussed the importance of protozoan infections in regulating insect populations. Unlike bacterial and viral pathogens, many protozoa have been isolated from the Coleoptera. As opposed to some of the highly virulent bacteria and viruses, protozoan infections are often more subtle and chronic. Symptoms and pathological effects may also vary considerably. Kellen and Lindegren (1973a), for example, demonstrated an inflammatory response in Lepidoptera infected by *Nosema invadens*. The severe inflammatory response included hemocytic encapsulation of infected areas including the fat body. Schwalbe et al. (1974) demonstrated a yellow-green fluorescing material in *Trogoderma glabrum* (Herbst) larvae infected with *Mattesia trogodermae* that could be visualized by examining larvae under an ultraviolet light. The fluorescing material is localized in the spore coat (Burkholder and Dicke 1964). This is evidenced by the fact that many laboratory colonies can survive and reproduce even though heavily infected. Several have been reported to have very broad host ranges but have not been exploited as microbial control agents. Many have been shown to be transmitted on or in the egg (Kellen and Lindegren 1968). The debilitative effect of these pathogens should not be taken for granted, particularly as they may relate to the possibilities for population suppression. Besides causing mortality to various stages of infected insects, protozoan infection may increase development times, reduce survival to the adult stage, increase adult deformities, and impair reproduction. Tissue tropisms vary considerably between protozoan species. Responses to various natural and artificial (i.e., insecticides) environmental factors may also be influenced by these organisms.

Jafri (1961, 1964) demonstrated that *Tribolium castaneum* adults infected by a protozoan were hypersensitive to radiation. Weiser (1963) indicated that the same parasite caused increased sensitivity to DDT. Rabindra et al. (1988) showed that *T. castaneum* infected with *Farinocystis tribolii* were significantly more susceptible to several pesticides. Susceptibility increased by 1.75–7.35-fold depending on the insecticide. *Nosema whitei* infection was shown to influence *T. castaneum* larval survival on vitamin-deficient diets (Armstrong 1978). However, George (1971) showed that dietary deficiencies did not increase susceptibility of *T. castaneum* to *N. whitei*. Nara et al. (1981) noted decreased growth of the fat body, slower development rates, and large variation in larval size. Seventy-five percent of beetles emerging from infected larvae had normal longevity and fecundity, but the remainder were almost sterile and died within 30 days of emergence. Khan and Selman (1984) reported a distinct level of synergism in increasing the larval period of *T. castaneum* when an insecticide and *N. whitei* spores were fed together. Dunkel and Boush (1969) showed that during short periods of chronic starvation, larvae of the black carpet beetle, *Attagenus unicolor* (Brahm) (= *Attagenus megatoma* = *Attagenus pieceus*) infected with *Pyxinia frenzeli* lose weight almost twice as fast as their gregarine-free counterparts. These studies show that in addition to the acute effects of protozoans, a number of more subtle interactions may occur with their hosts. Possibly these and other interactions could also be used to complement other new control technologies to provide general reductions of populations that might reduce the need or frequency of specific control measures.

Kellen and Lindegren (1971) demonstrated in the laboratory that *Nosema plodiae* may be transmitted, in addition to normal routes of infection, both transovarially and venereally. Infection of the ovaries reduces fecundity and 27% of progeny of diseased females acquired infections transovarially. Transmission was both on and in the egg.

Although there are many descriptive and basic studies on protozoans infectious to insects, few practical studies to utilize these organisms as microbial control agents have been conducted. Of the few studies reported, most have been conducted with protozoans infectious to Coleoptera and were often combined with another control component. Marzke and Dicke (1958) conducted studies with a schizogregarine infectious to a number of *Trogoderma* species. Previous experience with this organism showed that 95% of *T. glabrum* laboratory cultures and 55% of *Trogoderma granarium* Everts cultures were infected with this organism. Five other species, *Trogoderma inclusum* LeConte, *Trogoderma variabile* (Ballion) (= *Trogoderma parabile* Beal), *Trogoderma teukton* Marzke and Dicke, *Trogoderma simplex* Jayne, and *Trogoderma sternale* Jayne, were also exposed to dead larvae and diet from heavily diseased colonies. Varying degrees of infection were observed in the first three species, while *T. simplex* and *T. sternale* were virtually wiped out. The authors suggest that this schizogregarine may have potential as a control agent for important *Trogoderma* sp. Kellen and Lindegren (1969) showed that two previously undescribed microsporidians isolated from the Indianmeal moth, *Nosema heterosporum* and *Thelohania nana*, will infect the naval orangeworm, *Amyelois transitella* (Walker), and the greater wax moth, *Galleria mellonella*. The almond moth, *C. cautella*, the raisin moth, *C. figulilella*, the tobacco moth, *E. elutella*, the confused flour beetle, *Tribolium confusum* Jacquelin du Val, and the driedfruit beetle, *Carpophilus hemipterus* (L.) were not susceptible based on laboratory tests. *Nosema oryzaephili* sp was isolated from a rapidly declining population of *Oryzaephilus surinamensis* (Burges et al. 1971). Its host range was shown to include both Coleoptera and Lepidoptera from five families. The authors regarded *O. surinamensis* as highly susceptible to this pathogen. Of interest was the host range of *Nosema parasiticum* reported by Kellen and Lindegren (1973b), which provides some insight into the potential host range of protozoans. Representatives of the Coleoptera, Lepidoptera, Diptera, and Acarina were found susceptible to this pathogen, which included major storage pests. Ashford (1970) conducted research on the relationship of the red flour beetle (*T. castaneum*) to the neogregarine *Lymphotropa triboli*. He found an extended host range to many *Tribolium* species; however, *T. confusum* was completely refractive. Adult *T. castaneum* were not observed to have actively developed infections.

Several studies have been conducted on the storage life of entomogenous protozoa. Ashford (1970) showed that spores were stable for three months at either

15 or 30°C; a marked decrease in infectivity occurred between 3 and 9 months, particularly at the higher temperature. Milner (1972) reported that *N. whitei* spores did not lose any detectable loss in viability after 15 months in storage at 4°C. Storage temperature also discussed by Nara et al. (1981); survival was best at −19°C.

One of the best examples of the use of protozoans to control stored-product pests has been provided by Schwalbe et al. (1974) and Burkholder and Boush (1974). Initially they showed that *Mattesia trogodermae* could cause high levels of infection in *T. sternale*, *T. simplex*, *T. glabrum*, and *T. inclusum*. *Trogoderma grassmani* Beal and *T. variabile* were not infected under their conditions. They found that 1 infected larva/50 g medium reduced adult production by 25% while 25–50 infected larvae/50 g medium caused 100% infection of larvae and complete suppression of adult production. The high larval mortality combined with horizontal and vertical transmission, and cannibalism/cadaver feeding combined with the ability of healthy adults to transmit spores to other insects, led the investigators to develop autoinoculation devices to distribute the pathogen. Schwalbe et al. (1974) and Burkholder and Boush (1974) then devised an inoculation device treated with a synthetic sex pheromone ((Z)-14-methyl-8-hexadecen-1-ol). The presence of the pheromone in the devices led to 96% contamination of the test insects while only 56% were contaminated in the absence of pheromone. Such contaminated adults would return to their habitat and infect other susceptible species. Shapas et al. (1977) reported on the results of their studies. Population increases using the inoculation device were only fourfold in the first posttreatment generation, compared to 24-fold in the controls. Treated populations were below pretreatment levels in the second generation compared to a 100-fold increase in the controls. They concluded that spore transfer to subsequent generations was mainly by larval ingestion of dead contaminated adults or larval food that adults had contaminated by contact.

### *Rickettsia*

Only a few examples of these organisms have been isolated from stored product insects. They also are strict intracellular parasites, depending on live host cells for reproduction. Infections caused by these organisms are generally chronic, although they may have significant impacts on insect populations. To date they have not been explored for use as microbial control agents (Table 2).

Kellen and Lindegren (1970) and Kellen et al. (1972) reported on the isolation, ultrastructure, and life cycle of a *Rickettsiella* infectious to the naval orangeworm, *A. transitella*. Kellen et al. (1981) also reported on a *Wolbachia* sp. symbiont of the almond moth, *C. cautella*, which influenced fertility (reproductive cytoplasmic incompatibility). Gottlieb (1972) noted mycoplasma-like inclusions in *Ephestia kuehniella* and described their location and morphology. To our knowledge, none of the above organisms has been tested or manipulated to determine its potential as a microbial control agent.

## 2. Parasites of Insects

*Nematodes*

Most parasitic nematodes are harmful to plants, animals, or humans, but a small but significant number are parasites of insects and thus can be considered beneficial. Some of these nematodes are of considerable importance because of their potential as biological control agents of pest insects. Unfortunately nematodes have been viewed as having very little potential for control of stored-product pests because they usually require damp, if not wet, environments to infect their hosts. Nematodes can be reared on their insect hosts and in many cases in vitro on relatively simple media. Although their use in stored grain may be limited, their potential in specialized situations should not be overlooked.

Entomogenous nematodes (those that parasitize insects) are sometimes mass-reared on the larvae of Lepidoptera that fall in the realm of stored-product pests. Chief among these is the greater wax moth, *Galleria mellonella*, and also include the sunflower moth, *Homoesoma electellum* (Hulst) and the navel orangeworm, *Amyelois transitella*. The Mediterranean flour moth, *Ephestia kuehniella* is also quite susceptible to at least two species of nematodes (Morris 1985), and application techniques for this species are quite conceivable. A number of species of stored-product beetles also can be infected by entomogenous nematodes (Morris 1985), and many other species, both of pests and nematodes, need to be studied. Among the pests that are known to be susceptible to infection are three species of *Tribolium*, and the yellow mealworm, *Tenebrio molitor* (Morris 1985); the bean weevil, *Acanthoscelides obtectus* (Luca 1976); and the lesser mealworm, *Alphitobius diaperinus* (Panzer) (Geden et al. 1985).

The most promising application for nematodes is to reduce field infestations of commodities before they enter storage. A good example is the proposal by Agudelo-Silva et al. (1987) to control *A. transitella* in almond orchards by applying the nematode, *Neoaplectana carpocapsae* Weiser in a spray solution to "split" almonds still on the trees. Species such as the yellow and lesser mealworms that occur in damp habitats also could be infected by application of nematodes to their environment. The potential of applications of nematodes to grain crops in the field to reduce field infestations of stored-product pests is unknown but worthy of consideration.

*Hymenoptera*

Most of the parasitoids attacking stored-product insect pests are in the order Hymenoptera. The female parasitic wasps attack eggs, larvae, or pupae of their hosts, depending on the nature of the parasitoid. Discussion here will be limited to the better-known parasitoids of some of the principal stored-product pests. This section is organized into four units, based on important ecological characteristics: bulk whole grain, which is attacked by internal-feeding pests; bulk foodstuffs attacked by external-feeding pests; packaged foodstuffs in warehouses; and specialty commodities such as tobacco, woolens, and spices, which are attacked by specialists.

**Internal-Feeding Grain Pests.** Damage to whole-grain cereals and legumes is caused primarily by *Sitophilus* spp., *Rhyzopertha dominica* (F.), *Prostephanus truncatus*, *Sitotroga cerealella*, *Callosobruchus* spp., *Zabrotes subfasciatus* (Boheman), and *Caryedon serratus* (Oliver) (Mital 1969), whose larvae develop inside seeds. All these pests are beetles except for the gelechiid moth, *S. cerealella*. They are attacked by a guild of parasitoids that is able to crawl through the grain, detect infested kernels, drill into the seed, sting the immature host, and oviposit. The developing wasp larva consumes that host and emerges as an adult. Parasitoids that attack the larvae and pupae of these concealed hosts are often capable of parasitizing more than one species because the immature parasitoid develops outside the host, safe from immune defenses. Such parasitoids include the pteromalids: *Anisopteromalus calandrae* (Howard), *Lariophagus distinguendus*, *Pteromalus* (*Habrocytus*) *cerealellae* (Ashmead), and *Theocolax* (*Choetospila*) *elegans* (Westwood). These parasitoids appear to be "facultative" (able to switch host species), attacking detectable insects developing inside suitable types of seed. For example, *P. cerealellae* attacks a wide variety of hosts besides *S. cerealella* (Brower 1991a), and *P. truncatus* can be attacked by the four parasitoids listed above (Brower 1991b). *Lariophagus distinguendus* has been reported to parasitize the primary parasitoids *Habrobracon hebetor* and *Venturia canescens* (Kashef 1956), a condition referred to as hyperparasitism.

*Anisopteromalus calandrae* has been reported to attack all the above pests, as well as *Lasioderma serricorne* (Ahmed and Khatun 1988), *Stegobium paniceum* (L.), *Caulophilus oryzae* (Gyllenhal) (*Caulophilus latinasus* [Say]) (Peck 1963), *Zabrotes subfasciatus* (Fujii 1983), and *Callosobruchus chinensis* (L.) (Shimada 1985). However, some of these host records may be attributed to different parasitoid strains. The ease with which a parasitoid can switch from one host or commodity to another has received little attention in stored products, but this may have an important influence on the efficacy of commercially produced parasitoids used to control different target pests in different commodities. Furthermore, only a limited range of host sizes may be suitable for parasitism, such as third instar larvae to pupae for *A. calandrae* and *L. distinguendus* (Putters and van den Assem 1988; Yoo and Ryoo 1989; Smith 1993b). *Anisopteromalus calandrae* attacks insects in a variety of grains such as maize, wheat, rice, beans, and barley (Battu and Dhaliwal 1976; Fujii 1983; Yoshida and Kawano 1958; Arbogast and Mullen 1990). However, body size and foraging behavior may affect the efficacy of different parasitoids on different commodities. The larger-sized *A. calandrae* is more effective than *T. elegans* parasitizing *Sitophilus* in maize (Wen et al. 1994), but may be less effective in wheat, which has smaller interstitial spaces (Press 1992.) Differences in grain surface characteristics can also affect the efficiency of these parasitoids. For example, *A. calandrae* has a lower attack rate on hosts in rough rice than brown rice, from which the silicaceous hull has been removed (Robert R. Cogburn, personal communication). Also, the type of commodity may affect parasitoid fecundity or longevity indirectly via nutritional effects on host body size (Ryoo et al. 1991a).

*Dinarmus* species (Pteromalidae) may also be important larval-pupal parasitoids of *Callosobruchus* spp., *Bruchus* spp., *Bruchidius atrolineatus* (Pic), and *Acanthoscelides obtectus* in legume seeds (Peck 1963; Boucek and Rasplus 1991; Monge and Huignard 1991). Natural populations of *Dinarmus basalis* (Rondani) and the eupelmid *Eupelmus vuilleti* (Crawford) were found parasitizing *Callosobruchus maculatus* (F.) and *Bruchidius atrolineatus* in cowpeas in Niger, in both the field and storage (Monge and Huignard 1991). Many other parasitoids of bruchids are listed by van Huis (1991).

*Heterospilus prosopidis* Viereck (Braconidae) parasitizes larvae of *Callosobruchus chinensis* and *Zabrotes subfasciatus* on azuki, blackeye, and red kidney beans (Charnov et al. 1981; Fujii 1983; Kistler 1985) and is also reported from *Acanthoscelides* species (Marsh 1979). Utida (1955) speculated that *A. calandrae* is more efficient at high host density and *H. prosopidis* is better at low host density, based on a long-term interspecific competition study. This may indicate that the latter species would be a good candidate for maintaining the low levels of host populations needed for commercial pest control.

*Dibrachys cavus* (Walker) (Pteromalidae) is a facultative hyperparasitoid that can parasitize either primary parasitoids or their concealed hosts (Burks 1979a; Boucek and Rasplus 1991). Releasing such species should probably be avoided in biological control programs because they may reduce the effectiveness of primary parasitoids (Ayal and Green 1993); however, such a negative impact may not always be true (Barclay et al. 1985). Because *D. cavus* has been reported to parasitize many other beneficial insects in many habitats (Peck 1963), its use should be avoided lest it cause problems outside the storage system.

The eggs of *Sitotroga cerealella*, which are deposited on the substrate surface, are attacked by egg parasitoids such as *Trichogramma principium* Sugonjaev & Sorokina (Adashkevich and Umarova 1985) and *Trichogramma pretiosum* Riley (Bai et al. 1992). Egg parasitoids are discussed further in the following section.

**External-Feeding Pests.** A variety of hymenopterous parasitoids attack the different immature stages of external-feeding pests. The exposed eggs of many pest species can be parasitized by small hymenopterous parasitoids, particularly trichogrammatids. *Trichogramma* species are generally able to attack a wide range of hosts in the laboratory, although in nature they tend to be more habitat-specific, attacking any suitable hosts they find (Burks 1979b). *Trichogramma pretiosum* and *T. evanescens* Westwood attack eggs of Lepidoptera such as *Cadra cautella*, *C. figulilella*, *Ephestia elutella*, and *Plodia interpunctella* (Brower 1983a,b;

Brower 1988). *Trichogramma minutum* has been reported from *E. kuehniella*, *C. cautella*, and *P. interpunctella*, as well as from other non-stored-product pests (Peck 1963). Natural populations of *T. pretiosum* and *T. parkeri* Nagarakatti were found in some peanut storages in Georgia, where they presumably were attacking *P. interpunctella* and *C. cautella* (Brower 1984). Weekly releases of *T. pretiosum* suppressed *C. cautella* up to 42% and *P. interpunctella* up to 57%, relative to untreated controls, in 200 kg of inshell peanuts in small buildings (Brower 1988). However, it is important to realize that both *T. evanescens* and *T. pretiosum* are usually reported from herbaceous plants of open areas (Burks 1979b), so that behavioral conditioning or genetic selection of strains that stay indoors may be critical for stored-product applications.

*Uscana mukerjii* (Mani) (Trichogrammatidae) attacks eggs of *Callosobruchus maculatus* (Kapila and Agarwal 1991), and *Uscana lariophaga* Steffan attacks *Bruchidius atrolineatus* and *C. maculatus* eggs (van Huis et al. 1991). Several species of *Uscana* have been imported as classic biological control agents for bruchids (van Huis 1991; van Huis et al. 1992). For example, *Uscana semifumipennis* Girault was accidentally imported from Texas to Hawaii around 1913, resulting in parasitization of up to 90% of *Caryedon serratus*, and the pest population subsequently declined (Bridwell 1919). *Uscana semifumipennis* was also introduced to Japan in 1931 to control several bruchids (Ishii 1940). Van Huis et al. (1992) list host records of 11 species of *Uscana* and review strategies for their use. However, it should be noted that all these pests oviposit on the commodity in the field (although *C. maculatus* also reproduces in storage). So *Uscana* species may occur primarily outdoors, which would limit their potential for stored-product applications.

Egg parasitoids can complement other natural enemies such as larval parasitoids (*Habrobracon hebetor*; Brower and Press 1990) and predators (*Xylocoris flavipes* [Reuter]; Brower and Press 1988). Although *Trichogramma* and *Uscana* species tend to be fairly polyphagous, strains can differ in host preference (Hassan and Guo 1991; van Huis et al. 1991), and this may be influenced by the species on which they were reared (van Huis et al. 1991; Kaiser et al. 1989). Mass production, storage, and distribution of *Trichogramma* have been well developed because of its use in field crops (e.g., Bouse and Morrison 1985). The estimated cost of production is $0.0178 per 1000 parasitized host eggs (Morrison 1985). However, these species also attack outdoor hosts and have lower attack rates in darkness than in light (Brower 1991c), which suggests some limitations for stored-product applications, unless specialized strains are used.

Large larvae of most of the pyralid moths (*Ephestia kuehniella*, *Cadra cautella*, *Ephestia elutella*, *Plodia interpunctella*) are parasitized by the cosmopolitan braconid *Habrobracon* (*Bracon*) *hebetor* (Marsh 1979). *Sitotroga cerealella* has also been reported as a host (Marsh 1979); however, this may be atypical because the larvae are usually concealed inside grain kernels. Kairomones secreted from host mandibular glands are involved in host location (Strand et al. 1989), and the parasitoids apparently prefer to attack mature larvae wandering in search of pupation sites (Hagstrum and Smittle 1977). The female stings and paralyzes the host and deposits several eggs externally. *Habrobracon hebetor* is frequently known to sting and paralyze several hosts without ovipositing (Hagstrum 1983). Such hosts remain suitable for oviposition for several weeks, so this behavior has been called "self-provisioning." Artificial release of paralyzed host larvae improved control of *C. cautella* in inshell peanuts by increasing the longevity and fecundity of *H. hebetor* that had been released (Nickle and Hagstrum 1981). Releasing such hosts, which are unable to complete development, may be a cheaper alternative to releasing parasitoids, once the parasitoid population has been established. Natural populations of *H. hebetor* have been observed in commercially stored inshell peanuts (Keever et al. 1985) and shelled maize (R. T. Arbogast, unpublished data) infested with *P. interpunctella* and *C. cautella*, despite the use of malathion. This suggests the possibility of resistance to this insecticide. Reinert and King (1971) estimated that maximum suppression of a cohort of *P. interpunctella* larvae under laboratory conditions could be achieved by releasing a ratio of one 1-day-old female *H. hebetor* to 7.2 host larvae, which would cause 97% mortality.

*Venturia canescens* is a cosmopolitan ichneumonid parasitoid usually encountered in flour mills and stores of flour and grain products (Carlson 1979). It parasitizes the pyralid moths *P. interpunctella*, *C. cautella*, *Ephestia* species, the carob moth *Ectomyelois ceratoniae* (Zeller), and the tineid moth *Nemapogon granella* (L.). It has also been reared successfully from many other stored-product moths in the laboratory and from moths in orchards and vineyards (Salt 1976). Males are known only in Europe and rarely there. Females locate hosts with the aid of kairomones secreted from host mandibular glands (Corbet 1973). Females probe the substrate repeatedly with their long ovipositors attempting to strike a host. The host is not permanently paralyzed during oviposition, and eggs are deposited internally. Moth larvae of a variety of sizes are attacked; however, development in small hosts is delayed until the host reaches the final larval instar (Corbet 1968). Longevity of adults depends greatly on supplemental nourishment, despite the presence of suitable hosts. Females lived only 2.6 days in the absence of honey and 22.6 days with honey (Matsumoto 1974). When hosts were absent, but honey was present, the females lived even longer (35.0 days), which suggests that providing artificial food sources may be very helpful in conserving parasitoid populations in storage environments. This parasitoid has been the subject of many laboratory studies, such as on the effects of host density (e.g., Cook and Hubbard 1980, Trudeau and Gordon 1989), host patchiness (Waage 1979, Marris et al. 1986), host–parasitoid population models (Hassell and Rogers 1972), superparasitism (Huffaker and Matsumoto 1982), learning (Taylor 1974), and optimal foraging behavior (Hubbard et al. 1982).

Several studies have been conducted to evaluate the efficacy of *V. canescens* as a biological control agent. Press et al. (1982) released 1000 *C. cautella* larvae on a mixture of shelled peanuts, maize, and rolled oats scattered on the floor of a room. Biweekly releases of 50 *V. canescens* over 2.5 weeks provided 92% control, which was comparable to that by similar releases of *H. hebetor* (97% control). *Venturia canescens* is compatible with the hemipteran predator *Xylocoris flavipes*, in that combinations of the two always killed more *C. cautella* than either alone, although total mortality was less than the sum of mortalities caused by each species alone (Press 1989). However, *V. canescens* may not be compatible with *H. hebetor*. Petri dish experiments have shown that hosts previously parasitized by *H. hebetor* are not parasitized by *V. canescens*; however, *H. hebetor* will parasitize and eliminate *V. canescens* from hosts it has previously parasitized (Press et al. 1977).

*Mesostenus gracilis* Cresson is a little-studied ichneumonid that parasitizes *Ephestia elutella* pupae in stored tobacco (Bare 1942). Females prefer to parasitize host cocoons but will also attack naked pupae. They have delayed development or diapause and overwinter in the larval or pupal stage in host cocoons. Bare notes that females are rarely seen in warehouses, although they were collected in suction traps. They are easy to rear in the laboratory, and honey, water, and subdued light greatly extend their longevity. Other reported hosts that are potential stored-product pests include the pyralid moths *E. kuehniella*, *Cadra figulilella*, *Euzophera semifuneralis* (Walker), and the tortricid *Laspeyresia caryana* (Fitch) (Carlson 1979).

*Holepyris sylvanidis* (Bréthes) (Bethylidae) is a cosmopolitan ectoparasitoid of larvae of *Oryzaephilus surinamenis*, *Sitophilus oryzae*, *Tribolium castaneum*, *T. confusum*, and *Cryptolestes ferrugineus* (Stephens) (Krombein 1979). This parasitoid has been studied little except for life history measurements by Abdella et al. (1985) on *T. confusum* over a range of temperatures and humidities. However, Spitler and Hartsell (1975) noted that *H. sylvanidis* caused a significant reduction of *Oryzaephilus mercator* (Fauvel) when it invaded untreated drums of inshell almonds during a toxicological study.

*Cephalonomia waterstoni* Gahan (Bethylidae) is probably cosmopolitan and attacks several species of *Cryptolestes*, preferring *C. ferrugineus* (Finlayson 1950a; Krombein 1979). It follows kairomone trails left by wandering mature host larvae (Howard and Flinn 1990), stings the larva, permanently paralyzing it, and hides it before ovipositing. *Cephalonomia waterstoni* has been used to control *Cryptolestes* infesting laboratory cultures of *Trichogramma* (Finlayson 1950b). A natural population was reported to reduce *C. ferrugineus* populations in Kansas wheat bins (Hagstrum 1987). Some life history data are available in Finlayson (1950b), and Flinn (1991) reported temperature-dependent functional response data for the parasitoid on *C. ferrugineus* in wheat. Other species of *Cephalonomia* include *Cephalonomia gallicola* (Ashmead), which parasitizes *Stegobium pani-*

*ceum*, *Lasioderma serricorne*, *Ptinus* sp, and *Araecerus fasciculatus* de Geer (Krombein 1979) and *Cephalonomia tarsalis* (Ashmead), which parasitizes *Oryzaephilus surinamenis* and *Sitophilus oryzae* (Powell 1938; Krombein 1979).

**Pests of Specialty Commodities.** This section pertains to commodities such as tobacco, spices, and wool. The first two categories provide a special challenge to parasitoids and predators because of the presence of natural toxins and repellents (e.g., Su 1989a,b; Gunasena et al. 1990; Barbosa et al. 1991). *Anisopteromalus calandrae*, *Lariophagus distinguendus*, *Habrobracon hebetor*, and *Mesostenus gracilis* have been found in surveys of tobacco warehouses (Bare 1942). However, these parasitoids represent populations adapted to the toxins in their environment. Individuals from other commodities, as may be the case for commercial insectaries, are probably extremely sensitive to a product such as tobacco and must be selected for many generations for tolerance (Brower, unpublished data). Although wool does not have such properties, the dermestids that feed on it have morphological defenses that protect them from most natural enemies.

Immature cigarette beetles, *Lasioderma serricorne* and drugstore beetles, *Stegobium paniceum*, are parasitized by *A. calandrae* (Ahmed and Khatun 1988), *L. distinguendus* (Kashef 1956; Boucek and Rasplus 1991), *Theocolax elegans* (Burks 1979a), and *Cephalonomia gallicola* (Krombein 1979). *Ephestia elutella* has reportedly been parasitized by *A. calandrae* (Burks 1979a), *H. hebetor* (Marsh 1979), and *M. gracilis* (Bare 1942). However, commodities such as tobacco or hosts feeding on it may be too toxic for unadapted strains of these parasitoids to survive (Brower, unpublished data).

Surveys of dried spices and medicinal plants have recovered a variety of natural enemies. For example, *Habrobracon hebetor* has been reported on cacao (Richards and Herford 1930). *Holepyris hawaiiensis* Ashmead and *Cephalonomia tarsalis* were found on cacao and chilies (Richards and Herford 1930). *Anisopteromalus calandrae* and *Lariophagus distinguendus* were associated with *Stegobium paniceum* and *Rhyzopertha dominica* on coriander (Tawfik et al. 1987). *Habrobracon hebetor* was associated with a tineid moth on stored garlic (El-Nahal et al. 1985).

There are six species of *Laelius* (Bethylidae) in North America, and all parasitize dermestid larvae, usually in and around buildings (Krombein 1979). Howard (1908) has described the behavior of *Laelius trogodermatis* Ashmead, which jumps on the back of a dermestid larva and stings it in the thorax, causing instant paralysis. She then pulls at the legs as if to test for the degree of paralysis. Hairs are removed from the host, she then stings it several more times and lays one to six eggs externally on the ventral side. If other hosts are nearby, she will attack and paralyze them before returning to lay eggs, similar to the "self-provisioning" of *H. hebetor* mentioned above. The larvae develop together at the same site, then spin cocoons to pupate. Overwintering occurs in the cocoon, which is firmer and darker than those of summer.

*Laelius pedatus* (Say) attacks the varied carpet beetle *Anthrenus verbasci* (L.) (Mertins 1980), the warehouse beetle *Trogoderma variabile*, and *T. glabrum* (Klein and Beckage 1990). The host range of this species appears to be limited by efficacy of the venom (Mertins 1980; Klein and Beckage 1990) or by the efficacy of defensive hairs (*Anthrenus flavipes* LeConte) (Ma et al. 1978). *Anthrenus fuscus* Olivier often recovered from paralysis in 1–2 days, in which case the parasitoid larvae also died (Mertins 1980). *Laelius pedatus* was usually unsuccessful on four species of *Trogoderma*, was unable to attack *A. flavipes* due to its long spicisetae, and showed little or no interest in *Attagenus unicolor* (= *megatoma*) or *Thylodrias contractus* Motschulsky (Mertins 1985). *Laelius utilis* Cockerell parasitizes *T. inclusum*, *T. variabile*, *Anthrenus fuscus*, *A. verbasci*, and *A. flavipes* (Mertins 1985).

The clothes moths, *Tinea pellionella* (L.) and *Tineola bisselliella* (Hummel) are parasitized by the braconids *Chremylus elaphus* Haliday and *Apanteles carpatus* (Say) (Marsh 1979). However, there is no recent information on these species except that *A. carpatus* was reported from house fly puparia in a caged-layer poultry ranch (Rutz and Scoles 1989), although they may actually have been parasitizing *Niditinea spretella* Denis and Schiffenmüller (= *Tinea fuscipunctella*).

**Pests of Warehouses and Stores.** Parasitoids and predators may be used to help reduce the rate of infestation of packaged processed commodities by reducing ambient pest populations developing in spilled foodstuffs or nearby. For example, releases of *Anisopteromalus calandrae* reduced rice weevil infestation of spilled wheat by 90% in a simulated warehouse (Press et al. 1984). *Anisopteromalus calandrae* also was able to reduce infestation of bagged wheat by suppressing the rice weevil population in spilled wheat (Cline et al. 1985). Regarding possible contamination of packaged food, this parasitoid was able to penetrate 0% of cotton, 30% of burlap, and 47% of polypropylene bags. Failure to penetrate the bags is important to prevent contamination of processed foods with insect parts. Semiweekly releases of *Habrobracon hebetor* reduced the incidence of infestation by *Cadra cautella* (emerging from nearby food debris) of cornmeal sealed in kraft paper bags from 42.5 to 7.5% (Cline et al. 1984). *Habrobracon hebetor* also protected packages of raisins and cornmeal from *C. cautella* for at least 8 weeks, significantly reducing levels of infestation, regardless of whether the packages had been punctured (Cline and Press 1990). Similar releases of *Venturia canescens* reduced incidence of *C. cautella* infestation of cornmeal bags from 60.0 to 3.3% (Cline et al. 1986).

*Trichogramma* may be useful for parasitizing moth eggs; however, the fact that some species perform better in light than darkness may be important for warehouse applications (Brower 1991c). For commodities, such as animal feeds, which can tolerate the presence of some insect fragments, some of the previously mentioned parasitoids may be of use. For example, *Habrobracon hebetor* helped provide natural control of *Cadra cautella* in a citrus pulp warehouse (Hagstrum and Sharp 1975).

*Diptera*

Tachinid flies in the genus *Clausicella* have been reported to parasitize stored product pyralid moths. *Clausicella floridensis* (Townshend) and *Clausicella neomexicana* (Townshend) have been reported to parasitize *Plodia interpunctella* (Arnaud 1978). *Clausicella floridensis* was reared from *P. interpunctella* in dried fruits in California but was considered rare and of little importance for natural control (Hamlin et al. 1931; Simmons et al. 1931). Both these parasitoids have been reported in surveys of the sunflower moth, *Homoeosoma electellum*, which suggests that they have little affinity for storage conditions (Satterthwait and Swain 1946; Teetes and Randolph 1969). *Clausicella suturata* (Townshend) parasitizes *Ectomyelois ceratoniae* in carob pods in the Mediterranean (Kugler and Nitzan 1977). The fly oviposits near the opening of its host's tunnel. The eggs hatch immediately and the larvae crawl into the tunnel, following the host's silk. Normal development is completed only on fourth and fifth instar larvae, which spin a cocoon before being killed by the parasite. Only one parasite is produced per host, and it pupates inside the host cocoon.

## 3. Insect Predators

A wide variety of predators attack stored-product pests, but many of them do not seem to play a major role in regulating prey populations. Many of these predators are probably only occasional visitors to the storage situation and environmental conditions for their increase are seldom ideal. A few notable exceptions to this generalization are often found in storage situations where they survive and reproduce without any obvious impediments (except pesticides) and where they probably exert considerable control over their prey populations. Several of these species will be discussed in more detail below.

*Hemiptera as Predators*

Many of the predators of stored-product insects belong to the order Hemiptera or true bugs. This is surprising because the true bugs make up only a small proportion of the predatory insects in the outdoor situation (Sweetman 1958). One possible reason for this unusual distribution arises from the close association between predatory bugs and insects under bark. Many of the pests of stored products are thought to have arisen from species associated with the subcortical (under bark) habitat (Yoshida 1984). As the pest species adapted to the new habitat of human-made artificial storage environments, so too did their natural predators, and they are still often found in close association in commodity storages.

The most studied of the hemipteran predators is *Xylocoris flavipes* (Anthocoridae), the warehouse pirate bug. As the common name suggests, it is commonly found in peanut warehouses, grain bins, and other commodity storage situations (Arbogast 1978). It appears to be truly adapted to the storage environment (Arbogast 1978), and a variety of studies have shown that this predator has great potential for suppressing populations of stored-product pests. The developmental biology of *X. flavipes* is fairly typical of other small bugs and was described in detail (Arbogast et al. 1971) with all stages illustrated. In 1978 Arbogast reviewed the biology and possible usefulness of this species, and concluded that *X. flavipes* possesses attributes that make it an effective predator. This publication (Arbogast 1978) provided the basis for many applied studies that have been conducted since then. Initial laboratory tests with the warehouse pirate bug showed that it would attack almost any small stage of several beetles and moths (Jay et al. 1968). The eggs and larvae of a wide array of the smaller species of beetles appear to be the preferred hosts of *X. flavipes*, but those of other species are utilized as well (Table 3). The internal grain feeders are not particularly subject to predation because of their hidden location, but even they suffer a certain amount of egg predation (Jay et al. 1968; LeCato and Arbogast 1979). This broad range of prey species is a significant attribute for a biological control agent because this single agent can suppress many different pest species simultaneously and can switch to less abundant "alternate" hosts when prey are scarce.

Most early tests employed only one or two prey species and relatively small quantities of commodity. For example, Arbogast (1976) infested 35-quart lots of shelled corn with 20 pairs of sawtoothed grain beetles *O. surinamensis*, then added predators in different ratios. After 16 weeks the degree of population reduction ranged from 97 to 99% compared with the untreated control, depending on predator/prey ratio. A successful test using *X. flavipes* to suppress populations of the red flour beetle, *Tribolium castaneum*, in simulated warehouses was also reported (Press et al. 1975). Various numbers of the predator were introduced into plywood bins containing six bushels of inshell peanuts infested with *T. castaneum*. Fourteen weeks after predator introduction, random samples of the peanuts and insects were taken and counted. The number of *T. castaneum* was reduced 90.6% when 10 predators per bushel were released and 97.6% when 80 predators were used, and insect damage to peanut kernels was reduced 66.2 to 90.8% at the two release rates, respectively. Reductions in pest numbers and commodity damage in this short test were significant, and reductions tended to increase dramatically as the length of the test is extended.

Populations of *C. cautella*, *T. castaneum*, and *O. surinamensis* did not increase in a room containing grain debris when *X. flavipes* was released in small numbers, whereas all three populations increased greatly in the room when no predators were released (LeCato et al. 1977). The degree of control achieved continued to increase over the 100 day duration of the test. Populations of all three pests in the untreated room were 100-fold greater than in the treated room at the termination of the test (LeCato et al. 1977).

**Table 3** Species of Stored-Product Insects that are Known Prey of *Xylocoris flavipes*

| Species | Eggs | Larvae | Pupae | Adults | Reference |
|---|---|---|---|---|---|
| Coleoptera | | | | | |
| *Attagenus unicolor* | | X | | | LeCato 1976 |
| *Trogoderma granarium* | | | | | Arbogast 1978 |
| *Lasioderma serricorne* | | X | | | Abdel-Rahmen et al. 1978–9 |
| *Stegobium paniceum* | | X | | | Awadallah et al., unpublished |
| *Rhyzopertha dominica* | | X | | X | Abdel-Rahmen et al. 1978–9 |
| *Lophocateres pusillus* | | | | | Brower, unpublished |
| *Carpophilus dimidiatus* | | | | | Brower and Press 1992 |
| *Ahasversus advena* | | | | | Brower and Press 1992 |
| *Cathartus quadricollis* | | | | | Press et al. 1979 |
| *Cryptolestes minutus* | | | | | Wen and Deng 1988 |
| *Cryptolestes pusillus* | | | | | Brower and Press 1992 |
| *Oryzaephilus mercator* | | | | | Press et al. 1979 |
| *Oryzaephilus surinamensis* | | X | | | Abdel-Rahmen et al. 1978–9 |
| *Latheticus oryzae* | | | X | | Tawfik et al. 1982 |
| *Tribolium castaneum* | X | X | X | X | LeCato 1976 |
| *Tribolium confusum* | | X | | | LeCato 1976 |
| *Palorus ratzeburgi* | | | | | Wen and Deng 1988 |
| *Typhaea stercorea* | | | | | Brower and Press 1992 |
| *Acanthoscelides obtectus* | X | | | | Brower and Sing, unpublished |
| *Callosobruchus maculatus* | X | | | | Brower and Sing, unpublished |
| *Zabrotes subfasciatus* | | X | | | Brower and Sing, unpublished |
| *Sitophilus oryzae* | | X | | | Abdel-Rahmen et al. 1978–9 |
| *Sitophilus zeamais* | | | | | Wen and Deng 1988 |
| Lepidoptera | | | | | |
| *Sitotroga cerealella* | X | X | | | LeCato and Arbogast 1979 |
| *Galleria mellonella* | | | | | Arbogast 1978 |
| *Corcyra cephalonica* | X | X | | | Tawfik et al. 1982 |
| *Plodia interpunctella* | X | X | | | Abdel-Rahmen et al. 1978–9 |
| *Ephestia kuehniella* | | X | | | Abdel-Rahmen et al. 1978–9 |
| *Cadra cautella* | | | | | Press et al. 1974 |

A biological control test established in small buildings serving as experimental inshell peanut storages showed that the release of large numbers of *X. flavipes* could suppress stored-product moth populations (Brower and Mullen 1990). The degree of suppression of *C. cautella* and *P. interpunctella* depended on both the prey species and environmental conditions. Two releases of about 40,000 *X. flavipes* into each of two treatment storages were followed by 6 months of sampling moth populations. Release of the predator suppressed populations of *C. cautella* and *P. interpunctella* by as much as 78.8% and 71.4%, respectively, before a freeze in January eliminated the *C. cautella* population. Suppression of *P. interpunctella* lasted through the 7 month test period (Brower and Mullen 1990).

In a recent test (Brower and Press 1992), populations of stored grain pests infesting grain residues in empty corn bins were affected differently by the release of 50 pairs of the predatory warehouse pirate bug, depending on their size and niche. Large insects such as late instar pyralid moth larvae and adults were apparently unaffected, and species such as the *Sitophilus* weevils and *R. dominica* that develop within grain kernels were much less affected than small external feeders. However, populations of a wide variety of small beetle species including both direct grain feeders and secondary feeders were greatly reduced in abundance or even eliminated entirely by the predator. Reductions in the populations of these species at the termination of the test ranged from 70.0 to 100% (Table 4). This test showed that residual populations of several species of small beetles in empty grain bins can be greatly reduced by weekly releases of small numbers of a predator, and if specific parasites were also released for both moths and primary grain pests, the whole pest complex might be greatly reduced or eliminated before newly harvested grain is placed in storage. However, this concept has yet to be tested.

Several other species of predatory bugs appear to have good potential for suppressing pest populations if conditions are favorable. Unfortunately, only

**Table 4** Percentage of Population Reduction in Insects from 2 kg Corn Residues Removed from Grain Bins Found in *Xylocoris flavipes* Release Bins Compared to Check Bins

| | Percentage population reduction | |
|---|---|---|
| *Pest Species* | At removal | Incubated for 3 weeks |
| *Tribolium castaneum* | 77.1 | 84.3 |
| *Oryzaephilus surinamensis* | 99.5 | 99.2 |
| *Cryptolestes pusillus* | 69.7 | 90.4 |
| *Ahasverus advena* | 72.9 | 38.5 |
| *Carpophilus* sp. | 100 | 95.0 |
| *Typhaea stercorea* | 100 | 100 |
| Miscellaneous beetle spp | 64.7 | 97.6 |
| *Sitophilus* spp. | 40.3 | 81.2 |
| *Rhyzopertha dominica* | 82.0 | (+58.8) |
| Pooled beetles | 87.4 | 84.7 |
| *Plodia interpunctella* | (+153) | 68.6 |
| *Cadra cautella* | 50.7 | 18.4 |
| *Plodia + Cadra* | (+31) | 33.9 |
| Total | 85.7 | 77.0 |

*Source*: Brower and Press (1992).

recently have any serious studies been devoted to them and these have been confined to laboratory tests. *Xylocoris sordidus* (Reuter) is a species widespread in the western hemisphere (Arbogast et al. 1983), and it has been found several times in association with insect-infested inshell peanuts in Georgia. This species can also attack several species of storage pests including *P. interpunctella*, *C. cautella* (Arbogast et al. 1983), *E. kuehniella*, *Corcyra cephalonica* (Stainton), and *T. confusum* (Awadallah et al., unpublished). The biology of this species has been reported by Arbogast et al. (1983) and by Tawfik et al. (in press), and they report that this species prefers warm but humid conditions for survival. Arbogast et al. (1983) also noted that this species prefers to oviposit in a soft substrate such as pith. A related species, *Xylocoris galactinus* (Fieber), also appears to prefer hot, moist conditions where it may be primarily a predator of house fly larvae, *Musca domestica* L. (Hall 1951). However, this species also occurs in stored commodities where it attacks larvae of *L. serricorne*, *T. confusum*, and probably several other species (Afifi and Ibrahim 1991). Another species of predatory bug in the family Anthocoridae is *Dufouriellus ater* (Dufour), a species that is probably a subcortical dweller but one that occasionally inhabits peanut warehouses and beehives (Arbogast 1984). Arbogast (1984a) studied the life history of this species and reported that it preyed readily on the eggs of *P. interpunctella* and *C. cautella* in the laboratory. However, he concluded that because of its low reproductive potential, it may not be possible to mass produce this species at a reasonable cost. Awadallah et al. (1981) also reported rearing this species on eggs of the rice moth, *C. cephalonica*, and on larvae of *T. confusum*. Recent work has demonstrated the potential as a biological control agent of yet another anthocorid, *Lyctocoris campestris* (F.) (Parajulee and Phillips 1992). This bug is considerably larger than the species discussed above and consequently kills much larger prey, including the larger larvae of *Plodia* and *Cadra* (Parajulee and Phillips, 1993). This species, like most of the others, prefers a moist or wet site in which to lay its eggs and moisture is essential for good egg survival. Unfortunately, this may limit the usefulness of this species in grain storages because moisture sources are lacking. Larvae of most stored-product moths were suitable hosts, as were 20 species of beetles associated with stored products. This species also has a rather low reproductive potential (Parajulee and Phillips 1992), which may further limit its potential usefulness as a biological control agent.

A few other hemipterans have been found attacking stored-product pests. The most notable of these is a member of the predaceous family Reduviidae, *Allaeocranum biannulipes* (Montrouzier and Signoret). This predator is larger than the anthocorids (ca. 7.5 mm in length) and seems to prefer the larger moth larvae such as *E. kuehniella*, *Plodia interpunctella*, and the meal moth *Pyralis farinalis* L. (Tawfik et al. 1983). This predator occurs commonly in flour mills and silos in lower Egypt, where it preys on both flour beetle and moth larvae (Awadallah et al. 1984). Small-scale laboratory studies gave population reductions for *E. kuehniella* of up to 62.6% compared to the control, for *Corcyra cephalonica* of up to 59.8%, and for *T. confusum* of up to 95.5%. The authors did not offer any opinion as to the utility of this predator but their results were encouraging enough to warrant further research on this aspect. Another species of reduviid that showed promise in tests in India was *Amphibolus venator* Klug (Pingale 1954). Laboratory culturing was satisfactory although the reproductive rate was rather low. However, in a warehouse test with 1000 bags of wheat infested with *C. cautella* and *Alphitobius diaperinus* in a check warehouse and a warehouse where 100 adult predators were released, control of both pests was excellent at the end of 5 months. The moth population decreased 95% in the treatment, whereas it increased 197% in the check warehouse, and the *A. diaperinus* population was reduced 89% as opposed to an increase of 161% in the check population (Pingale 1954). A number of other species of reduviids are reported as predators of storage pests but their occurrence is sporadic.

A termatophilid bug, *Termatophylum insigne* Reuter, was reared in the laboratory in Egypt on larvae of *T. confusum*, *L. serricorne*, and *S. paniceum* (Awadallah et al. 1984). Although normally considered a predator of the thrips *Gynaikothrips ficorum* (Marchal), *T. insignae* had a higher reproductive potential on *T. confusum* and was also quite successful on the other storage pests as well. This predator has been found associated with storage pests in plant materials stored for use as drugs, and the aforementioned authors recommend storing these drug materials at low temperature (20°) to favor the development of the predator over that of the most common prey, *T. confusum*.

### *Coleoptera as Predators*

Quite a number of beetles are known to be facultative predators of stored-product pests, but many of these can also be pests themselves in the absence of prey (LeCato 1975). A few species, however, appear to be obligate predators such as some of the Histeridae and Cleridae. Several species of small histerids are commonly found associated with storage situations where they are probably predators of storage pests, but, with one exception, they have not been studied as predators. The exception is *Teretriosoma nigrescens*, a predator of the larger grain borer, *Prostephanus truncatus*. As reported by Rees (1987), *P. truncatus* has in recent years been introduced into parts of East and West Africa from Central America, probably with shipments of corn. Rees continues that local storage practices and ideal environmental conditions together with little predation or competition from other insects have led to its becoming a very important pest. Rees (1985) had reported earlier from studies done in Central America that *T. nigrescens* was an effective predator of *P. truncatus* in shelled corn in laboratory tests and that it was not present in Africa. Thus, he proposed that its importation and release into Africa might help to slow the spread and reduce the severity of the outbreak of *P. truncatus* in that continent. Rees (1987) reported that over a 14 week period at

27°C the mean rate of increase of three pairs of *P. truncatus* was reduced by a factor of 10 by the addition of four adult predators after 3 weeks. He also reported that the predator affected the buildup of *T. castaneum* populations, reducing them about 50%, but it had no effect on *Sitophilus zeamais* Motschulsky populations. *T. nigrescens* apparently has a very close evolutionary association with its host, and this assumption was supported by the work of Boeye et al. (1992) showing the predator responding to traps baited with the synthetic sex pheromone of *P. truncatus*. This intimate association between predator and prey is not obligate, since *T. nigrescens* has been shown to also attack other species of bostrichids that are stored-grain pests such as *Rhyzopertha dominica* and *Dinoderus minutus* (F.) (Rees 1991). Populations of *P. truncatus*, *R. dominica*, and *D. minutus* were reduced 83, 36, and 91%, respectively, by only five adult predators, and the weight loss of the corn was also reduced significantly in each case. Rees (1991) speculates that the introduction and release of *T. nigrescens* into Africa for the control of *P. truncatus* at the farmer storage level has merit (classic biological control) and that an additional benefit might be the suppression of the native bostrichid, *Dinoderus minutus*.

The family Cleridae contains species that are primarily predators; however, several species have adapted to feeding on stored commodities such as cheese, cured meat, and dried fish and are considered stored-product pests. One of the predatory species, *Thaneroclerus girodi* Chev., is recorded as occurring in tobacco storages as a predator of the cigarette beetle, *L. serricorne* (Reed and Vinzant, 1942).

Many other insects such as staphylinid beetles, scenopinid and scatopsid flies, and psocids may be occasional predators of stored-grain pests, but few reports of potential effects on prey populations exist. Most of these species occur so sporadically that conditions in the storage environment are probably not favorable for their success and the likelihood of their use as commercial biological control agents is slim.

### *Acarina (Mites)*

Mites are very often associated with stored commodities and most species are considered to be pests themselves. However, any study of the stored grain ecosystem finds a whole community of mites that function as predators (Sinha 1961; Sinha and Watters 1985). Most of these mite species interact primarily with each other, with their prey, and with the surrounding environment, but some of them function as predators (ectoparasites) of the insects present.

**Mites as Insect Predators.** To date, most of the research in this area has been devoted to the straw itch mite, *Pyemotes tritici* (Bruce 1983). It has long been known by stored-product entomologists that infestations of *P. tritici* in stored-product insect cultures would greatly debilitate them. Because of this, Bruce and LeCato (1979) studied the potential of this parasite as biological control agent for storage pests. They reported that it would attack all stages of *P. interpunctella*, *C. cautella*, *O. mercator*, and *L. serricorne* and all but the adults of *T. castaneum* (Bruce and LeCato 1979). It is known to attack a wide range of other storage pests and to be particularly destructive to *Sitotroga cerealella* cultures. Bruce and LeCato (1979) concluded that *P. tritici* has great potential as a biological control agent for stored-product insects and that it is an extremely efficient and effective parasite. Unfortunately its use as a commercial biological control agent may be limited because it also attacks humans (hence its name, straw itch mite or grocer's itch mite) (Moser 1975).

One species of mite, *Blattisocius tarsalis* (Berlese), is a very common egg predator of many species of storage pests, both moths and beetles (Darst and King 1969; Haines 1981). This mite was used in some early laboratory experiments to study the interactions and resultant population effects of predator (*B. tarsalis*)–prey (*E. kuehniella*) interactions in a closed system (Flanders and Badgley 1960, 1963; White and Huffaker 1969a,b). Graham (1970a,b) studied bagged corn in a warehouse in Africa and reported the natural control of an infestation of *C. cautella* because of egg predation by *B. tarsalis*. He also speculated (Graham 1970c) that the use of chemical control measures was differentially affecting the *B. tarsalis* more than the prey *C. cautella*, leading to a breakdown in this "natural control." Haines (1981) extended this work and showed that fumigation of stacks of bagged grain under tarpaulins reduces predator effectiveness due to greater effects on the predator and by removal of alternative prey (*T. castaneum* eggs). The available data on *B. tarsalis* are encouraging and some good controlled studies on the application of this predator in small scale or simulated storages are urgently needed.

There are quite a number of other species of mites that frequently or sporadically attack the eggs of stored-product insects. Other species of *Blattisocius* are known with their own preferred stored-product prey (Barker 1967), and many species in other groups are also reported as predators of stored-product pests (Gerson and Smiley 1990; Kumar and Naqi 1990; Steinkraus and Cross 1993). However, little is known about their choice of prey or their possible utility as biological control agents.

**Mites as Mite Predators.** Predatory mites appear to have the greatest potential for controlling other species of mites that are pests in stored commodities (McMurtry 1984). More research has been conducted on this use of predatory mites and the work has progressed to the application stage in stored grain commodities. At present the most promising of these predators is *Cheyletus eruditus* (Schrank), a member of a large predaceous family of mites (Cheylelidae). Much of the early work on its biology and effectiveness as a predator was summarized by Solomon (1969). Pulpan and Verner (1965) reviewed the few studies that considered possible application of *C. eruditus* before conducting the seminal study on control of mite infestations in grain using releases of predatory mites. These authors report the complete control of populations of *Acarus siro* L. and *Glycyphagus destructor* (Schrank) in two large-scale trials following the introduction of the predaceous mite, *Cheyletus eruditus*. Further tests demonstrated that the introduction of the predator into uninfested grain destined for long-term storage prevented the establishment of pest mite populations in 89% of the cases (Pulpan and Verner 1965). As a general principle, the authors proposed that the predator be collected from storages where complete eradication of harmful mites had been completed and that they be introduced on the surface of grain in the spring or fall. This work has been greatly extended since then by Zdarkova and her colleagues. Recent proposals have centered around the release of *C. eruditus* into empty granaries and storages to reduce or eliminate residual pest mite populations, thus preventing them from infesting newly stored grain (Zdarkova 1991; Zdarkova and Horak 1990).

Many predatory mites are known that attack particular species or groups of species of mites. Some excellent bionomic studies of these predators are available, yet little is known about their potential usefulness as biological control agents (see Barker 1991, 1992 and references therein). A thorough discussion of this area of research is outside the scope of this volume and interested readers are referred to work by Griffiths and Bowman (1984).

## IV. FIELD TESTS

Very few field-scale experiments to evaluate the efficacy of biological control agents of stored-product pests have been performed. Small-scale laboratory experiments simulating storage conditions are mentioned in the preceding sections and were reviewed by Nilakhe and Parker (1990). However, it is critical to evaluate the performance of biological control agents in the field because of the potentially unpredictable influences of scale, environmental conditions, pesticide residues, and interspecies interactions. The results from such trials are needed to provide confidence in the level of control expected from applying a prescribed biological control program.

Keever et al. (1986) released a total of 324,000 *Habrobracon hebetor* and 191,000 *Xylocoris flavipes* in a commercial warehouse in Georgia containing 36 metric tons of inshell peanuts over the storage period (October 5 to January 3). Seven releases were made at roughly 2 week intervals. Both the treatment warehouse and a control one (with a conventional application of malathion) were naturally infested with *Plodia interpunctella* and *Cadra cautella*. At the end of the study, the moth populations were 54% lower in the biological control warehouse than in the malathion-treated warehouse (Fig. 3). Peanut damage caused by insects was also correspondingly less in the biological control warehouse (2.1 ± 0.7% [SE] of intact-shell and 17.3 ± 1.6% of loose-shell kernels) than in the malathion warehouse (5.4 ± 2.0% of intact-shell and 35.0 ± 10.4% of loose-shell kernels). Although this study indicates the potential effectiveness of biological control in

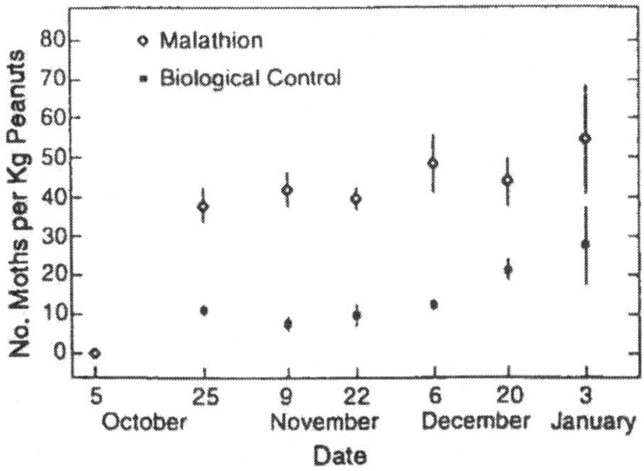


**Figure 3** Comparison of biweekly releases of *Habrobracon* (*Bracon*) *hebetor* and *Xylocoris flavipes* (biological control) to conventional application of malathion to control moths (*Cadra cautella* and *Plodia interpunctella*) in commercial warehouses of inshell peanuts (mean ± SE). (Data from Keever et al. 1986.)

inshell peanut storage, the experiment was not replicated, and it involved sites of different sizes 80 km apart.

The first commercial-scale experiment using replication was conducted by Parker and Nilakhe (1990) on grain sorghum in storage bins in Texas. Five treatments (Actellic, Reldan + Diacon, malathion, biological control, and untreated controls) were replicated three times using 2450 bushels per bin. The experiment started with fresh grain in September, 1988, and continued for 1 year. The biological control bins were managed by a commercial insectary (Biofac, Inc.) that released insects roughly every 9 days for a total of 12,700 *Anisopteromalus calandrae*, 1900 *Habrobracon hebetor*, 400 *Theocolax elegans*, 1,675,000 *Trichogramma pretiosum*, and 17,083 *Xylocoris flavipes* per bin. About 107,000 gravid female *Pyemotes tritici* were released during October and again in November. The principal pest species were *Cryptolestes pusillus* (Schönherr), *Cryptolestes ferrugineus*, *Rhyzopertha dominica*, *Sitophilus oryzae*, *Sitophilus zeamais* Motschulsky, *Tribolium castaneum*, *Typhaea stercorea* (L.), *Plodia interpunctella*, and *Latheticus oryzae* Waterhouse. All treatments failed to provide economic pest control due to constant migration of pests from nearby bins of sorghum that were heavily infested. However, the malathion and untreated bins suffered the most weight loss, and pest control in the biological control bins was comparable to that for Actellic and Reldan + Diacon (Table 5). The high cost of the biological control program (4.5¢/bu) is at least partly attributed to the release of excessive numbers of biological control agents. This is because the experiment was con-

**Table 5** Efficacy of Different Treatments for Protecting Grain Sorghum during 1 Year of Storage in Texas

| Treatment | Grain weight loss (%)[a] | Return relative to untreated (¢/bu) | Cost (¢/bu) | Added return over treatment costs (¢/bu) |
|---|---|---|---|---|
| Actellic | 2.71 a | 2.376 | 2.24 | 0.137 |
| Reldan + Diacon | 2.40 a | 3.120 | 3.48 | −0.360 |
| Malathion | 3.78 b | −0.002 | 0.90 | −0.900 |
| Biological control | 2.33 a | 3.288 | 4.5 | −1.212 |
| Untreated | 3.70 b | | | |

[a]Means within a column followed by the same letter are not significantly different at the 5% level (analysis of variance with Duncan's multiple range test).
*Source*: Parker and Nilakhe (1990).

ducted primarily to evaluate the potential efficacy of biological control, realizing that more work needs to be done to determine the species, numbers, and frequency of release required for control. Such information should greatly reduce costs.

Another test of biological control was conducted on long-grain rough rice stored in bins in Texas (Cogburn and Brower, 1990). Each of four metal bins was loaded with about 7.5 metric tons of uninfested U.S. #2 or #3 grade rice in March, 1988. Aeration fans were sealed, and openings were screened to obstruct rodents and insects. Two bins served as untreated controls and two received biological control agents. Each bin was infested with 100 unsexed adult *Sitophilus oryzae* and five pairs of *Plodia interpunctella*. Two weeks later weekly releases were begun of about 2200 *Anisopteromalus calandrae*, 2000 *Habrobracon hebetor*, and 13,000 *Trichogramma pretiosum* per bin. Five hundred *Xylocoris flavipes* were also released weekly per bin starting in November (when specimens became available). All biological control agents were shipped from the U.S. Department of Agriculture laboratory in Savannah, Georgia, except *X. flavipes*, which came from a commercial insectary in Texas.

The bins proved to be pervious to insects. The principal pests encountered in spear samples and pitfall traps were *Sitotroga cerealella*, *Rhyzopertha dominica*, *Sitophilus oryzae*, *Cryptolestes pusillus*, *Ahasverus advena* (Waltl), *Typhaea stercorea*, and *Tribolium castaneum*. Numbers of *R. dominica*, *C. pusillus*, and *T. stercorea* were substantially lower in the biological control bins than in the untreated bins (although no differences were statistically significant). *Sitophilus oryzae* and *T. castaneum* also tended to be lower, but *S. cerealella* tended to be higher in the biological control bins. During the 18 month course of this experiment all the bins were fumigated with phosphine six times because of excessive pest populations.

This experiment points out several important points. The important pests must be known in advance so that the appropriate biological control agents can be chosen. For example, the parasitoid *Pteromalus cerealellae* could be used against *S. cerealella*, *Cephalonomia waterstoni* parasitizes *C. pusillus*, and *Theocolax elegans* may be a better choice than *A. calandrae* for controlling *R. dominica*. The quantity and timing of releases may be critical. It is particularly important to establish natural enemies early before pest populations begin increasing exponentially. Thus, if pest invasion is a constant problem, massive initial releases for 2–3 weeks will establish a resident population of natural enemies that can attack incoming pests. Sustained weekly releases may be of value primarily when pest populations are too low to sustain a parasitoid or predator population. Biological control is not likely to be cost-effective in such a situation. Lastly, any test requires as much replication and as much sampling as can be afforded to improve statistical confidence in inferences made about the outcome. We are in the early stages of developing biological control for stored commodities and must design experiments that permit us to learn the most from our "mistakes."

## V. DESIGNING A BIOLOGICAL CONTROL PROGRAM

### 1. Using Demographic Data

One of the principal uncertainties of using biological control agents is knowing how many and how frequently to release in order to achieve economic control. The ability to make such predictions depends on the collection of extensive data on life history characteristics (survivorship, development time, fecundity, progeny sex ratio) and functional response (number of hosts killed in relation to the number available), particularly at different temperatures, with different commodities and hosts. Some of these data have been collected for *Lariophagus distinguendus* (e.g., Gonen and Kugler 1970; van den Assem 1971; Bellows 1985; Hong and Ryoo 1991; Ryoo et al. 1991b), *Anisopteromalus calandrae* (e.g., Ghani and Sweetman 1955; Heong 1981; Hassell et al. 1985; Smith 1992, 1993a,b), *Theocolax elegans* (van den Assem and Kuenen 1958), and *Habrobracon hebetor* (Nikam and Pawar 1993; Brower, unpublished). However, much more is needed to support the development of computer models to simulate population dynamics and help predict the effectiveness of parasitoids under changing environmental conditions.

### 2. Application Techniques

The concept behind the application of biological control in the stored-commodity ecosystem is different from the "classic" application of biological control. Biological controls usually are employed only after an exotic or native pest has increased to "outbreak" proportions; parasites or predators are released to restore a more natural balance that accompanies a greatly reduced pest abundance. This approach certainly is theoretically possible with stored-product pests, but from a practical standpoint it would probably be a case of "too little too late." Instead, the use of biological control for stored-product pests should be viewed as a preventive treatment. Several application approaches could be used, but inundative releases are the most likely to be effective. These are releases of large numbers of beneficials at intervals frequent enough not to miss any of the intended host or prey stage. Ideally these releases should start in the empty storages before the commodity is harvested and continue through the storage season, or until the danger of infestation from outside sources has passed. If some biological control agents already are present, this same approach is called augmentative; these agents can be augmented by releases of the same species or a complementary species. A strictly inoculative approach, in which relatively small numbers of beneficials are released early in the storage season, may provide adequate levels of control in storage situations where small, well-established populations of hosts are present. A good example is in farm-stored grain intended for use as animal feed. In this case, suppression, not elimination, of the pest populations is probably adequate, considering the small initial expenditure for inoculative releases and the higher tolerance for the presence of insects.

## 3. The Biological Control Strategy

Designing a biological control program for stored-product insect pests requires careful planning. It is not simply a matter of selecting parasitoids and predators from a directory (Hunter 1992). Because most of the natural enemies are host-specific, one must determine which pest species are causing the problem. In an ideal situation this can be based on which insects have been a problem in past years. A well-designed sampling program would indicate which pest species typically exceed economically damaging levels. Specific parasitoids and predators that attack these pests would then be selected, and because several commercial companies produce these insects, they could be purchased for release. In addition, the correct number of parasitoids and predators to release at the most appropriate time must be determined. Biological control is most effective when there is a high parasite to host ratio, such as 1:2. Each parasite can attack several hosts each day (Flinn 1991). For example, *Cephalonomia waterstoni* can paralyze up to 14 *Cryptolestes ferrugineus*/day, and lay two to three eggs per day. Low numbers of insects generally infest commodities initially; therefore parasite or predators need to be released early before pests reach high numbers. However, if they are released too early, suitable host stages may not be available. Sequential releases can add additional insurance, but each additional release will add to the control cost. Sampling the commodity can help you monitor how well the natural control methods are doing and whether additional releases are necessary.

Simulation models of parasitoids and predators can be invaluable in the design and evaluation of a biological control program. Kareiva (1990) has suggested that models should be used to develop biological control programs. Models are becoming available for stored-product insects, and will probably be used to design biological control programs. Models can be used to determine not only when and how often to release beneficials but also how effective the program will be. A simulation model already exists of a pathogen and its interactions with its host, *T. confusum* (Onstad and Maddox 1990). Simulations with a model for *C. waterstoni* parasitizing *C. ferrugineus* showed that changing the timing of a parasitoid release had a greater effect than releasing more parasitoids (Flinn and Hagstrum, unpublished). Adding four times as many parasitoids only decreased the maximum host population by 40%, but releasing parasitoids at day 25 instead of day 45 reduced the maximum host population by 75%. Models are in development for *Sitotroga cerealella* (Weaver et al., unpublished) and *Sitophilus zeamais* (Throne et al., unpublished).

## 4. Integrated Control

Integrating biological control with other control methods is critically important. Some methods will be compatible, whereas others will not. An example of a perfectly compatible control method for wheat is parasitoids plus fall aeration. In this system, parasitoids would be released in the grain about 3 weeks after the grain is binned (in June, July, or August, depending on location). Aeration would start in early October with the arrival of cool fall nights. The parasitoids would inhibit beetle populations from exceeding economically damaging levels during the warm summer months, until fall aeration could be used to cool the grain. Cooling the grain to 20°C almost completely inhibits beetle growth. Other controls, such as insecticidal protectants, are often not compatible because beneficials are typically more susceptible than their hosts (Croft 1990) (see below). Protectants are applied at binning, which precludes releasing parasitoids at this time, and they typically last for several months. Releases could be made after fumigation if sufficient time was allowed for the fumigant to dissipate (1–2 weeks). Many species of parasites and predators are probably able to overwinter in the grain, and thus would provide additional protection when the grain warms in the spring. This protection may carry through the marketing system until fumigants are used.

## 5. Integration with Pesticides

Insecticides have traditionally been considered incompatible with the application of biological control because they often affect parasitoids and predators more severely than the target pest (Croft 1990). For example, organophosphates, pyrethroids, and carbamates were found to be more toxic to *Anisopteromalus calandrae* than to *Callosobruchus maculatus* (Mohammed and Al-Jabery 1988). Control of *Sitophilus zeamais* and *Sitotroga cerealella* by *A. calandrae* in maize cribs was better in the absence of pyrethrin than when it was applied (Kockum 1965). Likewise, *Theocolax elegans* controlled *Sitophilus granarius* in wheat better in the absence of a pyrethrum–piperonyl butoxide mixture than in its presence (Loosjes 1957). One way to reduce the incompatibility of insecticides with biological control is to use formulations with high "selectivity," which are more toxic to the pest than to the biological control agent. For example, permethrin had a much lower impact on adult survival and fecundity of *Habrobracon hebetor* than did pyrethrin or pyrethrin–piperonyl butoxide (Press et al. 1981).

There is increasing interest in the genetic selection of insecticide resistance in parasitoids and predators (Hoy 1991). However, the natural occurrence of resistant strains should not be overlooked. *Trichogramma pretiosum*, *T. parkeri*, and *H. hebetor* were found naturally occurring in some peanut storages in Georgia, despite heavy use of insecticides (Brower 1984; Keever et al. 1985). Baker and Weaver (1993) recently discovered a strain of *Anisopteromalus calandrae* that is at least 90-fold more tolerant to malathion than its host, *Sitophilus oryzae* and Baker (1994) reported that it is cross-resistant to other protectants as well.

Microbial insecticides such as *Bacillus thuringiensis* (*Bt*) are often more selective than synthetic compounds; however, they may adversely affect parasitoids or predators. For example, *Bt*-infected *P. interpunctella* larvae reduced egg production and longevity of *H. hebetor* and reduced the attack rate and fecundity of *X. flavipes* (Salama et al. 1991). However, if such natural enemies preferentially attack hosts that received lower doses or that are less susceptible to an insecticide or pathogen, compatibility would be improved.

## VI. SUMMARY

Biological control promises to be an important component of integrated pest management strategies for many types of stored commodities. Biological control agents include pathogens, parasitoids, and predators, and are distinguished by the fact that they are capable of reproducing after release. In some situations, single inoculations may be sufficient to establish lasting control. In other cases, periodic inundative releases may be required to help maintain enough natural enemies to provide adequate pest control. As an alternative, supplemental food, factitious hosts, synthetic kairomones, or oviposition sites can be provided to boost populations of natural enemies. It is important to understand that biological control can only be used as a prophylactic, not remedial, strategy, and that it will be most effective when integrated with other control strategies such as sanitation, fumigation, aeration, and packaging.

Successful use of biological control depends on proper understanding of the ecology of each pest control problem. Because most biological control agents attack a limited number of pest species, it is important to identify which pest species are of interest and to match them with suitable agents. The type of commodity, environment, and storage conditions can affect decisions on what, when, and how many biological control agents to release. Quality control of biological control agents is a serious concern because their efficacy can be affected by conditions during production, genetics (strains, or inbreeding), rearing, storage, shipping, and handling conditions. Live material is generally more sensitive to mishandling than chemical compounds, and they will require modifications in the training and practice of pest control operators. However, some pathogens, such as *Bacillus thuringiensis* (*Bt*), or their toxins can be formulated and applied as conventional insecticides. But one limitation of such "biological" insecticides is that pests may develop resistance to them, as they have done to most chemical insecticides.

Biological control agents are now legal to use in the United States in most stored-food situations. *Bt* is commercially available, and at least one commercial insectary is providing a few species of parasitoids and predators appropriate for stored commodities. However, it should be realized that there is no system for registration of parasitoids or predators, establishing efficacy, safety, and application rates, as there is for chemical or "biological" insecticides. There is a great need for research to determine the number and frequency of beneficial species to release to provide specific levels of control. Such research is likely to benefit from the development of computer simulation models that will help to apply the results to changing conditions and to integrate biological control applications with other management options.

Biological control is certainly not a panacea for pest control in stored grain or other food commodities, but it is becoming an increasingly important part of an integrated pest management approach to controlling stored-product insects. At present, a great deal of information is accumulating on the behavior and basic biology of pathogens, parasitoids, and predators of stored-product pests and the magnitude of the interactions with their hosts. However, as yet, few data are available on the efficacy of these natural agents in the commercial stored-product environment. Basic research on grain storage systems and on the integration of biological controls with other alternative pest control methods is still urgently needed. Although biological control is not a panacea, the opportunity now exists to apply it to a much greater extent in many storage situations. We hope that developments now on the horizon will make biological control attractive from a philosophical point of view and also from an economic one.

## REFERENCES

Abdel-Rahman, H. A., Shaumrar, N. F., Soliman, A. A., and El-Agoze, M. M. (1978-9). Efficiency of the anthocorid predator *Xylocoris flavipes* Reut. in biological control of stored grain insects. *Bull. Entomol. Soc. Egypt, Econ. Series, 11*:27–34.

Abdella, M. M. H., Tawfik, M. F. S., and Awadallah, K. T. (1985). Biological studies on the bethylid parasitoid *Holepyris sylvanidis* Berthes. *Annals of Agric. Sc., Moshtohor, 23*:1355–1363.

Adams, J. R., and Wilcox, T. A. (1968). Histopathology of the almond moth, *Cadra cautella*, infected with a nuclear-polyhedrosis virus. *J. Invertebr. Pathol., 12*:269–274.

Adashkevich, B. P., and Umarova, T. M. (1985). Peculiarities of development of *Trichogramma principium* (Hymenoptera, Trichogrammatidae) in laboratory conditions. *Zool. Zh., 64*:1413–1417.

Afifi, A. I., and Ibrahim, A. M. A. (1991). Effect of prey on various stages of the predator *Xylocoris galactinus* (Fieber) (Hemiptera: Anthocoridae). *Bull. Fac. Agric. Univ. Cairo, 42*:139–149.

Agudelo-Silva, F., Lindegren, J. E., and Valero, K. A. (1987). Persistence of *Neoaplectana carpocapsae* (Kapow selection) infectives in almonds under field conditions. *Fl. Entomol., 70*:288–291.

Ahmed, K. N., and M. Khatun (1988). *Lasioderma serricorne* (F.), a possible alternate host of *Anisopteromalus calandrae* (Howard) (Hymenoptera: Pteromalidae) in Bangladesh. *Bangladesh J. Zool., 16*:165–166.

Aizawa, K., Shimazu, T., and Shimizu, S. (1976). Pathogenicity of microorganisms to stored-products insect. Proceedings of Joint U.S.–Japan Seminar on Stored Product Insects, Manhattan, Kansas, pp. 59–67.

Anonymous (1992). Parasitic and predaceous insects used to control insect pests; exemption from a tolerance. *Federal Register, 57*(78):14645–14646.

Arbogast, R. T. (1976). Suppression of *Oryzaephilus surinamensis* (L.) (Coleoptera: Cucujidae) on shelled corn by the predator *Xylocoris flavipes* (Reuter) (Hemiptera: Anthocoridae). *J. Georgia Entomol. Soc., 11*:67–71.

Arbogast, R. T. (1978). The biology and impact of the predatory bug *Xylocoris flavipes* (Reuter). Proceedings, 2nd International Conference on Stored-Product Entomology, Ibadan, Nigeria, September 10–16, 1978, pp. 91–105.

Arbogast, R. T. (1984a). Demography of the predaceous bug *Dufouriellus ater* (Hemiptera: Anthocoridae). *Environ. Entomol., 13*:990–994.

Arbogast, R. T. (1984b). Natural enemies as control agents for stored-product insects. Proceedings Third International Working Conference on Stored-Product Entomology, October 23–28, Kansas State University, Manhattan, KS, pp. 360–374.

Arbogast, R. T. (1984c). Biological control of stored-product insects: Status and prospects. In Insect Management for Food Storage and Processing (ed. Baur, F. J.). American Association of Cereal Chemists, pp. 226–238.

Arbogast, R. T., Carthon, M., and Roberts, J. R., Jr. (1971). Developmental stages of *Xylocoris flavipes* (Hemiptera: Anthocoridae), a predator of stored-product insects. *Ann. Entomol. Soc. Am., 64*:1131–1134.

Arbogast, R. T., Flaherty, B. R., and Press, J. W. (1983). Demography of the predaceous bug *Xylocoris sordidud* (Reuter). *Am. Midl. Nat., 109*:398–405.

Arbogast, R. T., and Mullen, M. A. (1990). Interaction of maize weevil (Coleoptera: Curculionidae) and parasitoid *Anisopteromalus calandrae* (Hymenoptera: Pteromalidae) in a small bulk of stored corn. *J. Econ. Entomol., 83*:2462–2468.

Armstrong, E. (1978). The effects of vitamin deficiencies on the growth and mortality of *Tribolium castaneum* infected with *Nosema whitei*. *J. Invertebr. Pathol., 31*:301–306.

Arnaud, P. H., Jr. (1978). Host-parasite catalog of North American Tachinidae (Diptera). U.S. Dept. Agriculture, Misc. Pub. 1319.

Arnott, H. J., and Smith, K. M. (1968a). An ultrastructural study of the development of a granulosis virus in the cells of the moth *Plodia interpunctella* (Hbn.). *J. Ultrastruct. Res.*, *21*:251–268.

Arnott, H. J., and Smith K. M. (1968b). Ultrastructure and formation of abnormal capsules in a granulosis virus of the moth *Plodia interpunctella* (Hbn.), *J. Ultrastruct. Res.*, 22:136–158

Ashford, R. W. (1970). Some relationships between the red flour beetle, *Tribolium castaneum* (Herbst) (*Coleoptera, Tenegrionidae*) and *Lymphotropha tribolii* Ashford (*Neogregarinida, Schizocystidae*). *Acta Protozoologica, 36*(*VII*):513–531.

Assem, J. van den (1971). Some experiments on sex ratio and sex regulation in the pteromalid *Lariophagus distinguendus*. *Neth. J. Zool.*, *21*:373–402.

Assem, J. van den, and Kuenen, D. J. (1958). Host finding of *Choetospila elegans* Westw. (Hym. Chalcid.) a parasite of *Sitophilus granarius* L. (Coleopt. Curcul.). *Entomol. Exp. Appl.*, *1*:174–180.

Awadallah, K. T., Tawfik, M. F. S., El-Husseini, M. M., and Ibrahim, A. M. A. (1981). The life history of *Dufouriellus ater* (Duf.). *Bull. Soc. Entomol. Egypt*, *63*:191–197.

Awadallah, K. T., Tawfik, M. F. S., and Abdellah, M. M. H. (1984). Suppression effect of the reduviid predator *Allaeocranum biannulipes* on populations of stored-product insect pests. *Z. Ang. Entomol.*, *97*:249–253.

Ayal, Y., and Green, R. F. (1993). Optimal egg distribution among host patches for parasitoids subject to attack by hyperparasitoids. *Am. Nat.*, *141*:120–138.

Bai, B., Luck, R. F., Forster, L., Stephens, B., and Janssne, J. A. M. (1992). The effect of host size on quality attributes of the egg parasitoid, *Trichogramma pretiosum*. *Entomol. Exp. Appl.*, *64*:37–48.

Baker, J. E. (1994). Sensitivities of laboratory and field strains of the parasitoid, *Anisopteromalus calandrae* (Hymenoptera: Pteromalidae) and its host, *Sitophilus oryzae* (Coleoptera: Curculionidae), the deltamethrin and cyfluthrin. *J. Entomol. Sci.*, *29*:100–109.

Baker, J. E., and Weaver, D. K. (1993). Resistance in field strains of the parasitoid *Anisopteromalus calandrae* (Hymenoptera: Pteromalidae) and its host, *Sitophilus oryzae* (Coleoptera: Curculionidae), to malathion, chlorpyrifos-methyl, and pirimiphos-methyl. *Biol. Control.*, *3*:233–242.

Baker, J. E., Weaver, D. K., Throne, J. E., and Zettler, J. L. (1995). Resistance to protectant insecticides in two field strains of the stored-product insect parasitoid *Bracon hebetor* (Hymenoptera:Braconidae). *J. Econ. Entomol.* (In Press).

Barbosa, P., Gross, P., and Kemper, J. (1991). Influence of plant allelochemicals on the tobacco hornworm and its parasitoid, *Cotesia congregata*. *Ecology*, *72*:1567–1575.

Barclay, H. J., Otvos, I. S., and Thomson, A. J. (1985). Models of periodic inundation of parasitoids for pest control. *Can. Ent.*, *117*:705–716.

Bare, C. O. (1942). Some natural enemies of stored-tobacco insects, with biological notes. *J. Econ. Entomol.*, *32*:185–189.

Barker, P. S. (1967). Bionomics of *Blattisocius keegani* (Fox) (Acarina: Ascidae), a predator on eggs of pests of stored grain. *Can. J. Zool.*, *45*:1093–1099.

Barker, P. S. (1991). Bionomics of *Cheyletus eruditus* (Schrank) (Acarina: Cheyletidae) a predator of *Lepidoglyphus destructor* (Schrank) (Acarina: Glycyphagidae) at three constant temperatures. *Can. J. Zool.*, *69*:2321–2325.

Barker, P. S. (1992). Bionomics of *Nodele calamondin* Muma (Acarina: Cheyletidae) fed on *Lepidoglyphus destructor* (Schrank) (Acarina: Glycyphagidae) at two constant temperatures. *Can. J. Zool.*, *70*:2333–2337.

Battu, G. S., and Dhaliwal, G. S. (1976). Seasonal fluctuations in the population of *Anisopteromalus calandrae* (Howard) in the laboratory culture of *Rhizopertha dominica* Fabricius. *Curr. Res. (Bangalore)*, *5*:106–107.

Bellows, T. S., Jr. (1985). Effects of host age and host availability on developmental period, adult size, sex ratio, longevity and fecundity in *Lariophagus distinguendus* Forster (Hymenoptera: Pteromalidae). *Res. Popul. Ecol.*, *27*:55–64.

Berliner, E. (1915). Über die Schlaffschut der Mehlmoffenraupe (*Ephestia kühniella* Zell.) und ihren Erreger *Bacillus thuringiensis* n. sp. *Z. Angew. Entomol.*, *2*:29–56.

Bigler, F. (1991). Quality control of mass-reared arthropods. Proceedings of the 5th Workshop of the IOBC Global Working Group, March 25–28, 1991, Wageningen.

Boeye, J., Laborius, G. A., and Schulz, F. A. (1992). The response of *Teretriosoma nigrescens* Lewis Col. Histeridae to the pheromone of *Prostephanus truncatus* Horn Col. Bostrichidae. *Anz. Schaedlingskd. Pflanzenschutz Unweltschutz*, *65*:153–157.

Boucek, Z., and Rasplus, J. Y. (1991). Illustrated Key to West-Palearctic Genera of *Pteromalidae* (Hymenoptera: Chalcidoidea). INRA Editions, Versailles, France.

Bouse, L. F., and Morrison, R. K. (1985). Transport, storage, and release of *Trichogramma pretiosum*. *Southwest. Entomol.*, *Suppl. 8*:36–48.

Bridwell, J. C. (1919). Some additional notes on Bruchidae and their parasites in the Hawaiian Islands. *Proc. Hawaii Ent. Soc.*, *4*:15–20.

Brooks, W. M. (1971). Protozoan infections of insects with emphasis on inflammation. Proceedings IVth Intl. Colloq. Insect Pathol. and Soc. Invertebr. Pathol., College Park, Maryland, pp. 11–27.

Brower, J. H. (1983a). Utilization of stored-product Lepidoptera eggs as hosts by *Trichogramma pretiosum* Riley (Hymenoptera: Trichogrammatidae). *J. Kans. Entomol. Soc.*, *56*:50–54.

Brower, J. H. (1983b). Eggs of stored-product Lepidoptera as hosts for *Trichogramma evanescens* (Hym. Trichogrammatidae). *Entomophaga*, *28*:355–362.

Brower, J. H. (1984). The natural occurrence of the egg parasite, *Trichogramma*, on almond moth eggs in peanut storages in Georgia. *J. Ga. Entomol. Soc.*, *19*:285–290.

Brower, J. H. (1988). Population suppression of the almond moth and the Indianmeal moth (Lepidoptera: Pyralidae) by release of *Trichogramma pretiosum* (Hymenoptera: Trichogrammatidae) into simulated peanut storages. *J. Econ. Entomol.*, *81*:944–948.

Brower, J. H. (1990). Pests of stored products. In Classical Biological Control in the Southern United States (ed. Habeck, D. H., Bennett, F. D., and Frank, J. H.). Southern Cooperative Series, Bulletin No. 355, pp. 113–122.

Brower, J. H. (1991a). Biologicals: insect diseases, insect parasites, and predators. In Management of Grain, Bulk Commodities, and Bagged Products (ed. Krischik, V., Cuperus, G., and Galliart, D.). USDA, Cooperative Extension Service, Circular E-912, pp. 195–200.

Brower, J. H. (1991b). Potential host range and performance of a reportedly monophagous parasitoid, *Pteromalus cerealellae* (Hymenoptera: Pteromalidae). *Entomol. News*, *102*:231–235.

Brower, J. H. (1991c). Biological control of *Prostephanus truncatus* (Horn) with parasitic Hymenoptera. In Proceedings of the Fifth International Working Conference on Stored-Product Protection, Bordeaux, France, September 9–14, 1990 (ed. Fleurat-Lessard, F., and Ducom, P.), pp. 1257–1258.

Brower, J. H. (1991d). Influence of light on dispersal of *Trichogramma pretiosum* in a warehouse. In *Trichogramma* and Other Egg Parasitoids (ed. Wajnberg, E., and Vinson, S. B.). Colloques de L'INRA, No. 56, Institut National de la Recherche Agronomique, Paris, France, pp. 55–58.

Brower, J. H., and Mullen, M. A. (1990). Effects of *Xylocoris flavipes* (Reuter) (Hemiptera: Anthocoridae) releases on moth populations in experimental peanut storages. *J. Entomol. Sci.*, *25*:268–276.

Brower, J. H., and Press, J. W. (1988). Interactions between the egg parasite *Trichogramma pretiosum* (Hymenoptera: Trichogrammatidae) and a predator, *Xylocoris flavipes* (Hemiptera: Anthocoridae) of the almond moth, *Cadra cautella* (Lepidoptera: Pyralidae). *J. Entomol. Sci.*, *23*:342–349.

Brower, J. H., and Press, J. W. (1990). Interaction of *Bracon hebetor* (Hymenoptera: Braconidae) and *Trichogramma pretiosum* (Hymenoptera: Trichogrammatidae) in suppressing stored-product moth populations in small inshell peanut storages. *J. Econ. Entomol.*, *83*:1096–1101.

Brower, J. H., and Press, J. W. (1992). Suppression of residual populations of stored-product pests in empty corn bins by releasing the predator, *Xylocoris flavipes* (Reuter). *Biol. Cont.*, 2:66–72.

Bruce, W. A. (1983). Mites as biological control agents of stored product pests. In Biological Control of Pests by Mites (ed. Hoy, M. A., Cunningham, G. L., and Knutson, L.). University of California, Division of Agriculture and Natural Resources, Special Publ. No. 3304, pp. 74–78.

Bruce, W. A., and LeCato, G. L. (1979). *Pyemotes tritici*: Potential biological control agent of stored-product insects. *Rec. Adv. Acarol.*, *1*:213–220.

Buchwald, J., and Berliner, E. (1910). *Habrobracon hebetor* Say in Bundesgenosse im Kampf gegen die Mehlmotte. *Z. Gas. Getreidewesen*, 2:1–4.

Burges, H. D. (1964). Possibilities of biological control of stored products insects. Proceedings XII International Congress of Entomology, London, England, Ind. Stored Products Entomology Section 9b:659.

Burges, H. D., Canning, E. U., and Hurst, J. A. (1971). Morphology, development, and pathogenicity of *Nosema oryzaephili* n. sp. in *Oryzaephilus surinamensis* and its host range among granivorous insects. *J. Invertebr. Pathol.*, *17*:419–432.

Burges, H. D., and Hurst, J. A. (1977). Ecology of *Bacillus thuringiensis* in storage moths. *J. Invertebr. Pathol.*, *30*:131–139.

Burkholder, W. E., and Dicke, R. J. (1964). Detection by ultraviolet light of stored-product insects infected with *Mattesia dispora*. *J. Econ. Entomol.*, *57*:818–819.

Burkholder, W. E., and Boush, G. M. (1974). Pheromones in stored product insect trapping and pathogen dissemination. *Bull. OEPP*, *4*:455–461.

Burkholder, W. E., and Faustini, D. L. (1991). Biological methods of survey and control.

Ecology and Management of Food Industry Pests (ed. Gorham, J. R.). *FDA Technical Bull.* 4. AOAC Press, pp. 361–372.

Burks, D. B. (1979a). Family Pteromalidae. In Catalog of Hymenoptera in America North of Mexico (ed. Krombein, K. V., Hurd, P. D., Jr., Smith, D. R., and Burks, D. B. Smithsonian Institute, Washington, D.C., pp. 768–834.

Burks, D. B. (1979b). Family Trichogrammatidae. In Catalog of Hymenoptera in America North of Mexico (ed. Krombein, K. V., Hurd, P. D., Jr., Smith, D. R., and Burks, D. B. Smithsonian Institute, Washington, D.C., pp. 1033–1043.

Cantwell, G. E. (1974a). *Insect Diseases*, Vol. I. Marcel Dekker, New York.

Cantwell, G. E. (1974b). *Insect Diseases*, Vol. II. Marcel Dekker, New York.

Carlson, R. W. (1979). Family Ichneumonidae. In Catalog of Hymenoptera in America North of Mexico (ed. Krombein, K. V., Hurd, P. D., Jr., Smith, D. R., and Burks, D. B. Smithsonian Institute, Washington, D.C., pp. 315–740.

Chapman, J. W., and Glaser, R. W. (1915). A preliminary list of insects which have wilt, with a comparative study of their polyhedra. *J. Econ. Entomol.*, *8*:140–149.

Charnov, E. L., Los-den Hartogh, R. L., Jones, W. T., and van den Assem, J. (1981). Sex ratio evolution in a variable environment. *Nature*, *289*:27–33.

Cline, L. D., and Press, J. W. (1990). Reduction in almond moth (Lepidoptera: Pyralidae) infestations using commercial packaging of foods in combination with the parasitic wasp, *Bracon hebetor* (Hymenoptera: Braconidae). *J. Econ. Entomol.*, *83*:1110–1113.

Cline, L. D., Press, J. W., and Flaherty, B. R. (1984). Preventing the spread of the almond moth (Lepidoptera: Pyralidae) from infested food debris to adjacent uninfested packages, using the parasite *Bracon hebetor* (Hymenoptera: Braconidae). *J. Econ. Entomol.*, *77*:331–333.

Cline, L. D., Press, J. W., and Flaherty, B. R. (1985). Suppression of the rice weevil, *Sitophilus oryzae* (Coleoptera: Curculionidae), inside and outside of burlap, woven polypropylene, and cotton bags by the parasitic wasp, *Anisopteromalus calandrae* (Hymenoptera: Pteromalidae). *J. Econ. Entomol.*, *78*:835–838.

Cline, L. D., Press, J. W., and Flaherty, B. R. (1986). Protecting uninfested packages from attack by *Cadra cautella* (Lepidoptera: Pyralidae) with the parasitic wasp *Venturia canescens (Hymenoptera: Ichneumonidae). J. Econ. Entomol.*, *79*:418–420.

Cogburn, R. R., and Brower, J. H. (1990). Biological control of stored-product insects in rice. In Proceedings of 23rd Rice Technology Working Conference, Biloxi, MS.

Cook, R. M., and Hubbard, S. F. (1980). Effect of host density on searching behaviour of *Nemeritis canescens* (Hymenoptera: Ichneumonidae). *Entomol. Exp. Appl.*, *27*:205–210.

Corbet, S. A. (1968). The influence of *Ephestia kuehniella* on the development of its parasite *Nemeritis canescens. J. Exp. Biol.*, *48*:291–304.

Corbet, S. A. (1973). Concentration effects and the response of *Nemeritis canescens* to a secretion of its host. *J. Insect Physiol.*, *19*:2119–2128.

Cowan, D. K., Vail, P. V., Kok-Yokomi, M. L., and Schreiber, F. E. (1986). Formulation of a granulosis virus of *Plodia interpunctella* (Hübner) (Lepidoptera: Pyralidae): Efficacy, persistence, and influence on oviposition and larval survival. *J. Econ. Entomol.*, 79:1085–1090.

Croft, B. A. (1990). Arthropod Biological Control Agents and Pesticides. John Wiley and Sons, Somerset, NJ.

Darst, P. H., and King, E. W. (1969). Biology of [Blattisocius] *Nekucgares tarsalis* in association with *Plodia interpunctella*. *Ann. Entomol. Soc. Am.*, *62*:747–749.

Davis, G. R. F., and Smith, J. D. (1977). Effect of temperature on production of fungal metabolites toxic to larvae of *Tenebrio molitor*. *J. Invertebr. Pathol.*, *30*:325–329.

Devauchelle, G. (1970). Inclusions cristalines et particules d'allure viral dans les noyaux des cellules de l'intestin moyen de Coléoptère *Tenebrio molitor* (L.). *J. Ultrastruct. Res.*, *33*:263–277.

Dick, K. (1988). A review of insect infestation of maize in farm storage in Africa with special reference to the ecology and control of *Prostephanus truncatus*. *Bull. Overseas Dev. Nat. Resources Inst.*, No. *18*:1–42.

Djerassi, C., Shih-Coleman, C., and Diekman, J. (1974). Insect control of the future: Operational and policy aspects. *Science*, *186*:596–606.

Dunkel, F. V., and Boush, G. M. (1969). Effect of starvation on the black carpet beetle, *Attagenus megatoma* infected with the eugregarine *Pyxinia frenzeli*. *J. Invertebr. Pathol.*, *14*:49–52.

El-Nahal, A. M., Tawfik, M. F. S., and Awadallah, K. T. (1985). Biology and ecology of the principal natural enemies of stored-product insects in Egypt, Final Rept EG-SEA-103, Cairo University, Giza, Egypt.

Ferron, P., and Robert, P. H. (1975). Virulence of entomopathogenic fungi (fungi imperfecti) for the adults of *Acanthoscelides obtectus* (Coleoptera: Bruchidae). *J. Invertebr. Pathol.*, 25:379–388.

Finlayson, L. H. (1950a). Host preference of *Cephalonomia waterstoni* Gahan, a bethylid parasitoid of *Laemophloeus* species. *Behaviour*, 2:275–316.

Finlayson, L. H. (1950b). The biology of *Cephalonomia waterstoni* Gahan (Hym., Bethylidae), a parasite of *Laemophloeus* (Col., Cucujidae). *Bull. Entomol. Res.*, *41*:79–97.

Flanders, S. E. (1930). Recent developments in *Trichogramma* production. *J. Econ. Entomol.*, *23*:837–841.

Flanders, S. E., and Badgley, M. E. (1960). A host–parasite interaction conditioned by predation. *Ecology*, *41*:363–365.

Flanders, S. E., and Badgley, M. E. (1963). Prey-predator interactions in self-balanced laboratory populations. *Hilgardia*, *35*:145–183.

Fletcher, L. W., and Long, J. S. (1971). A bacterial disease of cigarette beetle larvae. *Scientific Notes*, *64*:1559.

Flinn, P. W. (1991). Temperature-dependent functional response of the parasitoid *Cephalonomia waterstoni* (Gahan) (Hymenoptera: Bethylidae) attacking rusty grain beetle larvae (Coleoptera: Cucujidae). *Environ. Entomol.*, *20*:872–876.

Froggatt, W. W. (1912). Parasitic enemies of the Mediterranean flour moth. *Ephestia kuehniella*, Zeller. *Agric. Gaz. N.S.W.*, *23*:307–311.

Fujii, K. (1983). Resource dependent stability in an experimental laboratory resource–herbivore–carnivore system. *Res. Popul. Ecol.*, *Suppl. 3*:15–26.

Fuxa, J. R., and Tanada, Y. (1987). Epizootiology of Insect Disease. John Wiley and Sons, New York.

Geden, C. J., Axtell, R. C., and Brooks, W. M. (1985). Susceptibility of the lesser mealworm, *Alphitobius diaperinus* (Coleoptera: Tenebrionidae) to the entomogenous nematodes *Steinernema feltiae, S. glaseri* (Steinernematidae) and *Heterorhabditis heliothidis* (Heterorhabditidae). *J. Entomol. Sci.*, *20*:331–339.

George, C. R. (1971). The effects of malnutrition on growth and mortality of the red rust flour beetle *Tribolium castaneum* (Coleoptera: Tenebrionidae) parasitized by *Nosema whitei* (Microsporidia: Nosematidae). *J. Invertebr. Pathol.*, *18*:383–388.

Gerson, U., and Smiley, R. L. (1990). Acarine biocontrol agents: an illustrated key and manual. Chapman and Hall, New York.

Ghani, M. A., and Sweetman, H. L. (1955). Ecological studies on the granary weevil parasite. *Aplastomorpha calandrae* (Howard). *Biol. (Lahore)*, *1*:115–139.

Gonen, M., and Kugler, J. (1970). Notes on the biology of *Lariophagus distinguendus* (Foerster) (Hym. Pteromalidae) as a parasite of *Sitophilus oryzae* (L.) (Col. Curculionidae). *Israel J. Entomol.*, *5*:133–140.

Gordh, G. and Hartman, H. (1991). Hymenopterous parasites of stored-food insect pests. In Ecology and Management of Food-Industry Pests (ed. Gorham, J. R.). *Food and Drug Admin., Tech. Bull. 4*:217–227.

Gottlieb, F. J. (1972). A cytoplasmic symbion in *Ephestia (= Angasta) kuehniella*: Location and Morphology. *J. Invertebr. Pathol.*, *20*:351–355.

Graham, W. M. (1970a). Warehouse ecology studies of bagged maize in Kenya-I. The distribution of adult *Ephestia* (Cadra) *cautella* (Walker) (Lepidoptera: Phycitidae). *J. Stored Prod. Res.*, *6*:147–155.

Graham, W. M. (1970b). Warehouse ecology studies of bagged maize in Kenya-II. Ecological observations of an infestation by *Ephestia* (Cadra) *cautella* (Walker) (Lepidoptera: Phycitidae). *J. Stored Prod. Res.*, *6*:157–167.

Graham, W. M. (1970c). Warehouse ecology studies of bagged maize in Kenya-IV. Reinfestation following fumigation with methyl bromide gas. *J. Stored Prod. Res.*, *6*: 177–180.

Granados, R. A., and Federici, B. A. (1986a). The Biology of Baculoviruses. Vol. I. Biological Properties and Molecular Biology. CRC Press, Boca Raton, FL.

Granados, R. A., and Federici, B. A. (1986b). The Biology of Baculoviruses. Vol. II. Practical Application for Insect Control. CRC Press, Boca Raton, FL.

Griffiths, D. A., and Bowman, C. E. (ed.) (1984). Acarology VI. John Wiley and Sons, New York.

Gunasena, G. H., Vinson, S. B., and Williams, H. J. (1990). Effects of nicotine on growth, development, and survival of the tobacco budworm (Lepidoptera: Noctuidae) and the parasitoid *Campoletis sonorensis* (Hymenoptera: Ichneumonidae). *J. Econ. Entomol.*, *83*:1777–1782.

Hagstrum, D. W. (1983). Self-provisioning with paralyzed hosts and age, density, and concealment of hosts as factors influencing parasitization of *Ephestia cautella* (Walker) (Lepidoptera: Pyralidae) by *Bracon hebetor* Say (Hymenoptera: Braconidae). *Environ. Entomol.*, *12*:1727–1732.

Hagstrum, D. W. (1987). Seasonal variation of stored wheat environment and insect populations. *Environ. Entomol.*, *16*:77–83.

Hagstrum, D. W., and Sharp, J. E. (1975). Population studies on *Cadra cautella* in a citrus pulp warehouse with particular reference to diapause. *J. Econ. Entomol.*, *68*:11–14.

Hagstrum, D. W ., and Smittle, B. J. (1977). Host-finding ability of *Bracon hebetor* and its influence upon adult parasite survival and fecundity. *Environ. Entomol.*, *6*:437–439.

Haines, C. P. (1981). Laboratory studies on the role of an egg predator, *Blattisocius tarsalis* (Berlese) (Acari: Ascidae), in relation to the natural control of *Ephestia cautella* (Walker) (Lepidoptera: Pyralidae) in warehouses. *Bull. Entomol. Res.*, *71*:555–574.

Haines, C. P. (1984). Biological methods for integrated control of insects and mites in tropical stored products, III. The use of predators and parasites. *Trop. Stored Prod. Inf.*, *48*:17–25.

Hall, D. W. (1951). Observations on the distribution, habits and life history of the bug *Plezostethus galactinus* (Fieb.) (Hemiptera: Anthocoridae). *Ent. Mon. Mag., London*, *87*:45–52.

Hall, R. W., Ehler, L. E., and Bisabri-Ershadi, B. (1980). Rate of success in classical biological control of arthropods. *Bull. Entomol. Soc. Amer.*, *26*:111–114.

Hamlin, J. C., Reed, W. D., and Phillips, M. E. (1931). Biology of the Indian-meal moth on dried fruits in California. U.S. Dept. Agriculture, Tech. Bull. 242.

Harris, K. L., Nicholson, J. F., Randolph, L. K., and Trawick, J. L. (1952). An investigation of insect and rodent contamination of wheat and wheat flour. *J. Assoc. Off. Agric. Chemists*, *35*:115–158.

Harris, K. L., Trawick, J. L., Nicholson, J. F., and Weiss, W. (1953). An investigation of rodent and insect contamination of corn and corn meal. *J. Assoc. Off. Agric. Chemists*, *36*:1037–1064.

Harwood, R. F., and James, M. T. (1979). Entomology in Human and Animal Health, 7th ed. Macmillan, New York.

Hassan, S. A., and Guo, F. (1991). Selection of effective strains of egg parasites of the genus *Trichogramma* (Hymenoptera: Trichogrammatidae) to control the European corn borer *Ostrinia nubilalis* Hb. (Lepidoptera: Pyralidae). *J. Appl. Entomol.*, *111*:335–341.

Hassell, M. P., Lessells, C. M., and McGavin, G. C. (1985). Inverse density dependent parasitism in a patchy environment: A laboratory system. *Ecol. Entomol.*, *10*:393–402.

Hassell, M. P., and Rogers, D. J. (1972). Insect parasite responses in the development of population models, *J. Anim. Ecol.*, *41*:661–676.

Henry, J. E. (1981). Natural and applied control of insects by protozoa. *Annu. Rev. Entomol.*, *26*:49–73.

Heong, K. L. (1981). Searching preference of the parasitoid, *Anisopteromalus calandrae* (Howard), for different stages of the host, *Callosobruchus maculatus* (F.) in the laboratory. *Res. Popul. Ecol.*, *23*:177–191.

Hluchý, M., and Samšiňáková, A. (1989). Comparative study on the susceptibility of adult *Sitophilus granarius* (L.) (Coleoptera: Curculionidae) and larval *Galleria mellonella* (L.) (Lepidoptera: Pyralidae) to the entomogenous fungus *Beauveria bassiana* (Bals.) Vuill, *J. Stored Prod. Res.*, *25*:61–64.

Hodges, R. J. (1984). Biological methods for integrated control of insects and mites in tropical stored products, IV: The use of insect diseases, *Trop. Stored Prod. Inf.*, *48*:27–31.

Hong, Y. S., and Ryoo, M. I. (1991). Effect of temperature on the functional and numerical responses of *Lariophagus distinguendus* (Hymenoptera: Pteromalidae) to various densities of the host, *Sitophilus oryzae* (Coleoptera: Curculionidae). *J. Econ. Entomol.*, *84*:837–840.

Howard, L. O. (1908). The Insect Book. Doubleday, Page & Co., New York.

Howard, R. W., and Flinn, P. W. (1990). Larval trails of *Cryptolestes ferrugineus* (Coleoptera: Cucujidae) as kairomonal host-finding cues for the parasitoid *Cephalonomia waterstoni* (Hymenoptera: Bethylidae). *Ann. Entomol. Soc. Am.*, *83*:239–245.

Hoy, M. A. (1991). Genetic improvement of a parasitoid: response of *Trioxys pallidus* to laboratory selection with azinphosmethyl. *Biocontrol Sci. Technol.*, *1*:31–41.

Hoy, M. A., Nowierski, R. M., Johnson, M. W., and Flexner, J. L. (1991). Issues and ethics in commercial releases of arthropod natural enemies. *Am. Entomol.*, *37*:74–75.

Hubbard, S. F., Cook, R. M., Glover, J. G., and Greenwood, J. J. D. (1982). Apostatic selection as an optimal foraging strategy. *J. Anim. Ecol.*, *51*:625–633.

Huffaker, C. B., and Matsumoto, B. M. (1982). Group versus individual functional responses of *Venturia* [= *Nemeritis*] *canescens* (Grav.). *Res. Popul. Ecol.*, *24*:250–269.

Huffaker, C. B., Simmonds, F. J., and Laing, J. E. (1976). The theoretical and empirical basis of biological control. In Theory and Practice of Biological Control (ed. Huffaker, C. B., and Messenger, P. S.). Academic Press, New York.

Huis, van, A. (1991). Biological methods of bruchid control in the tropics: A review. *Insect Sci. Appl.*, *12*:87–102.

Huis, van, A., Wijkamp, M. G., Lammers, P. M., Klein, C. G. M., van Seeters, G. J. H., and Kaashoek, N. K. (1991). *Uscana lariophaga* (Hymenoptera: Trichogrammatidae), an egg parasitoid of bruchid beetles (Coleoptera: Bruchidae) storage pests in West Africa: Host-age and host-species selection. *Bull. Entomol. Res.*, *81*:65–75.

Huis, van, A., Kaashoek, N. K., and Maes, H. M. (1992). Biological control of bruchids (Col.: Bruchidae) in stored pulses by using egg parasitoids of the genus *Uscana* (Hym.: Trichogrammatidae): a review. In Proceedings of the Fifth International Working Conference on Stored-Product Protection, Bordeaux, France, September 9–14, 1990 (ed. Fleurat-Lessard, F., and Ducom, P.), pp. 99–108.

Hunter, C. D. (1992). Suppliers of beneficial organisms in North America, California Dept. of Food and Agriculture, Biological Control Services Program, Sacramento, CA.

Hunter, D. K. (1970). Pathogenicity of a granulosis virus of the Indian-meal moth. *J. Invertebr. Pathol.*, *16*:339–341.

Hunter, D. K., and Dexel, T. D. (1970). Observations on a granulosis of the almond moth, *Cadra cautella*. *J. Invertebr. Pathol.*, *16*:307–309.

Hunter, D. K., and Hoffmann, D. F. (1970). A granulosis virus of the almond moth, *Cadra cautella*. *J. Invertebr. Pathol.*, *16*:400–407.

Hunter, D. K., and Hoffmann, D. F. (1972). Cross infection of a granulosis virus of *Cadra cautella*, with observations on its ultrastructure in infected cells of *Plodia interpunctella*. *J. Invertebr. Pathol.*, *20*:4–10.

Hunter, D. K., and Hoffmann, D. F. (1973). Susceptibility of two strains of Indian meal moth to a granulosis virus. *J. Invertebr. Pathol.*, *21*:114–115.

Hunter, D. K., Collier, S. J., and Hoffmann, D. F. (1973). Effectiveness of a granulosis virus of the Indian meal moth as a protectant for stored inshell nuts: Preliminary observations. *J. Invertebr. Pathol.*, 22:481.

Ishii, T. (1940). The problems of biological control in Japan. Proceedings, 6th Pacific Science Congress, *4*:365–367.

Jacobs, S. E. (1951). Bacteriological control of the flour moth, *Ephestia kühniella* Z, *Proc. Soc. Appl. Bacteriol.*, *1*:83–91.

Jafri, R. H. (1961). Synergistic action of radiation and of *Bacillus thuringiensis* toxin on protozoan diseases of insects. Progress in Protozoology, Proceedings First Intl. Congr. Protozool., Prague 1961, Publ. House Czechoslov. Acad. Sci., Prague 1963, pp. 510–515.

Jafri, R. H. (1964). Influence of pathogens on the life span of irradiated insects. *Revue Patho. vég. Ent. agric. Fr.*, *43*:37–51.

Jay, E., Davis, R., and Brown, S. (1968). Studies on the predacious habits of *Xylocoris flavipes* (Reuter) (Hemiptera: Anthocoridae). *J. Georgia Entomol. Soc.*, *3*:126–130.

Johnson, D. E., Brookhart, G. L., Kramer, K. J., Barnett, B. D., and McGaughey, W. H. (1990). Resistance to *Bacillus thuringiensis* by the Indian meal moth, *Plodia interpunctella*: Comparison of midgut proteinases from susceptible and resistant larvae. *J. Invertebr. Pathol.*, *55*:235–244.

Kaiser, L., Pham-Delegue, M. H., and Masson, C. (1989). Behavioural study of plasticity in host preferences of *Trichogramma maidis* (Hym.: Trichogrammatidae). *Physiol. Entomol.*, *14*:53–60.

Kapila, R., and Agarwal, H. C. (1991). Biology of an egg parasite of *Callosobruchus maculatus* (Fab.) (Coleoptera: Bruchidae). In Proceedings of the Fifth International Working Conference on Stored-product Protection, Bordeaux, France, September 9–14, 1990 (ed. Fleurat-Lessard, F., and Ducom, P.), pp. 1265–1273.

Kareiva, P. (1990). Establishing a foothold for theory in biocontrol practice using models to guide experimental design and release protocols. *UCLA Symp. Mol. Cell Biol.*, *112*: 65–81.

Kashef, A. (1956). Ėtude biologique de *Stegobium paniceum* L. (Col. Anobiidae) et de son parasite: *Lariophagus distinguendus* Först. (Hym. Pteromalidae). *Ann. Soc. Ent. Fr.*, *124*:5–88.

Keever, D. W., Arbogast, R. T., and Mullen, M. A. (1985). Population trends and distributions of *Bracon hebetor* Say (Hymenoptera: Braconidae) and lepidopterous pests in commercially stored peanuts. *Environ. Entomol.*, *14*:722–725.

Keever, D. W., Mullen, M. A., Press, J. W., and Arbogast, R. T. (1986). Augmentation of natural enemies for suppressing two major insect pests in stored farmers stock peanuts. *Environ. Entomol.*, *15*:767–770.

Kellen, W. R. (1978). Microbial insecticides, Proceedings of Symposium on Prevention and Control of Insects in Stored-Food Products, pp. 223–230.

Kellen, W. R., Hoffmann, D. F., and Kwock, R. A. (1981). *Wolbachia* sp. (Rickettsiales: Rickettsiaceae) a symbiont of the almond month, *Ephestia cautella*: Ultrastructure and influence on host fertility. *J. Invertebr. Pathol.*, *37*:273–283.

Kellen, W. R., and Hoffmann, D. F. (1987). Laboratory studies on the dissemination of a granulosis virus by healthy adults of the Indianmeal moth, *Plodia interpunctella* (Lepidoptera: Pyralidae). *Environ. Entomol.*, *16*:1231–1234.

Kellen, W. R., and Lindegren, J. E. (1968). Biology of *Nosema plodiae* sp. n., and microsporidian pathogen of the Indian-meal moth, *Plodia interpunctella* (Hübner), (Lepidoptera: Phycitidae). *J. Invertebr. Pathol.*, *11*:104–111.

Kellen, W. R., and Lindegren, J. E. (1969). Host-pathogen relationships of two previously undescribed microsporidia from the Indian-meal moth, *Plodia interpunctella* (Hübner), (Lepidoptera: Phycitidae). *J. Invertebr. Pathol.*, *14*:328–335.

Kellen, W. R., and Lindegren, J. E. (1970). Previously unreported pathogens from the navel orangeworm, *Paramyelois transitella*, in California. *J. Invertebr. Pathol.*, *16*: 342–345.

Kellen, W. R., and Lindegren, J. E. (1971). Modes of transmission of *Nosema plodiae* Kellen and Lindegren, a pathogen of *Plodia interpunctella* (Hübner). *J. Stored Prod. Res.*, 7:31–34.

Kellen, W. R., and Lindegren, J. E. (1973a). *Nosema invadens* sp. n. (Microsporidia: Nosematidae), a pathogen causing inflammatory response in lepidoptera. *J. Invertebr. Pathol.*, *21*:293–300.

Kellen, W. R., and Lindegren, J. E. (1973b). New host records for *Helicosporidium parasiticum*. *J. Invertebr. Pathol.*, 22:296–297.

Kellen, W. R., Lindegren, J. E., and Hoffmann, D. F. (1972). Developmental stages and structure of a *Rickettsiella* in the navel orangeworm, *Paramyelois transitella* (Lepidoptera: Phycitidae). *J. Invertebr. Pathol.*, *20*:193–199.

Khan, A. R., and Selman, B. J. (1984). Effect of insecticide, microsporidian, and insecticide-microsporidian doses on the growth of *Tribolium castaneum* larvae. *J. Invertebr. Pathol.*, *44*:230–232.

Kinsinger, R. A., and McGaughey, W. H. (1976). Stability of *Bacillus thuringiensis* and a granulosis virus of *Plodia interpunctella* on stored wheat. *J. Econ. Entomol.*, *69*: 149–154.

Kinsinger, R. A ., and McGaughey, W. H. (1979). Susceptibility of populations of Indianmeal moth and almond moth to *Bacillus thuringiensis*. *J. Econ. Entomol.*, *72*:346–349.

Kistler, R. A. (1985). Host-age structure and parasitism in a laboratory system of two Hymenopterous parasitoids and larvae of *Zabrotes subfasciatus* (Coleoptera: Bruchidae). *Environ. Entomol.*, *14*:507–511.

Klein, J. A., and Beckage, N. E. (1990). Comparative suitability of *Trogoderma variabile* and *T. glabrum* (Coleoptera: Dermestidae) as hosts for the ectoparasite *Laelius pedatus* (Hymenoptera: Bethylidae). *Ann. Entomol. Soc. Am.*, *83*:809–816.

Kockum, S. (1965). Crib storage of maize a trial with pyrethrin and lindane formulations. *E. Afr. Agric. Forestry J.*, *31*:8–10.

Krombein, K. V. (1979). Superfamily Bethyloidea. In Catalog of Hymenoptera in America North of Mexico (ed. Krombein, K. V., Hurd, P. D., Jr., Smith, D. R., and Burks, D. B. Smithsonian Institute, Washington, D.C., pp. 1203–1251.

Kugler, J., and Nitzan, Y. (1977). Biology of *Clausicella suturata* [Dipt.: Tachinidae] a parasite of *Ectomyelois ceratoniae* [Lep.: Phycitidae}. *Entomophaga*, *22*:93–105.

Kumar, P., and Naqi, H. (1990). Study of host stage density effect on cannibalism in *Acaropsis sollers* (Acari: Cheyletidae) predatory mites and its role as a biological control agent. *Indian J. Helminthol.*, *42*:21–24.

Kumari, S. M., and Neelgund, Y. F. (1985). Preliminary infectivity tests using six bacterial formulations against the red flour beetle, *Tribolium castaneum*. *J. Invertebr. Pathol.*, *46*:198–199.

Kurstak, E. (1982). Microbial and Viral Pesticides. Marcel Dekker, New York.

LeCato, G. L. (1975). Red flour beetle: Population growth on diets of corn, wheat, rice, or shelled peanuts supplemented with eggs or adults on the Indianmeal moth. *J. Econ. Entomol.*, *68*:763–765.

LeCato, G. L. (1976). Predation by *Xylocoris flavipes* (Hem: Anthocoridae): Influence of stage, species, and density of prey and of starvation and density of predator. *Entomophaga*, *21*:217–221.

LeCato, G. L., and Arbogast, R. T. (1979). Functional response of *Xylocoris flavipes* to Angoumois grain moth and influence of predation on regulation of laboratory populations. *J. Econ. Entomol.*, *72*:847–849.

LeCato, G. L., Collins, J. M., and Arbogast, R. T. (1977). Reduction of residual populations of stored-product insects by *Xylocoris flavipes* (Hemiptera: Anthocoridae). *J. Kansas Entomol. Soc.*, *50*:84–88.

Lenteren, van, J. C., and Woets, J. (1988). Biological and integrated pest control in greenhouses. *Annu. Rev. Entomol.*, *33*:239–269.

Loosjes, F. E. (1957). Ervaringen met *Chaetospila elegans* (Westw.) (Hymenoptera, Pteromalidae), een parasiet van enige soorten voorraadinsecten. *Entomol. Ber. (Amsterdam)*, *17*:74–76.

Luca, Y. (1976). Destruction of imaginal forms of *Acanthoscelides obtectus* Say (Coleoptera: Bruchidae) by *Neoaplectana carpocapsae* Weiser. *Rev. Zool. Agric. Pathol. Veg.*, *75*:127–131.

Ma, M., Burkholder, W. E., and Carlson, S. D. (1978). Supra-anal organ: A defensive mechanism of the furniture carpet beetle, *Anthrenus flavipes* (Coleoptera: Dermestidae). *Ann. Entomol. Soc. Am.*, *71*:718–723.

Mardan, A. H., and Harein, P. K. (1984). Susceptibility of a malathion-resistant strain of Indianmeal moths (Lepidoptera: Pyralidae) to *Bacillus thuringiensis* Berliner. *J. Econ. Entomol.*, *77*:1260–1263.

Markham, R. H., and Herren, H. R. (1990). Biological control of the larger grain borer, Proceedings of an IITA/FAO Coordination Meeting, International Institute of Tropical Agriculture, Cotonou, Republic of Benin, 2–3 June, 1989.

Marris, G., Hubbard, S., and Hughes, J. (1986). Use of patchy resources by *Nemeritis canescens* (Hymenoptera: Ichneumonidae), I. Optimal solutions. *J. Anim. Ecol.*, *55*:631–640.

Marsh, P. M. (1979). Family Braconidae. In Catalog of Hymenoptera in America North of Mexico (ed. Krombein, K. V., Hurd, P. D., Jr., Smith, D. R., and Burks, D. B. Smithsonian Institute, Washington, D.C., pp. 144–295.

Marzke, F. O., and Dicke, R. J. (1958). Disease-producing protozoa in species of *Trogoderma*. *J. Econ. Entomol.*, *51*:916–917.

Matsumoto, B. M. (1974). On the adult longevity of the entomophagous parasite, *Venturia canescens* (Hymenoptera: Ichneumonidae). *Entomophaga*, *19*:325–329.

McGaughey, W. H. (1975). A granulosis virus for Indian meal moth control in stored wheat and corn. *J. Econ. Entomol.*, *68*:149–154.

McGaughey, W. H. (1978). Effects of larval age on the susceptibility of almond moths and Indianmeal moths to *Bacillus thuringiensis*, *J. Econ. Entomol.*, *71*:923–924.

McGaughey, W. H. (1982). Evaluation of commercial formulations of *Bacillus thuringiensis* for control of the Indianmeal moth and almond moth (Lepidoptera: Pyralidae) in stored inshell peanuts. *J. Econ. Entomol.*, *75*:754–757.

McGaughey, W. H. (1985). Insect resistance to the biological insecticide *Bacillus thuringiensis*. *Science*, *229*:193–195.

McGaughey, W. H. (1986). *Bacillus thuringiensis*: A critical review. In Proceedings of 4th

International Working Conference on Stored-Product Protection, Tel Aviv, Israel (ed. Donahaye, E., and Navarro, S.), pp. 14–23.

McGaughey, W. H., and Beeman, R. W. (1988). Resistance to *Bacillus thuringiensis* in colonies of Indianmeal moth and almond moth (Lepidoptera: Pyralidae). *J. Econ. Entomol.*, *81*:28–33.

McGaughey, W. H., and Kinsinger, R. A. (1978). Susceptibility of Angoumois grain moths to *Bacillus thuringiensis*, *J. Econ. Entomol.*, *71*:435–436.

McGaughey, W. H., and Whalon, M. E. (1992). Managing insect resistance to *Bacillus thuringiensis* toxins. *Science*, *258*:1451–1455.

McMurtry, J. A. (1984). A consideration of the role of predators in the control of acarine pests. In Acarology VI (ed. Griffiths, D. A., and Bowman, C. E.). John Wiley & Sons, New York, pp. 109–121.

Mertins, J. W. (1980). Life history and behavior of *Laelius pedatus*, a gregarious bethylid ectoparasitoid of *Anthrenus verbasci*. *Ann. Entomol. Soc. Am.*, *73*:686–693.

Mertins, J. W. (1985). *Laelius utilis* (Hym.: Bethylidae), a parasitoid of *Anthrenus fuscus* (Col.: Dermestidae) in Iowa. *Entomophaga*, *30*:65–68.

Métalnikov, S., and Métalnikov, S. S. (1935). Utilisation des microbes dans la lutte contre les insectes nuisibles. *Ann. Inst. Pasteur*, *55*:709.

Metcalf, R. L. (1982). Insecticides in pest management. In Introduction to Insect Pest Management, 2nd ed. (ed. Metcalf, R. L., and Luckmann, W. H.), John Wiley and Sons, New York, pp. 217–277.

Milner, R. J. (1972). The survival of *Nosema whitei* spores stored at 4°C. *J. Invertebr. Pathol.* [Notes] *20*:356–357.

Mital, V. P. (1969). *Anisopteromalus calandrae*, Howord (Hymenoptera: Chalcididae) a new record of pupal parasite of groundnut bruchid (*Caryedon gonagra*, F.) an serious pest of groundnut (*Arachis hypogea* L.) and tamarind (*Tamarindus indica* L.). *Bull. Grain Technol.*, *7*:234–235.

Mohammed, A. K. H., and Al-Jabery, I. A. R. (1988). Comparative toxicity of some insecticides in the laboratory to the southern cowpea weevil *Callosobruchus maculatus* (F.) and the parasite, *Anisopteromalus calandrae* (How.). *Mesopotamia J. Agric.*, *20*:289–306.

Monge, J. P., and Huignard, J. (1991). Population fluctuations of two bruchid species *Callosobruchus maculatus* F. and *Bruchidius atrolineatus* Pic (Coleoptera: Bruchidae) and their parasitoids *Dinarmus basalis* Rondani and *Eupelmus vuilleti* Crawford (Hymenoptera: Pteromalidae, Eupelmidae) in a storage situation in Niger. *J. Afr. Zool.*, *105*:187–196.

Morris, O. N. (1985). Susceptibility of 31 species of agricultural insect pests to the entomogeneous nematodes *Steinernema feltiae* and *Heterorhabditis bacteriophora*. *Can. Entomol.*, *117*:401–407.

Morrison, R. K. (1985). Mass production of *Trichogramma pretiosum* Riley. *Southwest. Entomol.*, *Suppl. 8*:21–27.

Moser, J. C. (1975). Biosystematics of the straw itch mite with special reference to nomenclature and dermatology. *Trans. R. Ent. Soc. Lond.*, *2*:185–191.

Nagy, C. G. (1986). Les femelles de *Laelius anthrenivorous trani* (Hymenoptera: Bethylidae) attaquent l'homme. *Riv. Parassitol.*, *29*:71–74.

Nara, J. M., Burkholder, W. E., and Boush, G. M. (1981). The influence of storage temperature on spore viability of *Mattesia trogodermae* (Protozoa: Neogregarinida). *J. Invertebr. Pathol.*, *38*:404–408.

Nickle, D. A., and Hagstrum, D. W. (1981). Provisioning with preparalyzed hosts to improve parasite effectiveness: A pest management strategy for stored commodities. *Environ. Entomol.*, *10*:560–564.

Nikam, P. K., and Pawar, C. V. (1993). Life tables and intrinsic rate of natural increase of *Bracon hebetor* Say (Hym., Braconidae) population on *Corcyra cephalonica* Staint. (Lep., Pyralidae) a key parasitoid of *Helicoverpa armigera* Hbn. (Lep., Noctuidae). *J. Appl. Ecol.*, *115*:210–213.

Nilakhe, S. S., and Parker, R. D. (1990). Implementation of parasites and predators for control of stored-product pests. In Proceedings of the III National Stored Grain Pest Management Training Conference, Kansas City, MO, Oct. 20–25, 1990, pp. 241–250.

Nwanze, K. F., Partida, G. J., and McGaughey, W. H. (1975). Susceptibility of *Cadra W. H. cautella* and *Plodia interpunctella* to *Bacillus thuringiensis* on wheat. *J. Econ. Entomol.*, *68*:751–752.

Onstad, D. W., and Maddox, J. V. (1990). Simulation model of *Tribolium confusum* and its pathogen, *Nosema whitei*. *Ecol. Model.*, *51*:143–160.

Parajulee, M. N., and Phillips, T. W. (1992). Laboratory rearing and field observations of *Lyctocoris campestris* (Heteroptera: Anthocoridae), a predator of stored-product insects. *Ann. Entomol. Soc. Am.*, *85*:736–743.

Parajulee, M. N., and Phillips, T. W. (1993). Effects of prey species on development and reproduction of the predator *Lyctocoris campestris* (Heteroptera: Anthocoridae). *Environ. Entomol.*, *22*:1035–1042.

Parker, R. D., and Nilakhe, S. S. (1990). Evaluation of predators and parasites and chemical grain protectants on insect pests of sorghum stored in commercial bins. In Proceedings of the 3rd National Stored Grain Pest Management Training Conference, Kansas City, MO, pp. 229–239.

Parrella, M. P., Heinz, K. M., and Nunney, L. (1992). Biological control through augmentative releases of natural enemies: A strategy whose time has come. *Am. Entomol.*, *38*:172–179.

Peck, O. (1963). A catalogue of the nearctic Chalcidoidea (Insecta: Hymenoptera). *Can. Entomol.*, suppl. no. 30.

Pingale, S. W. (1954). Biological control of some stored grain pests by the use of a bug predator, *Amphibolus venator* Klug, *Indian J. Entomol.*, *16*:300–302.

Powell, D. (1938). The biology of *Cephalonomia tarsalis* (Ash.), a vespoid wasp (Bethylidae: Hymenoptera) parasitic on the sawtoothed grain beetle. *Ann. Entomol. Soc. Am.*, *31*:44–48.

Powell, W. (1986). Enhancing parasitoid activity in crops. In Insect Parasitoids (ed. Waage, J. K., and Greathead, D. J.). Academic Press, London, pp. 319–340.

Press, J. W. (1989). Compatibility of *Xylocoris flavipes* (Hemiptera: Anthocoridae) and *Venturia canescens* (Hymenoptera: Ichneumonidae) for suppression of the almond moth, *Cadra cautella* (Lepidoptera: Pyralidae). *J. Entomol. Sci.*, *24*:156–160.

Press, J. W. (1992). Comparative efficacy in wheat between the weevil parasitoids *Anisopteromalus calandrae* and *Choetospila elegans* (Hymenoptera: Pteromalidae). *J. Entomol. Sci.*, 27:154–157.

Press, J. W., Cline, L. D., and Flaherty, B. R. (1982). A comparison of two parasitoids,

*Bracon hebetor* (Hymenoptera: Braconidae) and *Venturia canescens* (Hymenoptera: Ichneumonidae), and a predator *Xylocoris flavipes* (Hemiptera: Anthocoridae) in suppressing residual populations of the almond moth, *Ephestia cautella* (Lepidoptera: Pyralidae). *J. Kansas Entomol. Soc.*, *55*:725–728.

Press, J. W., Cline, L. D., and Flaherty, B. R. (1984). Suppression of residual populations of the rice weevil, *Sitophilus oryzae*, by the parasitic wasp, *Anisopteromalus calandrae*. *J. Ga. Entomol. Soc.*, *19*:110–113.

Press, J. W., Flaherty, B. R., and Arbogast, R. T. (1975). Control of the red flour beetle, *Tribolium castaneum*, in a warehouse by a predaceous bug, *Xylocoris flavipes*. *J. Georgia Entomol. Soc.*, *10*:76–78.

Press, J. W., Flaherty, B. R., and Arbogast, R. T. (1977). Interactions among *Nemeritis canescens* (Hymenoptera: Ichneumonidae), *Bracon hebetor* (Hymenoptera: Braconidae), and *Ephestia cautella* (Lepidoptera: Pyralidae). *J. Kansas Entomol. Soc.*, *50*:259–262.

Press, J. W., Flaherty, B. R., and Arbogast, R. T. (1979). Vertical dispersion and control efficacy of the predator *Xylocoris flavipes* (Reuter) (Hemiptera: Anthocoridae) in farmers' stock peanuts. *J. Kansas Entomol. Soc.*, *52*:561–564.

Press, J. W., Flaherty, B. R., and LeCato, G. L. (1974). Interactions among *Tribolium castaneum* (Coleoptera: Tenebrionidae), *Cadra cautella* (Lepidoptera: Pyralidae), and *Xylocoris flavipes* (Hemiptera: Anthocoridae). *J. Georgia Entomol. Soc.*, *9*:101–103.

Press, J. W., Flaherty, B. R., and McDonald, L. C. (1981). Survival and reproduction of *Bracon hebetor* on insecticide-treated *Ephestia cautella* larvae. *J. Ga. Entomol. Soc.*, *16*:231–234.

Pulpan, J., and Verner, P. H. (1965). Control of tryoglyphoid mites in stored grain by the predatory mite *Cheyletus eruditus* (Schrank). *Can. J. Zool.*, *43*:417–432.

Putters, F. A., and van den Assem, J. (1988). The analysis of partial preferences in a parasitic wasp. *Anim. Behav.*, *36*:933–935.

Rabindra, R. J., Jayaraj, S., and Balasubramanian, M. (1988). *Farinocystis tribolii*-induced susceptibility to some insecticides in *Tribolium castaneum* larvae. *J. Invertebr. Pathol.*, *52*:389–392.

Reed, W. D., and Vinzant, J. P. (1942). Control of Insects Attacking Stored Tobacco and Tobacco Products, Circular No. 635, USDA, Washington, D.C., p. 40.

Rees, D. P. (1985). Life history of *Teretriosoma nigrescens* Lewis (Coleoptera: Histeridae) and its ability to suppress populations of *Prostephanus truncatus* (Horn) (Coleoptera: Bostrichidae). *J. Stored Prod. Res.*, *21*:115–118.

Rees, D. P. (1987). Laboratory studies on predation by *Teretriosoma nigrescens* Lewis (Col.: Histeridae) on *Prostephanus truncatus* (Horn) (Col.: Bostrichidae) infesting maize cobs in the presence of other maize pests. *J. Stored Prod. Res.*, *23*:191–196.

Rees, D. P. (1991). The effect of *Teretriosoma nigrescens* Lewis (Coleoptera: Histeridae) on three species of storage Bostrichidae infesting shelled maize. *J. Stored Prod. Res.*, *27*:83–86.

Reinert, J. A., and King, E. W. (1971). Action of *Bracon hebetor* Say as a parasite of *Plodia interpunctella* at controlled densities. *Ann. Entomol. Soc. Am.*, *64*:1335–1340.

Richards, O. W., and Herford, G. V. B. (1930). Insects found associated with cacao, spices and dried fruits in London warehouses. *Ann. Biol.*, *17*:367–395.

Ridgway, R. L., and Vinson, S. B. (eds.) (1977). Biological Control by Augmentation of Natural Enemies. Plenum Press, New York.

Rutz, D. A., and Scoles, G. A. (1989). Occurrence and seasonal abundance of parasitoids attacking muscoid flies (Diptera: Muscidae) in caged-layer poultry facilities in New York. *Environ. Entomol.*, *18*:51–55.

Ryabov, M. A. (1926). The possibilities of applying the parasitic method of control in the case of granary pests. *Bull. N. Caucasian Plant Prot. Sta.*, *1*:19–54.

Ryoo, M., Yoo, C. K., and Hong, Y. S. (1991a). Influences of food quality for *Sitophilus oryzae* (Coleoptera: Curculionidae) on life history of *Lariophagus distinguendus* (Hymenoptera: Pteromalidae). In Proceedings of the Fifth International Working Conference on Stored-Product Protection, Bordeaux, France, September 9–14, 1990 (ed. Fleurat-Lessard, F., and Ducom, P.), pp. 211–219.

Ryoo, M. I., Hong, Y. S., and Yoo, C. K. (1991b). Relationship between temperature and development of *Lariophagus distinguendus* (Hymenoptera: Pteromalidae), an ectoparasitoid of *Sitophilus oryzae* (Coleoptera: Curculionidae). *J. Econ. Entomol.*, *84*:825–829.

Salama, H. S., El Moursy, A., Zaki, F. N., Aboul, R. Ela, and Razek, A. Abdel (1991). Parasites and predators of the meal moth *Plodia interpunctella* (Hbn.) as affected by *Bacillus thuringiensis*. *J. Appl. Entomol.*, *112*:244–253.

Salt, G. (1976). The hosts of *Nemeritis canescens*, a problem in the host specificity of insect parasitoids. *Ecol. Entomol.*, *1*:63–67.

Satterthwait, A. F., and Swain, R. B. (1946). The sunflower moth and some of its natural enemies. *J. Econ. Entomol.*, *39*:575–580.

Schesser, J. H. (1976). Commercial formulations of *Bacillus thuringiensis* for control of Indian meal moth. *Appl. Environ. Microbiol.*, *32*:508–510.

Schwalbe, C. P., Burkholder, W. E., and Boush, G. M. (1974). *Mattesia trogodermae* infection rates as influenced by mode of transmission, dosage and host species. *J. Stored Prod. Res.*, *10*:161–166.

Searle, T., and Doberski, J. (1984). An investigation of the entomogenous fungus *Beauveria bassiana* (Bals.) Vuill. as a potential biological control agent for *Oryzaephilus surinamensis* (L.). *J. Stored Prod. Res.*, *20*:17–23.

Shapas, T. J., Burkholder, W. E., and Boush, G. M. (1977). Population suppression of *Trogoderma glabrum* by using pheromone luring for protozoan pathogen dissemination. *J. Econ. Entomol.*, *70*:469–474.

Shimada, M. (1985). Niche modification and stability of competitive systems, II. Persistence of interspecific competitive systems with parasitoid wasps. *Res. Popul. Ecol.*, 27:203–216.

Simmonds, F. J., Franz, J. M., and Sailer, R. I. (1976). History of biological control. In Theory and Practice of Biological Control (ed. Huffaker, C. B., and Messenger, P. S.). Academic Press, New York.

Simmons, P., Reed, W. D., and McGregor, E. A. (1931). Fig insects in California, U.S. Dept. Agriculture, Circular 157.

Sinha, R. N. (1961). Insects and mites associated with hot spots in farm stored grain. *Can. Entomol.*, *93*:609–621.

Sinha, R. N., and Watters, F. L. (1985). Insect pests of flour mills, grain elevators, and feed mills and their control. *Res. Br. Agric. Canada*, Publ. #1776, Ottawa, Canada.

Smith, L. (1992). Effect of temperature on life history characteristics of *Anisopteromalus calandrae* (Hymenoptera: Pteromalidae) parasitizing maize weevil larvae in corn kernels. *Environ. Entomol.*, *21*:877–887.

Smith, L. (1993a). Effect of humidity on life history characteristics of *Anisopteromalus calandrae* (Hymenoptera: Pteromalidae) parasitizing maize weevil (Coleoptera: Curculionidae) larvae in shelled corn. *Environ. Entomol.*, 22:618–624.

Smith, L. (1993b). Host-size preference of the parasitoid *Anisopteromalus calandrae* [Hym.: Pteromalidae] on *Sitophilus zeamais* [Col.: Curculionidae] larvae with a uniform age distribution. *Entomophaga*, *38*:185–193.

Smith, R. F., and van den Bosh, R. (1967). Integrated control. In Pest Control: Biological, Physical and Selected Chemical Methods (ed. Kilgore, W. W., and Doutt, R. L.). Academic Press, New York, pp. 295–340.

Solomon, M. E. (1969). Experiments on predator–prey interactions of storage mites. *Acarologia*, *11*:484–503.

Spitler, G. H., and Hartsell, P. L. (1975). Pirimiphos-methyl as a protectant for stored inshell almonds. *J. Econ. Entomol.*, *68*:777–780.

Steinhaus, E. A., and Bell, C. R. (1953). The effect of certain microorganisms and antibiotics on stored-grain insects. *J. Econ. Entomol.*, *46*:582–598.

Steinkraus, C., and Cross, E. A. (1993). Descriptions and life history of *Acarophenaux mahunkai*, n. sp. (Acari, Tarsonemina: Acarophenacidae), an egg parasite of the lesser mealworm (Coleoptera: Tenebrionidae). *Ann. Entomol. Soc. Am.*, *86*:239–249.

Strand, M. R., Williams, H. J., Vinson, S. B., and Mudd, A. (1989). Kairomonal activities of 2-acylcylohexane-1,3 diones produced by *Ephestia kuehniella* Zeller in eliciting searching behavior by the parasitoid *Bracon hebetor* (Say). *J. Chem. Ecol.*, *15*:1491–1500.

Su, H. C. F. (1989a). Effects of *Myristica fragrans* fruit (Family: Myristicaceae) to four species of stored-product insects. *J. Entomol. Sci.*, *24*:168–173.

Su, H. C. F. (1989b). Laboratory evaluation of dill seed extract in reducing infestation of rice weevil in stored wheat. *J. Entomol. Sci.*, *24*:317–320.

Subramanyam, B., and Cutkomp, L. K. (1985). Moth control in stored grain and the role of *Bacillus thuringiensis*: An overview. *Residue Rev.*, *94*:1–47.

Summers, M., Engler, R., Falcon, L. A., and Vail, P. V. (1975). Baculoviruses for Insect Pest Control: Safety Considerations. American Society for Microbiology, Washington, D.C.

Sweetman, H. L. (1958). The Principles of Biological Control. Wm. C. Brown Co., Dubuque, IA.

Tawfik, M. F. S., Awadallah, K. T., and Abou-zeid, N. A. (1983). The biology of the reduviid *Allaeocranum biannulipes* (Montrouzier et Signoret), a predator of stored product insects. *Bull. Soc. Entomol. Egypt*, *64*:231–237.

Tawfik, M. F. S., Awadallah, K. T., Abou-zeid, N. A., and Abdella, M. M. H. (1982). Effect of feeding on various preys on the bio-cycle of *Xylocoris flavipes* (Reuter). Res. Bull. No. 623, Faculty of Agriculture, Zagazig University, 1–10.

Tawfik, M. F. S., Awadallah, K. T., El-Husseini, M. M., and Afifi, A. I. (1987). Survey on stored drug insect, mite pests and their associated natural enemies in Egypt. *Bull. Soc. Entomol. Egypt*, *65*:267–274.

Tawfik, M. F. S., Awadallah, K. T., El-Husseini, M. M., and Ibrahim, A. M. A. Effect of temperature and relative humidity on the biocycle of *Xylocoris sordidas*. *Dtsch. Entomol. Z.N.F.* (in press).

Taylor, R. L. (1974). Role of learning in insect parasitism. *Ecol. Monogr.*, *44*:89–104.

Teetes, G. L., and Randolph, N. M. (1969). Seasonal abundance and parasitism of the sunflower moth *Homoeosoma electellum* in Texas. *Ann. Entomol. Soc. Am.*, *62*:1461–1464.

Thompson, J. V., and Redlinger, L. M. (1968). Isolation of a nuclear-polyhedrosis virus from the almond moth, *Cadra cautella*. *J. Invertebr. Pathol.*10:441–444.

Trudeau, D., and Gordon, D. M. (1989). Factors determining the functional response of the parasitoid *Venturia canescens*. *Entomol. Exp. Appl.*, *50*:3–6.

Utida, S. (1955). Population fluctuation in the system of interaction between a host and its two species of parasite, Experimental studies on synparasitism, III. *Oyo-Kontyu*, *11*:43–48.

Vail, P. V. (1991). Novel virus composition to protect agricultural commodities from insects, U.S. Patent No. 07/212,641.

Vail, P. V., and Tebbets, J. S. (1990). Comparative biology and susceptibility of *Plodia interpunctella* (Lepidoptera: Pyralidae) populations to a granulosis virus. *Environ. Entomol.*, *19*:791–794.

Vail, P. V., Tebbets, J. S., Cowan, D. C., and Jenner, K. E. (1991). Efficacy and persistence of a granulosis virus against infestations of *Plodia interpunctella* (Hübner) (Lepidoptera: Pyralidae) on raisins. *J. Stored Prod. Res.*, *27*:103–107.

Vail, P. V., Hoffmann, D. F., and Tebbets, J. S. (1993). Autodissemination of *Plodia interpunctella* (Hübner) (Lepidoptera: Pyralidae) granulosis virus by healthy adults. *J. Stored Prod. Res.*, *29*:71–74.

Van Rie, J., McGaughey, W. H., Johnson, D. E., Barnett, D. B., and Van Mellaert, H. (1990). Mechanism of insect resistance to the microbial insecticide *Bacillus thuringiensis*. *Science*, *247*:72–74.

Waage, J. K. (1979). Foraging for patchily-distributed hosts by the parasitoid, *Nemeritis canescens*. *J. Anim. Ecol.*, *48*:353–371.

Waage, J. K. (1991). Biological control: the old and the new. In Proceedings of the Workshop on Biological Control of Pests in Canada, October 11–12, 1990 (ed. McClay, A. S.). Calgary, Alberta, Alberta Environmental Centre, Vegreville, Alberta, Canada, AECV91–P1, pp. 105–112.

Waters, W. E., Drooz, A. T., and Pschorn-Walcher, H. (1976). Biological control of pests of broad-leaved forests and woodlands. In Theory and Practice of Biological Control (ed. Huffaker, C. B., and Messenger, P. S.). Academic Press, New York, pp. 313–336.

Weiser, J. (1963). Sporozoan infections. Insect Pathology (ed. Steinhaus, E. A.). Academic Press Inc., New York, 2:291–334.

Wen, B., and Deng, W. (1988). Studies on the simulated control of several stored product insect pests by the predator *Xylocoris* sp. *Acta Phytophylact. Sin.*, *15*:273–276.

Wen, B., Smith, L., and Brower, J. H. (1994). Competition between *Anisopteromalus calandrae* and *Choetospila elegans* at different parasitoid densities on immature maize weevils in corn. *Environ. Entomol.*, *23*:367–373.

White, E. G., and Huffaker, C. B. (1969a). Regulatory processes and population cyclicity in laboratory populations of *Anagasta kuhniella* (Zeller) (Lepidoptera: Phycitidae), I-Competition for food and predation. *Res. Pop. Ecol. Kyoto Univ.*, *11*:57–83.

White, E. G., and Huffaker, C. B. (1969b). Regulatory processes and population cyclicity in laboratory populations of *Anagasta kuhniella* (Zeller) (Lepidoptera: Phycitidae), II-Parasitism, predation, competition and protective cover. *Res. Pop. Ecol. Kyoto Univ.*, *11*:150–185.

Yoo, C. K., and Ryoo, M. I. (1989). Host preference of *Lariophagus distinguendus* (Foerster) (Hymenoptera: Pteromalidae) for the instars of rice weevil (*Sitophilus oryzae* (L.) (Coleoptera: Curculionidae) and sex ratio of the parasitoid in relation to the host. *Korean J. Appl. Entomol.*, *28*:28–31.

Yoshida, T. (1984). Colonization by forest insects of stored barley or wheat in ancient times, in the near and middle east. *Biodeterioration*, *6*:660–663.

Yoshida, T., and Kawano, K. (1958). Seasonal fluctuation of the number of insects in the grains stored at farm-house, 1. The ecological studies of the pests infesting stored grains, Part 2. Mem. Fac. Liberal Arts Education, Miyazaki University. *Nat. Sci.*, *5*: 11–23.

Zdarkova, E. (1991). Application of the bio-preparation cheyletin in empty stores. In Modern Acarology, Vol. 1: VIII International Congress of Acarology, Ceske Budejovice, Czechoslovakia, August 6–11, 1990 (ed. Dusabek, F., and Bukvo, V.) SPB Academic Publishing by The Hague, Netherlands.

Zdarkova, E., and Horak, E. (1990). Preventive biological control of stored food mites in empty stores using *Cheyletus eruditus* (Schrank). *Crop Prot.*, *9*:378–382.

Zeikus, R. D., and Stienhaus, E. A. (1969). Teratology of the beetle *Tenebrio molitor*, V. Ultrastructural changes and viruslike particles in the foregut epithelium of pupal-winged adults. *J. Invertebr. Pathol.*, *14*:115–121.

# 7

# Chemical Control

**Noel D. G. White**

*Agriculture and Agri-Food Canada, Winnipeg, Manitoba, Canada*

**James G. Leesch**

*U.S. Department of Agriculture, Fresno, California*

Many insects are able to feed, develop, and multiply on whole seeds, damaged seeds, or processed cereals in storage (Pedersen 1992). When grain is stored at 25–35°C insects multiply rapidly (Sinha and Watters 1985) and can cause significant losses ranging from grain weight loss to quality loss, with decreased seed germination, biochemical and nutritional changes in seeds, and contamination by excreta, cast skins, and body parts (Pedersen 1992; Pomeranz 1992). The respiration by large populations of insects also produces heat and moisture, favoring mold growth in stored products. Some insects such as the weevils (*Sitophilus* spp) or lesser grain borer (*Rhyzopertha dominica* (F.)) feed directly on whole kernels, while others feed on damaged kernels or molds.

Storage insects can be found in grain and flour residues, old stored grain, processed and packaged food, and in nature feeding on seeds or fungus under bark of trees (Linsley 1944). Newly harvested grain can rapidly become infested by flying insects (Hagstrum 1989; Madrid et al. 1990). The chapter in this book on Ecology gives a good overview of how insect populations develop in stored grain.

Prevention and control of insects in stored products are important to farmers, grain handlers, grain processors, and consumers. Under many circumstances the easiest, most rapid, and economical method of controlling insects is with insecticides, often in the form of fumigants. All insecticides used on or near stored food must meet certain criteria and intense regulatory review. The ideal insecticide should rapidly kill pests but not harm humans or the environment, have a residual activity only as long as needed with acceptable product contamination, be inexpensive, easily handled and prepared, and produce nearly odorless protection (Bennett et al. 1988). No insecticides meet all of these criteria but these criteria guide the selection of appropriate chemicals for use near stored products.

There are many modes of actions for various types of insecticides ranging from neurological poisoning (acetylcholinesterase inhibitors), to desiccation (diatomaceous earth), to suffocation and other complex physiological abnormalities (modified atmospheres). Various insect species and developmental stages of each species respond differently to different insecticides (Snelson 1987; Harein and Davis 1992) (e.g., pyrethroids are more toxic to the lesser grain borer than are organophosphorus chemicals). Some insects, once established, are often very difficult to kill because they develop inside grain kernels (weevils). Classes of insecticide can affect insects differently as well: the synergized pyrethrins provide rapid knockdown of adult insects and provide a prolonged, repellent effect on insects whereas organophosphorus insecticides do not generally repel and produce a slower toxic effect (Harein and Davis 1992). All of these factors need to be weighed when choosing an insecticide for use on stored-grain insects.

Insecticide use should be considered carefully and grain managers should be aware that the presence of fungus-feeding insects indicates that grain is damp or wet and becoming moldy. Fumigation can kill the insects but will not prevent the fungal spoilage. Identification of insects in stored products is necessary for proper pest control and grain management.

## I. CONTACT INSECTICIDES

A relatively small number of contact insecticides are registered for use on or near stored cereals and their byproducts in the United States and Canada (Table 1). Discussions of use of insecticides in food processing areas are given by Zettler and Redlinger (1984), Heaps and Harein (1991), and Ostlie et al. (1994), Subramanyam et al. (1993), and Subramanyam (personal communication 1994) on residue tolerance limits by Grisamore et al. (1991) (Table 2). Insecticides cannot be applied in food processing plants during their operation.

Grain protectants, which are added directly to grain at the beginning of storage, are chlorpyrifos-methyl (Reldan) for barley, oats, rice, wheat, and sorghum, and pirimiphos-methyl (Actellic) for maize and sorghum in the United States. Until 1992, malathion was registered as a general stored-grain protectant but now registration is being canceled (Wienzierl 1992), although it may be relabeled for use as a top-dressing on the surface of grain masses. Manufacturers of malathion will not undertake expensive new toxicological studies because of a limited market and increasing resistance of insects to this chemical (Haliscak and Beeman 1983; Zettler and Cuperus 1990). The main grain protectant currently registered in Canada is malathion, although synergized pyrethrum or diatomaceous earth may be applied.

Chemicals currently used around the world to treat stored grain include bioresmethrin, bromophos, carbaryl, chlorpyrifos-methyl, dichlorvos, fenitrothion, lindane, malathion, pirimiphos-methyl, piperonyl butoxide plus pyrethrins.

Potential grain protectants include deltamethrin, diazinon, etrimphos, fenvalerate, iodofenphos, methacrifos, permethrin, phenothrin, phoxim, and tetrachlorvinphos (Snelson 1987). Some countries, such as Australia, typically apply a combination of grain protectants including an organophosphate such as fenitrothion and a pyrethroid such as bioresmethrin, which is needed specifically to control the lesser grain borer, *R. dominica*. Such applications can cost nearly $1 per ton of grain (Mills and White 1984), while the application of an organophosphate, such as malathion or chlorpyrifos-methyl alone, costs around 4 cents and 61 cents per ton, respectively (Reed et al. 1990).

In the United Kingdom, insecticides registered for use on or near stored grain and rapeseed are fenitrothion, fenitrothion + permethrin + resmethrin, and pirimiphos-methyl for flour beetles; fenitrothion + permethrin + resmethrin and pirimiphos-methyl for flour moths; etrimfos, fenitrothion, fenitrothion + permethrin + resmethrin, lindane, and pirimiphos-methyl for grain beetles; etrimfos and pirimiphos-methyl for grain storage mites; chloropicrin + methyl bromide, chlorpyrifos-methyl, etrimfos, lindane, methacriphos, permethrin, and pirimiphos-methyl for grain storage pests; etrimfos, fenitrothion, fenitrothion + permethrin + resmethrin, lindane, and pirimiphos-methyl for grain weevils; and pyrethrins for weevils (lesser grain borer) (Ivens 1989).

Grain protectants are generally used extensively in warm to hot climatic regions, where insect infestations in storage are severe and occur more or less continually.

The advantages of grain protectants over fumigants include their prolonged persistence for control of insects from months to years, generally safe application, they require little specialized equipment, they prevent establishment of pests, and they are effective in loosely constructed storage structures that cannot be fumigated effectively (Harein and Davis 1992). The main drawback of protectants are residues that remain on the food. However, residues generally degrade with storage time (Rowlands 1975) and food processing (Chamberlain 1981) and accumulate in localized regions of grain kernels (Rowlands and Bramhall 1977; Mensah et al. 1979). Other grain protectants usually used on seed grain are the inorganic dusts such as diatomaceous earth or silicon dioxide, which kill insects by abrasion and desiccation when fats are removed from the insect cuticle and the insect cannot regulate water loss. The insect growth regulator methoprene is registered for use as a grain protectant in the United States and for use near stored tobacco in the United States and Canada, and hydroprene is registered for use in food processing plants in the United States.

Some contact insecticides used in storage are for treating cracks and crevices, structural materials such as wood, concrete, or steel used in floors and walls, and localized or general treatment of packages and equipment to control residual insect populations. Presently malathion is used as an empty bin treatment for cereals in Canada. In the United States, cylfluthrin, diatomaceous earth, methoxy-

**Table 1** Contact Insecticides Registered for Use on or Near Stored Whole and Processed Cereals in the United States and Canada

| Country | Location | Chemical | | Use | Formulation Spray (S)/ Dust (D) | Long-term control | Toxicity ($LD_{50}$ rats-oral)[b] |
|---|---|---|---|---|---|---|---|
| | | Type | Name | | | | |
| United States | Farm | Organophosphate | Malathion[a] | Bin, protectant | S, D | Yes | 1000–2800 |
| | | | Chlorpyrifos-methyl | Bin, protectant | S, D | Yes | 1000–3700 |
| | | | Pirimiphos-methyl | Bin (pending), protectant | S, D | Yes | 2000 |
| | | | Dichlorvos | Space | S, Solid | Limited | 50 |
| | | Growth regulator | Methoprene | Bin, protectant | S | Yes | 34,600 |
| | | Botanical | Pyrethrins (synergized) | Bin, space | S | No | 1500 |
| | | Pyrethroid | Cyfluthrin | Bin | S | Yes | 900 |
| | | Bacteria | *Bacillus thuringiensis* | Top dressing, protectant | S, D | Some | Low toxicity |
| | | Chlorinated hydrocarbon | Methoxychlor | Bin, cracks | S, D | Yes | 6000 |
| | | Inorganic | Diatomaceous earth | Bin, cracks, protectant | D | Yes | >5000 |
| Canada | Farm | Organophosphate | Malathion | Bin, protectant | S, D | Yes | 1000–2800 |
| | | | Dichlorvos | Space | S | Limited | 50 |
| | | Botanical | Pyrethrins (synergized) | Bin, space | S | No | 1500 |
| | | Inorganic | Silicon dioxide | Cracks | D | Yes | >5000 |
| | Poultry barns | Organophosphate | Tetrachlorviphos | Bin | S | Yes | 4000–5000 |
| | | Carbamate | Carbaryl | Bin | S | Yes | 246 |
| | | Inorganic | Boric acid | Cracks | D | Yes | 500–5000 |

| | | | | | | | |
|---|---|---|---|---|---|---|---|
| | Seed grain | Organophosphate | Diazinon | Bin, protectant | | Yes | 300–400 |
| | | Chlorinated hydrocarbon | Lindane | Bin, protectant | S | Yes | 88–125 |
| | | | Methoxychlor | Bin, protectant | S, D | Yes | 6000 |
| | | Inorganic | Silicon dioxide | Protectant | D | Yes | >5000 |
| United States | Food processing plant | Organophosphate | Acephate | Cracks | S | Yes | 700–980 |
| | | | Chlorpyrifos | Cracks | S | Yes | 96–270 |
| | | | Diazinon | Cracks, local | S | Yes | 300–400 |
| | | Botanical | Pyrethrins (synergized) | Cracks, local | S | No | 1500 |
| | | | Microencapsulated pyrethins | Cracks, local | S | Limited | 1500 |
| | | Pyrethroids | D-*Trans* allethrin | | S | Limited | 860 |
| | | | Cyfluthrin | Cracks, local | | Yes | 900 |
| | | | S-fenvalerate | Cracks, local | S | Yes | 451 |
| | | | Lambda-cyhalothrin | Cracks, local | S | Yes | — |
| | | | Resmethrin | Cracks, local | S | Limited | 2500 |
| | | Carbamates | Propoxur | Cracks, local | S | Yes | 50–104 |
| | | | Bendiocarb | Cracks, local | S | Yes | 40–156 |
| | | Inorganic | Silica gel | Cracks | D | Yes | >5000 |
| | | | Boric acid | Cracks | D | Yes | 500–5000 |
| | | | Diatomaceous earth | Cracks | D | Yes | >5000 |
| | | Growth regulators | Methoprene | Cracks, local | S | Yes | 34,600 |
| | | | Hydroprene | Cracks, local | S | Yes | 34,000 |

**Table 1** Continued

| Country | Location | Chemical | | Use | Formulation Spray (S)/ Dust (D) | Long-term control | Toxicity ($LD_{50}$ rats-oral)[b] |
|---|---|---|---|---|---|---|---|
| | | Type | Name | | | | |
| Canada | Food processing plant | Organophosphate | Chlorpyrifos | Cracks, local | S | Yes | 96–270 |
| | | | Diazinon | Cracks, local | S | Yes | 300–400 |
| | | | Dichlorvos | Space | S | No | 50 |
| | | | Fenthion | Cracks, local | S, D | Yes | 250 |
| | | | Malathion | Cracks, local | S, D | Yes | 1000–2800 |
| | | | Naled | Space | S | Yes | 250 |
| | | Botanical | Pyrethrins (synergized) | Cracks, local, space | S | No | 1500 |
| | | Pyrethroid | Allethrin | Cracks, local, space | S | Limited | 685–1100 |
| | | | Cyfluthrin | (Warehouse, cracks, local) | S | Yes | 900 |
| | | | d-trans allethrin | Cracks, local, space | S | Limited | 860 |
| | | Carbamates | Bendiocarb | Cracks, local | S | Yes | 40–156 |
| | | | Propoxur | Cracks, local | S | Yes | 50–104 |
| | | Inorganic | Silicon dioxide | Cracks | D | Yes | >5000 |
| | | Growth regulators | Methoprene (tobacco) | Cracks, local | S | Yes | 34 600 |

[a]Malathion has not been widely used in the United States since 1992 because of unwillingness of manufacturers to obtain further toxicological data and widespread insect resistance.

[b]$LD_{50}$ = lethal dose of active ingredient for 50% of a test population.

chlor, and chloropicrin are registered for empty bin treatment. For bins receiving maize, chlorpyrifos-methyl may not be used. On farms, all residual grain should be removed from empty granaries and both the inside and outside of the granary sprayed several weeks before filling (Cink and Harein 1989).

General space treatments of empty storage structures or in headspaces above a product are occasionally done to control flying insects, often with synergized pyrethrins or dichlorvos. Occasionally bags holding a commodity may be treated with an insecticide such as permethrin, notably in tropical countries (Barakat et al. 1987). Synergized pyrethrins or silica gel on packages or between package layers can also be used to prevent insect infestations in some cases (Highland 1991).

## II. INSECTICIDE RESIDUES

The residue tolerances for insecticides used on or near stored products in the United States and Canada are given in Table 2. The maximum international limits for insecticide residues in grain, processed grain, and oilseeds that result from either preharvest or postharvest application set by the Food and Agricultural Organization of the United Nations and the World Health Organization are given in Table 3 (Smith 1990).

Most contact insecticides used in storage are lipophilic and accumulate in areas of high fat content such as the germ and bran of cereals (Mensah et al. 1979) and the oil in oilseeds (White and Nowicki, 1986).

## III. INSECTICIDE BREAKDOWN

The degradation of contact insecticides in stored grain is affected by grain moisture content and temperature (Snelson 1987). High temperatures generally result in rapid insecticide degradation but the chemicals are more toxic to insects at high temperature, although some pyrethroids are more toxic at cooler temperatures (Subramanyam and Cutkomp 1987). The presence of molds can also increase rates of insecticide breakdown to nontoxic compounds (Anderegg and Madisen 1983). The type of insecticide formulation does not generally affect rates of breakdown; however, these rates increase considerably when grain is in moisture equilibrium with 70% relative humidity (RH) or higher (Samson et al. 1988).

Desmarchelier and Bengston (1979) developed an equation to predict rates of degradation of grain protectants under various conditions:

$$\log t_{1/2} = \log (t_{1/2})_0 - BT^*$$

where $t_{1/2}$ = half-life of the chemical;
$(t_{1/2})_0$ = half-life of the chemical at 30°C, 50% RH;
B = temperature coefficient per week; and
$T^*$ = temperature in °C minus 30.

Values for $(t_{1/2})_0$ and B are given for 12 insecticides by Desmarchelier and Bengston (1979). Values for chlorpyrifos-methyl are $(t_{1/2})_0 = 19$, B = 0.04, mala-

**Table 2** Residue Tolerance (ppm) for Insecticides Used on or Near Stored Products in the United States and Canada

| Country | Type of material | Insecticide | Product | | | | |
|---|---|---|---|---|---|---|---|
| | | | Barley | Maize | Oat | Rye | Wheat |
| United States | Whole grain | Allethrin | — | 2 | — | — | — |
| | | Aluminum phosphide | 0.1 | 0.1 | 0.1 | 0.1 | 0.1 |
| | | Calcium cyanide | 25 | — | — | — | — |
| | | Carbaryl | 0 | 5 | 0 | 0 | 3 |
| | | Chlorpyrifos-methyl | 6 | — | 6 | — | 6 |
| | | Malathion | 8 | 8 | 8 | 8 | 8 |
| | | Magnesium phosphide | 0.1 | 0.1 | 0.1 | 0.1 | 0.1 |
| | | Methomyl | 1.0 | 0.1 | 1.0 | 1.0 | 1.0 |
| | | Methyl bromide | 50 | 50 | 50 | 50 | 50 |
| | | Methoxychlor | 2 | 2 | 2 | 2 | 2 |
| | | Piperonyl butoxide | 20 | 20 | 8 | 20 | 20 |
| | | Pirimiphos-methyl | — | 8 | — | — | — |
| | | Cereal | | Maize and meal | | | Wheat and macaroni |
| | Human food | Methoprene | 10 | 10 | 10 | 10 | 10 |
| | | Milled fractions | | | | | |
| | | Chlorpyrifos-methyl | 90 | — | 130 | | 30 |
| | | Pirimiphos-methyl | — | 40 | — | — | |

| | | | | | | | |
|---|---|---|---|---|---|---|---|
| | | Flour | | Maize grits | | | Maize malted beverages |
| | | Inorganic bromides | 125 | 125 | 125 | 125 | 25 |
| | | Milled fractions | | | | | |
| | Animal feed | Chlorpyrifos-methyl | 90 | — | 130 | — | 30 |
| | | Pirimiphos-methyl | — | 40 | — | — | — |
| | | Inorganic bromides | 125 | 125 | 125 | 125 | 125 |
| | | | Whole cereal seeds | | | | Oilseeds |
| Canada | Human food | Malathion | 8 (wheat, rye meal, flour) | | | | nr |
| | | | 2 | | | | |
| | | Methoxychlor | 2 | | | | nr |
| | | Naled | 0.5 (rice, soybean) | | | | nr |
| | | Piperonyl butoxide | 20 | | | | nr |
| | | Pyrethrins | 3 | | | | nr |
| | | Aluminum phosphide | no residue (nr, < 0.1 ppm) | | | | nr |
| | | Carbofuran | nr | | | | nr |
| | | Chlorfenvinphos | nr | | | | nr |
| | | Dichlorvos | nr | | | | nr |

US action levels for insecticides with no tolerance on cereal grains: aldrin (0.02 ppm), dieldrin (0.02 ppm), benzene hexachloride (0.05 ppm), chlordane (maize) (0.1 ppm), DDT (0.5 ppm), ethylene dibromide (cereal products to be cooked) 150 ppb, (cereal products ready to eat) 30 ppb, lindane (0.1 ppm).
nr, no residue.

**Table 3** Maximum Limits for Insecticide Residues in Grain, Processed Grain, or Oilseeds, Resulting from Preharvest or Postharvest Application, Set by the FAO/WHO of the United Nations

| Insecticide | Product | Maximum residue limit (ppm) |
|---|---|---|
| Aldicarb | Soybeans | 0.02 |
| | Maize, peanuts | 0.05 |
| | Cottonseed | 0.10 |
| | Sorghum | 0.20 |
| Aldrin/dieldrin | Rice in husk | 0.02 |
| | Cereal grains | 0.02 (extraneous sources) |
| Azinphos-methyl | Cereal grains, soybeans, sunflower seed | 0.20 |
| Bendiocarb | Rice (husked) | 0.02 |
| | Barley, maize, oat, wheat | |
| Bromophos | Rapeseed, rapeseed oil | 0.20 |
| | Cereal grains | 10.00 |
| | Wheat bran (unprocessed) | 20.00 |
| Carbaryl | Wheat flour | 0.20 |
| | Soybeans (dry) | 1.00 |
| | Peanuts (in shell), wholemeal flour | 2.00 |
| | Rice, rye, wheat | 5.00 |
| | Wheat bran | 20.00 |
| Carbofuran | Maize, oat, sorghum, wheat, oilseeds | 0.10 |
| Chlordane | Maize, oat, rice, rye, sorghum | 0.02 |
| | Wheat | 0.05 (extraneous sources) |
| Chlordimeform | Cottonseed | 2.00 |
| Chlorfenvinphos | Maize, peanuts (shelled), rice, wheat | 0.05 |
| Chlorpyrifos | Rice (with husk) | 0.10 |
| Chlorpyrifos-methyl | Flour, wholemeal bread | 2.00 |
| | Maize, sorghum, wheat | 10.00 |
| | Wheat (bran) | 20.00 |
| Cypermethrin | Maize, peanuts, soybeans | 0.05 |
| | Wheat, oilseeds | 0.20 |
| | Barley | 0.50 |
| DDT | Cereal grains | 0.10 (extraneous sources) |
| Deltamethrin | Oilseeds | 0.10 |
| Diazinon | Peanuts (shelled), rice (polished), wheat, sunflower seed | 0.10 |
| Dichlorvos | Bread, cake | 0.10 |
| | Milled, products from raw grain | 0.50 |
| | Cereal grains | 2.00 |
| Disulfoton | Peanuts (kernels) | 0.10 |
| | Cereal grains (except rice, maize) | 0.20 |
| Endosulfan | Cottonseed | 1.00 |

**Table 3** Continued

| Insecticide | Product | Maximum residue limit (ppm) |
|---|---|---|
| Endrin | Barley, cottonseed oil (edible), rice, sorghum, wheat | 0.02 |
| Ethiofencarb | Barley, oat, rye, wheat | 0.05 |
| Ethion | Maize | 0.05 |
| | Cottonseed | 0.50 |
| Etrimfos | Rapeseed oil | 0.50 |
| | Rapeseed | 10.00 |
| Fenitrothion | Soybeans (dry) | 0.10 |
| | Bread (white) | 0.20 |
| | Wheat flour (white), rice (polished) | 1.00 |
| | Processed wheat bran | 2.00 |
| | Wheat flour (wholemeal) | 5.00 |
| | Cereal grains | 10.00 |
| | Raw wheat bran, rice bran | 20.00 |
| Fenthion | Rice, wheat | 0.10 |
| Fenvalerate | Peanuts (whole), soybeans, sunflower seed | 0.10 |
| | Wheat flour | 0.20 |
| | Cereal grains | 2.00 |
| Heptachlor | Cottonseed, raw cereals | 0.02 (extraneous sources) |
| Hydrogen cyanide | Flour | 6.00 |
| | Cereal grains | 75.00 |
| Hydrogen phosphide | Breakfast cereals, flour, milled cereal products | 0.01 |
| | Cereal grains | 0.10 |
| Inorganic bromide | Cereal grains, wholemeal flour | 50.00 |
| Isofenphos | Rapeseed | 0.02 |
| Lindane | Cereal grains, rice | 0.50 |
| Malathion | Rye and wheat (wholemeal and flour) | 2.00 |
| | Cereal grains | 8.00 |
| | Rye (bran), wheat (bran) | 20.00 |
| Methidathion | Sorghum grain | 0.10 |
| | Cottonseed | 0.20 |
| | Cottonseed oil (crude) | 1.00 |
| Monocrotophos | Cottonseed oil, corn grain, soy beans | 0.05 |
| | Cottonseed | 0.10 |
| Parathion-methyl | Cottonseed oil | 0.05 |
| Permethrin | Beans | 0.10 |
| | Cereal grains | 2.00 |
| Phenthoate | Rice (hulled) | 0.05 |

**Table 3** Continued

| Insecticide | Product | Maximum residue limit (ppm) |
|---|---|---|
| Phosmet | Maize (kernels) | 0.20 |
| Phosphamidon | Cereal grains | 0.10 |
| Phoxim | Cereal grains, cottonseed | 0.05 |
| Piperonyl butoxide | Oilseeds | 18.00 |
| | Peanuts | 8.00 |
| | Cereal grains | 20.00 |
| Pirimiphos-methyl | Bread (wholemeal), rice (polished) | 1.0 |
| | Peanuts (kernels) | 2.00 |
| | Rye, wheat (wholemeal) flour | 5.00 |
| | Cereal grains, peanut oil | 10.00 |
| | Rice bran, wheat bran | 20.00 |
| | Peanuts (in shell) | 25.00 |
| Propoxur | Rice (hulled) | 0.10 |
| Pyrethrins | Oilseeds | 1.00 |
| | Cereal grains | 3.00 |
| Thiometon | Rapeseed, raw cereals | 0.05 |
| | Cottonseed oil | 0.10 |
| Trichlorfon | Cereal grains, cottonseed, peanuts (shelled), rapeseed | 0.10 |

thion $(t_{1/2})_0 = 12$, $B = 0.05$, and pirimiphos-methyl $(t_{1/2})_0 = 70$, $B$ = very small. The half-life for each chemical can be calculated for various relative humidities that will be in equilibrium with known moisture contents of the grain.

High temperatures and moisture contents cause rapid breakdown of many insecticides, such as malathion, by stimulating enzyme activity in seeds (Rowlands 1975), which degrades the insecticide mainly by hydrolysis (Orth and Minett 1975). An example of temperature effects on malathion residues (8 pm application) in stored wheat (12.5% moisture content) after 72 weeks was a decrease in malathion of 26% at −5°C, 41% at 5°C, 74% at 10°C, 95% at 20°C, and 96% at 27°C (Abdel-Kader et al. 1980). Chlorpyrifos-methyl degradation on stored wheat and maize at different temperatures and moisture contents was studied for 10 months and asymptotic equations derived (Arthur et al. 1991, 1992). The insecticide broke down rapidly as temperature and moisture increased and the need for rapidly cooling the grain in storage was emphasized.

The application of grain protectants is done while grain is moving at the auger or on a conveyor belt and it can be added as a dust or liquid spray or drip. The application rate of a known volume of a given concentration of insecticide on grain flowing at a known rate results in predetermined insecticide levels. Quinlan et al. (1979) and Minett et al. (1981) described cheap and easy-to-use drip applicators for adding insecticides to a grain stream at the auger or on conveyor belts. Applications using this method can often be crude. For example, White et al. (1986) applied malathion to wheat at calculated levels of 8 ppm but actual levels of insecticide varied from 5.8 to 19.0 ppm at the grain bulk surface and 2.8 to 3.6 ppm at the 1 m depth, although the average for the entire bulk was 8 ppm. However, since even partial treatment of a grain bulk can give effective control of insects (Minett and Williams 1971) uneven insecticide application is rarely a problem. Most of the organophosphate grain protectants control beetle populations with residue levels at 2 ppm or higher.

## IV. CLASSES OF INSECTICIDES

The classes of contact insecticides used on or near stored products include chlorinated hydrocarbon; botanical (i.e., synergized pyrethrins/pyrethroids); inert mineral dusts; carbamates; organophosphates; insect growth regulators; and bacteria.

Chlorinated hydrocarbon compounds are no longer used widely in developed countries because of bioaccumulation. Pyrethrins are often in scarce supply since they are produced from chrysanthemums grown in Ecuador and Kenya and are relatively expensive, with short residual activity. The new pyrethroids are also expensive and many are extremely irritating to applicators. Diatomaceous earth dust is messy, affects the flow characteristics and test weight of grain, is relatively ineffective at high grain moisture content (Le Patourel 1986), and blows out of structures, which may cause respiratory hazards in humans. It may be used to treat bags (Watters 1966) and is often applied as a slurry in water that sticks to surfaces. Because diatomaceous earth is relatively nontoxic to humans, it is receiving a great deal of attention as an alternative to traditional chemical insecticides. Carbamates are generally too toxic to mammals. Organophosphorus compounds are optimal, with suitable toxicity and residual activity.

Insect growth regulators disrupt a number of normal processes in the growth and development of insects leading to death, and have very little toxicity to mammals; examples are methoprene, hydroprene, and fenoxycarb. A bacterium, *Bacillus thuringiensis*, is often used to control moths at the surface of bulk grain.

## V. INSECTICIDE FORMULATIONS

Liquid formulations used as sprays include emulsifiable concentrates (EC) that are diluted with water, flowables (F) that are diluted with water, flowable microencapsulates (FM) that have active ingredients encased in very small plastic capsules suspended in water for spraying, and solutions (S) that are soluble in water or oil.

Dry formulations used as sprays include wettable powders (WP) that usually contain 50% or more active ingredient and remain suspended in water and soluble powders (SP) that dissolve in water. Dusts have the insecticide's active ingredients carried on dry particles such as wheat flour, inert clay, or talc, or inorganic dusts such as boric acid or silicon dioxide.

## VI. EFFECTS OF FORMULATION AND APPLICATION ON TOXICITY

Encapsulated formulations of malathion and fenitrothion are more effective on plywood than are emulsifiable concentrate (EC) formulations (LaHue and Kadoum 1979). Wettable powder formulations of fenitrothion, malathion, chlorpyrifos-methyl, and pirimiphos-methyl are more toxic to stored product beetles than emulsifiable concentrations on filter papers (Barson 1991).

Wettable powder solutions are best for treatment of absorbent materials such as bricks, cement, concrete, timber, and sacking since they are filtered by the surfaces; the carrier and insecticide remain on the surface and produce less contamination of materials in sacks than do emulsifiable concentrate solutions (Parkin 1966). Sprays are generally more economical, cleaner, and provide more controlled application than dusts. Contamination of stored products placed on concrete, wood, or metal treated with oil or wettable powder is greater from a wettable powder treatment (Watters and Grussendorf 1969). The application of synergized pyrethrins with a fog generator results in more insecticide on the floor than on the walls. There is a loss of insecticide during treatment and nonuniform distribution of deposits due to air currents (Watters 1968).

Commercial malathion dust consists of malathion on a wheat flour carrier and offers slightly longer activity on concrete than do sprays. Spruce sawdust treated with malathion has been tested on various surfaces. The size of sawdust particles does not affect residual activity on a given surface but the insecticide breaks down more quickly on concrete than on wood at 25°C, 50% RH. Sawdust with 0.8% active ingredient (AI) remains effective for insect control for 12 weeks on galvanized steel, and for 16 weeks on wood. A 2% AI sawdust mix is active for 4 weeks or longer on concrete, and at least 16 weeks on wood and steel (Mensah and White 1984).

Emulsions of malathion prepared in alkaline water break down rapidly. The use of emulsifiable concentrates is recommended on metal or wood, not on concrete or bricks or near electrical switches. Wettable powders should be used on brick and concrete. Oil solutions can be used on wood or metal and near electrical switches. Dusts can be used on floors or in wall spaces, since they are easily applied (Mills et al. 1990).

## VII. RESIDUAL ACTIVITY OF INSECTICIDES APPLIED IN STORAGE STRUCTURES

### 1. Effect of Surfaces

The residual activity of many insecticides is sharply decreased on surfaces such as concrete, which have a pH of about 10.5, because of hydrolysis; treated wood and steel remain effective for long periods because of a moderate pH near 6.0. This is important since floors in modern granaries, terminal elevators, and warehouses are concrete.

Methoxychlor and lindane are more toxic and persistent on wood than on metal or concrete (Watters and Grussendorf 1969) and both insecticides are more effective at controlling hairy spider beetles, *Ptinus villiger* (Reit.), in flour mills than is malathion or synergized pyrethrins. Methoxychlor, lindane, DDT, malathion, and bromophos on filter paper were tested against five species of stored-product beetles, and although methoxychlor was effective in the field it was ineffective in the laboratory. DDT and lindane were most toxic at 10 and 16°C; bromphos was effective at 27°C but not at 10 or 16°C; malathion was the most effective insecticide at all three temperatures (Iordanou and Watters 1969).

Malathion (0.42 g/m$^2$) degrades quickly on cement, whitewash, and unglazed tile but persists for several months on wood and sacking; fenthion is highly toxic on all surfaces to eight species of stored-product beetles with several days exposure but is ineffective on cement and whitewash by 8 months at 25°C. Chlorthion is similar to fenthion but degradation is more rapid once it begins. Diazinon fails almost immediately on cement and tile and is nontoxic after 2 months on whitewash and wood, or a little longer on sacking. Carbaryl is not effective on cement, sacking, or wood (Parkin 1966).

Alkaline surfaces can be treated to increase insecticide activity. A treatment of 1 g/m$^2$ malathion on concrete is ineffective in controlling insects after 4 days. If the concrete is washed continuously for 8 days the pH is lowered from 10.0 to 7.0 and the treated concrete remains toxic for 32 days (Okwelogu 1968). Treatment of concrete with waterproof silicone paint also lowers the pH and the presence of dust on floors increases the residual activity of malathion and bromophos (Watters 1970) or malathion and pirimiphos-methyl (White 1982) by lowering the pH from 8–10 to almost 6. Treatment with 1.5 g/m$^2$ malathion or bromophos applied to a dust-covered concrete terminal elevator floor produced 50–85% mortality of *Tribolium confusum* du Val for 33 weeks (Watters 1970). Malathion and fenthion at 1 g/m$^2$ remain effective for 17–20 weeks on concrete painted with silicone waterproofing but for less than 1 week with no paint. Painting increases the active life of the chemicals on masonite or glass to about 22 weeks (Burkholder and Dicke 1966). Pretreatment of concrete surfaces with starch paste increases the persistence of DDT or pyrethrum in oil (Parkin and Hewlett 1946), and talc or calcium carbonate extend the activity of malathion on concrete (Slominski and Gojmerac 1972).

Malathion applied at 0.25 and 0.50 g/m$^2$ on plywood or galvanized steel is slightly more effective against *Cryptolestes ferrugineus* (Stephens) and *Tribolium castaneum* (Herbst) than fenitrothion after 35 weeks (White et al. 1983). Malathion may remain active on wood and steel storage structures for up to a year. Factors affecting toxicity to insects are length of exposure, toxicity of the chemical, dosage, temperature, and type of surface.

Cyfluthrin applied at label rates on unpainted steel controlled *T. castaneum* and *T. confusum* for 235 days with 1 h exposure. Control on painted steel lasted only 3 weeks and required 24 h exposure. No difference was observed between EC and WP formulations (Arthur 1992).

In storage structures, insects usually have a refuge (cracks, etc.) to hide in and many factors affect refuge-seeking behavior (Cox et al. 1990). The presence of an artificial refuge results in a 10-fold increase in time required for 100% control of *Sitophilus granarius* L. and *Oryzaephilus surinamensis* (L.). All insects eventually die because they wander, but if an insecticide of low toxicity is used or rapid degradation occurs, incomplete control results. The use of an insecticide with flushing action that causes activity in insects (pyrethroids) might be used with organophosphorus compounds (Pinniger 1974). The vapors of pyrethroids generally have a repellent effect on insects, but the vapors of organophosphates do not, although they may be repellent on contact (Mondal 1984).

Many pyrethroids are of increasing interest because of their moderate mammalian toxicity. Aerosols of permethrin have been effective against the Dermestidae, but *T. confusum* are only knocked down and recover (Kirkpatrick and Gillenwater 1979). Cypermethrin, permethrin, and fenvalerate applied to plywood at 0.2, 0.5, and 1.0 $g/m^2$ were studied at 10, 20, and 30°C by bioassay with *T. castaneum* for 33 weeks. High knockdown in 24 h was observed after 33 weeks except with fenvalerate at 0.2 $g/m^2$. However, none of the insecticides gave 100% knockdown immediately after treatment, even at 1 $g/m^2$, and adults could often recover after 1 week of exposure to treated surfaces (Watters et al. 1983). If a granary is filled shortly after treatment there would be incomplete control. Synergized formulations may be more effective.

Dichlorvos is occasionally used but it has little residual activity because of its high volatility. In areas with little air movement dichlorvos strips can give good control of *Plodia interpunctella* (Hubner) (LaHue 1969) and other stored-product insects. Currently this chemical has had its food additive tolerance revoked and will probably no longer be used by the food industry in the United States.

## 2. Insecticide Uptake into Stored Products from Treated Surface

Factors that affect uptake are type of insecticide; type of grain (oil content, kernel size) or processed stored products; age of deposit; temperature; and interactions.

It is generally recommended that oilseeds not be stored in structures recently treated with insecticides since most are lipophilic and are rapidly taken up in the seed, which is about 45% oil (rapeseed, flax, sunflower, etc.). Since there are no legal tolerances established for insecticides in raw rapeseed in Canada the maximum allowable level for malathion is 0.1 ppm, which is easily exceeded. Rapeseed rapidly takes up malathion and fenitrothion from wood or steel, with fenitrothion being absorbed faster (White et al. 1983).

Rapeseed stored in a plywood-lined granary treated with 0.5 g bromophos/$m^2$ had residues of 0.1–1.6 ppm at 16 weeks and 0.4–3.5 ppm at 52 weeks, in excess of legal tolerances (Watters and Nowicki 1982). Laboratory studies with malathion or bromophos applied to wood or concrete showed malathion levels of 32.5 ppm (wood) and 0.01 ppm (concrete) in rapeseed, and 6.4 ppm (wood) and 0.2 ppm (concrete) in wheat after 16 weeks of storage. Bromophos levels of 21 ppm (wood) and 1.2 ppm (concrete) were found in rapeseed, and 11 ppm (wood) and 0.2 ppm (concrete) in wheat after 16 weeks. Bromophos uptake was faster than malathion uptake and insecticide uptake by both crops was faster from wood than concrete (Watters and Nowicki 1982).

Pirimiphos-methyl uptake into stored Stilton cheese was studied on wood treated with 0.44 g/$m^2$. The insecticide penetrated 20 mm into the cheese, mainly in the outer 6 mm. Levels of up to 10 ppm were found after 5 days but there was little increase in the next 44 days. A layer of cheesecloth between the board and cheese kept insecticide penetration to 8 mm and slowed uptake, with levels of 2–3 ppm in 14 days. The insecticide penetrated 1.2–1.5 mm into the board, and up to 5 mm in places where the cheese had sweated (Thomas 1980).

Wood or concrete treated with malathion, bromophos, and iodofenphos at 1 g/$m^2$ was placed under columns of wheat, barley, and maize. Insecticide uptake was greatest on wood, with bromophos giving best insect control in grain on concrete and malathion giving best control in grain on wood. Less iodofenphos moved into the grains than malathion or bromophos, and wheat and barley had more insecticides than maize. The persistence, and translocation of the insecticdes decreased with the age of deposit (Mensah et al. 1979).

Analyses of columns of dry barley placed on galvanized steel treated with 0.5 g/$m^2$ malathion indicated that little insecticide moved beyond 2 cm from the surface. At the 0–1 cm level from the steel surface malathion uptake at 20°C was 6.6 ppm at 2 weeks and 53.1 ppm at 7 months. Under ambient western Canadian storage conditions, levels were 6.1 ppm at 2 weeks and 24.4 ppm at 8 months. After 8 months of storage at 20 °C, malathion levels were 7.7 ppm (1–2 cm), 1.5 ppm (2–3 cm), and 0.8 ppm (3–4) cm), while at ambient storage conditions levels were 1.6 ppm (1–2 cm), 0.6 ppm (2–3 cm), and 0.9 ppm (3–4 cm) (White and Abramson 1984).

Rye, wheat, or triticale at 6.1–9.6% moisture content were stored for up to 8 months at 22°C in cylinders on fir plywood or galvanized steel panels treated with 0.5 g/$m^2$ malathion or pirimiphos-methyl, or on untreated panels, and the top (2–4 cm from the panel surface) and bottom (0–2 cm from the panel surface) samples were analyzed for residues. Insecticide residues did not exceed 0.8 ppm in top samples and generally declined between 1 and 8 months of storage, while residues in bottom samples usually increased with time. Insecticide residues in top samples on both surfaces showed no significant differences at 8 months. Maximum insecticide levels in bottom samples were 14.3 ppm malathion from wood, and 13.5 ppm pirimiphos-methyl from steel, in rye; 8.3 ppm malathion from steel, and 4.7 ppm pirimiphos-methyl from steel, in wheat; and 8.4 ppm malathion from steel, and 7.2 ppm pirimiphos-methyl from steel, in triticale. Insecticide levels were highest in rye and lowest in wheat (White 1985).

## VIII. PESTICIDE LABELS

In the United States, and with slight modifications in Canada, the label on a container is a legal document that must be followed and provides a great deal of information that must be read and understood before the product is used. Labels list the product name; company name and address; net contents; Environmental Protection Agency (EPA) pesticide registration number; EPA formulator manufacturer establishment number; ingredients statement; pounds/gallon statement (if liquid); front-panel precautionary statement; child hazard warning, signal word: Danger, Warning, or Caution; skull and cross-bones and the word Poison in red; statement of practical treatment (first-aid); referral statement (i.e., see side panel); side or back panel precautionary statements; hazards to humans and domestic animals; physical or chemical hazards; "restricted use pesticide" area (general or restricted); statement of pesticide classification; misuse statement warning; re-entry time statement; category of applicator; storage and disposal information; and directions for use.

## IX. SPECIFIC CONTACT INSECTICIDES COMMONLY USED IN STORAGE

General information on all pesticides registered in the United States is available in the *Farm Chemicals Handbook* (Sine 1992). The toxicity and general formulation of insecticides used on or near stored products in the United States and Canada are given in Table 1.

## X. FUMIGATION AND MODIFIED ATMOSPHERES

Only a few techniques are available to eliminate insects already present in raw or processed commodities. Two of the most important, fumigation and controlled atmospheres, employ gases to penetrate a commodity mass to kill insects hidden within. Other techniques such as heat/cold and radiation are discussed elsewhere. Fumigants and controlled atmospheres make up the most widely used methods and can be classed as a special type of pesticide/insecticide.

## 1. General Properties of Fumigants

*Diffusion/Molecular Size*

Fumigation is the process of killing insects by exposing them to a toxic gas or mixture of gases. Several properties of a gas suggest its potential success as a fumigant. The size of the molecule will indicate its ease in mixing with other gases and penetrating through materials. An important property of a fumigant is its ability to penetrate materials such as packaging films and into commodities. To kill insects in the center of a mass, the toxic gas must be able to get to the pest from the outside. Diffusion plays a critical role in a gas reaching the target pest and is related directly to the molecular weight of a gas and its density. Graham's law of diffusion says that the velocity of diffusion is inversely proportional to the square root of the density of the gas. Since the density of a gas is directly related to its molecular weight by Avogadro's principle, the diffusion of a gas is related to the molecular weight of that gas. In practice, diffusion is important when one is considering the behavior of a gas in space with only air as a mixing substrate. When a commodity or barrier (packaging film, corrugated container, etc.) becomes involved in the mixing and movement of gas, factors such as adsorption and absorption of the gas also play significant, sometimes dominant, roles in the penetration and movement of gas in the commodity storage ecosystem.

*Sorption*

Sorption is used to describe the phenomenon of the uptake of a gas that comes into contact with a solid material. This adherence of the gas to a solid material is the result of weak intermolecular forces called Van der Wall's forces. These forces can be classified at "absorption" or "adsorption," the difference being that adsorption is the holding of the gas on the surface of a solid while absorption is the entering of the gas molecule into the solid matrix where it is held by capillary forces that control the properties of solutions. Sorption is inversely related to temperature; more fumigant is sorbed at low temperature than at higher temperatures (Dumas and Bond 1979). In addition, the physical makeup of the commodity being fumigated can have a great influence on the sorption of fumigant. For example, gas may be more sorbed in commodities high in water or lipids, depending on whether the gas is hydrophilic or lipophilic (Berck 1964). Therefore, moisture content and lipid content can play very important roles in the sorption of fumigant. When large amounts of gas are sorbed during a fumigation, it is possible for the fumigation to fail because not enough gas was available for a long enough time to kill all the insects in the commodity.

The process of sorption described above is technically called "physical sorption." This should be differentiated from "chemisorption," which is the process of the fumigant gas entering a solid matrix and reacting to form a new chemical or chemicals. In the case of physical sorption, the process of the gas becoming attached to the surface or interior of the solid matrix is reversible and the gas can be removed from the solid by exhausting the free gas (e.g., aeration). With chemisorption this is not possible since the gas no longer has the same identity and is now chemically bonded to the solid or has formed another gas that may then be desorbed. This phenomenon was described for methyl bromide on wheat (Dennis et al. 1972). In theory, only chemisorption results in permanent residues on commodities after fumigation. Transient residues will be present while gas sorbed on the substrate is desorbed during aeration of that substrate. Permanent residues were found to occur when hydrogen cyanide was used to fumigate fruits containing reducing sugars, because the fumigant reacted with the reducing sugars to form cyanohydrins (Page and Blacklith 1956), or when methyl bromide is used to fumigate commodities with high sulfur-containing compounds (McLaine and Monro 1937). A thorough review of the phenomena of sorption and desorption of fumigants is provided by Banks (1986) and Banks and Sharp (1986).

*Concentration Times Time Product*

One concept that has been used in determining the efficacy of most fumigants is the concentration × time product (CXT; also CT or CTP). This product is obtained by measuring the concentration of fumigant during the fumigation and multiplying the mean concentration by the time of exposure. In other words, it is the area under the curve of concentration versus time plot. A basic explanation of CXT can be found in Bond (1984) and in the ASEAN Food Handling Bureau/Australian Centre for International Agricultural Research (ACIAR) book, *Suggested Recommendations for the Fumigation of Grain in the ASEAN Region* (Anonymous 1989), while a more rigorous treatment is presented by Banks et al. (1986) and Banks and Sharp (1986). CXT is very useful in constructing fumigation schedules for commodities because it provides for uniformity in quarantine treatments. The concept says that if you increase the concentration you can decrease the time of exposure to achieve the same CXT product. Of course, the method is not foolproof and the fumigator must keep in mind that factors such as temperature and moisture content may affect the response of the target insects (Whitney and Walkden 1961; Harein and Krause 1964; Estes 1965; Bell and Glanville 1973; Bell 1977, 1978). Most fumigants follow the CXT rule, with the exception of phosphine. The time factor is more important with phosphine, so the CXT product concept must be modified as suggested by Bond et al. (1969) due to the effect of high concentrations of phosphine causing insects to go into a protective narcosis. In addition, phosphine may not be taken up by the insect in direct proportion to its concentration. The protective narcosis that occurs in insects with phosphine has also been described for other fumigants (Lindgren 1938; Winks 1974).

**Table 4** Properties of the Fumigant Phosphine

| Property | Description |
|---|---|
| Formula | $PH_3$ |
| Molecular weight | 34.04 |
| Boiling point | −87.4°C |
| Specific gravity (air = 1) | 1.214 at 0°C |
| Lowest explosion point | 1.79% by volume in air |
| Odor | Carbide or garlic-like (probably due to impurities) |
| Solubility in water | 26 cc/100 ml at 17°C |
| Method of production (for fumigation) | From preparation of aluminum phosphide (AlP) or magnesium phosphide ($Mg_3P_2$) upon reacting with moisture in air |

Reaction of production:

$$AlP + 3H_2O \rightarrow PH_3\uparrow + Al(OH)_3$$

or

$$Mg_3P_2 + 6H_2O \rightarrow 2PH_3\uparrow + 3Mg(OH)_2$$

Useful concentration conversions (25°C at 760 mmHg) for phosphine in air:
1 mg/L = 0.0718% = 718 ppm
Threshold limit value[a] (TLV − TWA) = 0.3; TLV − STEL[b] = 1ppm
Alternative name: hydrogen phosphide

[a]ACGIH 1988.
[b]Short Term Exposure Limit (STEL).

*Factors Affecting Concentration*

Under practical conditions the major factors that determine the concentration of fumigant after application are temperature, sorption of the fumigant, relative humidity, moisture content of the commodity being fumigated, and leakage. Of course, in addition to these factors, when phosphine is the fumigant the time variable for generating the phosphine from aluminum or magnesium phosphide must be included in the calculation (Banks and Sharp 1986). In general, the warmer the commodity temperature, the faster the fumigant will kill insects. In the case of phosphine, it is also true that given similar moisture conditions the evolution of the phosphine from the solid formulation is in direct proportion to the temperature. Humidity and moisture content of the commodity will affect the sorption of fumigant and, in the case of phosphine, will affect the reaction that liberates the phosphine gas. Too often leakage is the one factor that gives the most trouble in obtaining a successful fumigation. A recent review of the behavior of gases in grain, and explanation of the leakage factor, is given by Banks (1990).

With any fumigation it is necessary to hold the fumigant gas in contact with the target pest for a certain amount of time for death to occur. If the material that holds the gas in contact with the commodity is leaky the gas will escape and the fumigation will be a failure. This is true whether the enclosure is a building, bag, or tarpaulin. The single most important consideration after safety is proper sealing of the enclosure in which the fumigation will take place. Even small leaks can cause failures, especially when pressure changes exist between the inside and the outside of the enclosure due to the wind blowing or the sun warming one side of the enclosure. The importance of adequate sealing was recognized by Australian researchers who adopted a leak standard for fumigation (Ripp 1985; Banks and Annis 1981; Banks and Ripp 1984; Newman 1990; Banks and Sticka 1981).

## XI. FUMIGANTS FOR TODAY

The most important fumigants used today in the control of insects in stored products are phosphine ($PH_3$) and methyl bromide ($CH_3Br$). Minor fumigants such as hydrogen cyanide (HCN) and chloropicrin are still used in limited applications (Storey and Davidson 1973; Quinlan and McGaughey 1983). In agriculture, methyl bromide is mostly (about 70%) used in the fumigation of soil to control nematodes, weeds, and disease pathogens prior to the planting of specialty crops such as tomatoes, strawberries, and peppers. A lesser amount of methyl bromide (about 30%) is used for the fumigation of commodities in postharvest marketing channels. Phosphine is used only in postharvest marketing channels. Because phosphine is corrosive to copper, silver, and gold (Bond et al. 1984), it has rarely been used in structural fumigation except in cases where a building is used as the fumitorium, such as the fumigation of tobacco in hogsheads stored in warehouses (Childs and Overby 1971; Keever 1990). In that situation, special techniques of sealing and protection of wiring and motors are needed to ensure that no damage is done to copper parts. However, we shall discuss here only the fumigation of commodities with methyl bromide and phosphine.

### 1. Phosphine

Phosphine is a colorless and tasteless gas that has a "garlic" or decayed fish smell when freshly generated from solid formulations of aluminum phosphide or magnesium phosphide. Some people who remember their elementary chemistry classes would probably say that its odor resembles that of the carbide used to generate acetylene. Other properties of phosphine are listed in Table 4. Of our present choices, phosphine comes closest to the ideal fumigant. It has a low molecular weight and low boiling point. It is only 1.2 times as heavy as air, so it mixes without stratifying or requiring fan circulation to mix it with the air. Because phosphine is a small molecule and somewhat nonpolar, it is a good penetrator of commodities and barriers such as packaging films. Penetration is an important consideration because often the commodity being fumigated is in large bulk volumes and a good penetration is required to kill target pests at the places in the mass furthest from the point of application. The major disadvantage of phosphine is the amount of time required to eliminate the target pest population completely. This may range from 3 to 7 days. This long period of time often precludes the use of phosphine, because time is an expensive part of the marketing channel. Short storage times are often encountered because of the rapid turnover of commodities and therefore phosphine cannot be used.

Phosphine acts on any target species, either insect or rodent, by interrupting respiration. It acts in much the same way as hydrogen cyanide does, by inhibiting the uptake or transfer of electrons from oxygen to the body (Kashi and Chefurka 1976; Chefurka et al. 1976; Price 1980a; Price and Walter 1987). The complex system responsible for electron transfer is referred to as "oxidative phosphorylation" and is a vital biochemical pathway found in all aerobic life forms including mammals, rodents, amphibians, and arthropods. Other research has identified additional biochemical pathways interrupted by phosphine (Price 1985; Price et al. 1982). Strong support for the inhibition of oxidative phosphorylation by phosphine is found in the fact that phosphine is nontoxic to insects in an environment lacking oxygen (Bond et al. 1967, 1969).

There are many advantages to using phosphine for fumigation. It is applied as an easily handled solid. It mixes readily with air to distribute itself and penetrates commodities more quickly than any known fumigant. Because it is a small molecule, it diffuses quickly and subsequently is quickly aerated from fumigated materials (Leesch et al. 1982). Phosphine leaves the least residue of any fumigant after fumigation and aeration, usually in the parts per billion range. Also phosphine does not interfere with germination and therefore can be used for seed stock (Sittisuang and Nakakita 1985).

### 2. Methyl Bromide

Methyl bromide is a colorless gas that is odorless at concentrations used for fumigation. At very high concentrations, methyl bromide has a sweet odor similar to that of chloroform. Other properties are listed in Table 5. In some cases methyl bromide is formulated with chloropicrin so that the presence of gas initially can be easily detected since chloropicrin is a very strong lacrimator. Methyl bromide can be used without the fear of explosion since it is nonflammable. Methyl bromide was first discovered to have insecticidal properties when LeGoupil (1932) attempted to mix it with other fumigants to obtain a better treatment than had previously been available. From that time methyl bromide has played a major part in the fumigation of commodities and in structural fumigation. Since we are interested in commodity fumigation, the following discussion will deal only with that subject. However, it should be kept in mind that methyl bromide has been the major fumigant used in structural fumigation and is still of major importance for that use.

The mode of action of methyl bromide on target pests or humans is not well understood. It has been observed that it acts on the central nervous system and that symptoms are often delayed for up to 2 days in the case of humans (Bond 1984). Even in insects delay in mortality has been observed and it is wise to wait for at least 24 h before determining the success of a methyl bromide fumigation (Bond 1984; Hole 1981). Lewis (1948) postulated that methylation of SH-containing enzymes could lead to impaired functions and Winteringham and Barnes (1955) postulated that the toxic action of methyl bromide was due to the methylation of proteins. In metabolism studies with radiolabeled methyl bromide Winteringham (1955) and his colleagues (Winteringham et al. 1955) found evidence that methyl bromide reacts with amino acids that have disulfide bonds. His work suggests that the mode of action of methyl bromide may indeed be the inhibition of some fraction containing these amino acids.

Due to the reaction of methyl bromide with disulfide bonds it cannot be used with some materials. Methyl bromide reacts with sulfur or sulfur-containing molecules. Material such as hair, natural rubber, feathers, and soybean flour cannot be fumigated with methyl bromide because of odors that remain after fumigation. Bills et al. (1969) describe an example of this reaction tainting the quality of nuts used in candy. In addition, care must be taken when using methyl bromide to ensure that it does not come into contact with aluminum in the absence of oxygen. In that case, it forms a complex molecule that reacts violently upon exposure to oxygen (as in the air). Because it is a powerful solvent, methyl bromide should never be allowed to come into contact with bituminous materials, which it will dissolve, or be used with some plastics such as polyvinyl chloride (PVC), which it may soften or dissolve.

The major advantages of methyl bromide are its high toxicity to pests, its ability to penetrate commodities at ambient temperatures and pressures, and its nonflammability. Because of these properties, relatively short periods of exposure are necessary to achieve an effective fumigation. Thus methyl bromide can be used when time is a critical factor, such as when large amounts of commodity must be fumigated in a short time. The shorter the time taken to fumigate a commodity with a short shelf life, the longer the commodity will be available for sale or processing. Often the marketing process will not allow the prolonged treatment of a commodity because it slows the movement of the product to its endpoint.

The disadvantages of methyl bromide are that it is a liquid and must be volatilized for application, it is much heavier than air and must be recirculated after application to prevent it from stratifying, it has recently been declared an ozone-depleting agent, and it leaves residues on commodities after aeration. Since methyl bromide is supplied in pressurized cylinders or cans, it is in a liquid state when applied. Either a long delivery hose or a heat exchanger is required to get the liquid into the gas phase. Because it is heavier than air, if uniform concentrations of the gas are expected, it is necessary to circulate and mix the gas with the air around commodities. Often this mixing is achieved by forced recirculation of the atmosphere around the commodity in places such as grain silos where air can be taken out of the bottom and pumped back into the top of the silo. Fumigation of some types of commodities is not advisable. For example, methyl bromide fumigation can cause a reduction in germination and therefore its use on seeds can be risky (Hanson et al. 1987; Powell 1975).

It has recently been found that methyl bromide can react with ozone, thus possibly contributing to a depletion of ozone in the upper atmosphere. Methyl bromide has been classified as an ozone depleter. The U.S. Clean Air Act (Section 602) as amended and the 1992 amendments to the Montreal Protocol of the Vienna Convention will mandate reduction in the production of methyl bromide back to 1991 levels and the elimination of all uses and production including fumigation in 2001 (Anon. 1993a, 1993b).

## XII. GENERAL PROPERTIES OF MODIFIED ATMOSPHERES

Humans have been using forms of modified atmospheres (MA) for centuries. The hermetic storage of grain is one form of MA by which grain itself through respiration creates an atmosphere rich in carbon dioxide and low in oxygen. This technique may have been used by the Egyptians centuries ago and persists today in Africa, where hermetic storage is still used to store grain in sealable vessels. Today atmospheres are modified to attain low oxygen environments by adding carbon dioxide ($CO_2$), nitrogen ($N_2$), or burning the storage facility atmosphere and recirculating the combustion products. The term "controlled atmosphere" usually refers to the process of changing the atmosphere of a facility artificially by introducing $CO_2$ or $N_2$.

Over the past few years, an abundant literature has accumulated and four major conferences have been organized to discuss the uses and effects of modified atmospheres in protecting the quality of commodities. These proceedings provide an excellent source of information that is recent and shows the possibilities for using modified atmospheres now and in the future (Navarro and Donahaye 1993; Champ et al. 1990; Ripp 1984; Shejbal 1980).

In changing the atmosphere of a storage facility, we expect to create an environment that will not support insects or microflora in the storage facility. Usually, in addition to the storage facility itself, we are interested in the stored commodity. In terms of the application and distribution of gas, the same factors affect the behavior of the gases that make up the modified atmosphere that affect the behavior of fumigants. Several types of modified atmospheres have been investigated over the years: high $CO_2$ and/or $N_2$ with low $O_2$, high $CO_2$ with reduced $O_2$, burner gas (high $CO_2$, low $O_2$ plus other gases), and hermetic storage. In addition, several methods exist for the modification of the atmospheres, including removal of $O_2$ and increasing $CO_2$ by burning the storage atmosphere, adding $CO_2$ either on the surface as dry ice, or adding $CO_2$ to the bottom of the storage as gas to displace the atmosphere out the top of the storage facility, adding $N_2$ as gas to the top or bottom of a facility, or adding $N_2$ to a facility by passing breathing air through a membrane or pulsating gas separator and displacing or recycling the facility's atmosphere. The application methods currently used are the ones that most easily and economically produce the required modification for the facility to be treated. For example, containers on shipboard can be maintained under modified atmosphere by using membrane-generated $N_2$ or displacement of the atmosphere by $CO_2$ before the container is loaded. In facilities near a source of liquid $CO_2$ it may be most advantageous to use liquid $CO_2$ to displace the atmosphere of the facility, while it may be advantageous to use a burner to lower the $O_2$ of a facility some distance from the source of $CO_2$. $CO_2$ or $N_2$ and combustion products have all been exempted from the requirement of establishing tolerances on raw agricultural products in the United States by the EPA (40 CFR180, Subpart D, Sec. 180.1049, 180.1050, and 180.1051; see Johnson 1980, 1981) and in Canada.

A large volume of literature exists that describes the uses of modified atmospheres on insects infesting stored grain. Most of the studies prior to 1970 were concerned with controlling only the adult of the insect species under laboratory conditions. Since that time studies have shown that insect species react differently to controlled atmospheres depending on the life stage of the insect, the temperature and moisture content of the grain, and the composition of the modified atmosphere to be used (Bailey and Banks 1980; Navarro and Jay 1987). From the early 1970s until 1976 research in Australia centered on the uses of $N_2$ to control stored product pests and after 1976 centered on the uses of $CO_2$ for this purpose (Davis and Jay 1983; Banks and Annis 1990). Guidelines for the uses of $N_2$ developed in Australia can be found in a paper by Banks and Annis (1977). In addition, a unique system for using $N_2$ was described by Zanon (1980).

Modified atmospheres have many advantages (Calderon and Barkai-Golan 1990). They provide a way to eliminate insects from stored commodities without polluting the atmosphere and are safer to apply than traditional fumigants such as methyl bromide. No harmful residues remain after the treatment of commodity with $N_2$ or $CO_2$ and the effects of modified atmospheres on organoleptic qualities of the commodities are minimal compared with the traditional fumigants. With new technology, the delivery systems for modified atmospheres will also bring the cost of application into the same range as those for the application of traditional fumigants such as phosphine. Even in 1985 the costs of applying controlled atmospheres to storage facilities in Australia were approaching those of phosphine and were lower than those of the grain protectants (Ripp et al. 1990).

The disadvantages of using modified atmospheres for insect control are the length of time required to obtain control as well as the cost of application and getting an adequate supply of gas to the treatment site. In addition, $CO_2$ cannot be used on some commodities because it forms carbonic acid, which then causes a flavor deterioration. Modified atmospheres usually require continuous monitoring during treatment, whereas once a fumigant is applied and possibly circulated for a short time, constant monitoring of the system is not necessary. The constant monitoring of modified atmospheres is required because of the stringent requirements of low $O_2$ or high $CO_2$ or $N_2$ that may have to be added to keep conditions within the desired ranges. As with phosphine, times required for the treatment process are long and often are incompatible with the marketing system.

## XIII. SAFETY CONSIDERATIONS FOR FUMIGANTS AND MODIFIED ATMOSPHERES

In dealing with materials that are as toxic as fumigants, adherence to good safety practices is essential. Knowing how to protect workers during the application and aeration process is only part of the picture. Safety begins with the planning of a fumigation and follows through the aeration phase of the job. A crucial element of safety is the ability to detect the fumigant gas. Several devices are currently available to detect methyl bromide, phosphine, and other fumigants.

The most commonly used device to detect specifically the presence of methyl bromide is the thermal conductivity meter. Several versions of this device will accurately determine levels of methyl bromide from very high concentrations to levels near the TLV of 5 ppm (ACGIH 1988). A device commonly used for the detection of leaks during methyl bromide fumigation is the halogen flame detector. This device is composed of a flame that draws a sample of fumigant-laden air into it and burns the sample while in contact with a copper ring. The presence of a halogenated hydrocarbon is indicated by the top portion of the flame turning light green (low concentrations) to bright blue (high concentrations). This device is neither quantitative nor qualitative. Halogenated hydrocarbons such as refrigerants (freons) will produce the same colors. A major disadvantage of the flame detector is that it can only be used in areas where open flames can be tolerated. The flame detector is useful in detecting leakage of methyl bromide occurring after application so that remedial action can be taken.

Other detectors exist for methyl bromide that are very specific and also very expensive. Gas chromatographs equipped with either a flame ionization detector or a thermal conductivity detector can be both qualitative and quantitative. Likewise, portable infrared spectrophotometers are available that also are both quantitative and qualitative. A less expensive way to determine methyl bromide is by using specific gas detection tubes. These tubes are very good for detecting low levels of methyl bromide; however, tubes to measure concentrations used during fumigation are not available.

For phosphine, two devices can be easily used in determining the concentrations both for safety and biological efficacy. The first is the detector tube, which is nothing more than a tube filled with a compound that reacts with phosphine as an air sample is drawn through the tube. As the phosphine–air mixture passes through the tube the phosphine reacts with the detector chemical, causing a color change that progresses down the tube in proportion to the concentration of phosphine in the original mixture. The tubes are graduated and the concentration of phosphine can be read directly. The advantages and disadvantages of using these tubes were discussed by Leesch (1982). Tubes are available that can be used to monitor the fumigants after application to decide whether a lethal concentration of gas was held and thus if the fumigation was successful. Low-level tubes will monitor for safety and determine the safety measures required at any time during the fumigation or aeration. In addition to gas detection tubes, other devices that can be used to detect phosphine are the Harris conductimetric detector (Harris, 1986); a gas chromatograph equipped with a flame photometric, photoionization, or thermal conductivity detector (Harris 1986; Ducom and Bourges 1986), or an infrared spectrometer equipped with a long-path gas cell.

Detection devices for $CO_2$ and $O_2$ consist of thermal conductivity cells either as a stand-alone device or as a detector on a gas chromatograph, and detector-tubes much like those described above for phosphine. The range of concentrations encountered is adequately handled by a variety of tubes. With $N_2$, detection is performed with a thermal conductivity device, usually as a detector attached to a gas chromatograph that will separate the injected air into its separate components.

Detection devices are essential because safety is of the highest priority and we cannot know if dangerous situations exist or occur if we are unable to detect the gas. In the United States, strict aeration procedures require that each fumigation be cleared for people to reoccupy areas that have been fumigated. According to labels used for phosphine formulations, no respiratory protection is required if the concentration is below 0.5 ppm. However, if the concentration is above this level, either a canister face respirator (0.5–15 ppm) or a full-face self-contained breathing apparatus (SCBA) (>15 ppm) must be worn upon entering a fumigated area. It is always best to wear a SCBA during any fumigation or controlled atmosphere treatment. The preferred type of SCBA is a pressure-demand device: it keeps a positive pressure inside the face mask at all times so that the chance of toxic gas intrusion into the mask is very slight. Of course, the first rule of fumigation or treatment with controlled atmospheres is *always to work so that one other person can see you.* This also means that you never work alone. This seems obvious, but experience has taught many people that failure to heed the obvious can be deadly.

## XIV. FUMIGATION APPROACHES

Fumigation of commodities takes two general forms. The first is a quarantine fumigation: the commodity being exported or imported must be fumigated to make sure that a pest associated with it is not transported to an area where the pest does not exist. These fumigations are the most rigorous type performed. They can be performed in the country of origin or the importing country. With the prospect of losing methyl bromide as a fumigant in the United States, much more of that country's quarantine fumigation will be performed in the exporting country. In quarantine control the objective is to kill 100% of the target pest population. To achieve this goal, the fumigation process is very rigidly controlled. Quarantine fumigations usually occur in specially built chambers designed to exacting standards so that concentrations of the fumigant and temperature of the commodity can be measured throughout the fumigation and the aeration process. Particular attention is given to sealing the chamber so that fumigant can be kept at a concentration that will kill insects at the commodity temperature. The aeration process is also closely monitored to determine the time when it is safe to enter the chamber and remove the treated commodity.

The other type of fumigation performed on commodities is that of control. This type of fumigation is used to kill any pests present which might damage the commodity and thus shorten its storage life. The objective is to increase the shelf or storage life of the commodity. Control fumigations are conducted in a variety of fumitoria on both bulk and packaged commodities as well as on raw and processed agricultural products. Several types of fumigation have been described in the literature: (1) all types of fumigations under tarpaulin (Brown 1959; Leesch et al. 1982; Leesch and Highland 1978; Banks and Sticka 1981); (2) fumigation of bulk raw agricultural products in permanently constructed fumitoria (Storey 1967, 1971; Leesch et al. 1978, 1986, 1990); and (3) fumigation of commodities, both raw and processed, in transportation carriers (Leesch et al. 1990; Banks et al. 1986; Jay et al. 1983).

From the time a fumigation or modified atmosphere treatment is anticipated, it is extremely important to formulate a plan to (1) prepare the commodity for treatment, which includes adequately sealing the enclosure; (2) conduct the fumigation or modified atmosphere treatment and measure the concentration of gas(es); and (3) aerate the commodity while monitoring its progress. In applying the fumigant, whether a methyl bromide/chloropicrin mixture or phosphine in the formulation of aluminum or magnesium phosphide, it is important to remember not to rely upon the odor of the fumigant because the agent responsible for the odor can be selectively sorbed from the gas mixture, leaving a deadly amount of odorless fumigant.

Fumigation and the use of modified atmospheres will remain the most important methods for eliminating live insects from large bulks of stored commodities. The effects of fumigants on the quality of commodities is well documented and depends upon the nature of the commodity and the conditions under which the fumigation is conducted. Residues from the application of fumigants have been documented by many research studies (Daft 1985; Rangaswamy 1985; Al-Omar and Al-Bassomy 1984; Dumas and Bond 1977; Rangaswamy and Muthu

1985; Rangaswamy and Sasikala 1986; Srinath and Ramchandani 1978; Starratt and Bond 1988; Lueck et al. 1984; Reeves et al. 1985). With phosphine, the important residue is phosphine itself that has not aerated from a fumigated commodity. Methyl bromide that reacts with components of the commodity leaves two types of residue, an inorganic bromide residue and an organic bromide residue, in addition to undersorbed methyl bromide itself.

## XV. RESISTANCE TO FUMIGANTS

Although methyl bromide has been in use for approximately 60 years, little or no resistance has developed to it. In 1976, the Food and Agriculture Organization (FAO) Global Survey of Pesticide Susceptibility showed that only 4.7% of the strains tested showed resistance to methyl bromide while 9.7% showed resistance to phosphine (Champ and Dyte 1976, 1977). Our dependence on phosphine will certainly grow in the near future as the elimination of methyl bromide becomes a reality. This will mean that techniques of application and sealing must become more stringent if we are to avoid severe resistance problems. Resistance to phosphine has already been demonstrated in several species of stored-product insects. The possibility of resistance to phosphine occurring in stored-product insects was reported by Monro et al. (1972) and the first serious resistance to phosphine by a wild strain was reported by Borah and Chahal (1979) in India. Shortly thereafter, Tyler et al. (1983) reported control failures in beetles from Bangladesh. Champ and Dyte (1976, 1977) reported that several strains of stored-product insects around the world were resistant to phosphine in varying degrees. Many of the investigations that followed pointed out that the probable reason for the resistance was repeated fumigations in leaky structures and poor fumigation technique (Attia and Greening 1981; Mills 1983; Mills et al. 1990; Banks and Annis 1981; Zettler 1982, 1990, 1991; Zettler and Cuperus 1990; Zettler et al. 1989). As of now, no resistance problem in the United States or Canada has been identified as the cause of a fumigation failure; however, if the practices of fumigating leaky storages and improper application and exposure continue, control failures due to phosphine resistance are inevitable. Therefore, investigations to identify new fumigants (Leesch and Sukkestad 1980, 1983) and new disinfestation techniques/methods (Leesch 1992; Calderon and Leesch 1983; Calderon et al. 1991) are becoming increasingly important if we are to protect stored commodities until they reach the consumer. Modified atmospheres certainly provide a partial solution to the dwindling number of fumigants available, but more technology needs to be developed.

Although no resistance to modified atmospheres has been reported under actual field conditions, the potential exists that such resistance might arise. Donahaye and Rindner (1993) have recently shown that laboratory strains of the red flour beetle *Tribolium castaneum* (Herbst) can be made less sensitive to hypoxia (reduced oxygen concentrations) and hypercarbia (increased carbon dioxide concentrations). It was found that the mechanisms for this increased tolerance to the two conditions differed from one another (Donahaye 1992).

## XVI. NEW RESEARCH

New technologies on the horizon include the use of flow-through fumigation, which, in theory, lessens the threat of severe resistance to phosphine from developing (Winks 1990). In this new technology, phosphine is bled upward through the commodity at very low concentrations in carbon dioxide. The treatment requires several days and purports to work on resistant insects as well as susceptible insects. Other new application methods have been described and are now in use to help the distribution of the fumigant phosphine. The first method is one of slowly recirculating phosphine and was first described by Cook (1980, 1983). The technique was put to practical use in the in-transit fumigation of grain in deep-draft ships (Leesch et al. 1990). The second method investigated showed that the distribution of phosphine in bulk grain could be enhanced if the phosphide formulation was applied with a small amount of carbon dioxide (Leesch 1990, 1992). The use of this technique will allow faster penetration of phosphine in large bulks without the need to install circulation equipment before the commodity is loaded into the storage facility.

In this age of concern for the environment, people are justifiably concerned about the effects of releasing fumigants into the atmosphere. With the discovery of methyl bromide in the stratosphere and its classification as an ozone depleter, we are now faced with the elimination of it from our arsenal of control tools. More research is needed on the amount and fate of fumigants in the atmosphere and on the amount of artificial methyl bromide contributing to ozone depletion in the stratosphere. Some research has begun, as evidenced by the investigations of Keever (1990) on ways to reduce the rate of phosphine emissions into the atmosphere during the aeration of a tobacco warehouse. More studies, both on the breakdown and the rate of emission of fumigants into the air, are needed to answer questions regarding the impact of fumigants on the environment. Health hazards have already arisen related to the use of phosphine. Garry et al. (1989) and Potter et al. (1991) demonstrated a possibility that prolonged exposure to low levels of phosphine may cause chromosome damage in pesticide applicators. Although much more work is needed to determine the implications of these aberrations, it appears that we must be more diligent in investigating the chronic human health hazards of phosphine.

Modified atmospheres are not in the same position as fumigants in relation to damaging the environment. Since they are viewed as naturally occurring compounds, they will probably escape the intense scrutiny received by the fumigants. However, do not be deceived into thinking that these compounds are not just as

**Table 5** Properties of the Fumigant Methyl Bromide

| Property | Description |
|---|---|
| Formula | $CH_3Br$ |
| Molecular weight | 94.95 |
| Boiling point | 3.6°C(38.5°F) |
| Specific gravity (air = 1) | 3.27 at 0°C |
| Flammability limits in air | Nonflammable |
| Odor | None at low concentrations; strong musty or sickly sweet at high concentrations |
| Solubility in water | 1.34 g/100 ml at 25°C = 345.6 cc/100 ml at 25°C |
| Method of production (fumigation) | From steel cylinders under pressure of from 1.5 lb cans under pressure |
| Useful concentration conversions (25°C at 760 mmHg) for methyl bromide in air: | |
| 1 mg/L = 0.0257% = 257 ppm | |
| Threshold limit value[a] (TLV-TWA) = 5 ppm | |
| Alternative name: monobromomethane, bromomethane | |

[a]ACGIH 1988.

dangerous to apply as fumigants. Both carbon dioxide and nitrogen are capable of killing humans (Mehler et al. 1992). Carbon dioxide is actually toxic and has a TLV value of 5% (v/v) in air. On the other hand, nitrogen is more insidious because as the concentration is increased above its natural level of about 78%, a person merely falls asleep as the oxygen is consequently reduced.

More research is needed to keep the price of modified atmospheres low enough to make them acceptable alternatives to fumigants and other forms of insect control. Research on membrane generation of nitrogen may provide means to disinfest commodities without the high cost of bottled gas. Although there is a relatively large amount of literature on the use of modified atmospheres on commodities, there is relatively little literature on the effects of modified atmospheres on all stages of many stored-product insects. Most research has been conducted on adult insects, and more knowledge is required on other insect stages. The long-term effects of modified atmospheres on commodities are also not known. The short-term effects of modified atmospheres on many commodities have been studied; the storage of commodities under continuous nitrogen, carbon dioxide, or burner gas has not been thoroughly researched.

In order to construct predictive models to use in the storage of various commodities, we need to plan experiments that monitor many variables so that each variable and its change during the course of the study can be put into models.

Computer models of grain storage will soon be available and since fumigants and modified atmospheres are integral parts of any storage regimen, it will be important to include them in the models. With that in mind, new research on fumigants and modified atmospheres must be directed toward incorporating the techniques into integrated pest management (IPM) programs for the storage and protection of stored products of all kinds.

## XVII. DEVELOPMENT OF AN APPROPRIATE CHEMICAL PREVENTION AND CONTROL PROGRAM

The management of stored grain is a complex task and insecticides are just one tool available to minimize grain damage and economic losses caused by insects. Chemicals should be used in conjunction with other storage practices in an integrated pest management system, as outlined in the chapter in this book on IPM. The use of sound, weatherproof granaries; keeping grain dry; and cooling it with aeration will minimize insect problems. Insecticides should be used to prevent insect infestations by the spraying of contact insecticide in and around empty granaries or the application of a grain protectant when prolonged warm storage makes it necessary. Chemicals chosen will be based on the small number of registered products, cost, effectiveness against the insect pests present, insecticide resistance, environmental conditions, formulation desired, and length of residual activity required. Fumigation will be used to control insects, occasionally in debris residues such as inaccessible grain under perforated floors in granaries, in empty mills or ships (space fumigation), or in grain bulks.

## XVIII. SUMMARY

Chemical control of insects in stored products relies on contact insecticides as space and structural treatments or as protectants on grain and on fumigants such as phosphine, methyl bromide gas, or modified atmospheres. Contact insecticides leave chemical residues in grain and grain by-products which break down with time in relation to moisture and temperature. Classes of contact insecticides included chlorinated hydrocarbons, botanicals, inert mineral dusts, carbamates, organophosphates, pyrethroids, insect growth regulators, and bacteria. Insecticide formulations (sprays, dust) and types of structural surfaces determine the residual activity of the chemicals and affect stored-product contamination. Information included on insecticide labels is listed. The properties of fumigants and fumigation practices, including safety guidelines, are discussed for modified atmospheres ($CO_2$, $N_2$, burner gases), phosphine, and methyl bromide. Insect resistance to fumigants is discussed and recent research on new application of fumigants is reviewed. The development of an appropriate chemical prevention and control program for stored-product insects is based on available chemicals, cost, effectiveness, insecticide resistance, air-tightness of storage structures, environmental conditions, insecticide formulation, and length of residual activity required.

## REFERENCES


Abdel-Kader, M. H. K., Webster, G. R. B., Loschiavo, S. R., and Watters, F. L. (1980). Low temperature degradation of malathion in stored wheat, *J. Econ. Entomol.*, *73*:654–656.

Al-Omar, M. A., and Al-Bassomy, M. (1984). Persistence of phosphine gas in fumigated Iraqi dates, *J. Food Safety*, *6*:253–260.

American Conference of Government Industrial Hygienists (ACGIH). (1988). TLV's, Threshold Limit Values and Biological Exposure Indices for 1987–1988. ACGIH, Cincinnati.

Anderegg, B. N., and Madisen, L. J. (1983). Degradation of $^{14}C$-malathion in stored corn and wheat inoculated with *Aspergillus glaucus*, *J. Econ. Entomol.*, *76*:733–736.

Anonymous (1989). Suggested Recommendations for the Fumigation of Grain in the ASEAN Region, Part 1: Principles and general practice, ASEAN Food Handling Bureau (AFHB), Australian Centre for International Agricultural Research (ACIAR), Kuala Lumpur/Canberra, July 1989.

Anonymous (1993a). Methyl Bromide Substitutes and Alternatives: A Research Agenda for the 1990s. USDA, ARS, Washington, D.C.

Anonymous (1993b). The Biological and Economic Assessment of Methyl Bromide. USDA, National Agricultural Pesticide Impact Assessment Program (NAPIAP), Washington, D.C.

Arthur, F. H. (1992). Cyfluthrin WP and EC formulations to control malathion resistant red flour beetles and confused flour beetles (Coleoptera: Tenebrionidae): Effects of paint on residual efficacy, *J. Entomol. Sci.*, *27*:436–444.

Arthur, F. H., Throne, J. E., and Simonaitis, R. A. (1991). Chlorpyrifos-methyl degradation and biological efficacy toward maize weevils (Coleoptera: Curculionidae) on corn stored at four temperatures and three moisture contents, *J. Econ. Entomol.*, *84*:1926–1932.

Arthur, F. H., Throne, J. E., and Simonaitis, R. A. (1992). Degradation and biological efficacy of chlorpyrifos-methyl on wheat stored at five temperatures and three moisture contents, *J. Econ. Entomol.*, *83*:1994–2002.

Attia, F. I., and Greening, H. G. (1981). Survey of resistance to phosphine in coleopterous pests of grain in stored products in New South Wales, *Gen. Appl. Entomol.*, *13*:93–97.

Bailey, S. W., and Banks, H. J. (1980). A review of recent studies on the effect of controlled atmospheres on stored product pests. In Controlled Atmosphere Storage of Grains, An International Symposium held 12–15 May 1980 at Castelgandolfo (Rome) Italy (ed. Shejbal, J.). Elsevier, Amsterdam, pp. 101–118.

Banks, H. J. (1986). Sorption and desorption of fumigants on grains: Mathematical descriptions. In Pesticides and Humid Tropical Grain Storage Systems: Proceedings of an International Seminar, Manila, Philippines, 27–30 May 1985 (ed. Champ, B. R., and Highley, E.). ACIAR Proceedings No. 14, pp. 291–298.

Banks, H. J. (1990). Behavior of gases in grain storage. In Fumigation and Controlled Atmosphere Storage of Grain: Proceedings of an International Conference, Singapore,

14–18 February 1989. (ed. Champ, B. R., Highley, E., and Banks, H. J. ). ACIAR Proceedings No. 25, p. 96–107.

Banks, H. J., and Annis, P. C. (1977). Suggested procedures for controlled atmosphere storage of dry grain, CSIRO Aust. Div. Entomol. Tech. Pap. No. 13, 23 pp.

Banks, H. J., and Annis, P. C. (1981). Conversion of existing storages for modified atmospheric use. In Controlled Atmosphere Storage of Grains (ed. Shejbal, J.). Elsevier, Amsterdam, pp. 207–224.

Banks, H. J., and Annis, P. C. (1990). Comparative advantages of high $CO_2$ and low $O_2$ types of controlled atmospheres for grain storage. In Food Preservation by Modified Atmospheres (ed. Calderon, M., and Barkai-Golan, R.). Boca Raton, FL, pp. 93–122.

Banks, H. J., and Ripp, B. E. (1984). Sealing of grain storages for use with fumigants and controlled atmospheres, Proceedings of the 3rd International Working Conference on Stored-Product Entomology, Manhattan, KS, pp. 375–390.

Banks, H. J., and Sharp, A. K. (1986). Influence of transport on gas loss from freight containers under fumigation, Part II, Descriptive model, *Pestic. Sci.*, *17*:221–229.

Banks, H. J., Sharp, A. K., and Irving, A. R. (1986). Influence of transport on gas loss from freight containers under fumigation, Part I, Experimental investigation, *Pestic. Sci.*, *17*:207–220.

Banks, H. J., and Sticka, R. (1981). Phosphine fumigation of PVC-covered earth-walled bulk grain storages: full scale trials using a surface application technique, CSIRO Aust. Div. Entomol. Tech., Paper No. 18, 1–45.

Barakat, A. A., Khan, P., and Alaldul Karim, A. M. (1987). The persistence and activity of permethrin and chlorpyrifos-methyl sprays on jute and woven polypropylene bags, *J. Stored Prod. Res.*, *23*:85–90.

Barson, G. (1991). Laboratory assessment of the residual toxicity of commercial formulations of insecticides to adult *Oryzaephilus surinamensis* (Coleoptera: Silvanidae) exposed for short time intervals, *J. Stored Prod. Res.*, *27*:205–211.

Bell, C. H. (1977). Tolerance of the diapausing stages of four species of Lepidoptera to methyl bromide, *J. Stored Prod. Res.*, *13*:199–227.

Bell, C. H. (1978). Effect of temperature on the toxicity of low concentrations of methyl bromide to diapausing larvae of the warehouse moth *Ephestia elutella* (Huber), *Pestic. Sci.*, *9*:529–534.

Bell, C. H., and Glanville, V. (1973). The effect of concentration and exposure in tests with methyl bromide and with phosphine on diapausing larvae of *Ephestia elutella*, *J. Stored Prod. Res.*, *9*:165–170.

Bennett, G. W., Owens, J. M., and Corrigan, R. M. (eds.) (1988). Truman's Scientific Guide to Pest Control Operations, 4th ed. Purdue University Press, Indiana.

Berck, B. (1964). Some parameters in the use of fumigants, *World Rev. Pest. Control*, *3*:196–174.

Bills, D. D., Reddy, M. C., and Lindsay, R. C. (1969). Fumigated nuts can cause off-flavor in candy, *Southeastern Peanut Assoc.*, *12*:39–40.

Bond, E. J. (1984). Manual of Fumigation for Insect Control. FAO Plant Prod. and Prot. Paper No. 54, F.A.O. of the U.N., Rome.

Bond, E. J., Dumas, T., and Hobbs, S. (1984). Corrosion of metals by the fumigant phosphine, *J. Stored Prod. Res.*, *20*:57–63.

Bond, E. J., Monro, H. A. U., and Buckland, C. T. (1967). The influence of oxygen on the toxicity of fumigants to *Sitophilus granarius*, *J. Stored Prod. Res.*, *3*:289–294.

Bond, E. J., Robinson, J. R., and Buckland, C. T. (1969). The toxic action of phosphine. Absorption and symptoms of poisoning in insects, *J. Stored Prod. Res.*, *5*:289–298.

Borah, B., and Chahal, B. S. (1979). Development of resistance in *Trogoderma granarium* Everts to phosphine in the Punjab, *FAO Plant Prot. Bull.*, *27*:77–80.

Brown, W. B. (1959). Fumigation with methyl bromide under gas-proof sheets. Pest Infestation Research Bulletin No. 1 (2nd ed.), Dept. of Scientific and Industrial Research, Her Majesty's Stationery Office, London.

Burkholder, W. E., and Dicke, R. J. (1966). The toxicity of malathion and fenthion to dermestid larvae as influenced by various surfaces, *J. Econ. Entomol.*, *59*:253–254.

Calderon, M., and Barkai-Golan, R. (eds.) (1990). Food Preservation by Modified Atmospheres. CRC Press, Boca Raton, FL.

Calderon, M, and Leesch, J. G. (1983). Effect of reduced pressure and $CO_2$ on the toxicity of methyl bromide to two species of stored-product insects, *J. Econ. Entomol.*, *76*:1125–1128.

Calderon, M., Leesch, J. G., and Jay, E. G. (1991). Toxicity of high oxygen atmospheres to *Tribolium castaneum* (Coleoptera: Tenebrionidae), *J. Entomol. Sci.*, *26*:197–204.

Chamberlain, S. J. (1981). Etrimofos residues in rapeseed oil during laboratory scale refining, *J. Stored Prod. Res.*, *17*:183–185.

Champ, B. R., and Dyte, C. E. (1976). Report of the FAO global survey of pesticide susceptibility of stored grain pests. FAO Plant Prot. Ser. No. 2.

Champ, B. R., and Dyte, C. E. (1977). FAO global survey of pesticide susceptibility of stored product pests, *FAO Plant Prot. Bull.*, *25*:49–67.

Champ, B. R., Highley, E., and Banks, H. J. (eds.) (1990). Fumigation and Controlled Atmosphere Storage of Grain: Proceedings of an International Conference, Singapore, 14–18 February 1989. ACIAR Proceedings No. 25.

Chefurka, W., Kashi, K. P., and Bond, E. J. (1976). The effect of phosphine on electron transport in mitochondria, *Pestic. Biochem. Physiol.*, *6*:65–84.

Childs, D. P., and Overby, J. E. (1971). Phosphine fumigation of tobacco in louvered warehouses, *Tobacco Sci.*, *14*:49.

Cink, J., and Harein, P. (1989). Stored Grain Pest Management, Burgess, Minneapolis, MN. p. 282.

Cook, J. S. (1980). U.S. Patent No. 4,200,657.

Cook, J.S . (1983). The use of controlled air to increase the effectiveness of fumigation of stationary grain storages. In Controlled Atmosphere and Fumigation in Grain Storages: Proceedings of an International Symposium Held from 11 to 22 April 1983 in Perth, Western Australia (ed. Ripp, E., et al.). Elsevier, Amsterdam, pp. 419–424.

Cox, P. D., Parish, W. E., and Ledson, M. (1990). Factors affecting the refuge-seeking behavior of *Cryptolestes ferrugineus* (Stephens) (Coleoptera: Cucujidae), *J. Stored Prod. Res.*, *26*:169–174.

Daft, J. L. (1985). Preparation and use of mixed fumigant standards for multiresidue level determination of gas chromatography, *J. Agric. Food Chem.*, *33*:563–566.

Davis, R., and Jay, E. J. (1983). An overview of modified atmospheres for insect control, *Assoc. Oper. Millers Bull.*, *March*:4026–4029.

Dennis, N. M., Eason, G., and Gillenwater, H. B. (1972). Formation of methyl chloride during methyl bromide fumigations, *J. Econ. Entomol.*, *65*:1753–1754.

Desmarchelier, J. M., and Bengston, M. (1979). The residual behaviour of chemicals on stored grain, Proceedings Second International Working Conference on Stored-Product Entomology, Ibadan, Nigeria, pp. 138–151.

Donahaye, E. (1992). Physiological differences between strains of *Tribolium castaneum* selected for resistance to hypoxia and hypercarbia, and the unselected strain, *Physiol. Entomol.*, *17*:219–229.

Donahaye, E., and Rindner, M. (1993). Specificity of induced resistance to hypoxia and hypercarbia in two strains of *Tribolium castaneum* (Herbst). In Proceedings International Conference on Controlled Atmospheres and Fumigation in Grain Storages, Winnipeg, Canada, June 1992 (ed. Navarro, S., and Donahaye, E.). Caspit Press Ltd. Jerusalem, pp. 441–447.

Ducom, P., and Bourges, C. (1986). A new technique for the measurement of phosphine concentrations. GASGA Seminar on Fumigation Technology in Developing Countries, Tropical Development and Research Institute, Slough, UK, pp. 68–73.

Dumas, T., and Bond, E. J. (1977). Penetration, sorption, and desorption of fumigant in the treatment of food materials with a methyl bromide-acrylnotrile mixture, *J. Agric. Food Chem.*, *25*:677–680.

Dumas, T., and Bond, E. J. (1979). Relation of temperature to ethylene dibromide desorption from fumigated wheat, *J. Agric. Food Chem.*, *27*:1206–1209.

Estes, P. M. (1965). The effect of time and temperature on methyl bromide fumigation of adults of *S. granarius* and *Tribolium confusum*, *J. Econ. Entomol.*, *58*:611–614.

Garry, V. F., Griffith, J., Danzi, T. J., Nelson, R. L., Wharton, E. B., Krueger, L. A., and Cervenka, J. (1989). Human genotoxicity: Pesticide applicators and phosphine, *Science*, *246*:251–255.

Grisamore, S. B., Hile, J. P., and Otten, R. J. (1991). Pesticide residues on grain products, *Cereal Foods World*, *36*:434–437.

Hagstrum, D. W. (1989). Infestation by *Cryptolestes ferrugineus* (Coleoptera: Cucujidae) in newly harvested wheat stored on three Kansas farms, *J. Econ. Entomol.*, *82*:655–659.

Haliscak, J. P., and Beeman, R. W. (1983). Status of malathion resistance in five genera of beetles infesting farm-stored corn, wheat, and oat in the United States, *J. Econ. Entomol.*, *76*:717–722.

Hanson, P. R., Wainman, H. E., and Chakrabarti, B. (1987). The effects of methyl bromide on the seed of some varieties of barley, *Seed Sci. Technol.*, *15*:155–162.

Harein, P. K., and Davis, R. (1992). Control of stored grain insects. In Storage of Cereal Grains and Their Products (ed. Sauer D. B.). American Association of Cereal Chemistry, St. Paul, MN, pp. 491–534.

Harein, P. K., and Krause, G. F. (1964). Dosage-time relationships between 80:20 ($CCl_4$ : $CS_2$) and rice weevils, *J. Econ. Entomol.*, *57*:521–522.

Harris, A. H. (1986). A conductimetric method for determining the concentration of phosphine during fumigation. GASGA Seminar on Fumigation Technology in Developing Countries, Tropical Development and Research Institute, Slough, UK, pp. 56–65.

Heaps, J., and Harein, P. K. (1991). Insecticide recommendations for food processing plants. *Assoc. Operative Millers Bull.*, *Oct.*:5960–5962.

Highland, H. A. (1991). Protecting packages against insects. In Ecology and Management of Food Industry Pests (ed. Gorham, J. R.). FDA Tech. Bull. 4, Association of Official Analytical Chemists, Arlington, VA, pp. 345–350.

Hole, B. D. (1981). Variation in tolerance of seven species of stored-product Coleoptera to methyl bromide and phosphine in strains from twenty-nine countries, *Bull. Entomol. Res.*, *71*:299–306.

Iordanou, N. T., and Watters, F. L. (1969). Temperature effects on the toxicity of five insecticides against five species of stored-product insects, *J. Econ. Entomol.*, *62*: 130–339.

Ivens, G. W. (ed.). (1989). The U. K. Pesticide Guide. C.A.B. Intern. British Crop Prot. Council, Wallingford, U.K.

Jay, E. G., Banks, H. J., and Keever, D. W. (1990). Recent developments in controlled atmosphere technology. In Fumigation and Controlled Atmosphere Storage of Grain: Proceedings of an International Conference Held at Singapore, 14–18 February 1989 (ed. Champ, B. R., Highley, E., and Banks, H. J.). ACIAR Proceedings No. 25, pp. 134–143.

Jay, E. G., Davis, R., and Zehner, J. M. (1983). In-transit fumigation of truck-ship containers with hydrogen phosphide, Adv. Agric. Tech., USDA, ARS, No. AAT-S-28.

Johnson, E. H. (1980). Tolerances and exemptions from tolerances for pesticide chemicals in or on raw agricultural commodities; carbon dioxide, nitrogen, and combustion product gases, *Fed. Reg.*, *45*:75663.

Johnson, E. H. (1981). Carbon dioxide, nitrogen, and combustion product gases: tolerances for pesticides in food administered by the Environmental Protection Agency, *Fed. Reg.*, *46*:32865.

Kashi, K. P., and Chefurka, W. (1976). The effect of phosphine on the absorption and circular dichroic spectra of cytochrome-c and cytochrome oxidase, *Pestic. Biochem. Physiol.*, *6*:350–362.

Keever, D. W. (1990). Modified tobacco warehouse aerations for lowered acute emissions of phosphine into the surrounding environs. In Vol. II, Proceedings Fifth International Working Conference on Stored Product Protection, Bordeaux, France, September 9–14, 1990 (ed. Fleurat-Lessard, F., and Ducom, P.). Imprimerie du Medoc, Bordeaux-Blanquefort, pp. 837–846.

Kirkpatrick, R. L., and Gillenwater, H. B. (1979). Toxicity of selected dusts and aerosols to three species of stored-product insects, *J. Georgia Entomol. Soc.*, *14*:334–339.

LaHue, D. W. (1969). Control of malathion-resistant Indian meal moths *Plodia interpunctella* (Hubner) with dichlorvos resin strips, *Proc. NC Branch Entomol. Soc. Am.*, *24*:117–119.

LaHue, D. W., and Kadoum, A. (1979). Residual effectiveness of emulsions and encapsulated formulations of malathion and fenitrothion against four stored grain beetles, *J. Econ. Entomol.*, *72*:234–237.

Leesch, J. G. (1982). Accuracy of different sampling pumps and detector tube combinations to determine phosphine concentrations, *J. Econ. Entomol.*, *75*:899–905.

Leesch, J. G. (1990). The effect of low concentrations of carbon dioxide on the penetration of phosphine through wheat. In Proceedings Fifth International Working Conference on Stored Product Protection, Bordeaux, France, September 9–14, 1990, Vol. II (ed.

Fleurat-Lessard, F., and Ducom, P). Imprimerie du Medoc, Bordeaux-Blanquefort, France, pp. 859–865.

Leesch, J. G. (1992). Carbon dioxide on the penetration and distribution of phosphine through wheat, *J. Econ. Entomol.*, *85*:156–161.

Leesch, J. G., and Highland, H. A. (1978). Fumigation of shrink-wrapped pallets, *J. Georgia Entomol. Soc.*, *13*:43–50.

Leesch, J. G., Arthur, F. H., and Davis, R. (1990). Three methods of aluminum phosphide application for the in-transit fumigation of grain aboard deep-draft bulk cargo ships, *J. Econ. Entomol.*, *83*:1459–1467.

Leesch, J. G., Davis, R., Zettler, J. L., Sukkestad, D. R., Zehner, J. M., and Redlinger, L. M. (1986). Use of perforated tubing to distribute phosphine during the in-transit fumigation of wheat, *J. Econ. Entomol.*, *79*:1583–1589.

Leesch, J. G., Redlinger, L. M., and Dennis, N. M. (1978). Methyl bromide fumigation of farmers' stock peanuts in flat storage, *Peanut Sci.*, *5*:40–43.

Leesch, J. G., Redlinger, L. M., Gillenwater, H. B., and Zehner, J. M. (1982). Fumigation of dates with phosphine, *J. Econ. Entomol.*, *75*:685–687.

Leesch, J. G., and Sukkestad, D. R. (1980). Potential of two perflourinated alcohols as fumigants, *J. Econ. Entomol.*, *73*:829–831.

Leesch, J. G., and Sukkestad, D. R. (1983). Some fumigant properties of hexamethyl distannane against stored-product insects, *J. Georgia Entomol. Soc.*, *18*:385–394.

LeGoupil, M. (1932). Les propriétés du bromure de methyle, *Rev. Veg. Ent. Agric. Fr.*, *19*:167–172.

LePatourel, G. N. J. (1986). The effect of grain moisture content on the toxicity of a sorptive silica dust to four species of grain beetle, *J. Stored Prod. Res.*, *22*:63–69.

Lewis, S. E. (1948). Inhibition of SH enzymes by methyl bromide, *Nature*, *161*:692–693.

Lindgren, D. L. (1938). The stupefaction of red scale *Aonidiella aurantii* by hydrocyanic acid, *Hilgardia*, *11*:213–225.

Linsley, G. E. (1944). Natural sources, habitats, and reservoirs of insects associated with stored food products, *Hilgardia*, *16*:18–233.

Lueck, H., Allemann, A., and Stockol, J. M. (1984). Control of insect infestation during storage of dairy products. Fumigation with hydrogen phosphide, *S. Afr. J. Dairy Technol.*, *16*:103–107.

Madrid, F. J., White, N. D. G., and Loschiavo, S. R. (1990). Insects in stored cereals, and their associations with farming practices in southern Manitoba, *Can. Entomol.*, *122*:515–523.

McLaine, L. S., and Monro, H. A. U. (1937). Developments in vacuum fumigation at the port of Montreal, *Ann. Rep. Entomol. Soc. Ont.*, *67*:15–17.

Mehler, L. N., O'Malley, M. A., and Krieger, R. I. (1992). Acute pesticide morbidity and mortality: California, *Rev. Env. Contamin. Toxicol.*, *129*:51–66.

Mensah, G. W. K., and White, N. D. G. (1984). Laboratory evaluation of malathion-treated sawdust for control of stored-product insects in empty granaries and food warehouses, *J. Econ. Entomol.*, *77*:202–206.

Mensah, G. W. K., Watters, F. L., and Webster, G. R. B. (1979). Insecticide residues in milled fractions of dry or tough wheat treated with malathion, bromophos, iodofenphos and pirimiphos-methyl, *J. Econ. Entomol.*, 72:728–731.

Mills, J. T. (ed.). (1990). Protection of farm-stored grains and oilseeds from insects, mites, and molds. Agric. Canada Publ. 1851/E, Commun. Branch, Agric. Canada, Ottawa.

Mills, J. T., and White, N. D. G. (1984). Foreign visit report on science and technology. Australia—Improved procedures for controlling insects in stored grains, Agric. Can. Res. Branch Publ. IST 84-13.

Mills, K. A. (1983). Resistance to the fumigant hydrogen phosphide in some stored-product species associated with repeated inadequate treatments, *Mitt. Dtsch. Ges. Allg. Angew. Entomol.*, *69*:241–244.

Mills, K. A., Clifton, A. L., Chakrabarti, B., and Savvidou, N. (1990). The impact of phosphine resistance on the control of insects in stored grain by phosphine fumigation. British Crop Protection Conference—Pests and Diseases. British Crop Protection Council, Brighton, England, pp. 1181–1187.

Minett, W., and Williams, P. (1971). Influence of malathion distribution on the protection of wheat grain against insect infestation, *J. Stored Prod. Res.*, *7*:233–242.

Minett, W., Williams, P., and Amos, T. G. (1981). Gravity feed application of insecticide concentrate to wheat in a commercial silo, *Gen. Appl. Entomol.*, *13*:59–64.

Mondal, K. A. M. S. H. (1984). Repellent effect of pirimiphos-methyl to larval *Tribolium castaneum* Herbst, *Int. Pest Control*, *26*:98–99.

Monro, H. A. U., Upitis, E., and Bond, E. J. (1972). Resistance of a laboratory strain of *Sitophilus granarius* (L.) (Coleoptera, Curculionidae) to phosphine, *J. Stored Prod. Res.*, *8*:199–202.

Navarro, S. and Donahaye, E. (eds.) (1993). Proceedings of the International Conference on Controlled Atmosphere and Fumigation in Grain Storages, Winnipeg, Canada, June 1992, Caspit Press Ltd., Jerusalem, p. 560.

Navarro, S., and Jay, E. G. (1987). Application of modified atmospheres for controlling stored grain insects. BCPC MONO. NO. 37, Stored Products Pest Control, pp. 229–236.

Newman, C. J. E. (1990). Specification and design of enclosures for gas treatment. In Fumigation and Controlled Atmosphere Storage of Grain: Proceedings of an International Conference, Singapore, 14–18 February 1989 (ed. Champ, B. R., Highley, E., and Banks, H. J.). ACIAR Proceedings No. 25, Canberra, pp. 108–130.

Okwelogu, T. N. (1968). The toxicity of malathion applied to washed concrete, *J. Stored Prod. Res.*, *4*:259–260.

Orth, R. A., and Minett, W. (1975). Iodometric analysis and shelf life of malathion in formulations, *Pestic. Sci.*, *6*:217–221.

Ostlie, K. R., Noetzel, D. M., Hutchison, W. D., and Subramanyam, Bh. (1994). Insecticide suggestions to control insect pests of field crops in 1994. Minnesota Ext. Serv., Univ. of Minnesota BU-0500-E, St. Paul, MN.

Page, A. P. B., and Blacklith, R. E. (1956). Developments in fumigation practice, *Rep. Progr. Appl. Chem.*, *41*:535–545.

Parkin, E. A. (1966). The relative toxicity and persistence of insecticides applied as water-dispersible powders against stored-product beetles, *Ann. Appl. Biol.*, *57*:1–14.

Parkin, E. A., and Hewlett, P. S. (1946). The formation of insecticide films on building materials. I. Preliminary experiments with films of pyrethrum and DDT in heavy oil, *Ann. Appl. Biol.*, *33*:381–386.

Pedersen, J. R. (1992). Insects: Identification, damage, and detection. In Storage of Cereal Grains and Their Products, 4th ed. (ed. Sauer, D. G.). American Association of Cereal Chemistry, St. Paul, MN, pp. 435–490.

Pinniger, D. B. (1974). A laboratory simulation of residual populations of stored product pests and an assessment of their susceptibility to a contact insecticide, *J. Stored Prod. Res.*, *10*:217–223.

Pomeranz, Y. (1992). Biochemical, functional, and nutritive changes during storage. In Storage of Cereal Grains and Their Products, 4th ed. (ed. Sauer, D. G.). American Association of Cereal Chemistry, St. Paul, MN, pp. 55–142.

Potter, W. T., Rong, S., Griffith, J., White, J., and Garry, V. F. (1991). Phosphine-mediated Heinz body formation and hemoglobin oxidation in human erythrocytes, *Toxicol. Lett.*, *57*:37–45.

Powell, D. F. (1975). The fumigation of seeds with methyl bromide, *Ann. Appl. Biol.*, *81*:425–431.

Price, N. R. (1980a). Some aspects of the inhibition of cytochrome-c oxidases by phosphine in susceptible and resistant strains of *Rhyzopertha dominica*, *Insect Biochem.*, *10*:147–150.

Price, N. R. (1980b). The effect of phosphine on respiration and mitochondrial oxidation in susceptible and resistant strains of *Rhyzopertha dominica*, *Insect Biochem.*, *10*: 65–71.

Price, N. R. (1985). The mode of action of fumigants, *J. Stored Prod. Res.*, *21*:157–164.

Price, N. R., Mills, K. A., and Humphries, L. A. (1982). Phosphine toxicity and catalyse activity in susceptible and resistant strains of the lesser grain borer (*Rhyzopertha dominica*), *Comp. Biochem. Physiol.*, *73C*:411–413.

Price, N. R., and Walter, C. M. (1987). A comparison of some effects of phosphine, hydrogen cyanide and anoxia in the lesser grain borer, *Rhyzopertha dominica* (F.) (Coleoptera: Bostrychidae), *Comp. Biochem. Physiol.*, *86C*:33–36.

Quinlan, J. K., and McGaughey, W. H. (1983). Fumigation of empty grain drying bins with chloropicrin, phosphine, and liquid fumigant mixtures, *J. Econ. Entomol.*, *76*:184–187.

Quinlan, J. K., White, G. D., Wilson, J. L., Davidson, L. I., and Hendricks, L. H. (1979). Effectiveness of chlorpyrifos-methyl and malathion as protectants for high moisture stored wheat, *J. Econ. Entomol.*, *72*:90–93.

Rangaswamy, J. R. (1985). Phosphine residue and its desorption from cereals, *J. Agric. Food Chem.*, *33*:1102–1106.

Rangaswamy, J. R., and Muthu, M. (1985). Spectrophotometric determination of phosphine residue in rice, *J. Assoc. Off. Anal. Chem.*, *68*:205–208.

Rangaswamy, J. R., and Sasikala, V. B. (1986). Kinetics of phosphine residue dissipation from wheat and its milled products in storage, *J. Food Sci. Technol.*, *23*:54–58.

Reed, C., Anderson, K., Brochschmidt, J., Wright, V., and Pedersen, J. (1990). Cost and effectiveness of chemical insect control measures in farm-stored Kansas wheat, *J. Kansas Entomol. Soc.*, *63*:351–360.

Reeves, R. G., McDaniel, B. A., and Ford, J. H. (1985). Organic and inorganic bromide residues in spices fumigated with methyl bromide, *J. Agric. Food Chem.*, *33*: 780–783.

Ripp, B. E. (ed.) (1984). Controlled Atmosphere and Fumigation in Grain Storages, Proceedings of an International Symposium, Practical Aspects of Controlled Atmosphere and Fumigation in Grain Storages, held from 11 to 22 April 1983 in Perth, Australia. Elsevier, Amsterdam.

Ripp, B. E. (1985). Surface coatings for the sealings of gas storages. In Surface Coatings. Australia, pp. 6–10.

Ripp, B. E., deLargie, T. A., and Barry, C. B. (1990). Advances in the practical application of controlled atmosphere for the preservation of grain in Australia. In Food Preservation by Modified Atmospheres (ed. Calderon, M., and Barkai-Golan, R.). CRC Press, Boca Raton, FL, pp. 151–184.

Rowlands, D. G. (1975). The metabolism of contact insecticides in stored grains, III, 1970–1974, *Residue Rev.*, *58*:113–155.

Rowlands, D. G., and Bramhall, J. S. (1977). The uptake and translocation of malathion by the stored wheat grain, *J. Stored Prod. Res.*, *13*:13–22.

Samson, P. R., Parker, R. J., and Jones, A. L. (1988). Comparative effect of grain moisture on the biological activity of protectants on stored corn, *J. Econ. Entomol.*, *81*: 949–954.

Shejbal, J. (ed.) (1980). Developments in Agricultural Engineering 1. Controlled Atmosphere Storage of Grains, An International Symposium held 12 to 15 May 1980 at Castelgandolfa (Rome) Italy. Elsevier, Amsterdam.

Sine, C. (ed.). (1992). Farm Chemicals Handbook '92. Meister Publishing Co., Willoughby, OH.

Sinha, R. N., and Watters, F. L. (1985). Insect pests of flour mills, elevators and feed mills and their control. Agr. Canada Publ. 1776E, Ottawa, ON.

Sittisuang, P., and Nakakita, H. (1985). The effect of phosphine and methyl bromide on germination of rice and corn, *J. Pestic. Sci.*, *10*:461–468.

Slominski, J. W., and Gojmerac, W. L. (1972). The effect of surfaces on the activity of insecticides. College Agric. Life Sci., Univ. Wisconsin, Res. Rep.

Smith, B. L. (ed.) (1990). Codex Alimentarius; abridged version. Food and Agriculture Organization of the United Nations/World Health Org., Rome, Italy.

Snelson, J. T. (1987). Grain Protectants. ACIAR Monograph No. 3. Canberra, Australia.

Srinath, D., and Ramchandani, N. P. (1978). Investigations on fumigation of walnuts with methyl bromide, *J. Food Sci. Tech.*, *15*:192–194.

Starratt, A. N., and Bond, E. J. (1988). Methylation of DNA of maize and wheat grains during fumigation with methyl bromide, *J. Agric. Food Chem.*, *36*:1035–1039.

Storey, C. L. (1967). Comparative study of methods of distributing methyl bromide in flat storages of wheat: Gravity penetration, single pass and closed recirculation. U. S. Department of Agriculture, Washington, D. C., Marketing Research Report No. 794.

Storey, C. L. (1971). Distribution of grain fumigants in silo-type elevator tanks by aeration systems. U. S. Department of Agriculture, Washington, D. C., Marketing Research Report No. 915.

Storey, C. L., and Davidson, L. I. (1973). Relative toxicity of chloropicrin, phosphine, EDC-$CCl_4$, and $CCl_4$-$CS_2$ to various life stages of the Indianmeal moth. USDA, Marketing Res. Report 7316. (7316).

Subramanyam, Bh., and Cutkomp, L. K. (1987). Influence of posttreatment temperature on toxicity of pyrethroids to five species of stored-product insects, *J. Econ. Entomol.*, *80*:9–13.

Subramanyam, Bh., Heaps, J., and Harein, P. (1993). Insecticide recommendations for food processing plants. Minnesota Ext. Serv., Univ. of Minnesota, FS-3921-A, St. Paul, MN.

Thomas, K. P. (1980). The uptake and degradation of pirimiphos-methyl by stilton cheese, *J. Stored Prod. Res.*, *16*:105–108.

Tyler, P. S., Taylor, R. W. D., and Rees, D. P. (1983). Insect resistance to phosphine fumigation in food warehouses in Bangladesh, *Int. Pest Control*, *25*:10–13, 21.

Watters, F. L. (1966). Protection of packaged food from insect infestation by the use of silica gel, *J. Econ. Entomol.*, *59*:146–149.

Watters, F. L. (1968). Pyrethrins–piperonyl butoxide applied as a fog in an empty grain bin, *J. Econ. Entomol.*, *61*:1313–1316.

Watters, F. L. (1970). Toxicity to the confused flour beetle of malathion and bromophos on concrete floors, *J. Econ. Entomol.*, *63*:1000–1001.

Watters, F. L., and Grussendorf, O. W. (1969). Toxicity and persistence of lindane and methoxychlor on building surfaces for stored-grain insect control, *J. Econ. Entomol.*, *62*:1101–1106.

Watters, F. L., and Nowicki, T. W. (1982). Uptake of bromphos by stored rapeseed, *J. Econ. Entomol.*, *75*:261–264.

Watters, F. L., White, N. D. G., and Cote, D. (1983). Effect of temperature on toxicity and persistence of three pyrethroid insecticides applied to fir plywood for the control of the red flour beetle (Coleoptera: Tenebrionidae), *J. Econ. Entomol.*, *76*:11–16.

White, N. D. G. (1982). Effectiveness of malathion and pirimiphos-methyl applied to plywood and concrete to control *Prostephanus truncatus* (Coleoptera: Bostrichidae), *Proc. Entomol. Soc. Ontario*, *113*:65–69.

White, N. D. G. (1985). Uptake of malathion and pirimiphos-methyl by rye, wheat, or triticale stored on treated surfaces, *J. Econ. Entomol.*, *78*:1315–1319.

White, N. D. G., and Abramson, D. (1984). Uptake of malathion from galvanized steel surfaces into stored barley, *J. Econ. Entomol.*, *77*:289–293.

White, N. D. G., and Nowicki, T. W. (1986). Persistence of malathion and pirimiphos-methyl residues in two species of rapeseed stored at various moisture contents and temperatures, *Sci. Aliments*, *6*:273–286.

White, N. D. G., Nowicki, T. W., and Watters, F. L. (1983). Comparison of fenitrothion and malathion for treatment of plywood or galvanized steel surfaces for control of the red flour beetle (Coleoptera: Tenebrionidae) and the rusty grain beetle (Coleoptera: Cucujidae), *J. Econ. Entomol.*, *76*:856–863.

White, N. D. G., Sinha, R. N., and Mills, J. T. (1986). Long-term effects of an insecticide on a stored-wheat ecosystem, *Can. J. Zool.*, *64*:2558–2569.

Whitney, W. K., and Walkden, H. H. (1961). Concentrations of methyl bromide lethal to insects in grain. U. S. Agricultural Marketing Service, Washington, D. C., Report No. 511.

Wienzierl, R. (1992). Grain protectants. Goodbye to malathion? *ACIAR Postharvest Newsletter*, *20*:4.

Winks, R. G. (1974). Characteristic response of grain pests to phosphine. Commonwealth

Scientific and Industrial Research Organization, Division of Entomology, Canberra, Australia, Annual Report 1973–74.

Winks, R. G. (1990). Recent developments in fumigation technology, with emphasis on phosphine. In Fumigation and Controlled Atmosphere Storage of Grain: Proceedings of an International Conference, Singapore, 14–18 February 1989 (ed. Champ, B. R., Highley, E., and Banks, H. J.). ACIAR Proceedings No. 25, pp. 144–151.

Winteringham, F. P. W. (1955). The fate of labelled insecticides in food products, 4, The possible toxicological and nutritional significance of fumigating wheat with methyl bromide, *J. Sci. Food Agric.*, *6*:269–274.

Winteringham, F. W. P., and Barnes, J. M. (1955). Comparative response of insects and mammals to certain halogenated hydrocarbons used as insecticides, *Physiol. Rev.*, *35*:701–739.

Winteringham, F. W. P., Harrison, A., Bridges, R. G., and Bridges, P. M. (1955). The fate of labelled insecticide residues in food products, 2, The Nature of methyl bromide residues in fumigated wheat, *J. Sci. Food Agric.*, *6*:251–261.

Zanon, K. (1980). Systems of supply of nitrogen for the storage of grains in controlled atmosphere. In Controlled Atmosphere Storage of Grains, An International Symposium held 12 to 17 May at Castelgandolfo (Rome) Italy (ed. Shejbal, J.). Elsevier, Amsterdam, pp. 507–516.

Zettler, J. L. (1982). Insecticide resistance on selected stored-product insects infesting peanuts in the southeastern United States. *J. Econ. Entomol.*, *75*:359–362.

Zettler, J. L. (1990). Phosphine resistance in stored product insects in the United States. In Proceedings of the Fifth International Working Conference on Stored Product Protection, Bordeaux, France, September 9–14, 1990, Vol. II (ed. Fleurat-Lessard, F., and Ducom, P.). Imprimère du Medoc, Bordeaux-Blanquefort, France, pp. 1075–1081.

Zettler, J. L. (1991). Pesticide resistance in *Tribolium castaneum* and *T. confusum* (Coleoptera: Tenebrionidae) from flour mills in the United States, *J. Econ. Entomol.*, *84*: 763–767.

Zettler, J. L., and Cuperus, G. W. (1990). Pesticide resistance in *Tribolium castaneum* (Coleoptera: Tenebrionidae) and *Rhyzopertha dominica* (Coleoptera: Bostrichidae) in wheat, *J. Econ. Entomol.*, *83*:1677–1681.

Zettler, J. L., and Redlinger, L. M. (1984). Arthropod pest management with residual insecticides. In Insect Management for Food Storage and Processing (ed. Baur, F. J.). American Association of Cereal Chemistry, St. Paul, MN, pp. 111–130.

Zettler, J. L., Halliday, W. R., and Arthur, F. H. (1989). Phosphine resistance in insects infesting stored peanuts in the southeastern United States, *J. Econ. Entomol.*, *82*:1508–1511.

# 8

# Resistance Measurement and Management

**Bhadriraju Subramanyam**
*University of Minnesota, St. Paul, Minnesota*

**David W. Hagstrum**
*U.S. Department of Agriculture, Manhattan, Kansas*

Insecticide products registered for control of stored-product insects vary from country to country, and the types and number of insecticides registered within a country are largely dependent on the need and existing pesticide regulations. For instance, pyrethroids, such as deltamethrin and bioresmethrin, are not registered for use on stored grains in the United States, but these chemicals are registered for use on stored wheat in Australia. The important features of insecticides used on stored commodities and empty storage facilities were reviewed in Chapter 7. Insecticides are commonly used for stored-product insect management because their use is simple and cost-effective. The frequent use of insecticides as a substitute rather than as a supplement to nonchemical insect management techniques (Storey et al. 1984; Harein et al. 1985; Kenkel et al. 1994) has resulted in the failure of these chemicals to effectively control stored-product insects. There are several reasons for field failures of an insecticide. Field failures can occur if an insecticide is applied at less than the labeled rate, resulting in exposure of insects to sublethal doses. Generally, this problem can be corrected by proper calibration of the insecticide application equipment and the grain flow rate when using grain protectants. In the case of fumigants, proper sealing of the facilities is important to maintain lethal levels of the gas concentration. Uneven coverage of the insecticide on stored commodities, or applying an insecticide after the appearance of a heavy infestation, may allow some insects to escape the treatment. The most important reason for possible field failures is the evolution of resistance in insects to insecticides. Before resistance is suspected, it is important to rule out factors other than resistance for the failure of an insecticide to effectively control insects.

In this chapter, the evolution of insecticide resistance in insects is discussed along with quantitative techniques for measuring resistance and resistance inheritance. Reports on resistance in stored-product insects to several insecticides are reviewed, including factors influencing resistance estimation and the mode of resistance inheritance. A majority of resistance reports covered in this chapter are from the United States, United Kingdom, Australia, with a few reports from India, Philippines, Indonesia, Canada, Rwanda, and Brazil. Our survey of resistance literature on stored-product insects is not exhaustive but comprehensive. The last section of this chapter deals with the biological and operational factors affecting resistance development, and the development of a resistance management program. An understanding of factors influencing resistance development is essential for formulating effective resistance management strategies.

## I. RESISTANCE

Resistance is the ability in individuals of a species to withstand doses of toxic substances that would be lethal to the majority of individuals in a normal population. In a population of insects, very few individuals possess the genes for resistance. In other words, a majority of the insect population is susceptible. When an insecticide is applied, a large portion of susceptible individuals (≥ 95%) are killed, and resistant individuals survive. These resistant individuals may breed with other resistant individuals or with susceptible individuals immigrating into the treated area. As a result, the offspring may have only resistant genes, susceptible genes, or a combination of resistant and susceptible genes. If resistance is governed by a single gene, then individuals with the resistant genes can be denoted by RR (homozygous resistant) genotype, the ones with susceptible genes by SS (homozygous susceptible) genotype, and the ones with both susceptible and resistant genotypes by RS or SR genotype (heterozygous resistant, represented as RS from now on). The relative proportion of RR, RS, SS genotypes in an insect population depends on the selection pressure exerted by the insecticide, and the frequency of mating among the genotypes. Without genetic analysis or genetic markers that impart a different body color to RR or RS genotypes, it would be difficult to differentiate between them. Therefore, if a dose kills 80% of the SS genotype, we would not know what portion of the surviving individuals contain the RS or RR genotype. Therefore, the remaining 20% of the individuals are said to be phenotypically resistant. Repeated use of an insecticide eliminates the susceptibles (SS), and gradually a majority of the insect population will be made up of the RS and RR genotypes. At this stage, control failures become apparent. Increasing the insecticide dose will eliminate individuals with the RS genotype, and the surviving insect population will be made up of the RR genotype.

The number of generations required to develop resistance varies with the insecticide selection pressure, the genetic make up of the insect, insect species, insect stage, and the environment. For example, in the carpet beetle, *Anthrenus flavipes* Casey, a 70-fold resistance to permethrin occurred after 22 generations of selection in the laboratory, but such selection for permethrin resistance could not be obtained in the clothes moth, *Tineola bisselliella* (Hummel) (Schmid 1987). Dowd et al. (1984) developed a microcomputer program to demonstrate how resistance develops in insects to insecticides. This program is suitable for teaching purposes. During the evolution of resistance (from SS to RS and RR), there may be a metabolic or reproductive cost to the insect. For example, the resistant females may not lay as many eggs as the susceptible females, or the resistant insects may take longer than the susceptible insects to develop from egg to adult under similar environmental conditions.

In an insect population, resistance can be documented using several quantitative criteria (Tabashnik 1994), especially in situations where insecticide treatments are repeatedly applied to control insects. Resistance is suspected if higher doses are required to achieve a constant mortality of insects (Table 1); if there is a significant decrease in insect susceptibility to a fixed amount of the insecticide (McGaughey and Beeman 1988); if it takes longer to obtain a fixed mortality of insects (Table 2), and if the mortality of field populations of a species frequently exposed to insecticides is significantly less than mortality of the same species that has little or no insecticide exposure.

Rapid biochemical tests have been developed for detecting specific insecticide-degrading enzymes in individual insects (e.g., mosquitoes) (Pasteur and Geor-

**Table 1** Resistance in Two Strains of the Larger Grain Borer, *Prostephanus truncatus*, Confirmed by Survival of Progeny Exposed to Increasing Doses of Permethrin

| Generation | Permethrin dose (mg/kg) | % Average survival of strain from | |
|---|---|---|---|
| | | Mexico | Tanzania |
| Parents | 0.25 | 81.7 | 90.0 |
| $F_1$ | 0.42 | 81.8 | 77.9 |
| $F_2$ | 1.01 | 78.4 | 75.5 |
| $F_3$ | 1.95 | 68.2 | 77.1 |
| $F_4$ | 3.60 | 68.4 | 60.6 |
| $F_5$ | 4.02 | 72.2 | 71.3 |
| $F_6$ | 4.60 | 69.9 | 80.1 |
| $F_7$ | 7.30 | 54.7 | 25.5 |

*Source*: Golob et al. (1991).

**Table 2** Phosphine Resistance in Adults of the Red Flour Beetle, *Tribolium castaneum*, as Indicated by an Increase in the Lethal Time ($LT_{50}$) Estimates (days required to kill 50% of the population) for Progeny Exposed to a Fixed Phosphine Dose[a]

| Generation | $LT_{50}$ (95% confidence limits) | Resistance ratio[b] |
|---|---|---|
| Parents (laboratory strain) | 5.2 (4.6–6.0) | |
| $F_2$ | 7.9 (7.2–8.5) | 1.5 |
| $F_4$ | 19.8 (17.4–22.4) | 3.8 |
| $F_6$ | 23.6 (21.1–26.3) | 4.5 |

[a]Adults were exposed to 28.3 μg/L of phosphine at 26°C and 60% RH.
[b]$LT_{50}$ of $F_2$, $F_4$, or $F_6$ generation ÷ $LT_{50}$ of parental generation.
*Source*: El-Lakwah et al. (1992).

ghiou 1989). These techniques have not been developed for detecting resistance in stored-product insects. Ball (1981) suggested that a practical assessment of resistance should be based on the expected degree of control in the field, and not solely on the laboratory assessment of the dose required to kill a certain percentage of the population.

There are several mechanisms by which resistant insects are able to detoxify insecticides (Georghiou 1972). Insecticides, especially the organophosphates, are detoxified by the action of enzymes such as the mixed function oxidases, hydrolases, and transferases. The mixed function oxidases also degrade carbamates and pyrethroids. In addition, mechanisms such as reduced penetration and transport of the insecticide to the target site (nervous system), insensitivity of the target site, and increased excretion also confer resistance. Besides biochemical and physiological mechanisms, behavioral aspects could also select insects for resistance. Behavioral resistance occurs if insects are irritated or repelled by the insecticide, and Georghiou (1972) refers to this type of resistance as stimulus-dependent behavioristic avoidance. For stimulus-dependent behavioristic resistance to occur, insects must come in contact with the insecticide or insecticide-treated substrates. Insects that have increased irritability will avoid insecticide treatments after initial contact with it. Repellency occurs if insects are able to detect the insecticide molecules in air before actually contacting the insecticide-treated substrates, and selection favors insects that are able to detect the toxic substance and avoid it (Gould 1984). Examples of this behavior include the reduced response of insects to alight on toxic baits (Georghiou 1972). Stimulus-independent behavioristic avoidance or resistance occurs if insects naturally avoid areas treated with insecticides. In stimulus-independent behavioral resistance, insects increase their chances of survival by moving to areas that are insecticide free. In this type of resistance, the avoidance by insects is not due to irritant or repellent effects of the insecticide. It is important to note that insects with behavioral resistance are susceptible to insecticides, but have altered their behavior to avoid insecticide-treated areas. If behavioral resistance depends on insects first contacting the insecticide, then physiological resistance would lead to greater fitness (i.e., increase chances of survival) than behavioral resistance (Daly et al. 1988). The rate of resistance development in insects may be determined from the relationship between behavioral and physiological resistance. If the two responses are negatively correlated (i.e., increase in one resistance results in a decrease in the other), then insects that are highly behaviorally resistant tend to be extremely susceptible and vice versa. In other words, survival due to behavioral resistance will increase the frequency of SS genotypes in a population. The presence and maintenance of SS genotypes in a population is important for managing resistance (see Section IX).

## II. METHODS FOR MEASURING RESISTANCE

### 1. Dose-Response Tests and Probit Analysis

Resistance in field populations of stored-product insects is determined by exposing insects in the laboratory to a series of insecticide doses and recording mortality of insects at each dose. Field insects are also exposed to a treatment that does not include any insecticide (control treatment) to determine natural mortality of insects in the absence of an insecticide. This method is generally referred to as the dose-mortality test. To use this technique, large numbers of insects are collected from the field and mass reared in the laboratory under controlled conditions. The method of exposure includes treating individual insects topically with a small volume (usually 0.5–1 μl) of the insecticide at several prescribed doses, exposing a known number of insects to insecticide-impregnated filter papers, exposing insects in a chamber containing a measured concentration of the fumigant, or exposing insects to insecticide-treated commodities. The susceptibility, or lack thereof, of insects to the insecticide varies with the method of exposure. The mortality of insects at each insecticide dose is corrected for mortality of insects in the control treatment using the Abbott's (1925) formula as

$$P_{corr} = \frac{T - C}{1 - C} \qquad (1)$$

where $P_{corr}$ is the proportion of dead insects in insecticide treatments corrected for control mortality, $T$ = proportion of dead insects in the insecticide treatment, and $C$ = proportion of dead insects in the control treatment. In dose-mortality tests, the control mortality should be < 10%, and if control mortality exceeds 10%, the reasons for high control mortalities must be determined. Rosenheim and Hoy (1989) modified the Abbott's (1925) formula to determine corrected mortality. The modified formula includes the variation in the morality of insects in the control treatment

$$P_{corr} = 1 - \frac{(1 - T)(1 - K)}{1 - C} \text{ and } K = \frac{\text{Var}(C)t^2}{(1 - C)^2 n_c} \tag{2}$$

where Var ($C$) is the variance associated with $C$, $n_c$ = number of replicates used for estimating $C$, and $t$ = value of $t$ distribution at $n_c - 1$ degrees of freedom (df) at $\alpha = 0.05$. The 95% confidence limits (±95% CL) for $P_{corr}$ in Eq. 2 can be estimated as follows

$$\frac{t(1 - K)}{(1 - C)}\left[(1 - K)\left(\frac{\text{Var}(T)}{n_t}\right) + \frac{(1 - T)^2\text{Var}(C)}{(1 - C)^2 n_c}\right]^{0.5} \tag{3}$$

where Var ($T$) is the variance associated with $T$, and $n_t$ is the number of replicates used for estimating $T$. The 95% CL can be used to judge differences in mortality among treatments. The treatments are considered significantly different if the 95% CL for $P_{corr}$ among treatments do not overlap.

The corrected mortality is regressed on the corresponding doses after transformation of the mortality to probit scale and the dose to $\log_{10}$ scale, and this type of statistical analysis of dose-mortality data is called probit analysis (Finney 1971). Table 1 of Finney's (1971) book *Probit Analysis* gives transformations of percentages to probits. The probit regression model is

$$P_i = F(\alpha + \beta x_i) \tag{4}$$

where $P_i$ is the probability of response (mortality), $x_i$ is the $i$th dose or log dose, $\alpha$ and $\beta$ are y-intercept and slope estimates of the regression, respectively, and $F$ is the standard normal distribution function. Robertson and Preisler (1992) discuss the design and analysis of dose-mortality tests. They also developed an IBM-PC program (POLO-PC) for performing probit or logit analysis of dose-mortality data. This software is available from LeOra Software, 1119 Shattuck Avenue, Berkeley, California. Figure 1 shows the probit analysis of the lesser grain borer, *Rhyzopertha dominica* (F.), adults exposed to filter papers impregnated with various doses of malathion (Subramanyam, unpublished data). From this linear relationship, the dose which produces 50% mortality (probit 5) of the insects is estimated. This dose is called the median lethal dose or $LD_{50}$. The $LD_{50}$ for the example shown in Fig. 1 is 218.95 μg per 7-cm-diameter filter paper. The doses that cause 90 and 95% mortality of the insects are called $LD_{90}$ and $LD_{95}$, respectively. Performing probit analysis using the POLO-PC program provides the lethal dose estimates and associated 95% confidence limits (CL) (Robertson and Preisler 1992). When fitting the probit model to the dose-mortality data, it is important to determine if the observed mortality data significantly departs from the probit model. A $\chi^2$ test for goodness-of-fit of the probit model to the data must be performed (Robertson and Preisler 1992). A large $\chi^2$ value (at $n - 2$ df, where $n$ =

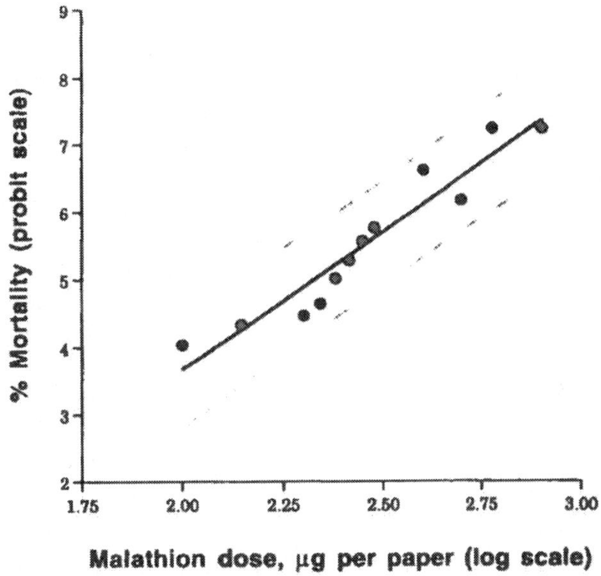


**Figure 1** Probit regression line for adults of the lesser grain borer, *Rhyzopertha dominica*, exposed to malathion-treated filter papers (Subramanyam, unpublished data). The dotted lines represent 95% confidence limits (CL) around the regression line. At each of the 12 malathion doses, 80 adults (20 adults × 4 replicates) were exposed to treated papers. The probit regression estimates are as follows: intercept ± SE = −4.46 ± 0.95; slope ± SE = 4.06 ± 0.39; $LD_{50}$ (95% CL) on linear scale = 218.95 µg per paper (192.63 − 244.56); $LD_{95}$ (95% CL) = 563.83 µg per paper (464.29 − 767.55). $\chi^2$ = 40.82; 10 df; $P$ <0.001.

number of doses) indicates poor fit of the probit model to the data. For the example in Fig. 1, the $\chi^2$ value of 40.82 at 10 df was highly significant ($P < 0.05$) indicating a poor fit of the probit model to *R. dominica* data. Therefore, when initiating dose-mortality tests, it is important to use insects of approximately similar age and weight, minimize errors in application of insecticide to insects or substrates, increase number of replicates used at each dose, use at least 120–240 insects per dose, randomize insects among treatments (across doses), use a healthy colony of insects, and conduct tests at controlled temperature and relative humidity.

Probit analysis performed on field strains is compared with similar analysis performed on strains maintained in the laboratory without any insecticide selection. The magnitude of resistance in field strains relative to an insecticide-susceptible laboratory strain is determined as a ratio of the $LD_{50}$ of the field and laboratory strains, and this ratio is commonly referred to as the resistance ratio (RR). The $LD_{50}$s are used for comparison because the 95% CL for mortality at

$LD_{50}$ are narrower than the 95% CL at the other lethal doses. In other words, the $LD_{50}$ estimates are more precise than other lethal dose estimates. The $LD_{50}$ of the field strain is considered to be significantly different from the $LD_{50}$ of the laboratory strain if the 95% CL for these lethal doses do not overlap. This comparison assumes that the probit regression slopes of the resistant and laboratory strains are similar (parallel). Because resistance ratios are affected by several factors (see Section V), Otto et al. (1992) discourage the use of resistance ratios and recommend the use of spline regression to evaluate resistance in the field.

As resistance evolves, the slope of the probit regression line becomes less steep, and slopes of the resistant and laboratory strains may not be parallel (Fig. 2). For example, Subramanyam et al. (1989) reported a slope of 10.70 for an insecticide-susceptible laboratory strain of the red flour beetle, *Tribolium castaneum* (Herbst), exposed to malathion-impregnated filter papers. In four field strains of the same species that were 5.9–40.6 times resistant than the laboratory strain at the $LD_{50}$, the slope values ranged from 1.79–2.66. The increasing frequency of RS genotypes relative to SS genotypes results in lower slope values (Fig. 2) because higher doses are required to kill the heterozygotes. Chilcutt and Tabashnik (1995) reported that the slope was not a good indicator of genetic variation in susceptibility of insects to insecticides, and they suggested that shift (decrease) in the slope values during resistance evolution could be due to genetic

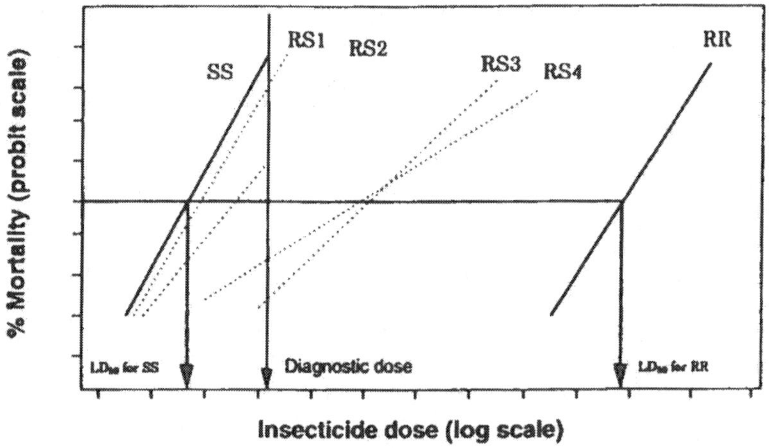


**Figure 2** A diagrammatic representation of the response of insects with SS, RS, and RR genotypes to a hypothetical insecticide. The probit regression slopes are steeper for SS and RR genotypes than for the RS genotypes. Most field-collected insects show a response represented by the RS lines. A diagnostic dose is the minimum effective dose of an insecticide that is able to detect RS individuals present at a low frequency in a random sample of insects.

and environmental factors and errors in estimating slopes from probit regressions. The $LD_{50}$ of the laboratory strain changes within a generation and across generations due to natural variation (Robertson et al. 1995), and as a result, resistance ratios for field strains tend to change with the $LD_{50}$ of the laboratory strain. Cogan (1982) showed that in a malathion-susceptible strain of the Indianmeal moth, *Plodia interpunctella* (Hübner), the dose for 50% knockdown ($KD_{50}$) of the moths in nine separate tests ranged from 12–20 parts per million (ppm), and the slopes ranged from 3.94–5.21. Because the resistance ratio does not have an error estimate, Robertson and Preisler (1992, p. 43–44) recommend constructing a 95% CL for the ratio based on probit regression y-intercept and slope, and variance–covariance matrixes of probit regression, all of which are produced by the POLO-PC program. If the 95% CL for the ratio includes a 1, then the $LD_{50}$s are not significantly different.

The survivors from the dose-mortality tests should be reared and the progeny tested similarly. If pure SS and RR strains are available, progeny resulting from crosses between these strains should be exposed to insecticides to determine lethal dose estimates. The progeny $LD_{50}$s are useful indicators of the progression, stability, and inheritance of resistance. These aspects are discussed in Section III. When insect response is measured over time at a fixed dose, probit analysis can be used to estimate the time for 50% mortality ($LT_{50}$). Valid estimates of $LT_{50}$ are obtained if independent samples of insects are observed at different time periods instead of repeated observations over time on the same cohort of insects. A complementary log–log model (CLL) is suitable for analyzing time-mortality data of insects (Robertson and Preisler 1992). Probit analysis is commonly used in quarantine entomology, where pest elimination procedures must be effective and prevent introduction of exotic pests. Robertson et al. (1993) criticized the use of $LD_{99.99}$ in quarantine entomology programs, and suggested alternative statistical methods for use in pest exclusion and eradication programs.

## 2. Diagnostic Dose Tests

Characterizing resistance using dose-mortality tests is time consuming, because insects have to be collected and reared in the laboratory, and large number of insects are required for testing. Generally, depending on the number of field strains tested, the process of collecting, rearing, and testing insects for resistance takes about one to two years. By this time, resistance levels of strains in the field may have increased. In this situation, laboratory assessments may underestimate field resistance. Another drawback of dose-mortality tests is that they fail to detect incipient resistance. For example, in Fig. 2, the SS line and RS1 line are close together, and a comparison of their $LD_{50}$s or $LD_{95}$ will fail to reveal low levels of resistant individuals present in the RS1 population because of overlapping 95% CL. In practical resistance management programs, it is important to develop techniques that can detect resistance at a phenotypic frequency of 1% (1 resistant individual in a random sample of 100 individuals), and these techniques should be quick, easy, and efficient (Roush and Miller 1986). Theoretical models predict that beyond the 1% resistance frequency, control failures could conceivably occur within one to six generations (Kable and Jeffrey 1980, Roush and Miller 1986). An alternative to the dose-mortality test is the use of a single dose, called the diagnostic dose.

The diagnostic dose is assumed to differentiate between susceptible (SS) and resistant (RS and RR) phenotypes (Table 3). Unlike the dose-mortality test, a diagnostic dose is useful in detecting incipient resistance (Roush and Miller 1986). Diagnostic doses are determined from probit regressions, and generally the $LD_{99}$, $2 \times LD_{99}$, or $LD_{99.99}$ is chosen as the diagnostic dose. The Food and Agricultural Organization (FAO) recommended methods for detection and measurement of resistance in stored-product insects to grain protectants (Champ and Dyte 1980), and to the fumigants methyl bromide and phosphine (Winks et al. 1980) uses the $LD_{99.99}$ as the diagnostic dose. Cogan (1982) described an apparatus for holding

**Table 3** Estimating Frequency of Phenotypically Resistant Adults of the Red Flour Beetle, *Tribolium castaneum*, from Mortality of Insects Exposed to a Diagnostic Dose of Malathion[a]

| Strain | *n*[b] | % Mortality (Mean ± SE) | % Phenotypically resistant individuals[c] |
|---|---|---|---|
| Laboratory[d] | 3 | 100.0 ± 0.0 | 0.0 |
| C-42[e] | 5 | 37.0 ± 12.5 | 63.0 |
| C-58 | 5 | 25.3 ± 12.8 | 74.7 |
| C-38 | 5 | 22.0 ± 7.7 | 78.0 |
| C-41 | 5 | 11.0 ± 4.3 | 89.0 |
| C-60 | 5 | 8.0 ± 3.0 | 92.0 |
| C-46 | 5 | 1.0 ± 1.0 | 99.0 |
| C-34 | 5 | 0.0 ± 0.0 | 100.0 |

[a]The diagnostic dose was 0.12 mg AI of malathion per 7-cm-diameter filter paper.

[b]Number of replicates. At each replicate, 20 adults of mixed ages and sexes were exposed.

[c]% Phenotypically resistant individuals = 100 − % mortality. The phenotypically resistant individuals may consist of RS, RR, or RS + RR genotypes.

[d]Strain reared continuously in the laboratory without insecticide exposure.

[e]Field strains collected during 1986 from farm-stored shelled maize in Minnesota, USA.

*Source*: Subramanyam and Harein (1990).

*P. interpunctella* adults, and a rapid method for detecting malathion resistance which involves exposing moths for 6 h to insecticide-coated 250 ml conical flasks. Subramanyam et al. (1989) have shown that the diagnostic doses of malathion, pirimiphos-methyl, and chlorpyrifos-methyl that consistently produced 100% mortality of the sawtoothed grain beetle, *Oryzaephilus surinamensis* (L.) and *T. castaneum* adults were 1.4–2.9 times lower than those estimated by doubling the $LD_{99}$ values. Therefore, experimentally obtained diagnostic doses are more effective in detecting insects with low levels of resistance. Halliday and Burnham (1990a) have shown that doses that produce less than 100% mortality of the susceptible strains are effective in detecting incipient resistance. They also provided two computer programs (BESTDOSE and RANGES) for determining the optimal diagnostic dose (Halliday and Burnham 1990b). Their methods are suitable when the probit regressions for SS and RS strains overlap.

Once a diagnostic dose is available, the number of individuals to be exposed to that dose is a function of the frequency of resistance and the probability of detecting resistance at that frequency (Fig. 3). The formula (Roush and Miller 1986) for calculating the number of individuals ($n$) required for the diagnostic dose test is

$$n = \frac{\log_{10}(1 - P)}{\log_{10}(1 - F)} \tag{5}$$

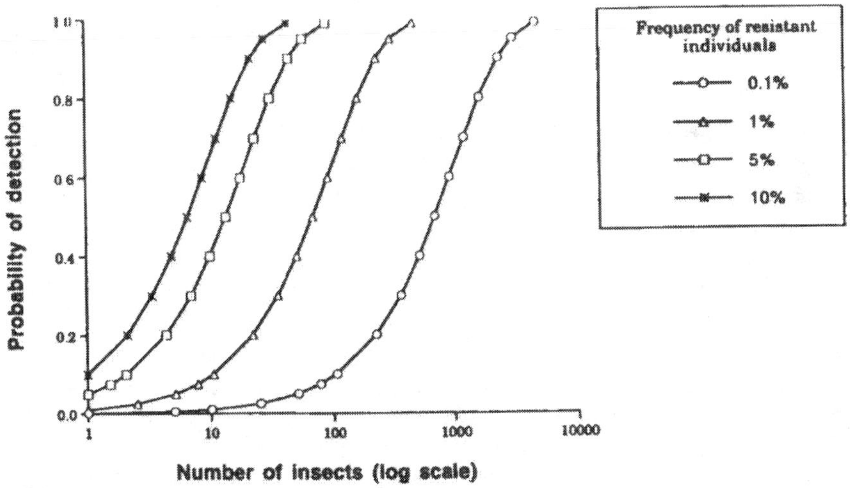


**Figure 3** Sample sizes required for detecting resistance in a random sample of insects. The number of insects required for testing is a function of the frequency of resistant individuals in the population and the probability (chance) of detecting at least one resistant individual.

where $P$ is the probability of detecting at least one resistant individual, and $F$ is the frequency of resistance in the population. Figure 3 shows the relationship between the probability of resistance detection and sample size at resistance frequencies of 0.1, 1, 5, and 10%. In general, a large number of individuals must be tested at low resistance frequencies and at higher detection probabilities. For example, if the resistance frequency in the population is 5% (i.e., 5 resistant individuals in a random sample of 100 individuals), at least 58 randomly selected individuals must be tested at the diagnostic dose to have a 95% chance of finding at least one resistant individual. The advantage of diagnostic dose tests is that field test kits can be developed for detecting and monitoring resistance quickly in the field. The test kits with instructions for use can be shipped to remote places for on-site evaluation of resistance. A field test kit containing filter papers treated with diagnostic doses of malathion, fenitrothion, and pirimiphos-methyl; plastic rings coated with polytetrafluoroethylene (Teflon®)*; and brush and forceps for handling insects, has been used for detecting resistance in field strains of the maize weevil, *Sitophilus zeamais* (L.) and *T. castaneum* in Zimbabwe, Mali, and Ghana (Jermannaud 1994).

## 3. Improved Resistance Detection Tests

Resistance evaluation procedures that simulate actual exposure of insects to insecticides in the field are needed. The evaluation of resistance in the laboratory using filter papers or topical treatment of insects (example, larvae of moths) with insecticides does not simulate exposure of insects to insecticides in the field. For example, dichlorvos resistance evaluations have been done by topically treating fully grown larvae of almond moth, *Cadra cautella* (Walker) and *P. interpunctella* (Zettler 1982). In the field, dichlorvos is used as a fog to kill the moths. Perhaps, larvae were chosen as the moths are too fragile to handle. Cogan's (1982) technique should enable testing the moths directly instead of fully grown larvae. Some species of insects associated with stored commodities feed exclusively on molds, and diets for mass rearing most of these species do not exist. The susceptibility of insects that feed on the molds and grain can be determined by treating the grain with an insecticide in the laboratory at a series of rates (including the labeled rate). Treated grain (100 grams) can be placed in pint size (0.47-liter) glass jars, along with jars containing untreated grain. The jars should be closed with lids and wire-mesh screens large enough to allow insect entry. These jars should be pushed into the top layers of the grain, such that the mesh screen is at the same level as the grain surface. A large number of jars containing insecticide-treated and untreated grain must be used, and a set of jars from each treatment removed at weekly intervals to determine number of live and dead insects in each jar. The grain

*To prevent insects from crawling on rings.

temperature and moisture should be recorded at the time of tests to correct insect mortalities for these environmental effects. This approach has great potential in studying the susceptibility or resistance to insecticides of insects that feed exclusively on molds.

Another approach is to treat grain in the laboratory at a series of doses, and expose insects collected from the field to the treated grain. A large number of insects (adults) can be collected by using probe traps (see Chapter 4). Weinzierl and Porter (1990) used this technique for detecting resistance in the hairy fungus beetle, *Typhaea stercorea* (L.), to malathion and pirimiphos-methyl. As no insecticide-susceptible strain was available, *T. stercorea* responses from treated grain (maize) were compared with responses of *T. stercorea* collected from insecticide-free maize (Table 4). In the same study, Weinzierl and Porter (1990) estimated the $LD_{50}$s and $LD_{95}$s for *T. castaneum* and *S. zeamias* collected from the grain. The practical advantage of grain assays is that the failure of insecticides to control insects can be predicted from assays on grain than from filter papers.

Collins and Wilson (1987) determined the relationship between filter paper and grain assays using the the probit regression estimates and resistance ratios. Table 5 shows the resistance ratios at the $LD_{50}$ for a field strain of *O. surinamensis* exposed to 12 different insecticides on filter papers and stored wheat. In general, resistance ratios on grain can be predicted from resistance ratios on filter papers (Collins and Wilson 1987). The slopes of the probit regressions were steeper for

**Table 4** Malathion and Pirimiphos-methyl Resistance in the Hairy Fungus Beetle, *Typhaea stercorea*, Determined by Directly Exposing Field-Collected Adults to Maize Treated with Insecticides

| Insecticide/Collection site | *n*[a] | $LD_{95}$ (95% Confidence Limits) (mg of AI per kg of maize) | Resistance ratio[b] |
|---|---|---|---|
| *Pirimiphos-methyl*[c] | | | |
| Livingston County[d] | 199 | 131.9 (71.5–503.0) | 87.9 |
| Franklin County[d] | 169 | 1.5 (1.1–3.7) | |
| *Malathion*[c] | | | |
| Livingston County | 233 | 909.0 (551.0–5,428.0) | 113.6 |
| Franklin County | 192 | 8.0 (6.1–15.3) | |

[a]Total number of adults used to generate the dose-mortality regression.
[b]Larger $LD_{95}$ ÷ smaller $LD_{95}$.
[c]The labeled rate of pirimiphos-methyl and malathion for use on maize is 6 and 10 mg/kg, respectively.
[d]These two counties are in the state of Illinois, USA.
*Source*: Weinzierl and Porter (1990).

**Table 5** Resistance Ratios for the Sawtoothed Grain Beetle, *Oryzaephilus surinamensis*, Exposed to Filter Papers and Wheat Treated With Insecticides

| | Resistance ratio[a] at $LD_{50}$ | | |
|---|---|---|---|
| Insecticide | Filter paper assay | Treated grain assay | Difference[b] |
| *Organophosphates* | | | |
| Fenitrothion | 172.3 | 92.0 | +80.3 |
| Chlorpyrifos-methyl | 3.1 | 5.7 | −2.6 |
| Methacrifos | 1.0 | 3.9 | −2.9 |
| Pirimiphos-methyl | 33.0 | 11.0 | −22.0 |
| Malathion | 16.0 | 31.0 | −15.0 |
| Dichlorvos | 3.9 | 3.0 | −0.9 |
| *Carbamate* | | | |
| Carbaryl | 150.0 | 200.0 | −50.0 |
| *Pyrethroids* | | | |
| Bioresmethrin | 3.0 | 3.5 | −0.5 |
| Permethrin | 3.2 | 3.9 | −0.7 |
| Cyfluthrin | 2.0 | 2.7 | −0.7 |
| Cypermethrin | 2.4 | 2.9 | −0.5 |
| Deltamethrin | 2.2 | 1.9 | +0.3 |

[a]Resistance ratio = $LD_{50}$ of resistant strain ÷ $LD_{50}$ of susceptible (laboratory) strain.
[b]Difference = resistance ratio of filter paper assay − resistance ratio of treated grain assay.
*Source*: Collins and Wilson (1987).

grain assays than for filter paper assays. However, the resistance ratios for pyrethroids from both assays were not correlated. Collins and Wilson (1987) compared the $LD_{99.9}$ obtained from grain assays for various insecticides with the degradation of the same insecticides on grain after nine months of storage (Table 6). These results showed that the recommended application rates of a majority of the insecticides (except for methacrifos and chlorpyrifos-methyl) would be ineffective in controlling the *O. surinamensis* field strain.

The mortality of strains of the confused flour beetle, *Tribolium confusum* Jacquelin du Val, at a diagnostic dose of 300 ppm was correlated with mortality of the same strains on galvanized steel and plywood treated with 1.63 g of active ingredient per 0.28 $m^2$ (Arthur and Zettler 1992). The mortality of insects on galvanized steel and plywood decreased as the resistance frequency increased. The mortality of *T. castaneum* strains at diagnostic doses did not correlate with mortality of strains exposed to insecticide-treated surfaces (Arthur and Zettler 1991).

In a majority of stored-product insects, the egg and pupal stages are more tolerant to phosphine than the adults (Hole et al. 1976). Practical doses, therefore,

**Table 6** Recommended Application Rate and Degradation of Insecticides on Wheat, and Insecticide Deposits on Wheat Required to Kill 99.9% Adults of the Sawtoothed Grain Beetle, *Oryzaephilus surinamensis*, Strains that Are Susceptible and Resistant to Fenitrothion

| Insecticide | Application rate for 9-month protection (mg/kg) | Residue remaining after 9 months of storage (mg/kg) | $LC_{99.9}$ (mg/kg) | |
|---|---|---|---|---|
| | | | Susceptible strain[a] | Resistant strain[b] |
| Fenitrothion | 12 | 4.5 | 0.6 | 24.0 |
| Chlorpyrifos-methyl | 10 | 3.5 | 0.7 | 2.5 |
| Methacrifos | 20 | 1.2 | 1.1 | 3.8 |
| Pirimiphos-methyl | 4 | 2 | 0.8 | 5.5 |
| Malathion | 18 | 6–7 | 0.9 | 51.0 |
| Carbaryl | 8 | 2.2 | 140.0 | 1000.0 |
| Bioresmethrin | 1[c] | 0.2–0.3 | 12.0 | 22.0 |
| Permethrin | 1[c] | 0.6 | 6.8 | 15.0 |
| Cyfluthrin | 2[c] | 1.2 | 2.2 | 6.3 |
| Cypermethrin | 4[c] | 3.5 | 2.6 | 6.1 |
| Deltamethrin | 1 | 0.9 | 0.9 | 1.2 |

[a]Strain VOS48.
[b]Strain QOS42.
[c]Insecticide mixed with 8 mg/kg of the synergist piperonyl butoxide (PBO).
*Source*: Collins and Wilson (1987).

should be designed to kill the most tolerant stages so that the adults are also eliminated. Thus, detecting resistance in adults to phosphine is of little practical value. Therefore, stage-specific resistance and correlation of resistance between stages (Bell et al. 1977) must be determined for detection and management of phosphine resistance. Pheromone-baited sticky traps laced with various doses or a diagnostic dose of an insecticide can provide a simple and effective means of monitoring resistance in the field, especially in actively flying moth pests (Haynes et al. 1986). This approach is gaining popularity for detection of resistance in stored-product moths. The resistance evaluation techniques must be ecologically relevant, and one must be able to predict field failures based on resistant risk assessment in the laboratory.

## III. STUDYING INHERITANCE OF RESISTANCE

### 1. Selecting Isogenic Lines

Resistance is an inherited phenomenon, and to study the mode of inheritance, field strains or insecticide-susceptible strains are selected in the laboratory with insecticides to obtain isogenic lines (SS and RR lines; see Fig. 2). This experimental selection for resistance is important for predicting the evolution of resistance to candidate insecticides, although laboratory selection experiments may not always accurately predict evolution of resistance in the field (Brown and Payne 1988). The rate of resistance development (RRD) can be predicted by the following equation:

$$\text{RRD} = \frac{\text{Resistant strain } \log_{10}(\text{LD}_{50}) - \text{Parent strain } \log_{10}(\text{LD}_{50})}{n} \tag{6}$$

where, $n$ = number of generations required to reach maximum resistance. RRD is an exponential rate constant, and number of generations selected ($n$) raised to this power indicates the magnitude of resistance development. For example, a value of 1.5 in 10 generations indicates that the resistance has increased 31.6-fold in 10 generations. Equation 6 is useful for comparing the rate of resistance development to different insecticides or insecticide mixtures. The same equation can be used to detect stability of resistance in the absence of insecticide selection (Tabashnik 1992). If the equation is used to predict the stability of resistance, $n$ is the number of generations without insecticide selection, and in the numerator, resistant strain is replaced by "final" and parent strain by "initial." The final $\log_{10}(\text{LD}_{50})$ is the resulting $\text{LD}_{50}$ in the absence of selection, and the initial $\log_{10}(\text{LD}_{50})$ is the $\text{LD}_{50}$ of the selected strain. The stability of resistance is determined by relaxing the selection pressure for 10 generations. Generally, selection work must be performed on field strains because they are genetically more adaptable than laboratory strains.

Brown and Payne (1988) outlined procedures for conducting laboratory selection experiments. Briefly, the procedure involves collecting a large number of insects from various field sites to increase the probability of detecting individuals with resistant genes. Initially, a dose capable of killing $\geq$ 80% of the exposed population must be used. In each generation, insects surviving exposure should be reared, and the progeny tested at the same or higher doses, until resistance stabilizes. During the selection process, data must be collected on the biological parameters of the unselected and selected insects to determine any changes in their fitness due to resistance development. Biological parameters of importance are the number of eggs laid, egg viability, developmental time, and longevity and survival of immature and mature stages.

The selection experiments are not necessarily limited to pest species. The natural enemies (parasitoids and predators) of the insect pests also can be selected to tolerate high insecticide doses, so that these beneficials are protected when released into insecticide-treated areas. Parasites and predators that are selected for insecticide resistance can be used in conjunction with insecticides for management of stored-product insects. This approach has not been explored with natural enemies of stored-product insects, although malathion resistance was documented in two hymenopterous parasitoids (Baker 1995), and an Anthocorid predator (Baker and Arbogast 1995). The protocols for selection of insecticide resistance in natural enemies are discussed by Johnson and Tabashnik (1994).

## 2. Dominance of Resistance

When the genotypes, SS, RS, and RR can be distinguished by phenotype or based on responses to an insecticide, then the alleles (a pair of genes) are said to be codominant, incompletely dominant, or partially dominant. If the allele R is completely dominant to S, then there are only two phenotypes, dominants and recessives. All recessives will be of SS genotype, and the dominant may be either RS or RR, the latter genotypes cannot be distinguished by their phenotypes (Doolittle 1986). Dipterans and coleopterans show a XY sex determination, with female individuals carrying the XX chromosomes and male individuals carrying the XY chromosomes. The X chromosome being larger carries some gene loci which do not occur in the Y chromosome. Such loci are said to be sex-linked. Therefore, females carry two copies of the genes at a sex-linked locus, and can be homozygous for any allele, or heterozygous for any pair of alleles, just as for any locus on an autosome (Doolittle 1986). An autosome is a chromosome other than X or Y. Males carry only one gene at the sex-linked locus. Therefore, males are haploid for sex-linked genes and females are diploid, and males and females differ in their inheritance of sex-linked traits.

Georghiou (1969) outlined a procedure for determining dominance as well as determining if resistance is associated with an autosome or a sex chromosome. First, the $F_1$ progeny are produced by crossing susceptible (SS) females or males with resistant (RR) males or females. The probit regression line for the $F_1$ progeny is determined. The $F_1$ progeny are mated to produce $F_2$ progeny, and a probit regression line is obtained for the $F_2$ progeny. Both the $F_1$ and $F_2$ progenies are backcrossed with the susceptible parent. If a single gene is involved in the resistance, the probit regression line plateaus at 25% and 75% mortality. In the $F_1$ backcross progeny, the regression will plateau at 50% mortality. Also, lack of change in the mortality of progeny resulting from repeated backcrosses to the susceptible parent shows that the resistance is monogenic. Polygenic resistance is weakened or lost by repeated backcrossing procedure (Crow 1957). The degree of dominance ($D$) can be calculated from the probit-regression analysis of susceptible, resistant, and $F_1$ progeny resulting from crosses between the susceptible and resistant parents (Stone 1968; Markwick 1994).

$$D = \frac{2X_2 - X_1 - X_3}{X_1 - X_3} \tag{7}$$

where $X_1$ is the $\log_{10}(\mathrm{LD}_{50})$ of the resistant parent, $X_2$ is the $\log_{10}(\mathrm{LD}_{50})$ of the $F_1$ progeny, and $X_3$ is the $\log_{10}(\mathrm{LD}_{50})$ of the susceptible parent. The dominance is recessive if $D = -1$, dominant if $D = 1$, and incomplete if $D$ is between $-1$ and 0 (incompletely recessive) or between 0 and 1 (incompletely dominant). Preisler et al. (1990) provided an equation for calculating the standard error and 95% CL associated with $D$. The mortality of progeny from repeated backcrosses or reciprocal crosses at a given insecticide dose depends on the genotype frequencies. If the progeny contain SS, RS, and RR genotypes, then at a given dose, 25, 50, and 25% mortality of these genotypes, respectively, is expected. Similarly, if the backcross results in SS and RS genotypes, then the expected mortality is 50% for each genotype. If the reciprocal crosses consist of RR and RS genotypes, 50% mortality of each genotype is expected. The observed mortality ($o$) of the progeny at a given insecticide dose is tested for departure (null hypothesis, $H_o$: The observed and expected mortality of insects are similar) from the expected mortality using a $\chi^2$ test as

$$\chi^2 = \frac{(o - np)^2}{np(1 - p)} \tag{8}$$

where, $o$ is the observed number of dead insects, $n$ is the total number of insects exposed, and $p$ is the expected proportion of dead insects. The $\chi^2$ test has 1 df. Preisler et al. (1990) presented a modified $\chi^2$ test that incorporates extra sources of variation between replicates at the same dose. Tabashnik (1991) provided guidelines for improving the design and interpretation of backcross experiments. In order to study the mode of resistance inheritance in stored-product insects, it is essential to sex insects in the adult stage. Halstead (1963) outlined external sex differences in stored-product coleoptera. Stored-product moths, especially the pyralids, can be sexed in the pupal stage using the characters described in Butt and Cantu (1962).

### 3. Realized Heritability

Estimates of heritability obtained by selection are called realized heritability ($h^2$), and $h^2$ estimates are useful in predicting the results of selection. The mode of inheritance of resistance and the number of genes involved in resistance are not considered when estimating $h^2$. Realized heritability is calculated as

$$h^2 = \frac{R}{S} \tag{9}$$

where $R$ is the response to selection and $S$ is the selection differential (Doolittle 1987), and $h^2$ is the proportion of phenotypic variance explained by the additive genetic variance (Firko and Hayes 1990). If the inheritance of a trait (for resistance) is strictly additive, and if dominance and environmental effects on the genotype were nil or negligible, then the $R/S$ ratio would be 1. However, dominance and environmental effects influence phenotypic values. Higher $h^2$ values indicate lower phenotypic variation and higher genetic variation. The $LD_{50}$ and slope values before and after selection, and percentage of the population surviving selection in each generation, are used for calculating $h^2$ (Tabashnik 1994). The response to selection (R) is calculated as

$$R = \frac{\log_{10}(\text{final LD}_{50}) - \log_{10}(\text{initial LD}_{50})}{n} \tag{10}$$

Initial $LD_{50}$ is the $LD_{50}$ of parents before insecticide selection, and final $LD_{50}$ is the $LD_{50}$ of the progeny after selection for $n$ generations. The selection differential (S) is the difference in mean phenotype between the selected parents and the entire parental generation (Tabashnik 1992), and is calculated as

$$S = i\sigma_p \tag{11}$$

where $i$ is the intensity of selection and $\sigma_p$ calculated as $[(\text{initial slope} + \text{final slope})/2]^{-1}$, is the phenotypic standard deviation. The intensity of selection is calculated as $p/z$, where $p$ is the percentage of insects surviving selection, and $z$ is the height of the ordinate of a normal distribution at the selection threshold. Tabashnik and McGaughey (1994) gave the following equation to calculate $i$ for $p$ between 10 and 80%

$$i = 1.583 - 0.0193336p + 0.0000428p^2 + \frac{3.65194}{p} \tag{12}$$

To compare the risk of resistance development among single insecticides, between mixtures, and sequences of two insecticides, $h^2$ estimates can be used (Tabashnik 1992, 1994).

## IV. RESISTANCE STATUS OF STORED-PRODUCT INSECTS

The extent of the resistance problem in stored-product insects can be measured by a number of criteria including number of species that are resistant, their geographical distribution, number of insecticides and fumigants to which resistance has developed, number of different types of storage and processing facilities in which resistance has been discovered, level of resistance, and extent to which the level of resistance is increasing. Several characteristics of stored-product ecosystems contribute to the slow development of resistance in field strains of stored-product insects (Parkin 1965; Dyte and Blackman 1970). Pesticide applications rarely cover all areas of the storage facility, and susceptible insects survive in untreated areas. The turnover of commodities in a facility results in susceptible insects being introduced into new commodities. Finally, fumigation, which is the most commonly used control measure, can eliminate insects that are resistant and susceptible to the grain protectants. Speirs et al. (1967) showed that a 9-fold resistance to malathion in field strains of *T. castaneum* developed slowly over a four-year period (Table 7), and the decrease in susceptibility of field strains correlated well with malathion use.

There are numerous reports from various countries on the widespread resistance in several stored-product insects to the organophosphate grain protectants

**Table 7** Relationship Between Malathion Use and Development of Resistance in Adults of the Red Flour Beetle, *Tribolium castaneum*

| Years of malathion use | No. strains | Average[a] $LD_{50}$ (μg/g/adult) | Resistance ratio[b] |
|---|---|---|---|
| 0 | 3 | 62.5 | 1.8 |
| 2 | 3 | 163.8 | 4.6 |
| 3 | 3 | 313.3 | 8.8 |
| 4 | 1 | 313.7 | 8.8 |

[a] $LD_{50}$s averaged across strains.
[b] Average $LD_{50}$ of field strains ÷ $LD_{50}$ of laboratory strain (35.6 μg/g/adult).
*Source*: Modified from Speirs et al. (1967).

and fumigants. A majority of the resistance reports, including the method of testing, have been summarized in Table 8. A brief discussion of the magnitude of resistance and cross resistance* problem in stored-product insects to insecticides is provided, along with notes on the factors affecting resistance, fitness of the resistant individuals, and inheritance of resistance. The resistance reports are discussed by country, and the literature from numerous countries is included to show similarities and differences in the resistance levels and resistance mechanisms of a given species collected from warehouses, farms, elevators, and food-processing plants.

### 1. Various Countries

During 1972–1973, stored-product insect samples from 85 countries were tested for resistance to malathion, lindane, phosphine, and methyl bromide using the FAO approved test methods (Champ and Dyte 1977). The insect species tested included the following: rice weevil, *Sitophilus oryzae* (L.), granary weevil, *Sitophilus granarius* (L.), *R. dominica*, *T. castaneum*, *T. confusum*, *O. surinamensis*, and merchant grain beetle, *Oryzaephilus mercator* (Fauvel). The type of malathion resistance (specific or nonspecific) was confirmed by the use of the synergist† triphenyl phosphate (TPP), which inhibits the carboxylesterase enzyme.

*Cross resistance involves cases where a single defense mechanism confers resistance against insecticides. Multiple resistance involves cases where different mechanisms confer resistances to various insecticides (Georghiou 1972). In the text, cross and multiple resistance are used interchangeably.
†Synergists are chemicals that are not toxic but increase the toxicity of insecticides to insects by suppressing enzymes responsible for detoxifying insecticides. Triphenyl phosphate (TPP) specifically suppresses the carboxylesterase enzyme, which is responsible for detoxifying malathion. Piperonyl butoxide (PBO) is a synergist that inhibits mixed function oxidases, and S,S,S, tributylphosphorotrithioate (TBPT or DEF) inhibits esterases and oxidases. Generally, in diagnostic dose tests, 5 parts of the synergist are added to 1 part of the insecticide (w:w).

Insects with carboxylesterase-dependent malathion resistance are said to possess malathion-specific resistance as these insects are susceptible to related insecticides. Resistance in insects due to factors other than carboxylesterase is said to be of the nonspecific type. Malathion resistance was detected in all species, with a large portion of the *T. castaneum* strains testing positive for malathion resistance followed by *T. confusum* and *R. dominica*. Nonspecific malathion resistance was also present in a few strains of *T. castaneum*, *T. confusum*, and *R. dominica* (Dyte 1991). Strains with nonspecific response to malathion tend to show resistance to other organophosphates (Champ and Campbell-Brown 1970; Pieterse et al. 1972). Lindane resistance was also widespread among the strains tested. Resistance to phosphine and methyl bromide was detected in only a few strains (Table 8). Research has shown that sublethal fumigations, such as in a leaky facility, predisposes insects to phosphine resistance. However, survivors of such fumigations are less fecund and fertile than the insecticide-susceptible adults. This has been documented with *T. castaneum*, *S. granarius*, Mediterranean flour moth, *Ephestia kuehniella* (Zeller), and *C. cautella* (Howe 1973; Saxena and Bhatia 1980; Al-Hakkak et al. 1985).

Table 9 shows the geographical distribution of malathion, lindane, phosphine, and methyl bromide resistance (Champ and Dyte 1977). Resistance to lindane and malathion was prevalent in several countries, and resistance to the fumigants methyl bromide and phosphine was detected in only a few countries.

Cogan (1982) tested 15 strains of adult *P. interpunctella* representing 15 countries, and found malathion resistance in five of the strains. These resistant strains were collected or obtained from Kenya, Nigeria, Switzerland, South Africa, and the United States.

Adults of five insect species obtained from 18 tropical countries were tested for phosphine resistance using the FAO diagnostic dose test method (Winks et al. 1980). Insects surviving the diagnostic dose were tested using higher phosphine doses and longer exposure durations. Increasing both the dose and exposure period is important because some insect species are able to regulate the amount of phosphine uptake by entering into a state of depressed metabolism, often described as narcosis. Nakakita and Kuroda (1986) found that the larval, pupal, and adult stages of susceptible *T. castaneum* absorb more phosphine than resistant strains under aerobic conditions. Similar mechanisms are found in *R. dominica*, *C. ferrugineus*, and *O. surinamensis*.

High levels of phosphine resistance were detected in *T. castaneum*, *R. dominica*, and *S. oryzae*, and the mortality of each of these species at the diagnostic dose varied within and among countries (Taylor and Halliday 1986). Phosphine resistant strains can be killed by increasing the duration of exposure. In a Karachi strain of *T. castaneum* that was not susceptible at a diagnostic dose of phosphine of 0.04 mg/L, a dose of 0.3 mg/L for 7 days caused 99% mortality. This study showed that the presence of phosphine resistant strains in locations where phosphine was not used was due to the movement of resistant strains through grain trade. For instance, strains of phosphine resistant *R. dominica* collected in Sri

**Table 8** Resistance Spectrum of Economically Important Stored-Product Insects

| Insecticide | Insect species | Test method[a] | Number of insects per replicate | Number of strains Tested | Number of strains Resistant | Resistance ratio[b] | Reference |
|---|---|---|---|---|---|---|---|
| *Bacillus thuringiensis* | *Cadra cautella* | DR | 50 (eggs) | 1 | 1 | 3.0–7.8 | McGaughey and Beeman (1988)[f] |
| | *Plodia interpunctella* | DR | 50 (eggs) | 2 | 1 | 141.4 | McGaughey and Johnson (1992)[f] |
| Bioresmethrin | *Oryzaephilus surinamensis* | DR | 40 | 2 | 2 | 3.8–3.9 | Attia and Frecker (1984)[g] |
| | | DR | 40 | 1 | 1 | 3.0 | Collins and Wilson (1987)[g] |
| | *Rhyzopertha dominica* | DD | 40 | 77 | 0 | —[c] | Herron (1990)[g] |
| | *Tribolium castaneum* | DR | 40 | 2 | 2 | 4.2; 12.0[d] | Collins (1990)[h] |
| Bioresmethrin + 8 mg/kg PBO[e] | *Rhyzopertha dominica* | DR | 30–50 | 1 | 1 | 2.0 | Collins et al. (1993)[h] |
| | | DD | 30–50 | 45 | 12 | — | Collins et al. (1993)[h] |
| Carbaryl | *Oryzaephilus surinamensis* | DR | 40 | 2 | 2 | >14.0 | Attia and Frecker (1984)[g] |
| | | DR | 40 | 1 | 1 | 150.0 | Collins and Wilson (1987)[g] |
| | *Rhyzopertha dominica* | DD | 40 | 83 | 0 | — | Herron (1990)[g] |
| | *Tribolium castaneum* | DR | 40 | 2 | 2 | 18.0; 86.0[d] | Collins (1990)[h] |
| Chlorpyrifos-methyl | *Cadra cautella* | DR | 10–20 | 7 | 7 | 1.8–4.1 | Arthur et al. (1988)[i] |
| | *Cryptolestes ferrugineus* | DD | 40 | 43 | 0 | — | Muggleton et al. (1991)[g] |
| | *Cryptolestes* spp. | DD | 25 | 15 | 0 | — | Beeman and Wright (1990)[g] |
| | *Oryzaephilus surinamensis* | DR | 40 | 2 | 2 | 2.2–2.9 | Attia and Frecker (1984)[g] |
| | | DD | 50 | 37 | 33 | — | Muggleton (1986)[g] |
| | | DR | 40 | 1 | 1 | 3.1 | Collins and Wilson (1987)[g] |
| | | DD | 20 | 6 | 0 | — | Subramanyam et al. (1989)[g] |
| | | DD | 25 | 22 | 3 | — | Beeman and Wright (1990)[g] |
| | | DD | 40 | 46 | 18 | — | Herron (1990)[g] |
| | | DD | 40 | 58 | 58 | — | Muggleton et al. (1991)[g] |
| | | DD | 30–50 | 34 | 17 | — | Collins et al. (1993)[h] |

| | | | | | | | |
|---|---|---|---|---|---|---|---|
| Chlorpyrifos-methyl (continued) | *Plodia interpunctella* | DD | 20 | 10 | 7 | — | Beeman et al. (1982)[i] |
| | | DR | 10–20 | 8 | 3 | 0.8–1.3 | Arthur et al. (1988)[i] |
| | *Rhyzopertha dominica* | DD | 25 | 18 | 3 | — | Beeman and Wright (1990)[g] |
| | | DD | 20 | 12 | 12 | — | Zettler and Cuperus (1990)[j] |
| | *Tribolium castaneum* | DD | 20 | 17 | 0 | — | Halliday et al. (1988)[j] |
| | | DD | 20 | 4 | 0 | — | Subramanyam et al. (1989)[g] |
| | | DD | 25 | 20 | 1 | — | Beeman and Wright (1990)[g] |
| | | DR | 40 | 2 | 2 | 3.4; 1.7[d] | Collins (1990)[h] |
| | | DD | 20 | 8 | 0 | — | Zettler and Cuperus (1990)[j] |
| | | DD | 100 | 28 | 10 | — | Zettler (1991b)[j] |
| | *Tribolium confusum* | DD | 100 | 17 | 9 | — | Zettler (1991b)[j] |
| Cyfluthrin | *Oryzaephilus surinamensis* | DR | 40 | 1 | 1 | 2.0 | Collins and Wilson (1987)[g] |
| | *Tribolium castaneum* | DR | 40 | 2 | 2 | 4.1; 270.0[d] | Collins (1990)[h] |
| Cyfluthrin + 8 mg/kg PBO | *Tribolium castaneum* | DR | 40 | 2 | 2 | 5.1; 310.0[d] | Collins (1990)[h] |
| Cyhalothrin | *Tribolium castaneum* | DR | 40 | 2 | 2 | 4.2; 310.0[d] | Collins (1990)[h] |
| Cypermethrin | *Oryzaephilus surinamensis* | DR | 40 | 1 | 1 | 2.4 | Collins and Wilson (1987)[g] |
| | *Tribolium castaneum* | DR | 40 | 2 | 2 | 3.2; 130.0[d] | Collins (1990)[h] |
| Cypermethrin + 8 mg/kg PBO | *Tribolium castaneum* | DR | 40 | 2 | 2 | 3.8; 110.0[d] | Collins (1990)[h] |
| DDT | *Oryzaephilus surinamensis* | DR | 40 | 2 | 2 | 6.2–6.4 | Attia and Frecker (1984)[g] |
| | *Sitophilus oryzae* | DR | 25 | 2 | 2 | 16.6–17.9 | Santhoy and Morallo-Rejesus (1984)[j] |
| Deltamethrin | *Oryzaephilus surinamensis* | DR | 40 | 1 | 1 | 2.2 | Collins and Wilson (1987)[g] |
| | *Sitophilus oryzae* | DR | 40 | 1 | 1 | 2633.7 | Guedes et al. (1994)[g] |
| | *Tribolium castaneum* | DR | 40 | 2 | 2 | 19.0; 950.0[d] | Collins (1990)[h] |
| Deltamethrin + 8 mg/kg PBO | *Tribolium castaneum* | DR | 40 | 2 | 2 | 4.3; 240.0[d] | Collins (1990)[h] |

**Table 8** Continued

| Insecticide | Insect species | Test method[a] | Number of insects per replicate | Number of strains | | Resistance ratio[b] | Reference |
|---|---|---|---|---|---|---|---|
| | | | | Tested | Resistant | | |
| Dichlorvos | *Cadra cautella* | DR | 10 | 8 | 4 | 1.1–1.3 | Zettler (1982)[j] |
| | | DR | 10–20 | 6 | 6 | 1.4–2.3 | Arthur et al. (1988)[j] |
| | *Oryzaephilus surinamensis* | DR | 40 | 2 | 2 | 7.8–8.2 | Attia and Frecker (1984)[g] |
| | | DR | 40 | 1 | 1 | 3.9 | Collins and Wilson (1987)[g] |
| | *Plodia interpunctella* | DR | 10 | 10 | 6 | 1.1–1.9 | Zettler (1982)[j] |
| | | DR | 10–20 | 16 | 16 | 2.1–9.3 | Arthur et al. (1988)[j] |
| | *Tribolium castaneum* | DR | 10 | 4 | 4 | 1.1–2.7 | Zettler (1982)[j] |
| | | DD | 20 | 16 | 8 | — | Halliday et al. (1988)[j] |
| | | DD | 100 | 28 | 18 | — | Zettler (1991b)[j] |
| | *Tribolium confusum* | DD | 100 | 17 | 4 | — | Zettler (1991b)[j] |
| | *Rhyzopertha dominica* | DD | 20 | 12 | 12 | — | Zettler and Cuperus (1990)[j] |
| Etrimfos | *Cryptolestes ferrugineus* | DD | 40 | 43 | 3 | — | Muggleton et al. (1991)[g] |
| | *Oryzaephilus surinamensis* | DD | 40 | 58 | 46 | — | Muggleton et al. (1991)[g] |
| Fenitrothion | *Cryptolestes ferrugineus* | DD | 40 | 43 | 1 | — | Muggleton et al. (1991)[g] |
| | *Cryptolestes* spp. | DD | 30–50 | 10 | 0 | — | Collins et al. (1993)[h] |
| | *Oryzaephilus surinamensis* | DR | 40 | 2 | 2 | 164.0–244.0 | Attia and Frecker (1984)[g] |
| | | DR | 50 | 1 | 1 | 145.0 | Collins (1986)[h] |
| | | DD | 50 | 31 | 2 | — | Muggleton (1986)[g] |
| | | DR | 40 | 1 | 1 | 172.3 | Collins and Wilson (1987)[g] |
| | | DD | 40 | 64 | 32 | — | Herron (1990)[g] |
| | | DD | 40 | 58 | 6 | — | Muggleton et al. (1991)[g] |
| | | DD | 30–50 | 34 | 34 | — | Collins et al. (1993)[h] |
| | *Rhyzopertha dominca* | DD | 40 | 20 | 2 | — | Pacheco et al. (1991)[g] |
| | *Sitophilus granarius* | DD | 40 | 8 | 0 | — | Herron (1990)[g] |

| | | | | | | | |
|---|---|---|---|---|---|---|---|
| Fenitrothion (continued) | *Sitophilus oryzae* | DD | 40 | 40 | 3 | — | Herron (1990)[g] |
| | | DD | 40 | 20 | 1 | — | Pacheco et al. (1991)[g] |
| | | DR | 30–50 | 1 | 1 | 14.3 | Collins et al. (1993)[h] |
| | | DD | 30–50 | 18 | 0 | — | Collins et al. (1993)[h] |
| | *Sitophilus zeamais* | DD | 40 | 8 | 1 | — | Pacheco et al. (1991)[g] |
| | | DD | 50 | 12 | 7 | — | Jermannaud (1994)[g] |
| | *Tribolium castaneum* | DR | 40 | 2 | 2 | 9.5; 8.2[d] | Collins (1990)[h] |
| | | DD | 40 | 110 | 0 | — | Herron (1990)[g] |
| | | DD | 40 | 25 | 10 | — | Pacheco et al. (1991)[g] |
| | | DD | 30–50 | 43 | 0 | — | Collins et al. (1993)[h] |
| | | DD | 50 | 16 | 9 | — | Jermannaud (1994)[g] |
| | *Tribolium confusum* | DD | 40 | 20 | 0 | — | Herron (1990)[g] |
| Fenvalerate | *Tribolium castaneum* | DR | 40 | 2 | 2 | 4.4; 29.0[d] | Collins (1990)[h] |
| Flucythrinate | *Tribolium castaneum* | DR | 40 | 2 | 2 | 4.4; 77.0[d] | Collins (1990)[h] |
| Fluvalinate | *Tribolium castaneum* | DR | 40 | 2 | 2 | 2.6; 9.0[d] | Collins (1990)[h] |
| Iodofenfos | *Alphitobius diaperinus* | DD | 10 | 42 | 17 | — | Wakefield and Cogan (1991)[g] |
| Lindane | *Oryzaephilus mercator* | DD | 25 | 34 | 17 | — | Champ and Dyte (1977)[g] |
| | *Oryzaephilus surinamensis* | DD | 25 | 152 | 112 | — | Champ and Dyte (1977)[g] |
| | | DR | 40 | 2 | 2 | >24.0 | Attia and Frecker (1984)[g] |
| | *Rhyzopertha dominica* | DD | 25 | 137 | 91 | — | Champ and Dyte (1977)[g] |
| | *Sitophilus granarius* | DD | 25 | 157 | 35 | — | Champ and Dyte (1977)[g] |
| | *Sitophilus oryzae* | DD | 25 | 235 | 176 | — | Champ and Dyte (1977)[g] |
| | *Sitophilus zeamais* | DD | 25 | 186 | 146 | — | Champ and Dyte (1977)[g] |
| | *Tribolium castaneum* | DD | 25 | 497 | 463 | — | Champ and Dyte (1977)[g] |
| | *Tribolium confusum* | DD | 25 | 94 | 38 | — | Champ and Dyte (1977)[g] |
| Malathion | *Alphitobius diaperinus* | DD | 10 | 42 | 16 | — | Wakefield and Cogan (1991)[g] |
| | *Cadra cautella* | DR | 10 | 6 | 6 | 1.2–7.2 | Zettler et al. (1973)[i] |
| | | DR | 10 | 9 | 9 | 3.0–>13.0 | Zettler (1982)[i] |
| | *Cryptolestes ferrugineus* | DD | 25 | 23 | 3 | — | White and Loschiavo (1985)[g] |
| | | DD | 40 | 43 | 5 | — | Muggleton et al. (1991)[g] |

**Table 8** Continued

| Insecticide | Insect species | Test method[a] | Number of insects per replicate | Number of strains Tested | Number of strains Resistant | Resistance ratio[b] | Reference |
|---|---|---|---|---|---|---|---|
| Malathion (continued) | *Cryptolestes* spp. | DD | 25 | 44 | 5 | — | Haliscak and Beeman (1983)[g] |
| | *Oryzaephilus mercator* | DD | 25 | 36 | 3 | — | Champ and Dyte (1977)[g] |
| | *Oryzaephilus surinamensis* | DD | 25 | 183 | 14 | — | Champ and Dyte (1977)[g] |
| | | DD | 25 | 12 | 0 | — | Haliscak and Beeman (1983)[g] |
| | | DR | 15 | 1 | 1 | 4.2 | Saleem and Wilkins (1983)[j] |
| | | DR | 40 | 2 | 2 | 15.4–16.6 | Attia and Frecker (1984)[g] |
| | | DD | 50 | 95 | 28 | — | Muggleton (1986)[g] |
| | | DR | 40 | 1 | 1 | 16.0 | Collins and Wilson (1987)[g] |
| | | DR | 20 | 6 | 0 | 0.7–1.0 | Subramanyam et al. (1989)[g] |
| | | DD | 20 | 6 | 0 | — | Subramanyam et al. (1989)[g] |
| | | DD | 25 | 22 | 15 | — | Beeman and Wright (1990)[g] |
| | | DD | 40 | 65 | 33 | — | Herron (1990)[g] |
| | | DD | 40 | 58 | 11 | — | Muggleton et al. (1991)[g] |
| | *Plodia interpunctella* | DR | 10 | 7 | 7 | 177.0–206.0 | Zettler et al. (1973)[i] |
| | | DD | 20 | 15 | 5 | — | Cogan (1982)[k] |
| | | DD | 20 | 44 | 42 | — | Beeman et al. (1982)[i] |
| | | DR | 10 | 12 | 12 | 45.7–>114.0 | Zettler (1982)[i] |
| | *Rhyzopertha dominica* | DD | 25 | 158 | 50 | — | Champ and Dyte (1977)[g] |
| | | DD | 25 | 13 | 11 | — | Haliscak and Beeman (1983)[g] |
| | | DD | 40 | 4 | 0 | — | Horton (1984)[g] |
| | | DD | 20 | 12 | 12 | — | Zettler and Cuperus (1990)[j] |
| | | DD | 40 | 79 | 79 | — | Herron (1990)[g] |
| | | DD | 40 | 11 | 9 | — | Sayaboc and Acda (1990)[g] |
| | | DD | 40 | 20 | 14 | — | Pacheco et al. (1991)[g] |

| | | | | | | | |
|---|---|---|---|---|---|---|---|
| Malathion (continued) | *Sitophilus granarius* | DD | 25 | 163 | 14 | — | Champ and Dyte (1977)[g] |
| | | DD | 40 | 2 | 0 | — | Horton (1984)[g] |
| | | DD | 40 | 7 | 0 | — | Herron (1990)[g] |
| | *Sitophilus oryzae* | DD | 25 | 257 | 33 | — | Champ and Dyte (1977)[g] |
| | | DD | 25 | 21 | 7 | — | Haliscak and Beeman (1983)[g] |
| | | DD | 40 | 9 | 0 | — | Horton (1984)[g] |
| | | DD | 40 | 39 | 16 | — | Herron (1990)[g] |
| | | DD | 40 | 20 | 14 | — | Pacheco et al. (1991)[g] |
| | *Sitophilus zeamais* | DD | 25 | 203 | 6 | — | Champ and Dyte (1977)[g] |
| | | DD | 40 | 25 | 0 | — | Horton (1984)[g] |
| | | DD | 40 | 38 | 0 | — | Sayaboc and Acda (1990)[g] |
| | | DD | 40 | 8 | 0 | — | Pacheco et al. (1991)[g] |
| | *Tribolium castaneum* | DR | 10 | 10 | 8 | 1.8–11.3 | Speirs et al. (1963)[j] |
| | | DD | 25 | 505 | 439 | — | Champ and Dyte (1977)[g] |
| | | DR | 20 | 1 | 1 | 44.0 | Bansode and Campbell (1979)[g] |
| | | DD | 50 | 56 | 51 | — | Osman and Morallo-Rejesus (1981)[g] |
| | | DR | 10 | 4 | 4 | 3.5–129.6 | Zettler (1982)[j] |
| | | DR | 30 | 1 | 1 | 73.0 | Beeman (1983)[g] |
| | | DD | 25 | 36 | 31 | — | Haliscak and Beeman (1983)[g] |
| | | DD | 40 | 28 | 16 | — | Horton (1984)[g] |
| | | DR | 30 | 3 | 3 | 3.6–111.0 | Beeman and Nanis (1986)[g] |
| | | DD | 20 | 15 | 13 | — | Halliday et al. (1988)[j] |
| | | DR | 40 | 1 | 1 | 405.6 | White and Bell (1988)[g] |
| | | DR | 20 | 4 | 4 | 5.9–40.6 | Subramanyam et al. (1989)[g] |
| | | DD | 20 | 4 | 4 | — | Subramanyam et al. (1989)[g] |
| | | DD | 25 | 22 | 22 | — | Beeman and Wright (1990)[g] |
| | | DR | 40 | 2 | 2 | 18.0; 15.0[d] | Collins (1990)[h] |

**Table 8** Continued

| Insecticide | Insect species | Test method[a] | Number of insects per replicate | Number of strains | | Resistance ratio[b] | Reference |
|---|---|---|---|---|---|---|---|
| | | | | Tested | Resistant | | |
| Malathion (continued) | | DD | 40 | 110 | 110 | — | Herron (1990)[g] |
| | | DD | 40 | 61 | 61 | — | Sayaboc and Acda (1990)[g] |
| | | DR | 20 | 7 | 7 | 6.1–46.3 | Subramanyam and Harein (1990)[g] |
| | | DD | 20 | 7 | 7 | — | Subramanyam and Harein (1990)[g] |
| | | DD | 20 | 8 | 8 | — | Zettler and Cuperus (1990)[j] |
| | | DD | 40 | 25 | 25 | — | Pacheco et al. (1991)[g] |
| | | DD | 50 | 15 | 14 | — | Jermannaud (1994)[g] |
| | | DD | 100 | 28 | 26 | — | Zettler (1991b)[j] |
| | *Tribolium confusum* | DD | 25 | 122 | 78 | — | Champ and Dyte (1977)[g] |
| | | DD | 40 | 3 | 0 | — | Horton (1984)[g] |
| | | DD | 40 | 20 | 7 | — | Herron (1990)[g] |
| | | DD | 100 | 17 | 14 | — | Zettler (1991b)[j] |
| | *Typaea stercorea* | DR | 10 | 1 | 1 | 65.7 | Weinzierl and Porter (1990)[l] |
| Methacrifos | *Oryzaephilus surinamensis* | DR | 40 | 1 | 0 | — | Collins and Wilson (1987)[g] |
| | | DD | 40 | 58 | 31 | — | Muggleton et al. (1991)[g] |
| | *Tribolium castaneum* | DR | 40 | 2 | 2 | 2.4; 2.3[d] | Collins (1990)[h] |
| Methyl bromide | *Oryzaephilus mercator* | DD | 25 | 14 | 0 | — | Champ and Dyte (1977) |
| | | DD | 25 | 60 | 1 | — | Champ and Dyte (1977) |
| | *Rhyzopertha dominica* | DD | 25 | 93 | 1 | — | Champ and Dyte (1977) |
| | *Sitophilus granarius* | DD | 25 | 98 | 0 | — | Champ and Dyte (1977) |
| | *Sitophilus oryzae* | DD | 25 | 140 | 1 | — | Champ and Dyte (1977) |
| | *Sitophilus zeamais* | DD | 25 | 109 | 6 | — | Champ and Dyte (1977) |
| | *Tribolium castaneum* | DD | 25 | 288 | 8 | — | Champ and Dyte (1977) |
| | *Tribolium confusum* | DD | 25 | 92 | 25 | — | Champ and Dyte (1977) |

| | | | | | | | |
|---|---|---|---|---|---|---|---|
| Permethrin | *Oryzaephilus surinamensis* | DR | 40 | 1 | 1 | 3.2 | Collins and Wilson (1987)[g] |
| | *Tribolium castaneum* | DR | 40 | 2 | 2 | 6.1; 21.0[d] | Collins (1990)[h] |
| d-Phenothrin | *Tribolium castaneum* | DR | 40 | 2 | 2 | 8.3; 25.0[d] | Collins (1990)[h] |
| Phosphine | *Cadra cautella* | DD | 10 | 18 | 3 | — | Zettler et al. (1989) |
| | | DR | 20 | 3 | 3 | 1.7–2.7 | Zettler (1991a) |
| | *Cryptolestes* spp. | DD | 40 | 6 | 6 | — | Sartori et al. (1991) |
| | *Lasioderma serricorne* | DD | 50 | 1 | 1 | — | Rajendran and Narasimhan (1994) |
| | *Oryzaephilus mercator* | DD | 25 | 13 | 0 | — | Champ and Dyte (1977) |
| | *Oryzaephilus surinamensis* | DD | 25 | 58 | 1 | — | Champ and Dyte (1977) |
| | | DD | 40 | 2 | 1 | — | Taylor and Halliday (1986) |
| | | DD | 40 | 64 | 1 | — | Herron (1990) |
| | | DD | 100 | 7 | 0 | — | White and Lambkin (1990) |
| | *Plodia interpunctella* | DD | 10 | 7 | 4 | — | Zettler et al. (1989) |
| | | DR | 20 | 4 | 2 | 1.4–2.7 | Zettler (1991a) |
| | *Rhyzopertha dominica* | DD | 25 | 94 | 22 | — | Champ and Dyte (1977) |
| | | DD | 50 | 112 | 3 | — | Attia and Greening (1981) |
| | | DD | 40 | 22 | 17 | — | Taylor and Halliday (1986) |
| | | DD | 40 | 82 | 9 | — | Herron (1990) |
| | | DD | 100 | 32 | 16 | — | White and Lambkin (1990) |
| | | DD | 20 | 12 | 8 | — | Zettler and Cuperus (1990) |
| | | DD | 40 | 16 | 16 | — | Sartori et al. (1991) |
| | *Sitophilus granarius* | DD | 25 | 85 | 8 | — | Champ and Dyte (1977) |
| | | DD | 40 | 7 | 1 | — | Herron (1990) |
| | *Sitophilus oryzae* | DD | 25 | 135 | 8 | — | Champ and Dyte (1977) |
| | | DD | 40 | 8 | 4 | — | Taylor and Halliday (1986) |
| | | DD | 40 | 38 | 7 | — | Herron (1990) |
| | | DD | 100 | 15 | 7 | — | White and Lambkin (1990) |
| | | DD | 40 | 6 | 5 | — | Sartori et al. (1991) |

**Table 8** Continued

| Insecticide | Insect species | Test method[a] | Number of insects per replicate | Number of strains | | Resistance ratio[b] | Reference |
|---|---|---|---|---|---|---|---|
| | | | | Tested | Resistant | | |
| Phosphine (continued) | *Sitophilus zeamais* | DD | 25 | 105 | 0 | — | Champ and Dyte (1977) |
| | | DD | 40 | 4 | 1 | — | Taylor and Halliday (1986) |
| | | DD | 40 | 16 | 5 | — | Sartori et al. (1991) |
| | *Tribolium castaneum* | DD | 25 | 267 | 15 | — | Champ and Dyte (1977) |
| | | DD | 50 | 90 | 11 | — | Attia and Greening (1981) |
| | | DD | 40 | 26 | 13 | — | Taylor and Halliday (1986) |
| | | DD | 10 | 23 | 8 | — | Zettler et al. (1989) |
| | | DD | 40 | 106 | 9 | — | Herron (1990) |
| | | DD | 100 | 36 | 7 | — | White and Lambkin (1990) |
| | | DD | 20 | 8 | 1 | — | Zettler and Cuperus (1990) |
| | | DD | 40 | 12 | 10 | — | Sartori et al. (1991) |
| | | DD | 100 | 28 | 13 | — | Zettler (1991b) |
| | *Tribolium confusum* | DD | 25 | 92 | 28 | — | Champ and Dyte (1977) |
| | | DD | 50 | 7 | 2 | — | Attia and Greening (1981) |
| | | DD | 40 | 20 | 1 | — | Herron (1990) |
| | | DD | 100 | 17 | 3 | — | Zettler (1991b) |
| | *Trogoderma granarium* | | | | | | |
| | First instar | DR | 10 | 1 | 1 | 40.7 | Borah and Chahal (1979) |
| | Second instar | DR | 10 | 1 | 1 | 13.7 | Borah and Chahal (1979) |
| | Third instar | DR | 10 | 1 | 1 | 11.7 | Borah and Chahal (1979) |
| | Fourth (final) instar | DR | 10 | 1 | 1 | 18.1 | Borah and Chahal (1979) |
| | Female pupae | DR | 10 | 1 | 1 | 2.6 | Borah and Chahal (1979) |
| | Male pupae | DR | 10 | 1 | 1 | 2.4 | Borah and Chahal (1979) |
| | Eggs | DR | 10 | 1 | 1 | 11.0 | Borah and Chahal (1979) |

| | | | | | | | |
|---|---|---|---|---|---|---|---|
| Pirimiphos-methyl | *Acanthoscelides obtectus* | DR | 10 | 2 | 2 | 49.4–149.8 | Sriharen et al. (1991)[g] |
| | *Cadra cautella* | DR | 10 | 8 | 8 | 1.7–3.5 | Zettler (1982)[j] |
| | | DR | 10–20 | 12 | 12 | 1.9–50.0 | Arthur et al. (1988)[j] |
| | *Cryptolestes ferrugineus* | DD | 40 | 43 | 0 | — | Muggleton et al. (1991)[g] |
| | *Cryptolestes* spp. | DD | 25 | 12 | 0 | — | Beeman and Wright (1990)[g] |
| | *Oryzaephilus surinamensis* | DR | 15 | 1 | 1 | 2.8 | Saleem and Wilkins (1983)[j] |
| | | DR | 40 | 2 | 2 | 22.0 | Attia and Frecker (1984)[g] |
| | | DD | 50 | 54 | 16 | — | Muggleton (1986)[g] |
| | | DR | 40 | 1 | 1 | 33.0 | Collins and Wilson (1987)[g] |
| | | DD | 20 | 6 | 0 | — | Subramanyam et al. (1989)[g] |
| | | DD | 25 | 8 | 1 | — | Beeman and Wright (1990)[g] |
| | | DD | 40 | 47 | 33 | — | Herron (1990)[g] |
| | | DD | 40 | 58 | 31 | — | Muggleton et al. (1991)[g] |
| | *Plodia interpunctella* | DR | 10 | 11 | 8 | 1.1–4.1 | Zettler (1982)[j] |
| | | DR | 10–20 | 11 | 11 | 1.4–3.2 | Arthur et al. (1988)[j] |
| | *Rhyzopertha dominica* | DD | 25 | 14 | 0 | — | Beeman and Wright (1990)[g] |
| | | DD | 40 | 20 | 1 | — | Pacheco et al. (1991)[g] |
| | *Sitophilus granarius* | DD | 40 | 7 | 0 | — | Herron (1990)[g] |
| | *Sitophilus oryzae* | DD | 40 | 23 | 0 | — | Herron (1990)[g] |
| | | DD | 40 | 20 | 3 | — | Pacheco et al. (1991)[g] |
| | | DR | 10 | 2 | 2 | 196.9–210.2 | Sriharen et al. (1991)[g] |
| | *Sitophilus zeamais* | DD | 40 | 8 | 0 | — | Pacheco et al. (1991)[g] |
| | | DD | 50 | 12 | 10 | — | Jermannaud (1994)[g] |
| | *Tribolium castaneum* | DR | 10 | 4 | 3 | 1.7–2.6 | Zettler (1982)[j] |
| | | DD | 20 | 17 | 1 | — | Halliday et al. (1988)[j] |
| | | DD | 20 | 4 | 0 | — | Subramanyam et al. (1989)[g] |
| | | DR | 40 | 2 | 2 | 6.7; 5.4[d] | Collins (1990)[h] |
| | | DD | 25 | 20 | 0 | — | Beeman and Wright (1990)[g] |
| | | DD | 40 | 81 | 1 | — | Herron (1990)[g] |
| | | DD | 40 | 25 | 9 | — | Pacheco et al. (1991)[g] |
| | | DD | 50 | 16 | 8 | — | Jermannaud (1994)[g] |

**Table 8** Continued

| Insecticide | Insect species | Test method[a] | Number of insects per replicate | Number of strains: Tested | Number of strains: Resistant | Resistance ratio[b] | Reference |
|---|---|---|---|---|---|---|---|
| Pirimiphos-methyl (continued) | *Tribolium confusum* | DD | 40 | 18 | 0 | — | Herron (1990)[g] |
| | *Typhaea stercorea* | DR | 10 | 1 | 1 | 38.2 | Weinzierl and Porter (1990)[l] |
| Pyrethrins, synergized | *Cadra cautella* | DR | 10 | 2 | 2 | 2.5–3.3 | Zettler et al. (1973)[i] |
| | | DR | 10–20 | 7 | 0 | 0.9–1.4 | Arthur et al. (1988)[i] |
| | *Oryzaephilus surinamensis* | DR | 40 | 2 | 2 | 2.3–2.4 | Attia and Frecker (1984)[g] |
| | *Plodia interpunctella* | DR | 10 | 2 | 1 | 2.5 | Zettler et al. (1973)[i] |
| | | DR | 10–20 | 20 | 18 | 1.4–1.8 | Arthur et al. (1988)[i] |
| | *Tribolium castaneum* | DD | 20 | 16 | 2 | — | Halliday et al. (1988)[j] |
| | | DD | 100 | 28 | 0 | — | Zettler (1991b)[j] |
| | *Tribolium confusum* | DD | 100 | 17 | 0 | — | Zettler (1991b)[j] |
| Resmethrin | *Tribolium confusum* | DD | 100 | 17 | 0 | — | Zettler (1991b)[j] |
| | | DD | 100 | 28 | 0 | — | Zettler (1991b)[j] |

[a]DR, Dose-response test. DD, Diagnostic-dose test. A strain showing 1–100% survival at the diagnostic dose was considered resistant.
[b]$LD_{50}$ of resistant field strain ÷ $LD_{50}$ of susceptible strain.
[c]Resistant ratios cannot be computed when a single (diagnostic) dose is used for determining resistance (see text for details).
[d]Resistance ratio of strains QTC285; QTC279 (both strains are resistant to cyfluthrin).
[e]PBO, Piperonyl butoxide, a synergist added to insecticides to enhance their activity.
[f]Egg-to-adult emergence on diet treated with *Bacillus thuringiensis* subsp. *kurstaki*.
[g]Adults exposed to insecticide-impregnated filter papers.
[h]Adults exposed to insecticide-treated wheat.
[i]Larvae (last instars) treated topically with insecticide solution(s).
[j]Adults treated topically with insecticide solution(s).
[k]Adults exposed to insecticide films on the inside of flasks.
[l]Adults exposed to insecticide-treated maize.

**Table 9** Geographical Distribution of Resistant Populations of Stored-Product Insects[a]

| Species | Insecticide | | | |
|---|---|---|---|---|
| | Lindane | Malathion | Methyl bromide | Phosphine |
| *Oryzaephilus mercator* | 11/18 | 3/19 | 0/8 | 0/10 |
| *Oryzaephilus surinamensis* | 6/52 | 10/53 | 1/23 | 1/23 |
| *Rhyzopertha dominica* | 41/51 | 23/50 | 1/45 | 11/46 |
| *Sitophilus granarius* | 14/28 | 4/28 | 0/24 | 8/23 |
| *Sitophilus oryzae* | 53/58 | 10/59 | 1/53 | 6/51 |
| *Sitophilus zeamais* | 46/48 | 5/52 | 6/37 | 0/37 |
| *Tribolium castaneum* | 75/76 | 75/78 | 6/69 | 13/69 |
| *Tribolium confusum* | 23/34 | 27/33 | 15/30 | 14/30 |

[a]The numerator shows the number of countries with resistant strains, and the denominator shows the number of countries from which the strains were collected for resistance testing.
*Source*: Champ and Dyte (1977).

Lanka originated in rice imported from Pakistan, where phosphine resistance in this species is widespread (Tyler et al. 1983). Movement of resistant strains through grain trade is a serious concern to importing countries. A strain of *C. ferrugineus* that was highly resistant to phosphine was intercepted in the United Kingdom on commodities imported from India (Dyte and Halliday 1985).

In Tanzania, where the larger grain borer, *Prostephanus truncatus* (Horn), is a devastating pest of maize and dried cassava, a combination of 3.3 ppm of permethrin and 16.6 ppm of pirimiphos-methyl is applied by farmers to control a wide variety of pests, including *P. truncatus*. In the laboratory, Golob et al. (1991) have shown that the Mexico, Tanzania, and Togo strains of *P. truncatus* can survive the recommended permethrin dose within three to four generations. After six generations of selection, survivors were apparent at the recommended pirimiphos-methyl dose. Dose-mortality tests indicated that the Mexico, Tanzania, and Togo strains were 3.5-, 10.5-, and 1.5-fold resistant to permethrin, respectively. Corresponding resistant ratios for the three strains selected with pirimiphos-methyl were 3.1, 3.1, and 11.1, respectively.

## 2. United States

Speirs et al. (1967) were the first to show malathion resistance in *T. castaneum* adults in the United States. In 8 of 10 field strains, 2 to 11-fold malathion resistance was detected, and there was a correlation between years of malathion use and resistance (Table 7). The resistance was 2 to 4-fold after two years of malathion use, and increased to 9-fold after three to four years of continuous malathion use.

Zettler et al. (1973) reported high levels of malathion resistance (> 177-fold) in *P. interpunctella* larvae, and low levels of resistance (< 8-fold) in *C. cautella* larvae. Low levels of resistance (< 4-fold) to synergized pyrethrins were detected in both species. Zettler (1974a) showed that five strains of *P. interpunctella* with 78 to >206-fold resistance to malathion were highly susceptible to pirimiphos-methyl.

In three of five field populations of *P. interpunctella* infesting almonds in California, Armstrong and Soderstrom (1975) found varying levels of resistance to malathion. Higher resistance levels were correlated with increased use of malathion. They also found that in the absence of malathion selection, the resistance levels were relatively stable. Continuous malathion pressure produced 110-fold resistance within four generations in a malathion susceptible (Fresno) strain.

Bansode and Campbell (1979) found a 44-fold malathion resistance in a composite field strain of *T. castaneum* from North Carolina that was collected from farm storage bins. The malathion resistance was TPP-suppressible, and this strain was not cross resistant to pirimiphos-methyl, fenitrothion, or bromophos. This TPP-suppressible malathion resistance, and lack of cross resistance to related organophosphates, has been independently reported by researchers in other countries (Dyte and Rowlands 1968; Dyte and Blackman 1972; Pasalu and Bhatia 1974).

Six insect species collected during 1979 and 1980 from stored grain in 34 of 46 counties of South Carolina were tested for resistance to malathion and pirimiphos-methyl (Horton 1984). Only malathion-specific resistance was detected in *T. castaneum*, and these *T. castaneum* strains were susceptible to pirimiphos-methyl. *Sitophilus oryzae*, *S. granarius*, *S. zeamais*, and *R. dominica* adults were susceptible to both malathion and pirimiphos-methyl.

All 39 of 44 *P. interpunctella* strains collected from farm-stored grain in northcentral states were 17-fold resistant to malathion, and the resistance was suppressed by TPP (Beeman et al. 1982). Elevated levels of the enzyme carboxylesterase were detected in the (larvae of) field strains when compared with the laboratory strain. Seven of 10 strains exposed to a diagnostic dose of chlorpyrifos-methyl survived, indicating a weak resistance to this compound. One *P. interpunctella* strain that was 200-fold resistant to malathion had low levels of carboxylesterase (Zettler 1974b). An explanation involving acetylcholinesterase was not tenable since the activity of this enzyme was similar in both the susceptible and resistant larvae. The physiological basis of malathion resistance in this strain appears to be due to enzymes other than carboxylesterase. Attia et al. (1980) reported on a similar nonspecific type of malathion resistance in certain *P. interpunctella* strains associated with stored grain in Australia.

Zettler (1982) tested fully grown larvae of *P. interpunctella* and *C. cautella*, and adults of *T. castaneum* against malathion, pirimiphos-methyl, and dichlorvos. The insect strains were collected during 1978 from peanut warehouses in Georgia,

Florida, and Alabama. High levels of malathion resistance were detected in *P. interpunctella* and *T. castaneum*, and low levels of resistance in *C. cautella*. Cross resistance to pirimiphos-methyl was evident in all three species. Field strains of all three species were susceptible to dichlorvos.

Haliscak and Beeman (1983) documented widespread malathion resistance in *T. castaneum* and *R. dominica*, collected from stored maize, wheat, and oats from 26 northcentral states. Malathion resistance in *T. castaneum* was suppressed by TPP, and resistance in *Cryptolestes* spp., *O. surinamensis*, and *Sitophilus* spp. was not widespread.

Resistance in *P. interpunctella* larvae to the bacterial insecticide, *Bacillus thuringiensis* subsp. *kurstaki* was documented for the first time by McGaughey (1985) through laboratory selection experiments. McGaughey (1985) also showed that *P. interpunctella* larvae collected from bins treated with this bacterial insecticide were less susceptible than those collected from untreated grain. This bacterial insecticide is not frequently used for controlling the lepidopterous pests associated with stored grain (Kenkel et al. 1994). This may be one of the reasons why field strains of moth larvae are not regularly monitored for resistance to *B. thuringiensis*. Subramanyam and Cutkomp (1985) discussed the mode of action and effectiveness of *B. thuringiensis* on moth larvae. Tabashnik (1994) reviewed the resistance and mechanisms conferring resistance in *P. interpunctella* to *B. thuringiensis*. The reduced binding of the bacterial toxin to the brush border membrane of the midgut epithelium is the primary mechanism of resistance to *B. thuringiensis* in *P. interpunctella* (Van Rie et al. 1990). A strain of *P. interpunctella* that was 141-fold resistant to *B. thuringiensis* subsp. *kurstaki* was cross resistant to 11 isolates of five serotypes of *B. thuringiensis* (McGaughey and Johnson 1987).

Four strains of *T. castaneum* and six strains of *O. surinamensis* collected from farm-stored barley in Minnesota during 1985 were tested for malathion resistance (Subramanyam et al. 1989). All *T. castaneum* strains were resistant to malathion. However, all *O. surinamensis* strains were highly susceptible to malathion. Although malathion resistance in *T. castaneum* strains was overcome by TPP, the increased mortality of adults in malathion + PBO treatments indicated that mixed function oxidases were involved to a lesser extent in malathion detoxification. Malathion resistant *T. castaneum* strains were not cross resistant to pirimiphos-methyl or chlorpyrifos-methyl. The lack of malathion resistance in *O. surinamensis* strains in spite of malathion use for three decades is interesting. All *O. surinamensis* strains were susceptible to pirimiphos-methyl, and four of six strains showed a low level of resistance to chlorpyrifos-methyl. Resistance to chlorpyrifos-methyl was suppressed by TBPT, and to a lesser degree by PBO, suggesting that nonspecific esterases and mixed-function oxidases were responsible for conferring resistance. A similar response to these three organophosphates was found in *T. castaneum* and *O. surinamensis* strains collected during 1985–

1986 from farm-stored maize (Subramanyam and Harein 1990). These are the first reports in the United States documenting resistance to chlorpyrifos-methyl in *O. surinamensis*, even before the approval of this insecticide for grain treatment. Therefore, field strains of *O. surinamensis* appear to be naturally tolerant to chlorpyrifos-methyl. Chlorpyrifos-methyl resistant strains of *O. surinamensis* were reported from the United Kingdom (Muggleton 1987; Muggleton et al. 1991) and Australia (Attia and Frecker 1984; Collins 1985).

Adults of *T. castaneum*, *O. surinamensis*, *R. dominica*, and *Cryptolestes* spp. collected in 1987 from farm-stored grain and grain stored in elevators throughout Kansas were screened for resistance to malathion, pirimiphos-methyl, and chlorpyrifos-methyl (Beeman and Wright 1990). *Cryptolestes* spp. were susceptible to both pirimiphos-methyl and chlorpyrifos-methyl. Among *O. surinamensis* populations, 14% were resistant to chlorpyrifos-methyl, 13% to pirimiphos-methyl, and 68.2% to malathion (see Table 8). About 17% of *R. dominica* and 5% of *T. castaneum* strains were resistant to chlorpyrifos-methyl. All *T. castaneum* strains were resistant to malathion and these strains were not cross resistant to pirimiphos-methyl.

*Plodia interpunctella* and *C. cautella* collected from 42 peanut warehouses and packaging plants in Georgia, Florida, and Alabama were tested for resistance to dichlorvos, malathion, pirimiphos-methyl, chlorpyrifos-methyl, and synergized pyrethrins (Arthur et al. 1988). Except for the lack of resistance to synergized pyrethrins in *C. cautella* larvae, larvae of both the species showed ≤ 13-fold resistance to all other insecticides (Table 8).

Halliday et al. (1988) tested *T. castaneum* strains collected from peanut warehouses in Georgia and Alabama with various organophosphate insecticides. About 87% of 15 strains were resistant to malathion. About 50% of 16 strains were resistant to dichlorvos. A previous survey (Zettler 1982) showed that *T. castaneum* strains were susceptible to dichlorvos. None of the strains showed resistance to chlorpyrifos-methyl, two strains showed resistance to synergized pyrethrins, and one to pirimiphos-methyl (Halliday et al. 1988).

Strains of *T. castaneum* and *T. confusum* originating from 42 flour mills in 21 states were collected during 1988–1989. These strains were evaluated for resistance to malathion, synergized pyrethrins, chlorpyrifos-methyl, dichlorvos, resmethrin, and phosphine. Strains of both the species were not resistant to synergized pyrethrins and resmethrin, but a few of the strains survived the diagnostic doses of phosphine, malathion, chlorpyrifos-methyl, and dichlorvos. Of the 28 *T. castaneum* strains tested, phosphine resistance was detected in 13 strains, while 26 strains were resistant to malathion, 10 to chlorpyrifos-methyl, and 18 to dichlorvos (see Table 8).

*Tribolium castaneum* and *R. dominica* adults were collected from 61 bins of stored wheat in 10 Oklahoma counties (Zettler and Cuperus 1990). Tests showed high levels of resistance in *T. castaneum* to malathion. Only one of the field strains was cross resistant to chlorpyrifos-methyl and phosphine. All 12 *R. dominica* populations showed resistance to malathion, and 8 of 12 populations showed phosphine resistance (Table 8).

Weinzierl and Porter (1990) documented a 38.2-fold resistance to pirimiphos-methyl in *T. stercorea* adults collected from pirimiphos-methyl-treated grain (maize) in Illinois. This strain showed 65.7-fold resistance to malathion. This is the first report in the United States of resistance in an insect species that feeds on storage molds. Because these insects are associated in nature with molds growing under the bark of trees (Linsley 1944), they may have physiological mechanisms to detoxify toxic mold metabolites (mycotoxins, which are complex molecules) as well as insecticides (Wright et al. 1982). The availability of suitable diets for rearing mold-feeding insects in the laboratory (Jacob 1988) will facilitate resistance evaluations on these species.

The diagnostic dose test method confirmed phosphine resistance in two of four *P. interpunctella* strains and two of two *C. cautella* strains. A 2.7-fold phosphine-resistant *P. interpunctella* strain had a probit regression slope of 7.57, which was similar to that of a susceptible strain (7.86). This indicated that resistance to phosphine in *P. interpunctella* was present at low levels. Three of 18 *C. cautella* strains, four of seven *P. interpunctella* strains, and eight of 23 *T. castaneum* strains, collected from peanut storage facilities in Southeastern United States, survived the diagnostic dose of phosphine indicating resistance (Zettler et al. 1989).

### 3. United Kingdom

Saleem and Wilkins (1983) found a low level of resistance in a field strain of *O. surinamensis* to malathion and pirimiphos-methyl. Muggleton (1987) showed a gradual increase in malathion resistance between 1971 and 1986 in field populations of *O. surinamensis*, *R. dominica*, *T. castaneum*, and *T. confusum*.

During 1987, *O. surinamensis* and *C. ferrugineus* strains were collected from 742 farms and central stores for resistance evaluations (Muggleton et al. 1991). All 30 *O. surinamensis* strains from farms, and 28 strains from central stores were classified as resistant to chlorpyrifos-methyl based on survivorship at the diagnostic dose. Resistance to methacrifos was more widespread in strains from farms than from central stores. The reverse was true for pirimiphos-methyl resistance. All strains from central stores were resistant to etrimfos, whereas about 77% of strains from farms showed resistance to this compound (Table 8). A majority of *O. surinamensis* strains were susceptible to malathion and fenitrothion, and resistance to these chemicals was seen in about ≤25% of the strains. Less than 12% of the 43 *C. ferrugineus* strains were resistant to malathion; 2.3% of the strains were resistant to fenitrothion and 7% to etrimfos. All *C. ferrugineus* strains were susceptible to pirimiphos-methyl and chlorpyrifos-methyl.

The lesser mealworm, *Alphitobius diaperinus* (Panzer), a pest in poultry houses, was sampled during 1983 and assayed for resistance to malathion and iodofenphos—two products approved for its control. Of the 42 strains tested, 40.5% were resistant to iodofenphos and 38.1% to malathion (Wakefield and Cogan 1991). Resistance in this species has serious implications for the feed mill and poultry industries as the larvae damage insulation and harbor disease organisms.

## 4. Australia

*Rhyzopertha dominica*, *T. castaneum*, and *T. confusum* collected between 1968–1980 from mills, produce stores, farms, grain elevators, and farm storages throughout New South Wales were tested for phosphine resistance (Attia and Greening 1981). Three of 112 *R. dominica* strains, 11 of 90 *T. castaneum* strains, and 2 of 7 *T. confusum* strains were resistant to phosphine. In *T. castaneum*, phosphine resistance was apparent since 1971, and phosphine resistance was detected in strains collected from 1971 through 1980.

Malathion resistance in stored grain moth pests is widespread in Australia. Malathion largely replaced DDT and lindane as a seed protectant, and dieldrin and lindane as a residual for treatment of structures. Attia (1976) reported more than 259-fold malathion resistance in two *C. cautella* strains. In a separate study, Attia et al. (1979) reported about 250-fold malathion resistance in field strains of *P. interpunctella*, *C. cautella*, and >244-fold resistance in a strain of *E. kuehniella*. Attia et al. (1979) have also observed that in highly resistant strains, TPP synergized malathion, indicating a specific-type of malathion resistance. However, in certain *P. interpunctella* strains, lack of synergism with TPP, and synergism with TBPT (Attia et al. 1980) or PBO (Attia 1977) suggested that esterases other than carboxylesterase, and oxidases played a role in conferring resistance.

During 1982–1985, field strains of *R. dominica*, *S. oryzae*, *O. surinamensis*, and *T. castaneum* were collected from farms, produce stores, seed merchants' premises, and grain silos in southern and central Queensland to determine phosphine resistance (White and Lambkin 1990). Dose-mortality tests on five *R. dominica* strains failed to show phosphine resistance, but strains collected after 1985 showed survival at the diagnostic dose of 0.03 mg/L (White and Lambkin 1990). Although, survival of insects at the diagnostic dose was low, continuous selection with phosphine showed that the insect response (mortality) was stable across generations. Phosphine resistance was common in *R. dominica* and *S. oryzae* strains, but not in *O. surinamensis* and *T. castaneum* strains. *Rhyzopertha dominica* strains selected with phosphine showed a 20-fold resistance, *O. surinamensis* showed a four-fold resistance, and *S. oryzae* showed a three-fold resistance. Winks (1982, 1984) criticized the FAO recommended method for detecting phosphine resistance because the 14-day holding period for exposed insects was inappropriate to determine end-point mortality. He also discouraged the use of dose-mortality regressions because the probit model may fit poorly to data on resistant strains and higher phosphine concentrations may induce narcosis in insects (Winks 1985). However, the resistance surveillance work by White and Lambkin (1990), conducted using the FAO-approved method for resistance detection, was found suitable for characterizing phosphine resistance in field strains.

Two highly fenitrothion-resistant (>160-fold) *O. surinamensis* strains showed less than 10-fold cross resistance to DDT, dichlorvos, chlorpyrifos-methyl, bioresmethrin, and pyrethrins, and less than 40-fold resistance to lindane, malathion, and pirimiphos-methyl (Attia and Frecker 1984). Both TBPT and PBO synergized these insecticides on resistant insects, and TPP failed to synergize malathion. These synergism studies suggested that mixed function oxidases and nonspecific esterases in resistant *O. surinamensis* were responsible for conferring cross resistance to a range of insecticides.

Six insect species (*T. castaneum*, *T. confusum*, *O. surinamensis*, *R. dominica*, *S. oryzae*, and *S. granarius*) collected during 1986 from 63 farms in New South Wales were tested for resistance to malathion, fenitrothion, carbaryl, bioresmethrin, pirimiphos-methyl, chlorpyrifos-methyl, and phosphine using the filter paper method (Herron 1990). Malathion resistance was detected in all species. About 70% of the *O. surinamensis* strains were resistant to pirimiphos-methyl, 50% to fenitrothion, and 39% to chlorpyrifos-methyl. Three *O. surinamensis* strains that were resistant to pirimiphos-methyl were also cross resistant to chlorpyrifos-methyl. Low levels of phosphine resistance were detected in all species.

A field strain of *O. surinamensis* (QOS42) from eastern Australia that was 170-fold resistant at the $LD_{50}$ to fenitrothion by treated filter-paper assay, was tested for resistance to 11 other insecticides (Collins and Wilson 1987). This strain exhibited anywhere from 2 to 150-fold resistance to the following insecticides: chlorpyrifos-methyl, pirimiphos-methyl, malathion, dichlorvos, carbaryl, bioresmethrin, permethrin, cyfluthrin, cypermethrin, and deltamethrin.

Collins (1990) determined the cross-resistance spectrum of two strains of *T. castaneum* exhibiting a low (QTC285) and high (QTC279) level of resistance to the pyrethroid cyfluthrin. These two strains showed high levels of resistance to other pyrethroids (cyhalothrin, cypermethrin, deltamethrin, permethrin, and phenothrin), carbaryl, and low levels of resistance to the organophosphate protectants (see Table 8). Tests with the synergists, PBO and TBPT, suggested target site insensitivity (*kdr* or knockdown resistance) and increased detoxification through oxidative and hydrolytic pathways. Reidy et al. (1990) have shown increased activity of the enzyme glutathione *S*-transferase in a strain of *T. castaneum* that was 71-fold resistant to cyfluthrin. This strain showed 9.5 and 37.1-fold cross resistance to fenitrothion and malathion, respectively.

Bioresmethrin + PBO is used to control organophosphate-resistant *R. dominica*. Fenitrothion is used to control all other species, and chlorpyrifos-methyl is used to control where fenitrothion-resistant *O. surinamensis* is present. Collins et al. (1993) tested populations of *O. surinamensis*, *R. dominica*, *Cryptolestes* spp.,

*T. castaneum*, and *S. oryzae* collected from central stores, grain merchants, and farmers in Queensland. About 27% of the 45 field strains of *R. dominica* were resistant to bioresmethrin + PBO. Fenitrothion resistance was present in all 34 field strains of *O. surinamensis*, and 50% of these strains also showed resistance to chlorpyrifos-methyl. Field strains of *Cryptolestes* spp., *S. oryzae*, and *T. castaneum* were susceptible to fenitrothion.

## 5. Canada

Incipient malathion resistance in 3 of 20 field strains of *C. ferrugineus* collected from farm storage bins, was detected on exposure to a diagnostic dose of 0.35% (w:v) (White and Loschiavo 1985). All insects from these strains died when exposed to a malathion dose of 0.5%. White and Bell (1988) produced 406-fold resistance to malathion in a susceptible strain of *T. castaneum* by rearing insects for 15 generations on wheat flour treated with 600 ppm of malathion.

## 6. Brazil

Mello (1970) reported DDT, lindane, and malathion resistance in *S. oryzae*, and Pacheco et al. (1991) reported organophosphate resistance in *S. oryzae*, *T. castaneum*, and *R. dominica*. In these three species, Sartori et al. (1991) documented phosphine resistance. Resistance to DDT and pyrethroids was detected in six strains of *S. zeamais* collected from four states in Brazil (Guedes 1993). Pacheco et al. (1991), using the diagnostic dose test method, evaluated malathion, pirimiphos-methyl, and fenitrothion resistance in *S. oryzae*, *S. zeamais*, *R. dominica*, and *T. castaneum*. The insect strains originated from grain storage areas, and were collected between 1986 and 1989. Malathion resistance was widespread in *S. oryzae*, *R. dominica*, and *T. castaneum*. *Sitophilus zeamais* was susceptible to all three organophosphates. Pirimiphos-methyl and fenitrothion resistance in *S. oryzae*, *R. dominica*, and *T. castaneum* was found in a few strains. Malathion resistance in *R. dominica* was predominantly of the specific type, and in *T. castaneum* it was of the nonspecific type. However, a few *T. castaneum* strains exhibited malathion-specific resistance. Unlike the Australian strains of *T. castaneum* (Collins 1990), strains showing nonspecific resistance to malathion were not cross resistant to pirimiphos-methyl and fenitrothion. Some *T. castaneum* strains resistant to pirimiphos-methyl were also resistant to fenitrothion (Pacheco et al. 1991).

All field strains of *R. dominica* and *Cryptolestes* spp. were resistant to phosphine, whereas 83.3% of *T. castaneum* and *S. oryzae* strains, and 31.3% of *S. zeamais* strains were resistant to phosphine (Sartori et al. 1991). The stability and severity of resistance was documented for a few strains by increasing both the phosphine dose and the duration of insect exposure to phosphine.

### 7. Philippines

DDT was used to protect stored seed from insect depredations. By 1966, DDT was rendered ineffective due to resistance in insect pests, especially in *S. oryzae* (Santhoy and Morallo-Rejesus 1984).

Sayaboc and Acda (1990) tested adults of 11 *R. dominica* strains, 38 *S. zeamais* strains, and 61 *T. castaneum* strains, collected from warehouses in 11 geographical regions, to malathion using the filter paper assays. All *S. zeamais* were susceptible to malathion. About 82% of *R. dominica* strains, and all *T. castaneum* strains were resistant to malathion. Resistance to malathion in 8 of 11 (72.7%) *R. dominica* strains was of the specific type, because TPP completely suppressed malathion resistance. In 79% of *T. castaneum* strains, malathion-specific resistance was detected, whereas in 21% of the strains the addition of TPP to malathion did not result in an increased mortality of the resistant strains, suggesting a nonspecific type of malathion resistance.

### 8. Rwanda

Pirimiphos-methyl has been in use since 1983 to control the bean bruchid, *Acanthoscelides obtectus* (Say) in dry edible beans, and to control *S. oryzae* and *R. dominica* in grain sorghum. These species collected in 1989 showed ≥50-fold resistance to pirimiphos-methyl. In areas where pirimiphos-methyl was not used, a high susceptibility in the insects was maintained.

### 9. India

In godowns containing bagged wheat, yearly treatments with phosphine resulted in the development of resistance in the Khapra beetle, *Trogoderma granarium* Everts (Borah and Chahal 1979). The resistance levels were highest in the first instars, followed by fourth instars, second instars, third instars, eggs, and pupae (Table 8). The use of ethylene dibromide (EDB) or EDB-carbon tetrachloride mixture (3:1) was effective in controlling these resistant insects.

Tests with *T. castaneum* and *R. dominica* strains that were 4.5 and 380-fold resistant to phosphine, respectively, at the $LD_{50}$, showed that maintaining a constant concentration of phosphine was important for killing both immature and adult stages of these species (Rajendran 1994). Exposure to changing concentrations of phosphine resulted in survival of these species.

Rajendran and Narasimhan (1994) reported field failures of phosphine against the cigarette beetle, *Lasioderma serricorne* (F.) in stored tobacco, caused by high levels of resistance, a result of repeated inadequate treatments (Mills 1983). Field test results showed that doubling both the phosphine dose and exposure duration from the recommended rate of 1 $g/m^3$ and five days, respectively, provided effective control of this pest (Rajendran and Narasimhan 1994).

### 10. Indonesia

Fifty-six strains of *T. castaneum* collected from all over Indonesia during 1978 showed survival of 1–99% at a diagnostic dose of 0.5% (w:v) malathion, and these strains were susceptible to pirimiphos-methyl (Osman and Morallo-Rejesus 1981).

## V. ENVIRONMENTAL AND BIOLOGICAL FACTORS AFFECTING RESISTANCE ESTIMATES

The choice of insecticide-susceptible laboratory strains that are used as standards for comparison against field strains can influence resistance ratio estimates. The $LD_{50}$ or response of laboratory strains to an insecticide may change within or between generations. For example, probit regression estimates on a susceptible strain of *T. castaneum* (QTC4) to fenitrothion on two separate occasions (Collins 1990; Collins et al. 1993) gave significantly different $LD_{50}$ and slope values. Collins (1990) originally reported an $LD_{50}$ (95% CL) and slope of 0.39 (0.37–0.41) mg of fenitrothion per kg of wheat and 11.9, respectively, for the QTC4 strain. In subsequent tests with fenitrothion on the same strain, Collins et al. 1993 reported an $LD_{50}$ (95% CL) and slope of 0.92 (0.82 − 1.10) and 5.5, respectively. Kinsinger and McGaughey (1979) found variation in response to *B. thuringiensis* among *P. interpunctella* strains. Unless the variation in susceptible strains is considered, both in the absence and presence of insecticides, the computation of resistance ratios could be misleading, indicating an increase or decrease in resistance levels of field strains when there is none. Bell (1976) reported that field (wild) populations of moth species were more tolerant to cold temperatures and longer phosphine exposures than laboratory strains of the same species. Bell (1976) also reported that the eggs of *C. cautella*, *E. kuehniella*, and *P. interpunctella* were more tolerant to phosphine than the other stages. In the egg stage, the tolerance to phosphine extends to 30–45% of the developmental period, and across these species, this tolerance was inversely related to temperature. Similar tolerance to fumigants was shown in the eggs of *T. confusum* (Lindgren and Vincent 1966), the cadelle, *Tenebroides mauritanicus* (L.) (Qureshi et al. 1965), *C. ferrugineus* (Barker 1969), and in three *Trogoderma* spp. (dermestids) (Vincent and Lindgren 1972).

The test method used to evaluate resistance can influence resistance ratio estimates. The resistance ratio of a lindane-selected strain of *T. castaneum* on lindane-treated diet was 55 times that of the unselected parental strain (Bhatia and Pradham 1971). The resistance ratio of the selected strain relative to the unselected parental strain was >87 and 13.1 when insects were exposed to lindane-treated films and direct spray, respectively.

The age of insects used in tests could influence $LD_{50}$ or resistance ratio estimates. Malathion resistance in *C. cautella* decreased with adult age (Wool and Kamin-

Belsky 1983). The adults of *C. cautella* that were ≤6 h old were resistant to malathion, and as the moths aged, their resistance to malathion decreased. Wool and Kamin-Belsky (1983) suggested that an increase in malathion susceptibility of older moths was due to abrasion and exhaustion caused by normal mating and flight activities. This reasoning was supported by the fact that moths held at 10°C for 96 h were not as susceptible as those held for the same time period at a higher temperature (28°C). Therefore, resistance ratio estimations based on mixed populations of insects of unspecified age, or on populations that contain a large portion of insects that are highly susceptible, may underestimate the true resistance level. In this respect, it would be useful to determine the age structure of population of insects so that different age groups are represented when screening field strains for resistance. This approach requires effective procedures for age-grading adults of field-collected insects. An alternative would be to rear field-collected insects in the laboratory and harvest cultures at specified time periods to obtain adults of various ages.

In stored-product lepidoptera, larvae are the only actively feeding stage. Therefore, larval fitness (weight, survival, and development time) is determined, to a certain extent, by the quality of larval diet. Consequently, adults produced from inferior larval diets tend to be less vigorous and more susceptible to insecticides (Wool and Kamin-Belsky 1984). Therefore, it is important to use healthy, unstressed insects for resistance evaluations.

Lethal dose estimates can change with environmental conditions, especially temperature. The increased susceptibility, or lower $LD_{50}$ values, at higher temperatures is a result of greater pickup of the insecticide by insects due to their increased activity. In an organophosphate-resistant strain of *O. surinamensis*, the $LD_{50}$ values for chlorpyrifos-methyl at 5, 10, 15, 20, 25, and 30°C and 70% relative humidity were 21.7, 6.1, 2.2, 1.6, 1.3, and 1.4 mg of active ingredient/$m^2$, respectively (Barson 1983). However, resistance ratios (obtained by testing a susceptible strain at the same temperatures) were similar across the temperatures. Barson (1983) also showed that *O. surinamensis* reared under fluctuating temperature and relative humidity conditions showed varying levels of susceptibility to chlorpyrifos-methyl.

In four malathion-resistant strains of *O. surinamensis*, the temperature experienced by adults affected their response to malathion (Muggleton et al. 1981). In this study, adults held for two weeks at 15, 20, 25, and 30°C were tested with a diagnostic dose of malathion at 25°C. Beetles held at lower temperatures showed higher mortality than did beetles reared at higher temperatures. For every 5°C drop in temperature, there was a 7.5% increase in mortality or knockdown. Muggleton et al. (1981) attributed the increased susceptibility as temperature decreased to increased uptake of malathion by beetles held at lower temperatures compared with beetles held at higher temperatures.

Most resistance reports on stored-product insects are based on tests with adults of beetles and larvae of moths. However, it is important to realize that in the field, insecticide selection pressure acts on eggs, larvae, pupae, and adults of insects. In the laboratory, Rajendran (1992) showed that resistance to the fumigant phosphine in the eggs, larvae, pupae, and adults of *T. castaneum* developed after one generation. After six generations, the highest amount of resistance was found in pupae, followed by eggs, adults, and larvae. A similar resistance pattern occurred among the insect stages exposed to the fumigant methyl bromide; however, the level of resistance to methyl bromide was less than the level of resistance to phosphine. There is a good correlation in phosphine resistance between *R. dominica* adults and eggs (Bell et al. 1977). In *T. castaneum*, early and mid-pupae are more tolerant to phosphine than the adults. In three *T. castaneum* strains, Binns (1986) showed that the final-stage larvae were twice as resistant as adults to malathion. However, the resistance in both larvae and adults was TPP suppressible, indicating a specific type of malathion resistance. In a fenitrothion-resistant strain of *O. surinamensis*, differences were detected between larvae and adults in the levels of an insecticide-degrading enzyme. In fenitrothion-resistant larvae, the level of cytochrome $P_{450}$ was three times higher than in larvae of the susceptible strain. However, in fenitrothion-resistant adults, the cytochrome $P_{450}$ level was 10–15 times greater than in susceptible adults (Rose and Wallbank 1986). Therefore, the rate of resistance evolution may vary with the insect stage and the insecticide, and finding resistance in adults implies that the immature stages could be resistant to varying degrees. It is important to recognize differences in susceptibility (or resistance) among insect stages, and adjust the insecticide dose, within reasonable limits, to control the most resistant stage.

Certain insect species harbor symbiotic microorganisms that produce amino acids, lipids, vitamins, or other nutrients for their hosts. Besides providing nutrition to the host, some intracellular or extracellular microorganisms are capable of detoxifying toxic chemicals encountered by the insect host. For instance, *L. serricorne* harbors intracellular yeasts in specialized tissues (mycetomes) at the junction of the foregut and midgut. Dowd (1989) reported a high level of hydrolytic activity in mycetomes excised from last instars and adults of *L. serricorne*. Shen and Dowd (1991) showed that the yeasts in *L. serricorne* were able to degrade various xenobiotics, including insecticides. Detoxifying enzymes produced by these symbiotic yeasts include hydrolases, glucosidase, phosphatase, and glutathione transferase (Shen and Dowd 1991). Insects harboring symbiotic microorganisms that detoxify insecticides are less susceptible to insecticides, whereas insects that contain pathogenic microorganisms tend to be more susceptible to insecticides. For example, *T. castaneum* larvae infected with the microsporidian *Nosema whitei* Weiser were more susceptible to methyl bromide than healthy larvae (M. V. Listov and V. A. Nesterov 1982, Altunion Research Institute of Applied Microbiology, Protvino, Moscow).

Errors while preparing insecticide solutions, and improper handling of insecticide-treated substrates before testing, can affect resistance test results. For example, malathion solubility in the carrier Risella oil is about 2%. A solution containing more than 2% malathion in Risella oil could result in uneven distribution of the insecticide concentrate in the oil. Serial dilutions of insecticides made from this stock solution could also be in error. The use of acetone as a solvent instead of Risella oil is recommended (White and Loschiavo 1985). However, it is important to minimize the evaporation of acetone. When using insecticide-impregnated filter papers for determining resistance, it is desirable to expose insects to papers immediately (≤24 h) after treatment, before the insecticide degrades. If tests are to be used at a later date, treated papers should be stored in a cooler set at 5°C or wrapped in aluminum foil for storage in a freezer (White and Loschiavo 1985). When exposing insects to insecticide-treated substrates, it is desirable to quantify residues on substrates so that insect response and resistance level can be related to the *actual* rather than to the *expected* insecticide level.

## VI. INHERITANCE OF RESISTANCE TO INSECTICIDES

DDT resistance in *S. oryzae* is incompletely dominant and is sex linked (Champ 1967). Resistance to lindane and cyclodiene in *T. castaneum* is associated with a major gene, and the resistance is due to target site (nervous system) insensitivity and not due to differences in penetration and metabolism of these two insecticides between susceptible and resistance individuals (Beeman and Stuart 1990).

In a 73-fold malathion-resistant strain of *T. castaneum* from Georgia, USA, Beeman (1983) showed that the resistance was inherited as a simple autosomal semidominant trait. A similar pattern of inheritance of malathion resistance in a Canadian strain of *T. castaneum* was reported by White and Bell (1988). In *T. castaneum* adults, malathion-specific resistance is controlled by closely linked alleles in the sixth linkage group at or near *R-mal* locus (Beeman and Nanis 1986), and the resistance levels among strains vary due to different *R-mal* alleles in each strain. Differences in probit regression lines and resistance levels among *T. castaneum* field strains to malathion (Subramanyam et al. 1989; Subramanyam and Harein 1990) and differences in mortality among strains at a diagnostic dose of malathion supports this view. In a Nigerian strain of *T. castaneum*, Wool et al. (1982) reported the presence of a Y-linked factor in addition to a major autosomal factor. This inheritance conferred enhanced malathion resistance in males compared with females, and $F_1$ males were generally more resistant than their female siblings.

Malathion resistance in *P. interpunctella* is due to an altered esterase, and resistance is inherited as a dominant factor that is controlled by a single autosomal gene or a set of closely linked genes (Attia et al. 1981; Beeman and Schmidt 1982).

Malathion resistance in *O. surinamensis* is controlled by a single dominant gene. However, fenitrothion resistance in *O. surinamensis* is controlled by two or more autosomally inherited factors, and the major genes involved are incompletely dominant (Collins 1986). The synergism studies with *O. surinamensis* strains (Attia and Frecker 1984), and biochemical work of Wallbank (1984), also support a multifactorial control of fenitrothion resistance. Insecticide resistance involving more than one factor has been reported in other beetle pests of stored products. These reports include resistance to malathion (Champ and Campbell-Brown 1976) and lindane (Champ and Campbell-Brown 1969) in *T. castaneum*, and resistance to lindane in *Sitophilus* spp. (Champ and Cribb 1965).

Ansell et al. (1991) reported that phosphine resistance in *R. dominica* and *T. castaneum* is semidominant and autosomally inherited.

Resistance to *B. thuringiensis* in *P. interpunctella* larvae is not sex linked and is partially to completely recessive (McGaughey 1985; McGaughey and Beeman 1988), because the susceptibility to *B. thuringiensis* of the $F_1$ progeny resembled that of the susceptible parent.

A strain of *S. zeamais* resistant to DDT and pyrethroids resembling knock-down resistance (*kdr*) was collected from stored-grain facilities in Brazil. This strain was 2,634 times more resistant than a susceptible strain to synergized deltamethrin. Guedes et al. (1994) found that in this strain, resistance was caused by a single recessive sex-linked gene. In *S. oryzae*, resistance to deltamethrin and permethrin was caused by a single sex-linked gene, but the factor was incompletely dominant (Heather 1986).

In general, recessive characters are harder to eliminate from populations. If resistance is recessive, and the dominant allele is susceptible, the heterozygotes will be killed by an insecticide application. Resistance that is sex linked and is associated with increased fitness of heterozygotes and susceptible individuals can spread within the insect population. In such instances, immigration of susceptible individuals into a resistant population may help decrease resistance levels of insects.

## VII. FITNESS OF RESISTANT INSECTS

Insects selected for resistance may be less fit than susceptible insects. Resistant insects may have reduced fecundity, decreased survival of immature and mature stages, or prolonged developmental time of certain life stages when compared with susceptible insects. A lindane-resistant strain of *T. castaneum* had lower fecundity than a susceptible strain (Bhatia and Pradhan 1971). However, the development of pre-adult stages was similar for both strains on untreated diet. Zettler (1977) did not find differences in fecundity and egg viability of four malathion-resistant strains of *P. interpunctella* compared with a malathion-susceptible strain. Ramsey and Farley (1978) reported that malathion-resistant males of *P. interpunctella* mated more readily than susceptible males, and resistant males stimulated greater fecundity of susceptible females. The resistant females laid more eggs than susceptible females, and the average number of eggs laid per female was less variable for resistant females than for susceptible females. White and Bell (1988) showed that homozygous resistant parents of *T. castaneum* produced the greatest number of progeny on malathion-treated wheat, whereas heterozygous parents were more productive than homozygous parents on untreated wheat. The mortality of larvae on malathion-treated diet was greater than mortality of adults on the same diet (White and Bell 1988). The susceptibility to malathion of two malathion-resistant strains of *O. surinamensis* increased in the absence of malathion pressure during a two-year period (Muggleton 1983). This disadvantage in fitness could be used to manage resistance (Muggleton 1982). There were differences in certain biological parameters of a four-fold phosphine-resistant *T. castaneum*, selected in the adult stage for four generations with a constant phosphine dose (El-Lakwah et al. 1992). Phosphine-resistant females laid twice as many eggs per day as the susceptible females (Fig. 4). However, only 76% of eggs laid by resistant females hatched, whereas 90% of eggs laid by susceptible females hatched. The time to hatch of eggs laid by resistant females was delayed by a day compared with the susceptible strain. In addition, 60% mortality occurred in the larval stage of resistant insects as opposed to 30% for susceptible insects. There were no differences between susceptible and resistant insects in the preoviposition period, instar duration, and total development period (El-Lakwah et al. 1992).

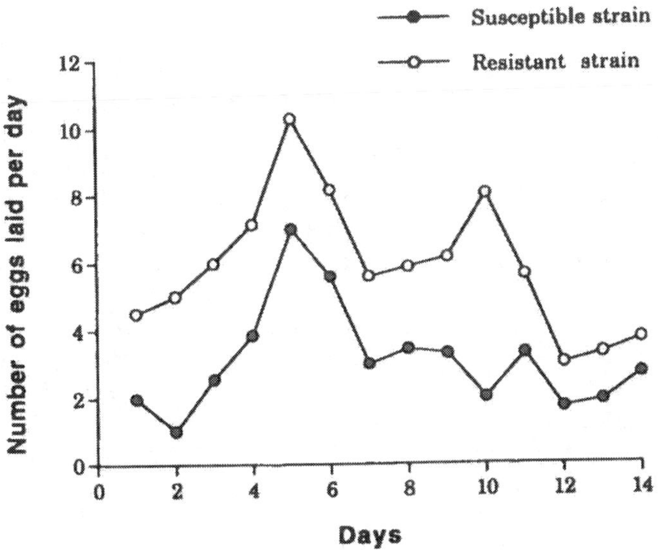


**Figure 4** Average number of eggs laid daily by phosphine-resistant and -susceptible females of the red flour beetle, *Tribolium castaneum*. The phosphine-resistant strain was selected for four generations with a constant dose of phosphine, and is four-fold resistant to phosphine (also see Table 2). (Redrawn from El-Lakwah et al. 1992.)

## VIII. EFFECTIVENESS OF INSECTICIDE ALTERNATIVES

As insects develop resistance to an insecticide, making it ineffective, newer insecticides or insecticide alternatives must be used. The use of newer insecticides may offer excellent insect control initially, but with time, certain insect species are capable of developing resistance to the newer products, rendering them ineffective. The widespread resistance to different classes of insecticides in an Australian strain of *T. castaneum* and *O. surinamensis* supports this view (Collins and Wilson 1987; Collins 1990). In the United States, widespread resistance to organophosphates is apparent, and pyrethroids such as cyfluthrin (Arthur 1992, 1994) and tralomethrin (Halliday 1992) are being explored as alternatives to control organophosphate-resistant insects. The pyrethroids are stable on the grain (Noble et al. 1982) and offer long-term protection against insects at low rates (≤2 ppm) (Arthur 1992, 1994; Halliday 1992). However, insects can develop high levels of resistance to pyrethroids (Collins and Wilson 1987; Collins 1990). Insect growth regulators such as methoprene and hydroprene do not provide effective control of adults but provide insect suppression by reducing fecundity and egg fertility and by disrupting development (Mian and Mulla 1982). A strain of *T. castaneum* that was resistant to at least 10 different insecticides was not resistant to methoprene (Collins 1990). In Australia, methoprene is recommended for control of *R. dominica* populations that are resistant to bioresmethrin + piperonyl butoxide (Collins et al. 1993). Certain insect species have been shown to develop resistance to methoprene. A low level of resistance to methoprene has been documented in *T. castaneum* (Dyte 1972) and *T. confusum* (Brown et al. 1978). In five of eight strains of *L. serricorne* infesting stored tobacco, Benezet and Helms (1994) reported low levels of methoprene resistance. In three of the five strains of *L. serricorne*, adults emerged from tobacco treated with a high methoprene dose of 14.1 ppm. The use of modified atmospheres for stored-product insect management has been discussed in Chapter 7. Donahaye (1991) reported that *T. castaneum* can develop low levels of resistance to low oxygen concentrations and high carbon dioxide concentrations. The granary weevils (*S. granarius*) developed a three-fold resistance to carbon dioxide after seven generations of selection in the laboratory (Bond and Buckland 1979). Similar results with *S. oryzae* were reported by Navarro et al. (1985). Limited tests with stored-product insects have shown that derivatives of L-tryptophan and proctolin have potential as feeding deterrents (Sobotka et al. 1992). These compounds are selective and nontoxic to vertebrates. More research is needed to determine the practical use of these compounds in stored-product insect management. Several plant materials have been tested on stored-product insects (Burroughs et al. 1988), and among the plant products, azadirachtin appears promising as a potential stored-product protectant.

The parasitoids and predators of stored-product insects may be naturally tolerant to insecticides (Baker 1995), and such insecticide-tolerant natural enemies can be selected, mass-reared, and released into storage environments to complement other insect management strategies (Johnson and Tabashnik 1994). Press et al. (1981) studied the survival and reproduction of the parasitic wasp *Habrobracon hebetor* on permethrin-treated *C. cautella* larvae. The topical treatment of permethrin at 0.013–0.53 mg/g/larva produced 20–97% mortality of *C. cautella* larvae. At the same rate, only 0–35% mortality of the parasite occurred. Inducing insecticide resistance in natural enemies through selection or genetic manipulation/biotechnology has potential for management of insects in stored commodities, where insecticides are frequently used as a substitute rather than as a supplement to nonchemical management strategies.

## IX. MANAGEMENT OF INSECTICIDE RESISTANCE

Models have been developed to examine the contribution of various factors to the rate of resistance development (Tabashnik 1990). Biological factors include initial frequency and dominance of the resistance gene, immigration rate, and number of generations of the insect species per year. Operational factors include the selection of insecticide, and the rate, coverage, timing, and frequency of insecticide applications. Operational factors are generally more easily manipulated than biological factors. Models have also been useful in exploring potential interactions between these factors.

### 1. Biological Factors

Resistance genes are generally considered monogenic, dominant, and initially rare. Computer simulations with polygenic models (Plapp et al. 1979; Via 1986; Uyenoyama 1986) have predicted responses similar to those with monogenic models. At high pesticide doses which kill heterozygotes, resistance genes become functionally codominant or recessive (Tabashnik 1986a). Immigration of susceptible insects into a population can slow the development of resistance, but generally immigration rates are not high enough for this to be significant (Tabashnik et al. 1992). Wool and Manheim (1980) and Wool and Noiman (1983) showed that the releases of insecticide-susceptible *T. castaneum* males slowed the development of resistance in populations treated with malathion. Therefore, resistance is more likely to be promoted by the immigration of resistant insects. Dyte and Blackman (1970) and Dyte (1979) reported the importation of insecticide-resistant strains of stored-product insects into the United Kingdom.

Fewer generations per year may slow the development of resistance because insects might be exposed to less insecticide selection pressure. Longstaff (1988) proposed that grain could be cooled by aeration to reduce the number of generations per year and, hence, the rate of resistance development. Analysis of field data of 682 species indicated that there was a poor correlation between number of generations per year and rate of resistance development (Tabashnik et al. 1992). The rate of resistance development was faster for species with an intermediate number of generations per year and was slower for species with less than five or greater than 12 generations per year. This might be explained by species with intermediate generation times being more likely to receive two pesticide applications per generation than those with shorter or longer generation time.

### 2. Operational Factors

The rate of selection for resistance increases with an increase in the dose, coverage, frequency of application, and persistence of an insecticide. Computer simulations by Tabashnik and Croft (1982) predicted that resistance development time decreased 10-fold in response to a 10-fold increase in dose. The extent to which the dose can be increased is severely limited by the level of residues that is acceptable on a commodity (Champ and Dyte 1977). Coverage is rarely complete and some susceptible insects survive in untreated areas. Muggleton (1986) developed a model for malathion resistance in *O. surinamensis*. This model predicted that resistance would be delayed the most by using the highest malathion dose and leaving the maximum acceptable proportion of the population untreated. Nonuniform applications of grain protectants might be used to provide high concentrations at some locations while leaving certain areas untreated (Amos et al. 1979).

Roush (1989) suggested that by carefully timing pesticide applications, a particular developmental stage of pest could be left untreated. The untreated stage act as a reservoir for susceptible genes. Insecticide applications might be timed to take advantage of differences in susceptibility between adults and larvae (Dittrich et al. 1980), and also differences in the susceptibility among larval instars (Daly et al. 1988). As frequent insecticide applications select for resistance (Tabashnik et al. 1990), the frequency of applications should be minimized by using chemical control only when nonchemical control methods are unavailable and only when insect densities have exceeded established economic thresholds. The extensive use of fumigants to control stored-product insects is a good example of choosing the least persistent chemical.

### 3. Prediction of Rate of Resistance Development

In addition to examining the relative influence of various factors on the development of resistance, models can be used to predict the rate of resistance development in the field. Tabashnik and Croft (1985) found that their model predicted the rate of resistant development for 12 pest and 12 beneficial species in apple orchards. Tabashnik (1986b) developed a model that correctly predicted the rate of resistance development of a cabbage pest (*Plutella xylostella* L.) in Taiwan. Simulations of 15 management strategies suggested that insecticide application frequency must be less than two treatments per cabbage crop or the application rate must be less than $LC_{75}$ to substantially delay the development of resistance.

Models can be useful in designing resistance management programs. Sinclair and Alder (1985) included resistance development in their model for stored-product insects associated with farm-stored grain. Their simulations showed that the development of resistance was delayed by not treating the grain intended for animal feed. The model also showed that for three insecticides, the rate of resistance development increased with increased insecticide persistence. As models are refined, they may be useful in scheduling the day-to-day operations of pest management programs designed to minimize resistance development.

## 4. Economics of Resistance

Insect pests and their susceptibility to insecticides are valuable property resources (Knight and Norton 1989) that should be preserved. Susceptibility is a valuable resource, because once insects become resistant, it is difficult to restore their susceptibility to insecticides (Metcalf 1980). Insects are mobile and can move among locations. They also move among locations through trade in infested commodities. Because of this mobility, insects and their susceptibility to insecticides are a shared resource. Therefore, pest management programs should consider resistance levels throughout the marketing system rather than at each facility.

Resistance reduces the effectiveness of insecticides and this results in commodity and treatment losses, which increase the price of commodities purchased by the consumer. The price of commodities may also increase if newer, expensive pesticides or a greater number of chemical applications are necessary to control resistant insects. The cost of replacing ineffective chemicals is also passed along to the consumer. Voss (1988) estimated that replacing an insecticide takes 8–10 years and costs about $50 million. Metcalf (1980) reported that the benefits of insecticides to field crop production was three to five dollars for every dollar spent, but the cost-benefit ratio is reduced to 1:2.4 if externalities such as the costs of government regulation, and health and environmental risks are considered.

Public intervention is needed to conserve insect susceptibility to insecticides, because optimal utilization of susceptibility can give farmers, elevator operators, food processors, and chemical manufacturers a short-term advantage (Miranowski and Carlson 1986). Dover and Croft (1984, 1986) suggested that public policies such as control of insecticide use by prescription and a federal end-user tax on chemicals to finance resistance-management programs might help solve the resistance problem. Resistance management programs can reduce the cost of pest management by reducing the number of insecticide applications and preventing the catastrophic losses due to control failures (Cox and Forrester 1992).

## 5. Resistance Management Program

The best way to slow the development of insecticide resistance is to use integrated pest management (IPM). IPM uses nonchemical methods instead of pesticides and applies pest control only when insect densities exceed the economic threshold. In apple orchards, natural enemies have been successfully used along with insecticides to manage resistance (Croft 1982). A more common approach has been to use multiple insecticides to manage resistance. Two insecticides can be used sequentially (the second after the first has failed), as mixtures, in rotation, or as mosaics (some areas treated with the first and other areas with the second) to delay the development of resistance (Roush 1989; Tabashnik 1989). Rotation is generally considered the preferred method, although, theoretically, use of mixtures is more effective. The higher cost of mixtures, however, is a disadvantage. Programs using two insecticides sequentially or application of insecticides in a mosaic pattern are considered least favorable. The use of insecticide mixtures can be the most effective strategy because few insects are likely to be resistant to two or more chemicals. Mixtures are most effective in delaying resistance when resistant alleles are rare, and rotation becomes a better strategy when the frequency of resistant alleles exceed 0.02%, according to the modeling study of Holloway and McCaffery (1988). The modeling study of Kable and Jeffery (1980) indicated that spray coverage had the greatest effect on the use of mixtures and there was little advantage to using mixtures when coverage was complete. Unequal rates of breakdown of residues of the two insecticides in a mixture can contribute to more rapid development of resistance, because sublethal doses allow heterozygotes to survive (Riddles and Nolan 1987). The higher cost of mixtures may be a disadvantage, although lower doses of two insecticides in a mixture may be more economical than the use of a high dose of a single pesticide. A procedure for calculating the optimal ratio of two insecticides in a mixture is given by Stone et al. (1988).

Rotation is considered better than sequential use of insecticides because susceptible genotypes generally have a reproductive advantage over resistant genotypes in the absence of an insecticide. The frequency of susceptible genotypes may increase during the periods when an insecticide is not used (Roush and Croft 1986). In the absence of the insecticide, Muggleton (1983) found that laboratory-selected malathion-resistant *O. surinamensis* adults were less fit than those with susceptible genotype at 25°C, but this effect was reversed above 30°C. Heather (1982) found that the population growth rates of resistant and susceptible *S. oryzae* populations in the absence of insecticides were similar. The reproductive disadvantage in resistant strains was found to be smaller for field-collected than laboratory-selected insects (Roush and Croft 1986) and, therefore, may be less effective in reducing resistance in the field.

The use of insecticide mosaics in delaying the development of resistance is not always better than rotating pesticides. In addition, mosaics may not delay resistance development, especially if there is extensive interbreeding of insects among areas treated with different insecticides. The use of different insecticides to which the insects are not cross resistant can contribute substantially to slowing the development of resistance (Metcalf 1980). Adults of *T. castaneum* populations in the United States with the malathion-specific type of resistance are not cross resistant to pirimiphos-methyl and chlorpyrifos-methyl, whereas Australian strains of *T. castaneum* and *O. surinamensis* show resistance to a wide variety of insecticides (see Section IV). Immaraju et al. (1990) found in field tests with thrips that the use of insecticide rotation was superior to using a mixture of insecticides. They indicated that the Environmental Protection Agency's policy of not registering newer pesticides until currently used pesticides have failed, encourages poor resistance management practices by users. Dover and Croft (1984, 1986) suggested that the risk of resistance be incorporated into pesticide registration requirements and that resistance management be used as justification for the registration of mixtures. Resistance management programs for stored-product insects should rely heavily on nonchemical methods because the number of insecticides available will limit the use of a multiple insecticide tactic for resistance management.

Just as monitoring insect populations is important for making pest management decisions, monitoring resistance is important for making resistance management decisions. Diagnostic tests that distinguish between resistant and susceptible individuals must be used (Roush and Miller 1986) instead of dose-response tests. Resistance must be detected before the resistant allele frequency reaches 1% because control failures may occur within one to six generations after resistance reaches this level (Kable and Jeffery 1980; Roush and Miller 1986). Therefore, at the diagnostic dose large numbers of insects must be tested to detect resistance. New enzyme and immunological test methods are needed to rapidly screen these large numbers of insects (Brown and Brogdon 1987). Alternatively, pheromone-baited sticky traps laced with insecticides can provide a simple and effective means of monitoring insecticide resistance in the field (Haynes et al. 1986). Monitoring can be used to choose an appropriate resistance management program, evaluate the success of resistance management programs, and improve models for predicting resistance development. Denholm and Rowland (1992) recommend that resistance management programs be flexible to incorporate new pest management approaches. Similarly, resistance management programs should be modified to respond to information from resistance monitoring.

## X. SUMMARY

Economically important stored-product insect species in many countries are resistant to most of the insecticides commonly used to protect stored commodities from insect infestation and damage. The high level of resistance in some insect species has resulted in field failures of several insecticides. Some species, especially *T. castaneum* and *O. surinamensis*, show resistance to a wide variety of insecticides. Techniques for detecting resistance in field strains of insects must be quick, reliable, and realistic. Generally, techniques must be able to detect incipient resistance in insect populations. Detection and accurate estimation of resistance and pattern of resistance inheritance are essential for developing a sound resistance management program. Resistance management programs must be designed to retain a high level of susceptible genes in the pest population and thus maintain the effectiveness of insecticides. The most effective resistance management program involves using insecticide alternatives and applying pest control measures or insecticides only when pest populations exceed economically damaging levels or thresholds. The development of resistance can be slowed by using lower insecticide application rates, fewer applications, and less persistent pesticides. The ratio of susceptible to resistant genes can be increased by eliminating resistant pests with a suitable noninsecticidal method. If two insecticides are available for insect management, they can be used sequentially, as mixtures, in rotation, or as mosaics. Susceptibility can be saved by providing untreated refugia for a portion of the pest population. Refugia may naturally occur as a result of incomplete insecticide coverage during treatment, or due to insecticide breakdown or dissipation (e.g., after fumigation and aeration). The level of insect susceptibility to insecticides can be increased by releasing susceptible insects or through increased immigration of susceptible insects into a population containing a high percentage of resistant individuals. Models have been used to examine the contribution of various factors to the rate of resistance development and to predict the rate of resistance development in the field. These theoretical models can be used in designing resistance management programs, and should be useful in scheduling insecticide applications within a given facility. A majority of stored-product insects are mobile, and this is one mechanism by which resistant insects spread within a geographical region. Insects are also transported in commodities through trade, and there are numerous instances of resistant strains moving through grain trade. Because of high insect mobility, insects and their susceptibility to insecticides are a shared resource. Therefore, integrated pest management programs should be designed to minimize the development of resistance and movement of resistant individuals throughout the commodity marketing system.

## ACKNOWLEDGMENTS

We thank Danielle Downey for assistance with the literature search and Sailaja Chandrapati for reviewing an earlier version of this chapter. Research of Bh. Subramanyam presented in this chapter was supported by grants from the United States Department of Agriculture to the Minnesota Pesticide Impact Assessment Program.

## REFERENCES

Abbott, W. S. (1925). A method of computing the effectiveness of an insecticide. *J. Econ. Entomol.*, **18**:265–267.

Al-Hakkak, Z. S., Murad, A. M. B., and Hussain, A. F. (1985). Phosphine-induced sterility in *Ephestia cautella* (Lepidoptera: Pyralidae). *J. Stored Prod. Res.*, *21*:119–121.

Amos, T. G., Williams, P., and Minett, W. (1979). Non-uniform application of grain protectants in commercial storages. In Proceedings, 2nd International Working Conference of Stored Product Entomology, Savannah, Georgia, pp. 344–349.

Ansell, M. R., Dyte, C. E., and Smith, R. H. (1991). The inheritance of phosphine resistance in *Rhyzopertha dominica* and *Tribolium castaneum*. In Proceedings, 5th International Working Conference of Stored Product Protection, INRA/SDPV, Bordeaux, France, pp. 961–970.

Armstrong, J. W., and Soderstrom, E. L. (1975). Malathion resistance in some populations of the Indianmeal moth infesting dried fruits and tree nuts in California. *J. Econ. Entomol.*, *68*:505–507.

Arthur, F. H. (1992). Cyfluthrin WP and EC formulations to control malathion-resistant red flour beetles and confused flour beetles (Coleoptera: Tenebrionidae): effects of paint on residual efficacy. *J. Entomol. Sci.*, 27:436–444.

Arthur, F. H. (1994). Residual efficacy of cyfluthrin applied alone, or in combination with piperonyl butoxide or piperonyl butoxide + chlorpyrifos-methyl as protectants for stored corn. *J. Entomol. Sci.*, *29*:276–287.

Arthur, F. H., and Zettler, J. L. (1991). Malathion resistance in *Tribolium castaneum*: differences between mortality caused by topical application and residues on treated surfaces. *J. Econ. Entomol.*, *84*:721–726.

Arthur, F. H., and Zettler, J. L. (1992). Malathion resistance in *Tribolium confusum* Duv. (Coleoptera: Tenebrionidae): correlating results from topical applications with residual mortality on treated surfaces. *J. Stored Prod. Res.*, *28*:55–58.

Arthur, F., Zettler, J. L., and Halliday, W. R. (1988). Insecticide resistance among populations of almond moth and Indianmeal moth (Lepidoptera: Pyralidae) in stored peanuts. *J. Econ. Entomol.*, *81*:1283–1287.

Attia, F. I. (1976). Insecticide resistance in *Cadra cautella* in New South Wales, Australia. *J. Econ. Entomol.*, *69*:773–774.

Attia, F. I. (1977). Insecticide resistance in *Plodia interpunctella* (Hübner) (Lepidoptera: Pyralidae) in New South Wales, Australia. *J. Aust. Entomol. Soc.*, *16*:149–152.

Attia, F. I., and Frecker, T. (1984). Cross-resistance spectrum and synergism studies in organophosphorous-resistant strains of *Oryzaephilus surinamensis* (L.) (Coleoptera: Cucujidae) in Australia. *J. Econ. Entomol.*, 77:1367–1370.

Attia, F. I., and Greening, H. G. (1981). Survey of resistance to phosphine in coleopterous pests of grain stored products in New South Wales. *Gen. Appl. Entomol.*, *13*:93–97.

Attia, F. I., Shanahan, G. J., and Shipp, E. (1979). Survey of insecticide resistance in *Plodia interpunctella* (Hübner), *Ephestia cautella* (Walker), and *Ephestia kuehniella* Zeller (Lepidoptera: Pyralidae) in New South Wales. *J. Aust. Entomol. Soc.*, *18*:67–70.

Attia, F. I., Shanahan, G. J., and Shipp, E. (1980). Synergism studies with organophosphorous resistant strains of the Indianmeal moth. *J. Econ. Entomol.*, *73*:184–185.

Attia, F. I., Shipp, E., and Shanahan, G. J. (1981). Inheritance of resistance to malathion, DDT, and dieldrin in *Plodia interpunctella* (Lepidoptera: Pyralidae). *J. Stored Prod. Res.*, *17*:109–115.

Baker, J. E. (1995). Stability of malathion resistance in two hymenopterous parasitoids. *J. Econ. Entomol.*, *88*:232–236.

Baker, J. E., and Arbogast, R. T. (1995). Malathion resistance in field strains of the warehouse pirate bug (Heteroptera: Anthocoridae) and a prey species *Tribolium castaneum* (Coleoptera: Tenebrionidae). *J. Econ. Entomol.*, *88*:241–245.

Ball, H. J. (1981). Insecticide resistance—a practical assessment. *Bull. Entomol. Soc. Am.*, *27*:261–262.

Bansode, P. C., and Campbell, W. V. (1979). Evaluation of North Carolina field strains of the red flour beetle for resistance to malathion and other organophosphorous compounds. *J. Econ. Entomol.*, *72*:331–333.

Barker, P. S. (1969). Susceptibility of eggs and young adults of *Cryptolestes ferrugineus* and *C. turcicus* to hydrogen phosphide. *J. Econ. Entomol.*, *62*:363–365.

Barson, G. (1983). The effects of temperature and humidity on the toxicity of three organophosphorous insecticides to adult *Oryzaephilus surinamensis* (L.) *Pestic. Sci.*, *14*:145–152.

Beeman, R. W. (1983). Inheritance and linkage of malathion resistance in the red flour beetle. *J. Econ. Entomol.*, *76*:737–740.

Beeman, R. W., and Nanis, S. M. (1986). Malathion resistance alleles and their fitness in the red flour beetle (Coleoptera: Tenebrionidae). *J. Econ. Entomol.*, *79*:580–587.

Beeman, R. W., and Schmidt, B. A. (1982). Biochemical and genetic aspects of malathion-specific resistance in the Indianmeal moth (Coleoptera: Pyralidae). *J. Econ. Entomol.*, *75*:945–949.

Beeman, R. W., and Stuart, J. J. (1990). A gene for lindane + cyclodiene resistance in the red flour beetle (Coleoptera: Tenebrionidae). *J. Econ. Entomol.*, *83*:1745–1751.

Beeman, R. W., Speirs, W. E., and Schmidt, B. A. (1982). Malathion resistance in Indianmeal moths (Lepidoptera: Pyralidae) infesting stored corn and wheat in northcentral United States. *J. Econ. Entomol.*, *75*:950–954.

Beeman, R. W., and Wright, V. F. (1990). Monitoring for resistance to chlorpyrifos-methyl, pirimiphos-methyl and malathion in Kansas populations of stored product insects. *J. Kansas Entomol. Soc.*, *63*:385–392.

Bell, C. H. (1976). The tolerance of developmental stages of four stored product moths to phosphine. *J. Stored Prod. Res.*, *12*:77–86.

Bell, C. H., Hole, B. D., and Evans, P. H. (1977). The occurrence of resistance to phosphine in adult and egg stages of *Rhyzopertha dominica* (F.) (Coleoptera: Bostrichidae). *J. Stored Prod. Res.*, *13*:91–94.

Benezet, H. J., and Helms, C. W. (1994). Methoprene resistance in the cigarette beetle, *Lasioderma serricorne* (F.) (Coleoptera: Anobiidae) from tobacco storages in southeastern United States. *Resist. Pest. Mgmt.*, *6*:17–19.

Bhatia, S. K., and Pradhan, S. (1971). Studies on resistance to insecticides in *Tribolium castaneum* (Herbst)-III. Selection of a strain resistant to lindane and its biological characteristics. *J. Stored Prod. Res.*, *6*:331–337.

Binns, T. J. (1986). The comparative toxicity of malathion, with and without the addition of triphenyl phosphate, to adults and larvae of a susceptible and resistant strains of *Tribolium castaneum* (Herbst) (Coleoptera: Tenebrionidae). *J. Stored Prod. Res.*, *22*: 97–101.

Bond, E. J., and Buckland, C. T. (1979). Development of resistance to carbon dioxide in the granary weevil. *J. Econ. Entomol.*, *72*:770–771.
Borah, G., and Chahal, B. S. (1979). Development of resistance in *Trogoderma granarium* Everts to phosphine in the Punjab. *FAO Plant Prot. Bull.*, *27*:77–80.
Brown, T. M., and Brogdon, W. G. (1987). Improved detection of insecticide resistance through conventional and molecular techniques. *Annu. Rev. Entomol.*, *32*:145–162.
Brown, T. M., and Payne, G. T. (1988). Experimental selection for insecticide resistance. *J. Econ. Entomol.*, *81*:49–56.
Brown, T. M., DeVries, D. H., and Brown, A. W. A. (1978). Induction of resistance to insect growth regulators. *J. Econ. Entomol.*, *71*:223–229.
Burroughs, R., Schenck-Hamlin, D., and Wright, V. F. (1988). A bibliography of plant materials tested for their activity against stored-product insects. Research Report No. 9, Kansas State University, Manhattan, Kansas, p. 38.
Butt, B. A., and Cantu, E. (1962). Sex determination of lepidopterous pupae. United States Deparatment of Agriculture, Agricultural Research Service Report-33-75, pp. 1–7.
Champ, B. R. (1967). The inheritance of DDT resistance in *Sitophilus oryzae* (L.) (Coleoptera: Curculionidae) in Queensland. *J. Stored Prod. Res.*, *3*:321–324.
Champ, B. R., and Campbell-Brown, M. (1969). Genetics of lindane resistance in *Tribolium castaneum*. *J. Stored Prod. Res.*, *5*:399–406.
Champ, B. R., and Campbell-Brown, M. (1970). Insecticide resistance in Australian *Tribolium castaneum* (Hbst.) (Coleoptera: Tenebrionidae)—II. Malathion resistance in eastern Australia. *J. Stored Prod. Res.*, *6*:111–131.
Champ, B. R., and Campbell-Brown, M. (1976). The inheritance of resistance in stored grain insects, *Tribolium castaneum*. In *Report of the Global Survey of Pesticide Susceptibility of Stored Grain Pests*, FAO Plant Prot. Ser. No. 5, pp. 232–233.
Champ, B. R., and Cribb, J. N. (1965). Lindane resistance in *Sitophilus oryzae* (L.) and *Sitophilus zeamais* Motsch. (Coleoptera: Curculionidae) in Queensland. *J. Stored Prod. Res.*, *1*:9–24.
Champ, B. R., and Dyte, C. E. (1977). FAO global survey of pesticide susceptibility of stored grain pests. *FAO Plant Prot. Bull.*, *25*:49–67.
Champ, B. R., and Dyte, C. E. (1980). Method for adults of some major beetle pests of stored cereals with malathion or lindane-FAO Method No. 15. In *Recommended Methods for Measurement of Pest Resistance to Pesticides* (ed. Busvine, J. R. ). FAO Plant Prod. and Prot. Paper 21, F.A.O. Rome, pp. 77–89.
Chilcutt, C. F., and Tabashnik, B. E. (1995). Evolution of pesticide resistance and slope of the concentration-mortality line: are they related? *J. Econ. Entomol.*, *88*:11–20.
Cogan, P. M. (1982). A method for rapid detection of malathion resistance in *Plodia interpunctella* (Hübner) (Lepidoptera: Pyralidae) with further records of resistance. *J. Stored Prod. Res.*, *18*:121–124.
Collins, P. J. (1985). Resistance to grain protectants in field populations of the sawtoothed grain beetle in southern Queensland. *Aust. J. Exp. Agric.*, *25*:683–686.
Collins, P. J. (1986). Genetic analysis of fenitrothion resistance in the sawtoothed grain beetle, *Oryzaephilus surinamensis* (Coleopter: Cucujidae). *J. Econ. Entomol.*, *79*:1196–1199.
Collins, P. J. (1990). A new resistance to pyrethroids in *Tribolium castaneum* (Herbst). *Pestic. Sci.*, *28*:101–115.

Collins, P. J., and Wilson, D. (1987). Efficacy of current and potential grain protectant insecticides against a fenitrothion-resistant strain of the sawtoothed grain beetle, *Oryzaephilus surinamensis* L. *Pestic. Sci.*, *20*:93–104.

Collins, P. J., Lambkin, T. M., Bridgeman, B. W., and Pulvirenti, C. (1993). Resistance to grain-protectant insecticides in coleopterous pests of stored cereals in Queensland, Australia. *J. Econ. Entomol.*, *86*:239–245.

Cox, P. G., and Forrester, N. W. (1992). Economics of insecticide resistance management in *Heliothis armigera* (Lepidoptera: Noctuidae) in Australia. *J. Econ. Entomol.*, *85*: 1539–1550.

Croft, B. A. (1982). Arthropod resistance to insecticides: a key to pest control failures and successes in North American apple orchards. *Entomol. Exp. Appl.*, *31*:88.

Crow, J. F. (1957). Genetics of insect resistance to chemicals. *Annu. Rev. Entomol.*, *2*: 227–246.

Daly, J. C., Fisk, J. H., and Forrester, N. W. (1988). Selective mortality in field trials between strains of *Heliothis armigera* (Lepidoptera: Noctuidae) resistant and susceptible to pyrethroids: functional dominance of resistance and age class. *J. Econ. Entomol.*, *81*:1000–1007.

Denholm, I., and Rowland, M. W. (1992). Tactics for managing pesticide resistance in arthropods: theory and practice. *Annu. Rev. Entomol.*, *37*:91–112.

Dittrich, V., Luetkemeier, N., and Voss, G. (1980). OP-resistance in *Spodoptera littoralis*: inheritance, larval and imaginal expression, and consequence for control. *J. Econ. Entomol.*, *76*:356–362.

Donohaye, E. J. (1991). The potential for stored product insects to develop resistance to modified atmospheres. In Proceedings, 5th International Working Conference of Stored Product Protection, INRA/SDPV, Bordeaux, France, pp. 989–997.

Doolittle, D. P. (1986). *Population Genetics: Basic Principles*. Springer-Verlag, New York.

Dover, M., and Croft, B. A. (1984). *Getting Tough: Public Policy and the Management of Pesticide Resistance*. World Resources Institute, Washington, D.C.

Dover, M., and Croft, B. A. (1986). Pesticide resistance and public policy: resistance management could become the key to continuing effective pest control. *Bioscience*, *36*:78–85.

Dowd, P. F. (1989). In situ production of hydrolytic detoxifying enzymes by symbiotic yeasts in the cigarette beetle (Coleoptera: Anobiidae). *J. Econ. Entomol.*, *82*:396–400.

Dowd, P. F., Sparks, T. C., and Mitchell, F. L. (1984). A microcomputer simulation program for demonstrating the development of insecticide resistance. *Bull. Entomol. Soc. Am.*, *30*:37–41.

Dyte, C. E. (1972). Resistance to synthetic juvenile hormone in a strain of the flour beetle, *Tribolium castaneum*. *Nature* (London), *238*:48–49.

Dyte, C. E. (1979). The importation of insecticide-resistant strains of stored-product pests. *Ann. Appl. Biol.*, *91*:414–417.

Dyte, C. E. (1991). Living with resistant strains of storage pests. In Proceedings, 5th International Working Conference of Stored Product Protection, INRA/SDPV, Bordeaux, France, pp. 947–959.

Dyte, C. E., and Blackman, D. G. (1970). The spread of insecticide resistance in *Tribolium castaneum* (Herbst) (Coleoptera: Tenebrionidae). *J. Stored Prod. Res.*, *6*:255–261.

Dyte, C. E., and Blackman, D. G. (1972). Laboratory evaluation of organophosphorous insecticides against susceptible and malathion-resistant strains of *Tribolium castaneum*. *J. Stored Prod. Res.*, *8*:103–109.

Dyte, C. E., and Halliday, D. (1985). Problems of development of resistance to phosphine by insect pests of stored grain. *Bull. OEPP/EPPO Bull.*, *15*:51–57.

Dyte, C. E., and Rowlands, D. G. (1968). The metabolism and synergism of malathion in resistant strains of *Tribolium castaneum*. *J. Stored Prod. Res.*, *4*:157–173.

El-Lakwah, S. M., Ahmed, S. M., Khaltab, M. M., and Abdel-Latief, A. M. (1992). Selection of the red flour beetle (*Tribolium castaneum* Herbst) for resistance to phosphine in the laboratory and biological observations on the resistant strain. In *Insecticides: Mechanism of Action and Resistance* (ed. Otto, D., and Weber, B.), Andover, United Kingdom, pp. 409–426.

Finney, D. J. (1971). *Probit Analysis*, 3rd edition. Cambridge University Press, London.

Firko, M. J., and Hayes, J. L. (1990). Quantitative genetic tools for insecticide resistance risk assessment: estimating the heritability of resistance. *J. Econ. Entomol.*, *83*: 647–654.

Georghiou, G. P. (1969). Genetics of resistance to insecticides in houseflies and mosquitoes. *Exp. Parasitol.*, *26*:224–255.

Georghiou, G. P. (1972). The evolution of resistance to pesticides. *Annu. Rev. Ecol. Syst.*, *3*:133–168.

Golob, P., Broadhead, P., and Wright, M. (1991). Development of resistance to insecticides by populations of *Prostephanus truncatus* (Horn) (Coleoptera: Bostrichidae). In Proceedings, 5th International Working Conference of Stored Product Protection, INRA/SDPV, Bordeaux, France, pp. 999–1007.

Gould, F. (1984). Role of behaviour in the evolution of insect adaptation to insecticides and resistant host plants. *Bull. Entomol. Soc. Am.*, *30*:34–41.

Guedes, R. N. C. (1993). *Detecçao e herança de resistência ao DDT e a piretroides em Sitophilus zeamais Motschulsky (Coleptera: Curculionidae)*. M.Sc. Entomology Thesis, Universidade Federal de Viçosa, Viçosa, Minas Gerais, Brasil, p. 67.

Guedes, R. N. C., Lima, J. O. G., Santos, J. P., and Cruz, C. D. (1994). Inheritance of deltamethrin resistance in a Brazilian strain of maize weevil (*Sitophilus zeamais* Mots.). *Int. J. Pest Mgmt.*, *40*:103–106.

Haliscak, J. P., and Beeman, R. W. (1983). Status of malathion resistance in five genera of beetles infesting farm-stored corn, wheat, and oats in the United States. *J. Econ. Entomol.*, *76*:717–722.

Halliday, W. R. (1992). Tralomethrin as a long-term protectant of stored corn and wheat. *J. Agric. Entomol.*, *9*:145–163.

Halliday, W. R., and Burnham, K. P. (1990a). Choosing the optimal diagnostic dose for monitoring insecticide resistance. *J. Econ. Entomol.*, *83*:1511–1559.

Halliday, W. R., and Burnham, K. P. (1990b). BESTDOSE and RANGES: two computer programs for determining the optimal dose in diagnostic dose tests. *J. Econ. Entomol.*, *83*:1160–1169.

Halliday, W. R., Arthur, F. H., and Zettler, J. L. (1988). Resistance status of red flour beetle (Coleoptera: Tenebrionidae) infesting stored peanuts in the Southeastern United States. *J. Econ. Entomol.*, *8*:74–77.

Halstead, D. G. H. (1963). External sex differences in stored-products Coleoptera. *Bull. Entomol. Res.*, *54*:119–134.

Harein, P. K., Gardner, R. D., and Cloud, H. (1985). 1984 review of Minnesota stored grain management practices. University of Minnesota Agricultural Experiment Station Bulletin AD-SB-2705.

Haynes, K. F., Miller, T. A., Staten R. T., Li, W. G., and Baker, T. C. (1986). Monitoring insecticide resistance with insect pheromones. *Experientia*, *42*:1293–1295.

Heather, N. W. (1982). Comparison of population growth rates of malathion resistant and susceptible populations of the rice weevil, *Sitophilus oryzae* (Linnaeus) (Coleoptera: Curculionidae). *Queensland J. Agric. Anim. Sci.*, *39*:61–68.

Heather, N. W. (1986). Sex-linked resistance to pyrethroids in *Sitophilus oryzae* (L.) (Coleoptera: Curculionidae). *J. Stored Prod. Res.*, 22:15–20.

Herron, G. A. (1990). Resistance to grain protectants and phosphine in coleopterous pests of grain stored on farms in New South Wales. *J. Aust. Entomol. Soc.*, *29*:183–189.

Hole, B. D., Bell, C. H., Mills, K. A., And Goodship, G. (1976). The toxicity of phosphine to all developmental stages of thirteen species of stored-product beetles. *J. Stored Prod. Res.*, *12*:235–244.

Holloway, G. J., and McCaffery, A. R. (1988). Reactive and preventative strategies for the management of insecticide resistance. In Proceedings British Crop Protection Conference—Pests and Diseases, BCPC Publications, Surrey, United Kingdom, pp. 465–469.

Horton, P. M. (1984). Evaluation of South Carolina field strains of certain stored-product Coleoptera for malathion resistance and pirimphos-methyl susceptibility. *J. Agric. Entomol.*, *1*:1–5.

Howe, R. W. (1973). The susceptibility of the immature and adult stages of *Sitophilus granarius* to phosphine. *J. Stored Prod. Res.*, *8*:241–262.

Immaraju, J. A., Morse, J. G., and Hobza, R. F. (1990). Field evaluation of insecticide rotation and mixtures as strategies for citrus thrip (Thysanoptera: Thripidae) resistance management in California. *J. Econ. Entomol.*, *83*:306–314.

Jacob, T. A. (1988). The effect of temperature and humidity on the development period and mortality of *Typhaea stercorea* (L.) (Coleoptera: Mycetophagidae). *J. Stored Prod. Res.*, *24*:221–224.

Jermannaud, A. (1994). Field evaluation of a test kit for monitoring insecticide resistance in stored-grain pests. In Proceedings, 6th International Working Conference of Stored Product Protection, Canberra, Australia, pp. 795–797.

Johnson, M. W., and Tabashnik, B. E. (1994). Laboratory selection for pesticide resistance in natural enemies. In *Application of Genetics to Arthropods of Biological Control Significance* (eds. Narang, S. K., Bartlett, A. C., and Faust, R. M.). CRC Press, Boca Raton, Florida, pp. 91–105.

Kable, P. F., and Jeffery, H. (1980). Selection for tolerance in organisms exposed to sprays of biocide mixtures: a theoretical model. *Phytopathology*, *70*:8–12.

Kenkel, P., Criswell, J. T., Cuperus, G. W., Noyes, R. T., Anderson, K., Fargo, W. S., Shelton, K., Morrison, W. P., and Adam, B. (1994). Current management practices and impact of pesticide loss in the hard red wheat post-harvest system. Oklahoma Cooperative Extension Service, Oklahoma State University Publication No. E-930.

Kinsinger, R. A., and McGaughey, W. H. (1979). Susceptibility of populations of Indianmeal moth and almond moth to *Bacillus thuringiensis*. *J. Econ. Entomol.*, *72*:346–349.

Knight, A. L., and Norton, G. W. (1989). Economics of agricultural pesticide resistance in arthropods. *Annu. Rev. Entomol.*, *34*:293–313.

Lindgren, D. L., and Vincent, L. E. (1966). Relative toxicity of hydrogen phosphide to various stored-product insects. *J. Stored Prod. Res.*, 2:141–146.

Linsley, E. G. (1944). Natural sources, habitats, and reservoirs of insects associated with stored food products. *Hilgardia*, *16*:187–222.

Longstaff, B. C. (1988). Temperature manipulation and the management of insecticide resistance in stored grain pests: a simulation study for the rice weevil, *Sitophilus oryzae*. *Ecol. Model.*, *43*:303–313.

Markwick, N. P. (1994). Genetic analysis of resistance of two strains of *Typhlodromus pyri* to synthetic pyrethroid insecticides. In *Applications of Genetics to Arthropods of Biological Control Significance* (ed. Narang, S. K., Bartlett, A. C., and Faust, R. M.). CRC Press, Boca Raton, Florida, pp. 107–120.

McGaughey, W. H. (1985). Insect resistance to the biological insecticide *Bacillus thuringiensis*. *Science*, *229*:193–195.

McGaughey, W. H., and Johnson, D. E. (1987). Toxicity of different serotypes and toxins of *Bacillus thuringiensis* to resistant and susceptible Indianmeal moth (Lepidoptera: Pyralidae). *J. Econ. Entomol.*, *80*:1122–1126.

McGaughey, W. H., and Beeman, R. W. (1988). Resistance to *Bacillus thuringiensis* in colonies of Indianmeal moth and almond moth (Lepidoptera: Pyralidae). *J. Econ. Entomol.*, *81*:28–33.

Mello, E. J. R. (1970). Constataçao de resistência ao DDT e lindane em *Sitophilus oryzae* (L.) em milho armazenado, no localidade de Capinopilis, Minas Gerais. In *Reuniao Brasileira Milho*, Porto Alegre, PIP pp. 130–131.

Metcalf, R. L. (1980). Changing role of insecticides in crop protection. *Annu. Rev. Entomol.*, *25*:219–256.

Mian, L. S., and Mulla, M. S. (1982). Residual activity of insect growth regulators against stored-product beetles in grain commodities. *J. Econ. Entomol.*, *75*:599–603.

Mills, K. A. (1983). Resistance to the fumigant hydrogen phosphide in some stored product species associated with repeated inadequate treatments. In Proceedings, 2nd European Congress of Entomology, Kiel, 1982. Mitt. Dtsch., *Ges. Allg. Angew. Entomol.*, *4*: 98–101.

Miranowski, J. A., and Carlson, G. A. (1986). Economic issues in public and private approaches to preserving pest susceptibility. In *Pesticide Resistance: Strategies and Tactics for Management*, National Academy of Sciences, Washington, D.C, pp. 436–448.

Muggleton, J. (1982). A model for the elimination of insecticide resistance using heterozygous disadvantage. *Heredity*, *49*:247–251.

Muggleton, J. (1983). Relative fitness of malathion-resistant phenotypes of *Oryzaephilus surinamensis* L. (Coleoptera: Silvanidae). *J. Appl. Ecol.*, *20*:245–254.

Muggleton, J. (1986). Selection for malathion resistance in *Oryzaephilus surinamensis* (L.) (Coleoptera: Silvanidae): fitness values of resistant and susceptible phenotypes and their inclusion in a general model describing the spread of resistance. *Bull. Entomol. Res.*, *76*:469–480.

Muggleton, J. (1987). Insecticide resistance in stored product beetles and its consequences for their control. In Proceedings, 1986 British Crop Protection Conference—Pests and Diseases, BCPC Monograph No. 37, Stored Products Pest Control, pp. 177–186.

Muggleton, J., Duran, J. E., and Rowlands, D. G. (1981). The effect of holding adult *Oryzaephilus surinamensis* (L.) (Coleoptera: Silvanidae) at different temperatures on their subsequent susceptibility to malathion. *J. Stored Prod. Res.*, *17*:125–130.

Muggleton, J., Llewellin, J. A., and Prickett, A. J. (1991). Insecticide resistance in populations of *Oryzaephilus surinamensis* and *Cryptolestes ferrugineus* from grain stores in the United Kingdom. In Proceedings, 5th International Working Conference of Stored Product Protection, INRA/SDPV, Bordeaux, France, pp. 1019–1028.

Nakakita, H., and Kuroda, J. (1986). Differences in phosphine uptake between susceptible and resistant strains of insects. *J. Pestic. Sci.*, *11*:21–26.

Navarro, S., Dias, R., and Donahaye, E. (1985). Induced tolerance of *Sitophilus oryzae* adults to carbon dioxide. *J. Stored Prod. Res.*, *21*:207–213.

Noble, R. M., Hamilton, D. J., and Osborne, W. J. (1982). Stability of pyrethroids on wheat in storage. *Pestic. Sci.*, *13*:246–252.

Osman, N., and Morallo-Rejesus, B. (1981). Evaluation of Indonesian strains of *Tribolium castaneum* (Herbst) for resistance to malathion and pirimiphos-methyl. *Philipp. Ent.*, *4*:405–414.

Otto, D., Moli, E., and Richter, P. (1992). A critical comment on the evaluation of the resistance level in field populations be the resistance index Ri. In *Insecticides: Mechanism of Action and Resistance* (ed. Otto, D., and Weber, B.). Intercept, Andover, United Kingdom, pp. 363–375.

Pacheco, I. A., Sartori, M. R., and Bolonhezi, S. (1991). Resistance to malathion, pirimiphos-methyl and fenitrothion in Coleoptera from stored grains. In Proceedings, 5th International Working Conference of Stored Product Protection, INRA/SDPV, Bordeaux, France, pp. 1029–1037.

Parkin, E. A. (1965). The onset of insecticide resistance among field populations of stored-product insects. *J. Stored Prod. Res.*, *1*:3–8.

Pasalu, I. C., and Bhatia, S. K. (1974). Specific nature of malathion-resistant strain of *Tribolium castaneum* (Hbst.) in India. *Bull. Grain Technol.*, *12*:229–231.

Pasteur, N., and Georghiou, G. P. (1989). Improved filter paper test for detecting and quantifying increased esterase activity in organophosphate-resistant mosquitoes (Diptera: Culicidae). *J. Econ. Entomol.*, *82*:347–353.

Pieterse, A. H., Schulten, G. G. M., and Kuyken, W. (1992). A study on insecticide resistance in *Tribolium castaneum* (Hbst.) in Malawi (Central Africa). *J. Stored Prod. Res.*, *8*:183–191.

Plapp, F. W., Jr., Browning, C. R., and Sharpe, P. J. H. (1979). Analysis of rate of development of insecticide resistance based on simulation of a genetic model. *Environ. Entomol.*, *8*:494–500.

Press, J. W., Flaherty, B. R., and McDonald, L. L. (1981). Survival and reproduction of *Bracon hebetor* on insecticide-treated *Ephestia cautella* larvae. *J. Georgia Entomol. Soc.*, *16*:231–234.

Preisler, H. K., Hoy, M. A., and Robertson, J. L. (1990). Statistical analysis of modes of inheritance for pesticide resistance. *J. Econ. Entomol.*, *83*:1649–1655.

Qureshi, A. H., Bond, E. J., and Monro, H. A. U. (1965). Toxicity of hydrogen phosphide to the granary weevil, *Sitophilus granarius* and other insects. *J. Econ. Entomol.*, *58*: 324–331.

Ramsey, P. R., and Farley, T. K. (1978). Mating and fecundity in malathion-resistant and susceptible strains of the Indian meal moth, *Plodia interpunctella*. *Ann. Entomol. Soc. Am.*, *71*:513–516.

Rajendran, S. (1992). Selection for resistance to phosphine or methyl bromide in *Tribolium castaneum* (Coleoptera: Tenebrionidae). *Bull. Entomol. Res.*, *82*:119–124.

Rajendran, S. (1994). Responses of phosphine-resistant strains of two stored-product insect pests to changing concentrations of phosphine. *Pestic. Sci.*, *40*:183–186.

Rajendran, S., and Narasimhan, K. S. (1994). Phosphine resistance in the cigarette beetle *Lasioderma serricorne* (Coleoptera: Anobiidae) and overcoming control failures during fumigation of stored tobacco. *Intl. J. Pest Mgmt.*, *40*:207–210.

Reidy, G. F., Rose, H. A., Visetson, S., and Murray, M. (1990). Increased glutathione *S*-transferase activity and glutathione content in an insecticide resistant strain of *Tribolium castaneum* (Herbst). *Pestic. Biochem. Physiol.*, *36*:269–276.

Riddles, P. W., and Nolan, J. (1987). Prospects for the management of arthropod resistance to pesticides. *Intl. J. Parasitol.*, *17*:679–688.

Robertson, J. L., and Preisler, H. K. (1992). *Pesticide Bioassays with Arthropods*. CRC Press, Boca Raton, Florida.

Robertson, J. L., Preisler, H. K., Frampton, E. R., and Armstrong, J. W. (1993). Statistical analyses to estimate efficacy of disinfestation treatments. In *Quarantine Treatments for Pests of Food Plants* (ed. Sharp, J. L., and Hallman, G. J.). Westview Press, San Francisco, California, pp. 47–65.

Robertson, J. L., Preisler, H. K., Ng, S. S., Hickle, L. A., and Gelernter, W. D. (1995). Natural variation: a complicating factor in bioassays with chemical and microbial pesticides. *J. Econ. Entomol.*, *88*:1–10.

Rose, H. A., and Wallbank, B. E. (1986). Mixed-function oxidase and glutathione *S*-transferase activity in a susceptible and a fenitrothion-resistant strain of *Oryzaephilus surinamensis* (Coleoptera: Cucujidae). *J. Econ. Entomol.*, *79*:896–899.

Rosenheim, J. A., and Hoy, M. A. (1989). Confidence intervals for the Abbott's formula correction of bioassay data for control response. *J. Econ. Entomol.*, *82*:331.

Roush, R. T. (1989). Designing resistance management programs: how can you choose? *Pestic. Sci.*, *26*:423–441.

Roush, R. T., and Croft, B. A. (1986). Experimental population genetics and ecological studies of pesticide resistance in insects and mites. In *Pesticide Resistance: Strategies and Tactics for Management*. National Academy of Sciences, Washington, D.C., pp. 257–270.

Roush, R. T., and Miller, G. L. (1986). Considerations for designing of insecticide resistance monitoring programs. *J. Econ. Entomol.*, *79*:293–298.

Saleem, M. A., and Wilkins, R. M. (1983). Toxicity of primiphos-methyl against a malathion resistant and susceptible strain of the sawtoothed grain beetle *Oryzaephilus surinamensis* (L.). *Pakistan J. Zool.*, *15*:89–93.

Santhoy, O., and Morallo-Rejesus, B. (1984). Toxicity of six organophosphorous insecticides to field-collected DDT-resistant strains of rice weevil, *Sitophilus oryzae* (L.) and red flour beetle, *Tribolium castaneum* (Herbst). *Philipp. Ent.*, 2:283–290.

Sartori, M. R., Pacheco, I. A., and Vilar, R. M. G. (1991). Resistance to phosphine in stored grain insects in Brazil. In Proceedings, 5th International Working Conference of Stored Product Protection, INRA/SDPV, Bordeaux, France, pp. 1041–1050.

Saxena, J. D., and Bhatia, S. K. (1980). Reduction in fecundity of *Tribolium castaneum* due to fumigation and phosphine resistance. *India. J. Ent.*, *42*:796–798.

Sayaboc, P. D., and Acda, M. A. (1990). Resistance of major coleopterous pests of stored grain to malathion and pirimiphos-methyl. *Philipp. Ent.*, *8*:653–660.

Schmid, V. W. (1987). Permethrin-resistenz beim Teppichkäfer *Anthrenus flavipes* Casey (Col., Dermestidae). *Anz. Schädlingskde., Pflanzenschutz, Umweltschutz*, *60*:15–18.

Shen, S. K., and Dowd, P. F. (1991). Detoxification spectrum of the cigarette beetle symbiont *Symbiotophrina kochii* in culture. *Entomol. Exp. Appl.*, *60*:51–59.

Sinclair, E. R., and Alder, J. (1985). Development of a computer simulation model of stored product insect populations on grain farms. *Agric. Syst.*, *18*:95–113.

Sobotka, W., Konopinska, D., and Nawrot, J. (1992). A new class of antifeedants against stored product insects. In *Insecticides: Mechanism of Action and Resistance* (ed. Otto, D., and Weber, B.). Intercept, Andover, United Kingdom, pp. 117–124.

Speirs, R. D., Redlinger, L. M., and Boles, H. P. (1967). Malathion resistance in the red flour beetle. *J. Econ. Entomol.*, *60*:1373–1374.

Stone, B. F. (1968). A formula for determining the degree of dominance in cases of monofactorial inheritance of resistance to chemicals. *Bull. W.H.O.*, *38*:325–326.

Stone, N. D., Makela, M. E., and Plapp, F. W. (1988). Nonlinear optimization analysis of insecticide mixtures for the control of the tobacco budworm (Lepidoptera: Noctuidae). *J. Econ. Entomol.*, *81*:989–994.

Storey, C. L., Sauer, D. B., and Walker, D. (1984). Present use of pest management practices in wheat, corn, and oats stored on the farm. *J. Econ. Entomol.*, *77*:784–788.

Sriharen, S., Dunkel, F., and Nizeyimana, E. (1991). Status of Actellic resistance in three stored product insects infesting sorghum and beans in Rwanda. In Proceedings, 5th International Working Conference of Stored Product Protection, INRA/SDPV, Bordeaux, France, pp. 1051–1060.

Subramanyam, Bh., and Cutkomp, L. K. (1985). Moth control in stored grain and the role of *Bacillus thuringiensis*: An overview. *Residue Rev.*, *94*:1–47.

Subramanyam, Bh., and Harein, P. K. (1990). Status of malathion and primiphos-methyl resistance in adults of red flour beetle and sawtoothed grain beetle infesting farm-stored corn in Minnesota. *J. Agric. Entomol.*, *7*:127–136.

Subramanyam, Bh., Harein, P. K., and Cutkomp, L. K. (1989). Organophosphate resistance in adults of red flour beetle (Coleoptera: Tenebrionidae) and sawtoothed grain beetle (Coleoptera: Cucujidae) infesting barley stored on farms in Minnesota. *J. Econ. Entomol.*, *82*:989–995.

Tabashnik, B. E. (1986a). Computer simulation as a tool for pesticide resistance management. In *Pesticide Resistance: Strategies and Tactics for Management*. National Academy of Sciences, Washington, D.C., pp. 194–206.

Tabashnik, B. E. (1986b). Model for managing resistance to fenvalerate in the diamondback moth (Lepidoptera: Plutellidae). *J. Econ. Entomol.*, *79*:1447–1451.

Tabashnik, B. E. (1989). Managing resistance with multiple pesticide tactics: theory, evidence, and recommendations. *J. Econ. Entomol.*, *82*:1263–1269.

Tabashnik, B. E. (1990). Modeling and evaluation of resistance management tactics. In *Pesticide Resistance in Arthropods* (ed. R. T. Roush and B. E. Tabashnik). Chapman and Hall, New York, pp. 153–182.

Tabashnik, B. E. (1991). Determining the mode of inheritance of pesticide resistance with backcross experiments. *J. Econ. Entomol.*, *84*:703–712.

Tabashnik, B. E. (1992). Resistance risk assessment: realized heritability of resistance to *Bacillus thuringiensis* in diamond back moth (Lepidoptera: Plutellidae), tobacco budworm (Lepidoptera: Noctuidae), and Colorado potato beetle (Coleoptera: Chrysomelidae). *J. Econ. Entomol.*, *85*:1551–1559.

Tabashnik, B. E. (1994). Evolution of resistance to *Bacillus thuringiensis*. *Annu. Rev. Entomol.*, *39*:47–79.

Tabashnik, B. E., and Croft, B. A. (1982). Managing pesticide resistance in crop-arthropod complexes: interactions between biological and operational factors. *Environ. Entomol.*, *11*:1137–1144.

Tabashnik, B. E., and Croft, B. A. (1985). Evolution of pesticide resistance in apple pests and their natural enemies. *Entomophaga*, *30*:37–49.

Tabashnik, B. E., and McGaughey, W. H. (1994). Resistance risk assessment for single and multiple insecticides: responses of Indianmeal moth (Lepidoptera: Pyralidae) to *Bacillus thuringiensis*. *J. Econ. Entomol.*, *87*:834–841.

Tabashnik, B. E., Croft, B. A., and Rosenheim, J. A. (1990). Spatial scale of fenvalerate resistance in pear psylla (Homoptera: Psyllidae) and its relationship to treatment history. *J. Econ. Entomol.*, *83*:1177–1183.

Tabashnik, B. E., Rosenheim, J. A., and Caprio, M. A. (1992). What do we really know about management of insecticide resistance? In *Resistance '91: Achievements and Developments in Combating Pesticide Resistance* (ed. I. Denholm, A. L. Devonshire, and Hollomon, D. W.). Elsevier, New York, pp. 124–135.

Taylor, R. W. D., and Halliday, D. (1986). The geographical spread of resistance to phosphine by coleopterous pests of stored products. In Proceedings, 1986 British Crop Protection Conference—Pests and Diseases, BCPC Monograph No. 37, Stored Products Pest Control, pp. 607–613.

Tyler, P. S., Taylor, R. W. D., and Rees, D. P. (1983). Insect resistance to phosphine fumigation in food warehouses in Bangladesh. *Intl. Pest Control*, *25*:10–13.

Uyenoyama, M. K. (1986). Pleiotropy and the evolution of genetic systems conferring resistance to pesticides, In *Pesticide Resistance: Strategies and Tactics for Management*. National Academy of Sciences, Washington, D.C., pp. 207–221.

Van Rie, J., McGaughey, W. H., Johnson, D. E., Barnett, B. D., and Van Mellaert, H. (1990). Mechanism of insect resistance to the microbial insecticide *Bacillus thuringiensis*. *Science*, *247*:72–74.

Via, S. (1986). Quantitative genetic models and the evolution of pesticide resistance. In *Pesticide Resistance: Strategies and Tactics for Management*. National Academy of Sciences, Washington, D.C., pp. 222–235.

Vincent, L. E., and Lindgren, D. L. (1972). Toxicity of phosphine to the life stages of four species of dermestids. *J. Econ. Entomol.*, *65*:1429–1431.

Voss, G. (1988). Insecticide/Acaricide resistance: industry's efforts and plans to cope. *Pestic. Sci.*, *23*:149–156.

Wakefield, M. E., and Cogan, P. M. (1991). Resistance to iodofenphos and malathion in the lesser mealworm *Alphitobius diaperinus*. In Proceedings, 5th International Working Conference of Stored Product Protection, INRA/SDPV, Bordeaux, France, pp. 1065–1073.

Wallbank, B. E. (1984). Fenitrothion resistance in *Oryzaephilus surinamensis* (L.) saw-toothed grain beetle, in New South Wales. In Proceedings, 4th Australian Applied Entomology Resistance Conference, Adelaide, Australia, pp. 507–511.

Weinzierl, R. A., and Porter, R. P. (1990). Resistance of hairy fungus beetle (Coleoptera: Mycetophagidae) to pirimiphos-methyl and malathion. *J. Econ. Entomol.*, *83*: 325–328.

White, G. G., and Lambkin, T. A. (1990). Baseline responses to phosphine and resistance status of stored-grain beetle pests in Queensland, Australia. *J. Econ. Entomol.*, *83*:1738–1744.

White, N. D. G., and Bell, R. J. (1988). Inheritance of malathion-resistance in a strain of *Tribolium castaneum* (Coleoptera: Tenebrionidae) and effects of resistance genotypes on fecundity and larval survival in malathion-treated wheat. *J. Econ. Entomol.*, *81*:381–386.

White, N. D. G., and Loschiavo, S. R. (1985). Testing for malathion resistance in field-collected populations of *Cryptolestes ferrugineus* (Stephens) and factors affecting reliability of the tests. *J. Econ. Entomol.*, *78*:511–515.

Winks, R. G. (1982). The toxicity of phoshine to adults of *Tribolium castaneum* (Herbst): time as a response factor. *J. Stored. Prod. Res.*, *18*:159–169.

Winks, R. G. (1984). The toxicity of phosphine to adults of *Tribolium castaneum* (Herbst): time as a dosage factor. *J. Stored. Prod. Res.*, *20*:45–56.

Winks, R. G. (1985). The toxicity of phosphine to adults of *Tribolium castaneum* (Herbst): phosphine induced narcosis. *J. Stored Prod. Res.*, *21*:25–29.

Winks, R. G., Champ, B. R., Dyte, C. E., Bond, E. J., Barker, P. S., Lindgren, D. L., and Davis, R. (1980). Method for adults of some major pest species of stored cereals with methyl bromide and phosphine-FAO Method No. 16. In *Recommended Methods for Measurement of Pest Resistance to Pesticides* (ed. J. R. Busvine). FAO Plant Production and Protection Paper 21, F.A.O. Rome, pp. 91–102.

Wool, D., and Kamin-Belsky, N. (1983). Age-dependent resistance to malathion in adult almond moths, *Ephestia cautella* (Walker). *Z. Ang. Ent.*, *96*:386–391.

Wool, D., and Kamin-Belsky, N. (1984). Effects of diet and larval density on adult sensitivity to malathion and on ecological parameters in *Ephestia cautella* (Walker) (Lepidoptera: Phyticidae). *Z. Ang. Ent.*, *98*:58–62.

Wool, D., and Manheim, O. (1980). Genetically-induced susceptibility to malathion in *Tribolium castaneum* despite selection for resistance. *Entomol. Exp. Appl.*, *28*: 183–190.

Wool, D., and Noiman, S. (1983). Integrated control of insecticide resistance by combined genetic and chemical treatments: a warehouse model with flour beetles (*Tribolium*; Tenebrionidae, Coleoptera). *Z. Ang. Ent.*, *95*:22–30.

Wool, D., Noiman, S., Manheim, O., and Cohen, E. (1982). Malathion resistance in *Tribolium* strains and their hybrids: inheritance patterns and possible enzymatic mechanisms (Coleoptera: Tenebrionidae). *Biochem. Genet.*, *20*:621–636.

Wright, V. F., Vesonder, R. F., and Ciegler, A. (1982). Mycotoxins and other fungal metabolites as insecticides. In *Microbial and Viral Pesticides* (ed. Kurstak, E.). Marcel Dekker, New York, pp. 559–583.

Zettler, J. L. (1974a). PP 511: toxicity to malathion-resistant strains of the Indianmeal moth. *J. Econ. Entomol.*, *67*:450–451.

Zettler, J. L. (1974b). Esterases in a malathion-susceptible and a malathion-resistant strain of *Plodia interpunctella* (Lepidoptera: Phycitidae). *J. Georgia Entomol. Soc.*, *9*: 207–213.

Zettler, J. L. (1977). Fecundity of malathion-resistant *Plodia interpunctella* and *Ephestia cautella* (Lepidoptera: Phycitidae). *J. Ga. Entomol. Soc.*, *12*:333–336.

Zettler, J. L. (1982). Insecticide resistance in selected stored-product insects infesting peanuts in the Southeastern United States. *J. Econ. Entomol.*, *75*:359–362.

Zettler, J. L. (1991). Phosphine resistance in stored product insects in the United States. In Proceedings, 5th International Working Conference of Stored Product Protection, INRA/SDPV, Bordeaux, France, pp. 1075–1081.

Zettler, J. L. (1991). Pesticide resistance in *Tribolium castaneum* and *T. confusum* (Coleoptera: Tenebrionidae) from flour mills in the United States. *J. Econ. Entomol.*, *84*: 763–767.

Zettler, J. L., and Beeman, R. W. (1991). Resistance to chemicals. In *FGIS Handbook on Management of Grain, Bulk Commodities, and Bagged Products*. Oklahoma State Univ. Coop. Ext. Serv. Circ. E-912, pp. 145–148.

Zettler, J. L., and Cuperus, G. W. (1990). Pesticide resistance in *Tribolium castaneum* (Coleoptera: Tenebrionidae) and *Rhyzopertha dominica* (Coleoptera: Bostrichidae) in wheat. *J. Econ. Entomol.*, *83*:1677–1681.

Zettler, J. L., Halliday, W. R., and Arthur, F. H. (1989). Phosphine resistance in insects infesting stored peanuts in the Southeastern United States. *J. Econ. Entomol.*, *82*:1508–1511.

Zettler, J. L., McDonald, L. L., Redlinger, L. M., and Jones, R. D. (1973). *Plodia interpunctella* and *Cadra cautella* resistance in strains to malathion and synergized pyrethrins. *J. Econ. Entomol.*, *66*:1049–1050.

# 9

# Integrated Pest Management

**David W. Hagstrum and Paul W. Flinn**
*U.S. Department of Agriculture, Manhattan, Kansas*

Integrated pest management (IPM) is an approach to pest control that uses cost-benefit analysis in making decisions. Metcalf and Luckmann (1982) provide a general discussion of insect pest management. In IPM programs, control is cost-effective when the cost of control is less than the reduction in market value due to pests. The alternative to IPM has been the use of pesticides on a regular schedule without determining whether insect control was needed. IPM can reduce the use of pesticides by avoiding unnecessary chemical applications and by using nonchemical control methods whenever possible. Biological or physical control methods are beneficial in delaying the development of insecticide resistance and in reducing health risk from human exposure to pesticides. The development of IPM programs has been considered by the food industry for both raw (Hagstrum and Flinn 1991, 1992; Cuperus et al. 1993) and processed (Troller 1982; Marriott 1989; Mills and Pedersen 1990; Jones and Thompson 1991) commodities. The food industry will need to use IPM programs more extensively in the future to satisfy the increased demands of consumers and regulatory agencies for reduced use of pesticides.

Previous chapters have discussed elements of IPM programs, which include sampling, insect ecology, control methods, and insecticide resistance management. This chapter will discuss how these elements fit together in an IPM program, review recent research on the IPM concept, and examine the extent to which IPM decision-making methods developed for field crops are applicable to stored-product insects.

## I. ACTION THRESHOLDS

Action thresholds are insect densities or damage levels at which managers need to control insect populations. These thresholds must be established to decide when pest control measures are cost-effective. Cost–benefit analysis for IPM programs is based on an economic injury level (EIL). An EIL is the insect density that causes a reduction in market value greater than the cost of insect control (Onstad 1987). Control costs also include the costs of sampling programs, insecticide resistance management programs, and managing the risk to human health and the environment. Nyrop et al. (1986) provide a method for estimating the value of sampling information for pest control decision-making. A second threshold, the economic threshold (ET), is the insect density at which control must be applied to prevent damage or contamination from reaching the EIL (Fig. 1). The ET allows for delays in making control decisions and applying control measures, and the time required for control measures to reduce insect density (Onstad 1987). The critical control points (CCP) concept of the hazard analysis/critical control points (HACCP) approach to food safety used by the food industry (Mills and Pedersen 1990) is similar to the ET and EIL concepts in IPM programs. Both emphasize regular monitoring and taking corrective action when established thresholds are exceeded. A major difference between the HACCP and IPM approaches is that HACCP emphasizes the identification of points in the system where hazards are likely to be introduced and the development of programs to prevent their introduction, while IPM emphasizes the use of cost–benefit analysis to determine whether control is cost effective. Dunaif and Krysinski (1992) suggest that the food industry needs to use both approaches to reduce the level of pesticide residues in food. They indicate that testing for residues is expensive and seem to indicate that cost–benefit analysis is needed to evaluate the contribution of any residue testing to the overall goal of pesticide residue control program.

### 1. Quality or Weight Losses

The EIL can be developed for commodities using either the loss of quality or weight. However, in the food industry the loss of quality due to insect infestation generally becomes an issue well before measurable weight losses occur. Quality loss is generally considered more difficult to quantify and assign a monetary value than weight loss (Olkowski 1974; Zungoli and Robinson 1984; Raupp et al. 1987). The term *aesthetic injury level* has been used instead of EIL to differentiate a perceived loss of quality from a loss of weight. Traditional EILs for stored wheat include the designations "infested" given when insect densities are two or more live insects per kilogram of wheat and "sample grade" given when insect damaged kernels (IDK) exceed 32/100 g (Hagstrum and Flinn 1992). These thresholds might generally be considered EIL because of the monetary loss associated with them. The penalties for delivering "infested" wheat to Kansas elevators ranged from \$0 to \$0.6 per bushel and were often less than the cost of control (Reed et al. 1989). For densities of 0.1–5 insects/kg, the penalty averaged \$0.02 per bushel. The monetary loss due to "sample grade" designation is much greater than

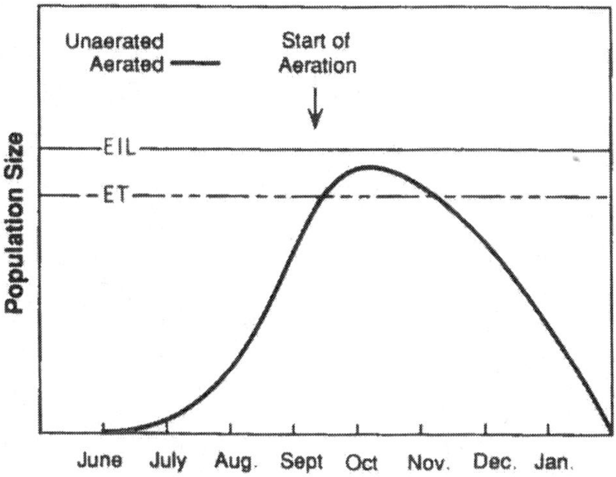


**Figure 1** Economic threshold concept shows when aeration to cool grain needs to begin to prevent insect population from reaching economic injury level. (Redrawn from Hagstrum and Flinn, 1991.)

that due to "infested" designation. Also, the "infested" designation can be removed by fumigating to kill insects but the "sample grade" designation cannot be removed. The "infested" designation might therefore be considered an ET that allows managers to prevent grain from reaching "sample grade." For flour, an EIL is 75 insect fragments per 50 g, a level at which flour cannot be sold for human consumption. Flour mills often use a lower threshold of 5 IDK for rejecting wheat and keep fragment counts in flour below 15 insect fragments per 50 g. These lower thresholds used by mills might be considered ETs.

For many packaged commodities, consumer complaint levels are the EIL. High levels of consumer complaints can be an indicator of customer dissatisfaction that might result in a loss of sales due to customers buying other brands. Studies by Loschiavo and Sabourin (1982) and Turner and Maude-Roxby (1989) found that 13–19% of households had stored-product insect infestations, but that consumers in 20–78% of these households discarded infested food without contacting the store where it was purchased or the manufacturer. When consumer complaints are received, the insect problems are traced back to the source so that insect infestation can be controlled. Similar trace-back procedures are recommended in flour mills when insects are found in the final sieving (American Institute of Baking 1979). These trace-back procedures are part of a good IPM program: sampling program or consumer complaints are used to identify areas that need spot treatments rather than treating an entire facility.

### 2. Sampling Rate and Action Threshold

The pest management programs of food processors that consider the ET to be any density above zero insects are actually allowing the ET to be determined by the sampling rate. Higher sampling rates would increase the chances of detecting insects. The sampling rate must be high enough to avoid having insects or excess numbers of insect fragments in processed food. In the flour industry, all of the flour can be checked for insects by examining the tailings from the final sieving (American Institute of Baking 1979) and the use of this high sampling rate in making pest management decisions can result in an ET very close to zero insects in flour. The lower sampling rate used in flour mill inspection programs for residual insect populations gives the appearance of having a zero insect ET. However, insect eradication is extremely difficult and expensive. The characteristics of insect populations that make them difficult to eradicate were discussed in the chapter on ecology. Low densities of insects in a facility are probably tolerated, and insects only become a problem when the sanitation program or other control measures are relaxed and insect densities increase.

## II. DECISION MAKING

IPM programs use action thresholds and cost–benefit analysis to make pest control decisions. Onstad (1987) provides equations that compare the cost–benefit analysis for different control methods. The higher cost of integrating more than one control method will generally make using a single method preferable. However, insecticides degrade and fumigants escape leaving the commodity unprotected. Thus, several methods may need to be used to control insects in commodities or facilities throughout the year. A practical IPM program for farm-stored grain includes cleaning bins before storing grain, periodically sampling grain for insects, and aerating to cool grain early in the fall. For mills, a simple IPM program includes not purchasing wheat with high insect infestations or pesticide residue levels, use of impact-aspiration machines to remove insects from inside wheat kernels, a sanitation program to reduce residual insect infestations in facilities, and monitoring the tailings from the final sieving of the flour for insects. Both of these simple IPM programs will slow the development of insect resistance to pesticides because using nonchemical control methods is the most effective resistance management program.

### 1. Predictive Models

Predictions of future insect populations (and the economic loss they can cause) are important because only future losses can be prevented. The growth rate of insect populations is determined by the temperature and moisture of the stored grain (Flinn and Hagstrum 1990a; Hagstrum and Flinn 1990). Computer simulation models have been developed to predict the population growth of five of the most important insect pests of stored wheat. These models provide a means of using our knowledge of insect ecology in pest management programs. With these models, the effect of different control strategies on insect numbers can be compared. These models should be particularly useful in predicting the effectiveness of nonchemical control methods, such as parasites and aeration, that generally depend more on an understanding of insect ecology than does chemical control. Hagstrum and Flinn (1990) showed that the timing of fall aeration to cool grain had a large effect on the population growth of five species (Fig. 2). Each month that aeration was delayed resulted in a 4–24-fold increase in insect density at the end of the year. These models have been used to develop an expert system for stored grain management that makes recommendations for insect control (Flinn and Hagstrum 1990b).

Computer simulation models are useful in making pest management decisions because they can help us consider the uncertainty about expected market value, sampling and control costs, and preventable economic losses. Uncertainty about the accuracy of insect density estimates from a sampling program also needs to be considered (Nyrop et al. 1986). The EIL changes with market value of the commodity, sampling costs, and control costs. The ET is also dynamic and may vary between facilities, control methods, and times of the year. Computer simulation models that predict dynamic EIL and ET have been shown to provide better control decisions than fixed economic thresholds or extension recommendations (Nordh et al. 1988; Stone and Schaub 1990).

## 2. Cost–Benefit Calculations

The economics of insect problems in stored wheat have been investigated by Anderson et al. (1990), Reed et al. (1988, 1989, 1990), and Reed and Pedersen (1987). We will illustrate the use of cost–benefit analysis in an IPM program using farm-stored wheat as an example. A control measure is cost effective if the value of wheat at the time of sale minus the cost of sampling and control is greater than the value of wheat without insect control. If wheat is sold at harvest on July 1st, there are no insect control costs. However, the value of wheat typically increases between one harvest and the next (Reed and Pedersen 1987) and the value of wheat will be lower than if wheat is stored and sold later. Let us assume that the wheat was sampled on August 15th and again on September 15th. Sampling costs were $0.002 per bushel ($5.00/h for 2h on each sampling date). Sampling shows that the average density is 2 insects per kilogram and, if detected at the time of sale, the elevator operator would charge a penalty of $0.02 per bushel. If wheat is sold on September 25th, the value of wheat would have had to increase by $0.022 per bushel to cover these costs. As an alternative, the wheat could be stored for another 6 months. If the farmer did not sell the wheat, he or she would need to have

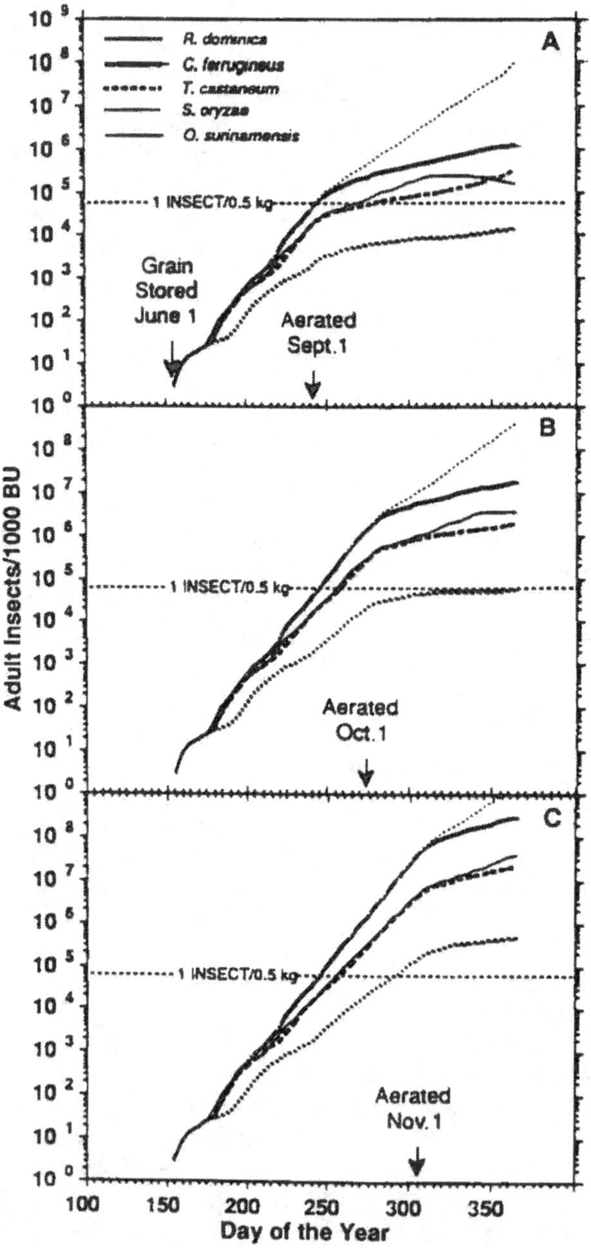


**Figure 2** Effect of time of aeration on population growth of five species of stored grain insects. (Redrawn from Hagstrum and Flinn, 1990.)

the wheat fumigated at a cost of $0.04 per bushel. During this period, the wheat would be sampled again on October 15th and cooled by aeration to reduce insect population growth ($0.01 per bushel). If the wheat was stored another 6 months, the value of wheat would have had to increase by $0.052 per bushel to cover sampling and control costs.

### 3. Preventive or Responsive Control

A primary difference between preventive and responsive control is that with responsive control management decisions are based on estimated insect density or damage. Many IPM programs will include both preventive and responsive elements (Vandermeer and Andow 1986). Exclusion, sanitation, impact machines (Entoleter), and stock rotation are generally considered preventive (Mills and Pedersen 1990; Jones and Thompson 1991) because these control measures tend to be used regularly rather than in response to insect density or damage estimates. Cost–benefit analysis should be used when developing an IPM program to decide which preventive control measure(s) to use and the amount of effort to give each control measure. Information on the relationship between the costs of increased insect control effort and the benefits of increased insect suppression will be needed for these analyses because, with some methods, such as sanitation, each increase in effort will be less beneficial as remaining food residues and insects become harder to find. The pest management program may need to be reevaluated periodically to determine whether these insect control methods and the amount of effort given to each method are still cost-effective. Most other control methods can be preventative or responsive (i.e., insect control is sometimes used as insurance against incorrect pest management decisions even when cost–benefit analysis would indicate that it is unnecessary). For example, space or crack and crevice treatments with insecticides can be used in response to finding insects, but are often used on a regular schedule as a preventive control measure. In the food industry, the probability of exceeding the EIL must be very low. However, the use of chemical control as insurance against incorrect management decisions should become less common when we are able to make reliable pest management decisions.

### 4. Control in Response to Sampling

The cost of chemical control at grain elevators might be reduced by fumigating individual bins only when the sampling program finds insect densities above an ET. Cuperus et al. (1990) found that elevators in Oklahoma averaged 2.6 fumigations of the entire facility per year. They indicated that wheat stored in Oklahoma had a higher risk of insect damage than wheat stored in states further north and therefore more insect pest management was needed in Oklahoma. Storage in Oklahoma provides a good example of the potential savings achieved by shifting from a preventive to a responsive control program. For an elevator storing 2 million bushels of wheat, fumigation costs at a rate of $0.02 per bushel would equal $80,000.00 for two fumigations per year. This includes the cost of moving grain to a new bin so that the fumigant can be added and the cost of the fumigant. If a sampling program found that during the year insect densities exceeded an ET in only 20% of bins, fumigation costs could be only $16,000.00, a savings of $64,000.00. In the future, cables with acoustical sensors that can automatically monitor insect populations throughout the grain mass (Hagstrum et al. 1991) will make it easier to fumigate only the bins in which insect densities exceed the ET.

### 5. Effectiveness of Control

Sampling programs need to evaluate the effectiveness of insect control measures. In many tropical countries, the surfaces of warehouses and stacks of bagged commodities are sprayed with a contact insecticide after fumigation and then at monthly intervals (Hodges et al. 1992). The first spraying reduces the insect populations outside the gas-proof sheets and the rate of reinfestation of the bagged commodities. The ET for fumigation was 10 insects per kilogram of rice. However, additional spraying at monthly intervals after fumigation was found to be ineffective in increasing the interval between fumigations, and these workers recommended that respraying be discontinued.

## III. SUMMARY

Integrated pest management (IPM) is an approach to pest control that uses cost–benefit analysis in making decisions. In IPM programs, control is cost-effective when the cost of control is less than the reduction in market value due to pests. Control is applied when insect densities reach an economic threshold (ET) to prevent damage or contamination from reaching the economic injury level (EIL). Computer models can be used to predict when control will be needed and to recommend which control method(s) to use. These models allow knowledge of insect ecology to be used in pest management programs. IPM will reduce the use of pesticides because control measures will be used only when sampling indicates that insect densities have exceeded the ET. Further reductions of chemical control can be achieved by substituting biological and physical for chemical control methods. Decreased reliance on chemical control methods is the best resistance management program. Sampling programs need to be used more frequently to evaluate the effectiveness of insect control measures. In the future, automation of insect monitoring may help to lower the cost of sampling programs. The IPM approach is complex because many factors influence pest management decisions. Fully utilizing the IPM approach will require more effort than other types of control programs, but the IPM approach can provide more reliable pest management decisions.

## REFERENCES

American Institute of Baking. (1979). Searching bulk flour systems for obscure insect sources. In Basic Food Plant Sanitation Manual. American Institute of Baking, Manhattan, KS, pp. 195–197.

Anderson, K., Schurle, B., Reed, C., and Pedersen, J. (1990). An economic analysis of producers' decisions regarding insect control in stored grain. *N. Central J. Agric. Econ.*, *12*:23–29.

Cuperus, G., Noyes, R. T., Fargo, W. S., Clary, B. L., Arnold, D. C., and Anderson, K. (1990). Management practices in a high-risk stored-wheat system in Oklahoma. *Amer. Entomol.*, *36*:129–134.

Cuperus, G., Noyes, R. T., Fargo, W. S., Kenkel, P., Criswell, J. T., and Anderson, K. (1993). Reducing pesticide use in wheat postharvest systems. *Cereal Foods World*, *38*: 199–203.

Dunaif, G. E., and Krysinski, E. P. (1992). Managing the pesticide challenge: A food processor's model. *Food Technology*, *46*:72–76.

Flinn, P. W., and Hagstrum, D. W. (1990a). Simulations comparing the effectiveness of various stored-grain management practices used to control the lesser grain borer, *Rhyzopertha dominica* (F.) (Coleoptera: Bostrichidae). *Environ. Entomol.*, *19*:725–729.

Flinn, P. W., and Hagstrum, D. W. (1990b). Stored Grain Advisor: A knowledge-based system for management of insect pests of stored grain. *AI Appl. Nat. Resource Manage.*, *4*:44–52.

Hagstrum, D. W., and Flinn, P. W. (1990). Simulations comparing insect species differences in response to wheat storage conditions and management practice. *J. Econ. Entomol.*, *83*:2469–2475.

Hagstrum, D. W., and Flinn, P. W. (1991). IPM in grain storage and bulk commodities. In FGIS Handbook on Management of Grain, Bulk Commodities, and Bagged Products. Oklahoma State Univ. Coop. Ext. Serv. Circ. E-912, pp. 183–187.

Hagstrum, D. W., and Flinn, P. W. (1992). Integrated pest management of stored-grain insects. In Storage of Cereal Grains and Their Products (ed. Sauer, D. B.). American Association of Cereal Chemists, St. Paul, MN, pp. 535–562.

Hagstrum, D. W., Vick, K. W., and Flinn, P. W. (1991). Automatic monitoring of *Tribolium castaneum* populations in stored wheat with computerized acoustical detection system. *J. Econ. Entomol.*, *84*:1604–1608.

Hodges, R. J., Sidik, M., Halid, H., and Conway, J. A. (1992). Cost efficiency of respraying store surfaces with insecticide to protect bagged milled rice from insect attack. *Tropical Pest Manage.*, *38*:391–397.

Jones, D. F., and Thompson, E. G. (1991). Integrated pest management for the food industry. In Food and Drug Administration Technical Bulletin 4: Ecology and Management of Food-Industry Pests (ed. Gorham, J. R.), pp. 551–556.

Loschiavo, S. R., and Sabourin, D. (1982). The merchant grain beetle, *Oryzaephilus mercator* (Silvanidae: Coleoptera), as a household pest in Canada. *Can. Entomol.*, *114*:1163–1169.

Marriott, N. G. (1989). Principles of Food Sanitation. Van Nostrand Reinhold, New York.

Metcalf, R. L., and Luckmann, W. H. (1982). Introduction to Insect Pest Management. John Wiley and Sons, New York.

Mills, R., and Pedersen, J. (1990). A Flour Mill Sanitation Manual. Eagan Press, St. Paul, MN.

Nordh, M. B., Zavaleta, L. R., and Ruesink, W. G. (1988). Estimating multidimensional economic injury levels with simulation models. *Agric. Syst.*, *26*:19–33.

Nyrop, J. P., Foster, R. E., and Onstad, D. W. (1986). Value of sample information in pest control decision making. *J. Econ. Entomol.*, *79*:1421–1429.

Olkowski, W. (1974). A model ecosystem management program. *Proc. Tall Timbers Conf. Ecol. Anim. Control Habitat Manage.*, *5*:103–117.

Onstad, D. W. (1987). Calculation of economic-injury levels and economic thresholds for pest management. *J. Econ. Entomol.*, *80*:297–303.

Raupp, M. J., Davidson, J. A., Koehler, C. S., Sadof, C. S., and Reichelderfer, K. (1987). Decision-making considerations for aesthetic damage caused by pests. *Bull. Entomol. Soc. Amer.*, *34*:27–32.

Reed, C., and Pedersen, J. (1987). Farm-stored wheat in Kansas: facilities, conditions, pest control, and cost comparisons. *Kans. Agric. Exp. Stn. Bull.* 652

Reed, C., Anderson, K., Brockschmidt, J., Wright, V., and Pedersen, J. (1990). Cost and effectiveness of chemical insect control measures in farm-stored Kansas wheat, *J. Kansas Entomol. Soc.*, *63*:351–360.

Reed, C., Schurle, B., and Fleming, R. (1988). Stored wheat and the Kansas country elevator. *Kans. Agric. Exp. Stn. Rep. Prog.* 535.

Reed, C., Wright, V. F., Pedersen, J. R., and Anderson, K. (1989). Effects of insect infestation of farm-stored wheat on its sale price at country and terminal elevators. *J. Econ. Entomol.*, *82*:1254–1261.

Stone, N. D., and Schaub, L. P. (1990). A hybrid expert system/simulation model for the analysis of pest management strategies. *AI Appl. Nat. Resource Manage.*, *4*:17–26.

Troller, J. A. (1982). Sanitation in Food Processing. Academic Press, New York.

Turner, B., and Maude-Roxby, H. (1989). The prevalence of the booklice *Liposcelis bostrychophilus* Badonnel (Liposcelidae, Psocoptera) in British domestic kitchens. *Int. Pest Control*, *31*:93–97.

Vandermeer, J., and Andow, D. A. (1986). Prophylactic and responsive components of an integrated pest management program. *J. Econ. Entomol.*, *79*:299–302.

Zungoli, P. A., and Robinson, W. H. (1984). Feasibility of establishing an aesthetic injury level for German cockroach pest management programs. *Environ. Entomol.*, *13*:1453–1458.

# Index